Fields Institute Communications

VOLUME 75

The Fields Institute for Research in Mathematical Sciences

The Fields Institute is a centre for research in the mathematical sciences, located in Toronto, Canada. The Institute's mission is to advance global mathematical activity in the areas of research, education and innovation. The Fields Institute is supported by the Ontario Ministry of Training, Colleges and Universities, the Natural Sciences and Engineering Research Council of Canada, and seven Principal Sponsoring Universities in Ontario (Carleton, McMaster, Ottawa, Queen's, Toronto, Waterloo, Western and York), as well as by a growing list of Affiliate Universities in Canada, the U.S. and Europe, and several commercial and industrial partners.

More information about this series at http://www.springer.com/series/10503

Philippe Guyenne • David Nicholls
Catherine Sulem
Editors

Hamiltonian Partial Differential Equations and Applications

FIELDS The Fields Institute for Research in the Mathematical Sciences

Springer

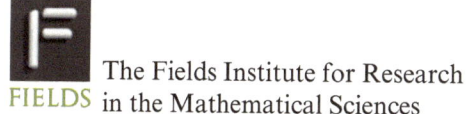

Editors
Philippe Guyenne
Department of Mathematical Sciences
University of Delaware
Newark, DE, USA

David Nicholls
Department of Mathematics, Statistics, and Computer Science
University of Illinois at Chicago
Chicago, IL, USA

Catherine Sulem
Department of Mathematics
University of Toronto
Toronto, ON, Canada

ISSN 1069-5265 ISSN 2194-1564 (electronic)
Fields Institute Communications
ISBN 978-1-4939-2949-8 ISBN 978-1-4939-2950-4 (eBook)
DOI 10.1007/978-1-4939-2950-4

Library of Congress Control Number: 2015947929

Mathematics Subject Classification (2010): 35-XX, 37-XX, 46-XX, 76-XX, 83-XX

Springer New York Heidelberg Dordrecht London

Cover illustration: Drawing of J.C. Fields by Keith Yeomans

Printed on acid-free paper

Springer Science+Business Media LLC New York is part of Springer Science+Business Media (www.springer.com)

Preface

Partial differential equations (PDEs) are a fundamental tool in the modeling of phenomena arising in the physical sciences. PDEs with Hamiltonian structure are a distinguished subset, which not only model systems with conserved quantities (e.g., energy and momentum), but also possess an array of special techniques for their analysis and simulation. They constitute an active area of research where major innovations have been and continue to be made from both the mathematical and computational sides. Not only do these innovations benefit the field itself but also contribute to progress in a vast range of other scientific areas. Applications of Hamiltonian PDEs are numerous in fluid mechanics, plasma physics, and nonlinear optics with such notable examples as the Korteweg–de Vries equation and the nonlinear Schrödinger equation.

In the last few decades, significant progress has been achieved in the mathematical study of these evolutionary PDEs by adopting the "dynamical systems" approach, extending refined analytical techniques of Hamiltonian dynamical systems to the setting of PDEs. This point of view has led to the consideration of the global behavior of orbits for a Hamiltonian PDE in an appropriate phase space, the pursuit of the mathematical technology of normal forms, the study of stable orbits and Kolmogorov–Arnold–Moser (KAM) tori, and a number of results analogous to Nekhoroshev stability and Arnold diffusion. In particular, building on the experience gained from the qualitative study of finite-dimensional dynamical systems, the search for periodic and quasi-periodic solutions has been regarded as a first step towards better understanding the complicated flow evolution of Hamiltonian PDEs. A central tool is transformation theory including Birkhoff normal form transformations. In the broad picture, the goal is to understand some of the important structures of infinite-dimensional phase spaces in which these evolutionary equations are naturally posed, such as periodic orbits, embedded invariant tori, center manifolds, and the different effects of resonances in the non-compact versus compact cases. Techniques from transformation theory for Hamiltonian PDEs with a small parameter have also been successfully used in recent work on water waves, allowing for the systematic derivation of Hamiltonian models in various asymptotic limits.

On 10–12 January 2014, a conference on "Hamiltonian PDEs: Analysis, Computations and Applications" was held at the Fields Institute for Research in Mathematical Sciences in Toronto, bringing together a group of world-class researchers to present and discuss the latest developments in this field. Given the wide range of applications and mathematical tools, a motivating theme of this event was the interaction of specialists in dynamical systems, KAM theory, normal form theory, PDE theory and variational methods, as well as applied and numerical analysts, and experts in water waves. The program consisted of eighteen lectures by distinguished faculty, together with three shorter presentations by junior speakers including two graduate students. The participants came from Canada, Europe, and the USA.

This conference was also an opportunity to honor our friend and colleague Walter Craig, who has made significant contributions to this field, on the occasion of his 60th birthday. Walter obtained his Ph.D. degree from the Courant Institute of Mathematical Sciences (NYU) in 1981. He has held faculty positions at CalTech, Stanford University and Brown University before joining McMaster University as a Professor and Canada Research Chair of Mathematical Analysis and its Applications. He has received a number of prestigious awards including an Alfred P. Sloan Fellowship, an NSF Presidential Young Investigator Award, a Killam Research Fellowship and is a Fellow of the AMS, the AAAS, the Fields Institute and the Royal Society of Canada. He has served on the editorial boards of several journals including the Philosophical Transactions of the Royal Society, the Proceedings of the AMS, and the SIAM Journal on Mathematical Analysis. Walter is a world-renowned mathematical analyst with interests in nonlinear PDEs, Hamiltonian dynamical systems and their physical applications. He has authored more than 100 research articles.

This special volume presents a unique selection of both survey and original research papers by experts who participated in that conference. The various topics discussed in this volume are representative of the wide scope covered by Hamiltonian PDEs, and the results range from mathematical modeling to rigorous analysis and numerical simulation. These topics also reflect Walter Craig's breadth in research interests and his influence in this field. This book will be of particular interest to graduate students as well as researchers in mathematics, physics, and engineering, who wish to learn more about the powerful and elegant analytical techniques for Hamiltonian PDEs.

The editors would like to thank the Fields Institute for Research in Mathematical Sciences and the Department of Mathematics & Statistics at McMaster University for their generous support. In particular, we are grateful to Alison Conway, Drs. Matheus Grasselli and Hans Boden for their assistance with the organization of the conference, as well as to Debbie Iscoe, Dr. Carl Riehm, and the Springer team for their assistance with the publication of this special volume. We are also thankful to the authors for contributing such excellent articles and to the referees for their invaluable help during the review process. Finally, we dedicate this book to Walter Craig who has been a constant source of inspiration, and whose enthusiasm and friendship have never waned. We would like to extend to him our warmest wishes for many more happy events to come.

Newark, DE, USA
Chicago, IL, USA
Toronto, ON, Canada
April 2015

Philippe Guyenne
David Nicholls
Catherine Sulem

Contents

Hamiltonian Structure, Fluid Representation and Stability for the Vlasov–Dirac–Benney Equation

Claude Bardos and Nicolas Besse

Pour Walter Craig, en remerciement pour ses contributions scientifiques et son amitié.

Abstract This contribution is an element of a research program devoted to the analysis of a variant of the Vlasov–Poisson equation that we dubbed the Vlasov–Dirac–Benney equation or in short V–D–B equation. As such it contains both new results and efforts to synthesize previous observations. One of main links between the different issues is the use of the energy of the system. In some cases such energy becomes a convex functional and allows to extend to the present problem the methods used in the study of conservation laws. Such use of the energy is closely related to the Hamiltonian structure of the problem. Hence it is a pleasure to present this article to Walter Craig in recognition to the pioneering work he made for our community, among other things, on the relations between Hamiltonian systems and Partial Differential Equations.

1 Introduction

This article extends a program (cf. [1, 2]) devoted to the mathematical analysis of an avatar of the Vlasov–Poisson equation, where the "Coulomb potential" is replaced by the Dirac mass. Since it was proposed by Benney [3] and Zakharov [28] for the description of water waves, it is dubbed Vlasov–Dirac–Benney equation (or in short V–D–B equation). Therefore the V–D–B equation reads

---

C. Bardos (✉)
Laboratoire Jacques-Louis Lions, Université Denis Diderot, Paris, France
e-mail: claude.bardos@gmail.com

N. Besse
Département Physique de la Matière et des Matériaux, Institut Jean Lamour UMR CNRS 7198, Université de Lorraine, BP 70239, 54506 Vandoeuvre-lès-Nancy Cedex, France
e-mail: nicolas.besse@univ-lorraine.fr

P. Guyenne et al. (eds.), *Hamiltonian Partial Differential Equations and Applications*, Fields Institute Communications 75,
DOI 10.1007/978-1-4939-2950-4_1

$$\partial_t f + v \cdot \nabla_x f - \nabla_x \rho_f \cdot \nabla_v f = 0\,, \quad \text{with} \quad \rho_f(t,x) = \int_{\mathbb{R}^d} f(t,x,v)dv. \tag{1}$$

And the classical conservation of mass and energy turn out to be given by the formula,

$$\partial_t \int_{\mathbb{R}^d_v} f(t,x,v)dv + \nabla_x \cdot \int_{\mathbb{R}^d_v} vf(t,x,v)dv = 0,$$

$$\begin{aligned}
\partial_t \int_{\mathbb{R}^d_x}\int_{\mathbb{R}^d_v} \frac{|v|^2}{2} f(t,x,v)dxdv &= -\int_{\mathbb{R}^d_x} \nabla_x \rho_f(t,x) \cdot \int_{\mathbb{R}^d_v} \frac{|v|^2}{2} \nabla_v f(t,x,v)dxdv \\
&= \int_{\mathbb{R}^d_x} \rho_f(t,x) \nabla_x \cdot \int_{\mathbb{R}^d_v} vf(t,x,v)dxdv \\
&= -\int_{\mathbb{R}^d_x} \rho_f(t,x)\partial_t \rho_f(t,x)dx,
\end{aligned}$$

or eventually

$$\frac{d\mathscr{E}}{dt} = \frac{d}{dt}\left(\frac{1}{2}\int_{\mathbb{R}^d_x} dx \left(\int_{\mathbb{R}^d_v} dv \,\frac{|v|^2}{2} f(t,x,v) + (\rho_f(t,x))^2\right)\right) = 0.$$

1.1 *Some Physical Motivations for the Introduction of the Dirac Potential*

One of the many physical motivations for the introduction of this equation is the description of a plasma constituted of ions in a background of "adiabatic" electrons which instantaneously reach a thermodynamical equilibrium (i.e. electrons follow a Maxwell–Boltzmann distribution). Therefore the charge density of electrons is given in term of the electrical potential Φ_ϵ by the formula

$$\rho_- = \rho_0 e^{-\frac{e\Phi_\epsilon}{k_B T_e}},$$

with k_B the Boltzmann constant, e the electron charge and T_e the equilibrium temperature of electrons. Finally the parameter ϵ represents the Debye length. Hence the "Coulomb law" couples the electrical potential Φ_ϵ to the charge density such that

$$-\epsilon^2 \Delta\Phi_\epsilon = \rho_\epsilon - \rho_0 e^{-\frac{e\Phi_\epsilon}{k_B T_e}}, \quad \text{with} \quad \rho_\epsilon = \int_{\mathbb{R}^d_v} f_\epsilon(t,x,v)dv.$$

Now since the electrical potential energy $e\Phi_\epsilon$ is supposed to be small in comparison to the kinetic energy $k_B T_e$, i.e $|e\Phi_\epsilon/(k_B T_e)| \ll 1$, after linearization on the exponential function, at first order we get

$$-\epsilon^2 \Delta \Phi_\epsilon = \rho_\epsilon - \rho_0 + \frac{e\rho_0}{k_B T_e} \Phi_\epsilon .$$

Setting ϵ to zero (quasineutrality assumption) and since ρ_0 and T_e are supposed to be constant, we obtain for the electric field E_ϵ the expression

$$E_\epsilon = -\nabla_x \Phi_\epsilon = \frac{k_B T_e}{e\rho_0} \nabla_x \int_{\mathbb{R}^d_v} f_\epsilon(t,x,v)dv,$$

which appears in the Vlasov equation (1).

1.2 Some Mathematical Motivations for this Analysis

Since in the Eq. (1) the electric field E is given in term of the electrons density by an operator of order 1, while in the classical Vlasov–Poisson case it is given by an operator of degree -1, the solution is much more dependent on the initial data. Therefore, while for the classical Vlasov–Poisson equation the issue is the large time asymptotic behavior, here what is at stake is the well-posedness of the problem in term of the initial data. On the other hand since the electrical potential is given by a purely local operator there exists a strong connection between the dynamics of hyperbolic systems of conservation laws and the V–D–B equation. This connection appears even more clearly when one uses for the Vlasov equation a kinetic representation of the form (cf. Sect. 3.2)

$$f(t,x,v) = \int_M \rho(t,x,\sigma)\delta(v - u(t,x,\sigma))d\sigma, \tag{2}$$

which leads to non local "operator type" conservation laws.

For such conservation laws the invariants play an essential role and as expected, they coincide (cf. Theorem 6) with the Lax–Godunov conserved quantities. When such invariants turn out to be convex (with respect to the parameters of the dynamics) they play the role of convex entropies and ensure the local-in-time stability and well-posedness of the Cauchy problem.

As this is the case for the most general Vlasov equations (as explained for instance in [22]) the present V–D–B equation can be viewed as a Hamiltonian system related to the minimization of an energy. Moreover the same point of view can be used to formalize the relations between classical and quantum mechanics via semi-classical (WKB) limits and Wigner measure (cf. Sect. 6). Such convergence will be always true at the formal level, or with analytical initial data. However, as expected, proofs in the Distributions (or Sobolev) setting will be available only when the limit enjoys the same stability i.e. mostly in the case where a convex entropy is present. Even if the analyticity hypothesis is not "physical", conclusions that follow are important, and especially in the case of the one-dimensional space variable. Then the cubic nonlinear Schrödinger equation and its generalization as infinite systems of coupled nonlinear Schrödinger equations (cf. [28]) are integrable systems with a rich algebraic structure including in particular the construction of infinite family of conservation laws. In the semi-classical limits these structures (at least for analytic initial data) do persist and make the one-dimensional-space-variable V–D–B equation a quasi-integrable system in the sense of [28].

The paper is then organized as follows. First the emphasis is put on the one-dimensional space variable which as quoted above contains more mathematical structure and provides also more explicit examples. To underline the dimension one in the corresponding equations, the symbol ∇_x and ∇_v are replaced by the symbol ∂_x and ∂_v. In Sect. 2, the analysis of the linearized problems turns out to be (and this should not be a surprise) in full agreement with the properties of the fully nonlinear systems. Moreover this produces also a natural tool for the study of nonlinear perturbations which is the object of the next section.

In the Sect. 3, the Hamiltonian structure and the fluid representation of the kinetic V–D–B equation are described. In this setting, under strong analyticity hypothesis a local-in-time stability result can be proven and this is the object of the Theorem 5. To obtain stability results with finite order regularity, the entropies have to be introduced and compared with the classical invariants of the Hamiltonian system. This is the object of the Sect. 4 and Theorem 6. The next Sect. 5 is devoted to several examples of application.

For the discussion of the semi-classical limits in the Sect. 6, we follow similar route. First formal computations are given. Then there are validated with analyticity hypothesis (cf. Theorem 9). Such results are compared with a theorem of Grenier which is valid in any space variable, with Sobolev type regularity hypothesis, but which concerns only the Wigner limit of "pure states" i.e. mono-kinetic solutions of the V–D–B equation.

As a conclusion we return to the relation between Wigner limit and inverse scattering.

2 Properties of the Linearized Problem and Consequences

Long time ago, it has been observed that x-independent solutions

$$v \mapsto G(v) \geq 0 \quad \text{with} \quad \int_{\mathbb{R}} G(v)dv = 1,$$

are stationary solutions of the Vlasov–Poisson equation. Same simple observation is also valid for the V–D–B equation. Writing

$$f(t,x,v) = G(v) + \tilde{f}(t,x,v),$$

retaining only the linear terms in $\tilde{f}$ and omitting henceforth the tilde notation, one obtains the evolution equation

$$\partial_t f + v\nabla_x f - G'(v)\partial_x \rho_f(t,x) = 0.$$

It then turns out that in one space dimension the spectral analysis, hence the stability, can be described in term of the shape of the stationary profile $v \mapsto G(v)$. In particular:

i) One can prove for the classical Vlasov–Poisson equation (this goes back to Kruskal [19]) that if the profile has only one maximum (one bump profile) then the solution is described in a convenient Hilbert space by a unitary group and therefore is stable. This remark can be adapted to the V–D–B equation and is shortly described below.
ii) In the presence of several extrema, a criterion due to Penrose [25] for the original Vlasov–Poisson equation, gives the existence (resp. the non-existence) of unstable generalized eigenvalues which may imply large time linear instabilities (cf. [10] for this point of view). However for V–D–B equation, due to the homogeneity of the dispersion relation, unstable modes whenever they exist are of the form $\omega(k) = \omega^* k$ with $\Im\omega^* \neq 0$. Hence the relation to prove their existence is not a simple adaptation of the Penrose criterion.

Below explicit examples given in [1] and [2] are recalled to show that the well-posedness of the Cauchy problem depends on the structure of the function $v \mapsto G(v)$. For some initial data the linearized problem may have no solution even in the sense of distributions. This remark extends to the nonlinear case which illustrates the natural connections between the stability of the linearized and the full nonlinear system.

2.1 *The Stability for the One Bump Profile*

To emphasize the role of the "bumps" in the stationary profile we recall below the following

Theorem 1. *Assume that the stationary profile $v \mapsto G(v)$ satisfies for some $a \in \mathbb{R}$, the relation*

$$G'(v) := -H(v)(v-a) \text{ with } \quad H(v) > 0. \tag{3}$$

Then any smooth solution $f(t,x,v)$ of the linearized V–D–B equation

$$\partial_t f(t,x,v) + v\partial_x f(t,x,v) - G'(v)\partial_x \rho_f(t,x) = 0, \tag{4}$$

satisfies the energy identity,

$$\frac{d}{dt}\left(\int_{\mathbb{R}\times\mathbb{R}} H^{-1}(v)(f(t,x,v))^2 dxdv + \int_{\mathbb{R}} (\rho(t,x))^2 dx\right) = 0.$$

The proof (cf. [19] and [1]) follows from the basic conservation laws of mass and energy combined with the formula (3). From the above Theorem 1 one deduces the following

Corollary 1. *With the function $v \mapsto G(v)$ as in the Theorem 1, and denoting by $\mathscr{H}$ the Hilbert space of functions f such that*

$$\int_{\mathbb{R}\times\mathbb{R}} H^{-1}(v)(f(t,x,v))^2 dxdv + \int_{\mathbb{R}} (\rho(t,x))^2 dx < \infty,$$

the solutions of the Cauchy problem with initial data $f_0(x,v) \in \mathscr{H}$ are described by a unitary group of operators.

2.2 Synthesis of Plane Waves and Unstable Modes

Plane waves of the form,

$$e_k(t,x,v) = A(k,v) \exp\left(i(kx - \omega(k)t)\right),$$

are solutions of the Eq. (4) whenever they satisfy the dispersion relation

$$(-i\omega(k) + ikv)A(k,v) - ik\left(\int_{\mathbb{R}} A(k,v)dv\right) G'(v) = 0,$$

or with $\omega(k) = \omega^* k$,

$$1 - \int_{\mathbb{R}} \frac{G'(v)}{v-\omega^*} dv = 0. \tag{5}$$

Then for any ω^* solution of (5), the functions

$$f(t,x,v) = \int_{\mathbb{R}} \frac{G'(v)}{v-\omega^*}\, \hat{\rho}(k)\, e^{i(kx-k\omega^* t)}\, dk,$$

are (if they exist) the unique solutions of the linear Cauchy problem with initial data

$$f(0,x,v) = \int_{\mathbb{R}} \frac{G'(v)}{v-\omega^*}\, \hat{\rho}(k)\, e^{ikx}\, dk.$$

As a consequence if there exists a ω^* solution of (5) with $\Im\omega^* \neq 0$, the Cauchy problem is ill-posed in any Sobolev space (because polynomial decreasing of Fourier modes with any speed does not compensate exponential growth).

The following statement, which is an adapted version of the Penrose criterion [25], illustrates the relation between multiple bumps and instabilities.

Theorem 2. *Assume that the stationary profile*

$$v \mapsto G(v) \geq 0, \quad \textit{and} \quad \int_{\mathbb{R}} G(v)dv = 1,$$

has a minimum for $v = 0$ and is even (i.e. $G(v) = G(-v)$), then for some $\epsilon > 0$ small enough, there is at least one non oscillatory unstable mode $\omega^* = i\beta$ for the equation linearized near $G_\epsilon = \epsilon^{-1}G(\epsilon^{-1}v)$.

Proof. Introducing the ϵ-dependent continuous function,

$$\beta \mapsto I_\epsilon(\beta) = \int_{\mathbb{R}} \frac{G'_\epsilon(v)}{v - i\beta}dv = \int_{\mathbb{R}} \frac{G'_\epsilon(v)(v + i\beta)}{v^2 + \beta^2}dv = \int_{\mathbb{R}} \frac{G'_\epsilon(v)v}{v^2 + \beta^2}dv,$$

for which one has,

$$I_\varepsilon(\infty) = 0, \quad \text{and} \quad I_\varepsilon(0) = \int_{\mathbb{R}} \frac{G'_\epsilon(v)}{v}dv = \int_{\mathbb{R}} \frac{G'_\epsilon(v) - G'_\epsilon(0)}{v}dv = \int_{\mathbb{R}} \frac{G_\epsilon(v)}{v^2}dv.$$

The last integration by part is justified by the fact that $G'(0) = 0$ and the convergence at $\pm\infty$ of the integral of v^{-2}. Eventually one has

$$I_\varepsilon(0) = \int_{\mathbb{R}} \frac{G_\epsilon(v)}{v^2}dv = \frac{1}{\epsilon^2}\int_{\mathbb{R}} \frac{G(v)}{v^2}dv > 1, \quad \text{for } \epsilon \text{ small enough.}$$

Therefore by continuity there exists at least one $\omega^* = i\beta^*$ solution of the dispersion relation (5). □

2.3 Consequences for the Original Nonlinear V–D–B Equation with General Initial Data

Theorem 3. *Let $\dot{H}^m$ be the space of functions $f \in L^\infty(\mathbb{R}_x, L^1(\mathbb{R}_v))$ with, for $1 \leq \ell \leq m$, derivatives $\partial_x^\ell f \in L^2(\mathbb{R}_x; L^1(\mathbb{R}_v))$.*
For every m, the Cauchy problem for the dynamics $S(t)$ defined by the V–D–B equation is not locally ($\dot{H}^m \mapsto \dot{H}^1$) well-posed.

Proof. Let be

$$\phi(t,x,v) = \int \frac{G'(v)}{v-\omega^*}\, \hat{\rho}(k)\, e^{i(kx-\omega^* kt)}\, dk.$$

A solution $f(t,x,v,s)$, with the initial data

$$f(0,x,v,s) = G(v) + s\phi(0,x,v),$$

for the nonlinear V–D–B equation gives

$$\tilde{f}(t,x,v) = \frac{d}{ds} f(t,x,v,s)_{|s=0},$$

as a solution for the linearized one. But we obtain a contradiction since $\tilde{f}$ is not well defined even in the distributional sense. □

From the structure of the solution

$$f(t,x,v) = \int_{\mathbb{R}} \frac{G'(v)}{v-\omega(k)/k}\, \hat{\rho}(k)\, e^{i(kx-\omega(k)t)}\, dk,$$

with $w(k) = w{*}k$ and $\Im w^* \neq 0$ one observes that the linear problem (and a fortiori the nonlinear one) will be well-posed if the Fourier transform of initial data are exponentially decreasing, which by use of Paley–Wiener Theorem [24] means that initial data must be analytic in a strip. And this is in agreement with the following forerunner result of Jabin and Nouri [17]:

Theorem 4 (Jabin-Nouri 2011). *For any (x,v) analytic function $f_0(x,v)$ with*

$$\forall \alpha,\, m,\, n, \quad \sup_x |\partial_x^m \partial_v^n f_0(x,v)|(1+|v|)^\alpha = C(m,n)o(|v|),$$

there exists, for a finite time T, an analytic solution of the Cauchy problem for the V–D–B equation.

3 Hamiltonian Structure of the Vlasov Equation and Application to Some Examples

3.1 Hamiltonian Structure

Below we just recall what would be essential for the present contribution. As it is the case for the classical Vlasov–Poisson equation, one may (cf. [22]) start from the conservation of the energy

$$\mathcal{E} = \int_{\mathbb{R}_x^d}\int_{\mathbb{R}_v^d} \frac{|v|^2}{2} f(t,x,v)dxdv + \frac{1}{2}\int_{\mathbb{R}_x^d} (\rho_f(t,x))^2 dx, \quad \rho_f(t,x) = \int_{\mathbb{R}_v^d} f(t,x,v)dv.$$

With the introduction of the Gâteaux derivative of this energy

$$\frac{\delta\mathcal{E}}{\delta f} = \frac{|v|^2}{2} + \rho_f(t,x),$$

and of the Poisson bracket

$$\{g,f\} = \nabla_v g \cdot \nabla_x f - \nabla_x g \cdot \nabla_v f,$$

the V–D–B equation is equivalent to the "Hamiltonian system"

$$\partial_t f = \left\{ f, \frac{\delta\mathcal{E}}{\delta f} \right\}. \tag{6}$$

Remark 1. Following Benney [3], we can obtain a new family of invariants for the one-dimensional Vlasov–Dirac–Benney and Vlasov–Poisson equations. To this purpose, let us define the velocity moments of the distribution function f such that

$$\partial_x A_0 = E(t,x) = -\partial_x \phi(t,x), \quad \text{and for} \quad n \geq 1, \quad A_n[f](t,x) = \int_{\mathbb{R}} v^n f(t,x,v)dv.$$

Velocity integration of V–D–B or V–P equations against polynomial in velocity leads to the moment hierarchy

$$\partial_t A_n + \partial_x A_{n+1} + n\partial_x A_0(x)\partial_x A_{n-1} = 0. \tag{7}$$

Defining the generating function $f(t,x;z)$ such that

$$f(t,x;z) := \sum_{n\geq 0} A_n(t,x) z^n,$$

it can be easily shown that the moment hierarchy (7) is equivalent to an equation on the generating function f, given by

$$z\partial_t f + \partial_x f = (1 - L(zf))\partial_x f(z=0), \tag{8}$$

with $L = z^2\partial_z$. Now rescaling the differential operator L as $\varepsilon z^2\partial_z$ in (8), using a recursive procedure which involves recursive multiplication of (8) by εz^2 and its z-differentiation, gathering terms of same power in ε and making some summation manipulations (cf. [3]), we obtain the following infinite system of conservation laws

$$\frac{\partial}{\partial t}\left(\sum_{n\geq 0} L^n\left(\frac{z^{n+1}f^{n+1}}{(n+1)!}\right)\right)+\frac{\partial}{\partial x}\left(\sum_{n\geq 0} L^n\left(\frac{z^n f^{n+1}}{(n+1)!}\right)-f(0)\right)=0. \tag{9}$$

In (9) each power n of z yields a distinct conservation laws and invariant. These new invariants for the one-dimensional V–D–B and V–P equations are thus polynomials of ϕ and velocity moments A_n, $n \geq 1$.

3.2 Zakharov–Grenier Representation and Benney Equation

Observe also that it is always possible to write the solutions of the Vlasov equation on the form (cf. [14, 28])

$$f(t,x,v)=\int_M \rho(t,x,\sigma)\delta(v-u(t,x,\sigma))d\sigma, \tag{10}$$

with $(M, d\sigma)$ a probability space. These notations are consistent with the macroscopic definition of density and momentum, according to the formulas,

$$\rho(t,x)=\int_{\mathbb{R}^d_v} f(t,x,v)dv=\int_M \rho(t,x,\sigma)d\sigma,$$

$$\rho(t,x)u(t,x)=\int_{\mathbb{R}^d_v} vf(t,x,v)dv=\int_M u(t,x,\sigma)\rho(t,x,\sigma)d\sigma.$$

Such decomposition is not unique and depends in particular on the form of this decomposition at time $t = 0$. Moreover a distribution function $f(t, x, v)$ given by (10) is a distributional solution of the V–D–B equation if and only if the functions $\rho(t, x, \sigma)$ and $u(t, x, \sigma)$ are solutions of the system

$$\begin{aligned}&\partial_t\rho(t,x,\sigma)+\nabla_x\cdot(\rho(t,x,\sigma)u(t,x,\sigma))=0,\\&\partial_t\left(\rho(t,x,\sigma)u(t,x,\sigma)\right)+\nabla_x\cdot(\rho(t,x,\sigma)u(t,x,\sigma)\otimes u(t,x,\sigma))\\&\quad+\rho(t,x,\sigma)\nabla_x\int_M \rho(t,x,\sigma)d\sigma=0.\end{aligned} \tag{11}$$

In one space dimension with $(M, d\sigma)$ being respectively the interval $(0, 1)$ and the Lebesgue measure, the system (11) turns out to be the Benney system

$$\begin{aligned}&\partial_t\rho(t,x,\sigma)+\partial_x(\rho(t,x,\sigma)u(t,x,\sigma))=0,\\&\partial_t u(t,x,\sigma)+u(t,x,\sigma)\partial_x u(t,x,\sigma)+\partial_x\int_0^1 \rho(t,x,\sigma)d\sigma=0,\end{aligned} \tag{12}$$

which has been derived by Zakharov from the original Benney equation [3] by using a Lagrangian parametrization (cf. [28]) as a model of water-waves for long waves. Hence the name "Benney" in the title of this contribution.

3.3 Kinetic Representations

Replacing the Lebesgue measure by the counting measure on the discrete set

$$M = \{1, 2, \dots, N \leq \infty\},$$

the formula (10) becomes the multi-kinetic representation

$$f(t, x, v) = \sum_{1 \leq n \leq N} \rho_n(t, x)\delta(v - u_n(t, x)).$$

In particular for $N = 1$ and in any space dimension, the mono-kinetic distribution

$$f(t, x, v) = \rho(t, x)\delta(v - u(t, x)),$$

is a solution of the V–D–B equation if and only if the moments $\rho(t, x)$ and $u(t, x)$ are solutions of an isentropic fluid equations,

$$\partial_t \rho + \nabla_x \cdot (\rho u) = 0\,, \qquad \partial_t(\rho u) + \nabla_x \cdot (\rho u \otimes u) + \nabla_x\left(\frac{\rho^2}{2}\right) = 0, \tag{13}$$

while in a one-dimensional space variable the multi-kinetic distribution function $f(t, x, v)$ will be a solution of the V–D–B equation if and only if the unknowns $U = ((\rho_1, \rho_2, \dots, \rho_N), (u_1, u_2, \dots, u_N))$ are solutions of the system of conservation laws,

$$\partial_t \rho_n + \nabla_x(\rho_n u_n) = 0\,,$$
$$\partial_t(\rho_n u_n) + \partial_x(\rho_n u_n^2) + \rho_n \partial_x\Big(\sum_{1 \leq l \leq N} \rho_l\Big) = 0.$$

3.4 Waterbag Representation and Equations

Assume that the density profile $v \mapsto f(t, x, v)$, with $0 \leq f(t, x, v) \leq 1$, has only one bump (say for $v = \nu(t, x)$). Then with

$$f_+(t, x, v) = f(t, x, v) \text{ for } v \geq \nu(t, x), \quad \text{and}$$
$$f_-(t, x, v) = f(t, x, v) \text{ for } v \leq \nu(t, x),$$

one defines on $]0,1[$, two functions $\sigma \mapsto v_\pm(t,x,\sigma)$ according to the formulas

$$f_\pm(t,x,v_\pm(t,x,\sigma)) = \sigma, \quad \text{if } \sigma \le \sup_v f_\pm(t,x,v),$$
$$v_\pm(t,x,\sigma) = 0, \quad \text{otherwise.}$$

As it was observed in [5] the density profile f can be reconstructed according to the standard formula (with Y denoting the Heaviside function)

$$\begin{aligned} f(t,x,v) = & \int_0^1 \delta(v - v_+(t,x,\sigma))\mathrm{Y}(v - \nu(t,x))|\partial_\sigma v_+(t,x,\sigma)|\sigma d\sigma \\ & + \int_0^1 \delta(v - v_-(t,x,\sigma))\mathrm{Y}(\nu(t,x) - v)|\partial_\sigma v_-(t,x,\sigma)|\sigma d\sigma, \end{aligned}$$

or also as

$$f(t,x,v) = \int_0^1 (\mathrm{Y}(v_+(t,x,\sigma) - v) - \mathrm{Y}(v_-(t,x,\sigma) - v))d\sigma,$$

which is an exact weak solution of the V–D–B equation if and only if

$$\partial_t v_\pm + v_\pm \partial_x v_\pm + \partial_x \int_0^1 (v_+(t,x,\sigma) - v_-(t,x,\sigma))d\sigma = 0. \tag{14}$$

Remark 2. From the formulas (14) one deduces the relations

$$\partial_t(v_+ - v_-) + \frac{(v_+ + v_-)}{2}\partial_x(v_+ - v_-) + (v_+ - v_-)\partial_x \frac{(v_+ + v_-)}{2} = 0,$$
$$\partial_t(\partial_\sigma v_\pm) + v_\pm \partial_x \partial_\sigma v_\pm + (\partial_\sigma v_\pm)\partial_x v_\pm = 0,$$

which imply that the following properties,

$$\begin{aligned} & \forall \sigma \in (0,1) \quad v_-(x,t,\sigma) \le v_+(x,t,\sigma), \\ & \sigma \mapsto v_+(x,t,\sigma) \text{ is decreasing and } \sigma \mapsto v_-(x,t,\sigma) \text{ is increasing,} \end{aligned} \tag{15}$$

are preserved by the dynamics [1, 4, 8].

With the infinite set of σ-dependent densities ρ and velocities u,

$$\rho(t,x,\sigma) = v_+(t,x,\sigma) - v_-(t,x,\sigma), \quad u(t,x,\sigma) = \frac{1}{2}(v_+(t,x,\sigma) + v_-(t,x,\sigma)),$$

the (v_-, v_+)-system (14) is equivalent to the fluid type system,

$$\partial_t \rho(t,x,\sigma) + \partial_x(\rho(t,x,\sigma)u(t,x,\sigma)) = 0,$$

$$\partial_t u(t,x,\sigma) + \partial_x\left(\frac{1}{2}u^2(t,x,\sigma) + \frac{1}{8}\rho^2(t,x,\sigma)\right) + \partial_x \int_0^1 \rho(t,x,a)da = 0. \quad (16)$$

3.5 Hamiltonian Formulation of Fluid Representations

In fact fluid representations of the V–D–B equation, such as "mono-kinetic" model (13), the Zakharov–Benney model (12) and the waterbag model (16), inherit of the Hamiltonian structure of the V–D–B equation (6), with the energy $\mathscr{E}$ specified by the fluid representation that we choose for the distribution function f. To this purpose we introduce the matrix $\mathscr{J}$ defined by

$$\mathscr{J} = -\begin{pmatrix} 0 & 1 \\ 1 & 0 \end{pmatrix}.$$

For the "mono-kinetic" model (13), setting $\mathfrak{m}(t,x) = (\rho(t,x), u(t,x))^T$, we obtain the Hamiltonian formulation

$$\partial_t \mathfrak{m} = \{\mathfrak{m}, \mathscr{E}\}_{\text{MoK}} := \mathscr{J}\,\partial_x \frac{\delta \mathscr{E}}{\delta \mathfrak{m}},$$

leading to the Poisson bracket structure [22, 23],

$$\{F(\mathfrak{m}), G(\mathfrak{m})\}_{\text{MoK}} = \int_{\mathbb{R}} dx \frac{\delta F}{\delta \mathfrak{m}} \mathscr{J}\,\partial_x \frac{\delta G}{\delta \mathfrak{m}}.$$

For the Zakharov–Benney model (12), setting $\mathfrak{m}(t,x,\sigma) = (\rho(t,x,\sigma), u(t,x,\sigma))^T$, we obtain the Hamiltonian formulation

$$\partial_t \mathfrak{m} = \{\mathfrak{m}, \mathscr{E}\}_{\text{ZB}} := \mathscr{J}\,\partial_x \frac{\delta \mathscr{E}}{\delta \mathfrak{m}},$$

leading to the Poisson bracket structure

$$\{F(\mathfrak{m}), G(\mathfrak{m})\}_{\text{ZB}} = \int_0^1 d\sigma \int_{\mathbb{R}} dx \frac{\delta F}{\delta \mathfrak{m}} \mathscr{J}\,\partial_x \frac{\delta G}{\delta \mathfrak{m}}.$$

For the waterbag model (16), setting $\mathfrak{m}(t,x,\sigma) = (\rho(t,x,\sigma), u(t,x,\sigma))^T = (v_+ - v_-,\ [v_+ + v_-]/2)^T$, we obtain the Hamiltonian formulation

$$\partial_t \mathfrak{m} = \{\mathfrak{m}, \mathscr{E}\}_{\text{WB}} := \mathscr{J}\,\partial_x \frac{\delta \mathscr{E}}{\delta \mathfrak{m}},$$

leading to the Poisson bracket structure

$$\{F(\mathfrak{m}), G(\mathfrak{m})\}_{\mathrm{WB}} = \int_0^1 d\sigma \int_{\mathbb{R}} dx \frac{\delta F}{\delta \mathfrak{m}} \mathscr{J} \partial_x \frac{\delta G}{\delta \mathfrak{m}}.$$

3.6 *Analytic Well-Posedness for Solutions in Fluid Representations*

Following Safonov [26], one introduces the Hardy type spaces H^s of x-analytic vector-valued functions $U(t,z,\sigma) = (\rho(t,x+iy,\sigma), u(t,x+iy,\sigma))$ defined on the tube $\{(x+iy,\sigma) \in \mathbb{C}^d \times M \, : \, |y_i| < s, \, i = 1,\ldots,d\}$ of the d-dimensional complex plane $\mathbb{C}^d$ with norm,

$$\|U\|_s^2 = \sup_{0 \le |y| \le s, \, \sigma \in M} \left(\int_{\mathbb{R}^d} \left| \left(\mathrm{I} + (-\Delta_x)^{1/2}\right)^{\frac{d}{2}+1} U(x+iy,\sigma) \right|^2 dx \right),$$

and the Banach space X_{s0}^γ equiped with the norm,

$$\|U\|_{s_0}^\gamma = \sup_{0 \le s + \lambda t < s_0} (s_0 - s - \lambda t)^\gamma \|U(t)\|_s,$$

where $\gamma \ge 0$, $s_0 > 0$ and $\lambda > 0$. Eventually one denotes by $\mathscr{B}_{s_o}^\gamma(r)$ the ball of radius r in such space, i.e.

$$\mathscr{B}^\gamma s_o(r) = \left\{ U \in X_{s_0}^\gamma ; \sup_{0 \le s + \lambda t < s_0} (s_0 - s - \lambda t)^\gamma \|U(t)\|_s < r \right\}. \tag{17}$$

Observe that, for all the examples in a one-dimensional space variable, from the "general Benney equation" (cf. Sect. 3.2) to the "waterbag" (cf. Sect. 3.4), the Cauchy problem can be written in the form,

$$f(t,x,v) = \int_M \rho(t,x,a) \delta(v - u(t,x,\sigma)) d\sigma,$$

$$U(t,x,\sigma) = (\rho(t,x,\sigma), u(t,x,\sigma)), \quad U(t,\sigma) = U(0,\sigma) + \int_0^t \mathscr{F}(U)(\tau,\sigma) d\tau, \tag{18}$$

where $\mathscr{F}$ is an operator which satisfies the hypothesis of the Safonov version of the Cauchy–Kowalewski Theorem (namely Assumptions 1.1 in [26]). Indeed one has $\mathscr{F}(0) = 0$; For $r > 0$, the correspondence $U \mapsto \mathscr{F}(U)$ is a continuous mapping of $\{U \in \mathrm{H}^s \, : \, \|U\|_s^2 < r\}$ into $\mathrm{H}^{s'}$ with $0 < s' < s < s_0$; and for any $0 < s' < s < s_0$, and for all $U, \, V \in \mathrm{H}^s$, with $\|U\|_s < r$, $\|V\|_s < r$, we have

$$\|\mathscr{F}(U) - \mathscr{F}(V)\|_{s'} \leq \frac{C(r)}{s - s'}\|U - V\|_s.$$

Finally this leads to the following

Theorem 5. *For solutions given by the formula (18) there exists $\lambda > 0$ depending only the dimension d, and the constant parameters $s_0 > 0$, $r > 0$, and $\gamma \in (0, 1)$ such that for any initial data*

$$U(0, x, \sigma) = (\rho(0, x, \sigma), u(0, x, \sigma)) \in \mathscr{H}^{s_0}$$

with $\|U(0)\|_{s_0} < r$, one has on the time interval $(0, \frac{s_0}{\lambda})$ a solution $U(t, x, \sigma) \in \mathscr{H}^{s_0 - \lambda t}$, with $\|U(t)\|_{s_0} - \lambda t<r$ of the corresponding Cauchy problem.

Remark 3. In agreement with the representation formula (18), the Theorem 5 concerns (at variance with the Jabin–Nouri Theorem 4) solutions which are analytic with respect to x and t but which can exhibit singularities in the v variable (Dirac masses, sum of Dirac masses, step or Heaviside functions, etc ...).

4 Entropy and Local-in-Time Stability in Sobolev Spaces

4.1 Energy, Conserved Quantities and Entropies

The energy takes the form

$$\mathscr{E}(f) = \frac{1}{2}\int_{\mathbb{R}\times\mathbb{R}} |v|^2 f(t, x, v)dxdv + \frac{1}{2}\int_{\mathbb{R}} (\rho_f(t, x))^2 dx,$$

which is the basic conserved quantity of the V–D–B equation written in the Hamiltonian formalism according to the formula

$$\partial_t f + \left\{\frac{\delta\mathscr{E}}{\delta f}, f\right\} = 0. \tag{19}$$

Obviously the energy $\mathscr{E}(f)$ is not a convex function of f. However with the representation

$$f(t, x, v) = \int_M \rho(t, x, \sigma)\delta(v - u(t, x, \sigma))d\sigma,$$

this energy may become a convex functional of the variable $U(t, x, \sigma) = (\rho(t, x, \sigma), u(t, x, \sigma))$ solution of the system

$$\partial_t U + \partial_x F(U) = 0, \tag{20}$$

where the application $U \mapsto F(U)$ is a twice continuously Gâteaux-differentiable nonlinear unbounded operator in $\mathbb{L}^2(M, d\sigma) := L^2(M, d\sigma) \times L^2(M, d\sigma)$, with domain $\mathscr{D}(F) = \mathbb{L}^2 \cap \mathbb{L}^\infty(M, d\sigma)$.

More generally, the invariants $\eta(f) = \eta(U)$ of the dynamics given by (19) or (20) are characterized by the relation

$$0 = \frac{d}{dt}\int_{\mathbb{R}} \eta(U)dx = \int_{\mathbb{R}} dx\, D_U\eta(U)\partial_t U = \int_{\mathbb{R}} dx\, D_U\eta(U)D_UF(U)\partial_x U, \tag{21}$$

where the symbol D_U denotes the differential with respect to the variable U. In the classical theory of conservation laws, solutions of (21) are called "conserved quantities" and are associated to the notion of flux according to the formula $D_U\eta(U)D_UF(U) = D_UQ(U)$ which implies the relation

$$D_U^2\eta(U)D_UF(U) = (D_UF(U))^T D_U^2\eta(U),$$

i.e. the fact that $D_U^2\eta(U)$ is a symmetrizer for the conservation law and a positive definite symmetrizer when $u \mapsto \eta(U)$ is a convex function.

Extension of these considerations to the system (20) is the object of the next theorem,

Theorem 6. *Let us consider solutions $(t, x, \sigma) \mapsto U(t, x, \sigma)$ of the system,*

$$\partial_t U + \partial_x F(U) = 0,$$

where the application $U \mapsto F(U)$ is a twice Gâteaux-differentiable local operator (with respect to the variables (t, x) which can be considered as fixed parameters) in $\mathbb{L}^\infty(M, d\sigma)$. For a twice Gâteaux-differentiable function $U \mapsto \eta(U)$ defined on $H^s(\mathbb{R}_x; \mathbb{L}^\infty(M, d\sigma))$ with value in $\mathbb{R}$, the following assertions are equivalent:

i) $D_U\eta(U)D_UF(U) = D_UQ(U)$ is a flux.
ii) $\eta(U)$ is a conserved quantity.
iii) $D_U^2\eta(U)D_UF(U)$ is a symmetric (self-adjoint) operator.

Proof. ii) follows from i) by integration of the relation

$$\partial_t\eta(U) = -\partial_x Q(U).$$

If $\eta(U)$ is a conserved quantity, one has

$$0 = \frac{d}{dt}\int \eta(U(t,x))dx = -\int_{\mathbb{R}} D_U\eta(U)D_UF(U)\partial_x U. \tag{22}$$

Using Gâteaux-derivative of (22), we get for any vector-valued function V,

$$
\begin{aligned}
0 &= \frac{d}{ds}\left(\int_{\mathbb{R}} D_U\eta(U+sV)D_U(F(U+sV))\partial_x(U+sV)dx\right)_{|s=0}\\
&= \int_{\mathbb{R}} D_U^2\eta(U)[V, D_UF(U)\partial_xU]dx + \int_{\mathbb{R}} D_U\eta(U)D_U^2F(U)[V,\partial_xU]dx\\
&+ \int_{\mathbb{R}} D_U\eta(U)D_UF(U)\partial_xVdx,\\
&= \int_{\mathbb{R}}(\partial_xU)^T(D_UF(U))^TD_U^2\eta(U)Vdx + \int_{\mathbb{R}}\{D_U(D_U\eta(U)D_UF(U))\\
&-D_U^2\eta(U)D_UF(U)\}[V,\partial_xU]dx + \int_{\mathbb{R}} D_U\eta(U)D_UF(U)\partial_xVdx,\\
&= \int_{\mathbb{R}}(\partial_xU)^T(D_UF(U))^TD_U^2\eta(U)Vdx - \int_{\mathbb{R}}(\partial_xU)^TD_U^2\eta(U)D_UF(U)Vdx\\
&+ \int_{\mathbb{R}}\partial_x(D_U\eta(U)D_UF(U))Vdx + \int_{\mathbb{R}} D_U\eta(U)D_UF(U)\partial_xVdx,\\
&= \int_{\mathbb{R}}(\partial_xU)^T(D_UF(U))^TD_U^2\eta(U)Vdx - \int_{\mathbb{R}}(\partial_xU)^TD_U^2\eta(U)D_UF(U)Vdx,
\end{aligned}
$$

which implies the Lax condition

$$D_U^2\eta(U)D_UF(U) = (D_UF(U))^TD_U^2\eta(U), \tag{23}$$

and shows that ii) implies iii). The proof of the assertion "iii) implies i)" is a direct adaptation of the same property for functions depending of a finite number of variables and is done as follows. Assume that $U \mapsto \mathscr{R}(U)$ is a linear operator in $\mathbb{L}^\infty(M, d\sigma)$, with $\mathscr{R}(U) = D_U\eta(U)D_UF(U)$, then define $Q(U)$ by the formula

$$Q(U) = \int_0^1 \mathscr{R}(sU)(U)ds,$$

and show with one integration by part and self-adjointness of $D_U\mathscr{R}(U)$, that one has for any vector-valued function V,

$$\frac{d}{d\tau}Q(U+\tau V)_{|\tau=0} = \mathscr{R}(U)(V),$$

which explicitly means that $U \mapsto \mathscr{R}(U)$ is the Gâteaux derivative of $U \mapsto Q(U)$. It remains to show that $D_U\mathscr{R}(U)$ is self-adjoint, which follows from the Lax condition (23), the obvious relation: for any vector-valued functions V and W one has

$$(D_U^2F(U)V)W = (D_U^2F(U)W)V,$$

and the equation

$$D_U\mathscr{R}(U) = D_U(D_U\eta(U)D_UF(U)) = D_U^2\eta(U)D_UF(U) + D_U\eta(U)D_U^2F(U). \quad \square$$

Remark 4. If we denote by the bracket $\{\cdot,\cdot\}_{\mathscr{S}}$, a generic Poisson bracket with $\mathscr{S} \in \{\text{MoK, ZB, WB}\}$ and whose corresponding definitions are established in Sect. 3.5, from the proof of Theorem 6 we get that

$$\frac{d}{dt}\int_{\mathbb{R}} \eta(U)\,dx = \{\eta(U), \mathscr{E}(U)\}_{\mathscr{S}} = 0,$$

which means that the invariant $\eta(U)$ is an involution.

4.2 Stability of Mono-Kinetic and Multi-Kinetic Solutions

In any space dimension d, the energy

$$\mathscr{E}(f) = \frac{1}{2}\int_{\mathbb{R}_x^d} (|u(t,x)|^2 + \rho(t,x))\rho(t,x)dx,$$

of the isentropic system

$$\begin{aligned} &\partial_t\rho + \nabla_x\cdot(\rho u) = 0, \\ &\partial_t(\rho u) + \nabla_x\cdot(\rho u\otimes u) + \nabla_x\left(\frac{\rho^2}{2}\right) = 0, \end{aligned} \tag{24}$$

(i.e. for a mono-kinetic solution $f(t,x,v) = \rho(t,x)\delta(v - u(t,x))$), is strictly convex near any constant state $U_0 = (\rho_0, u_0)$ with $\rho_0 > 0$. Following the classical theory of hyperbolic systems of conservation laws [9, 21], this implies the

Theorem 7. *The Cauchy problem for the system (24) and initial data of the form* $U_0+\tilde{U}_0(x)$ *with* $\tilde{U}_0(x) \in H^s(\mathbb{R}^d)$ *and* $s > d/2+1$, *has for a finite time* $(0 < t < T*)$, *a unique solution of the form* $U_0(t,x) + \tilde{U}(t,x)$ *with* $\tilde{U}(t,x) \in C(0,T;H^s(\mathbb{R}^d))$.

On the other hand in a one-dimensional space variable, the parameters of the multi-kinetic representation $U = ((\rho_1,u_1),(\rho_2,u_2),\ldots,(\rho_N,u_N))$ are also solutions of a system of $2N$ conservation laws,

$$\begin{aligned} &\partial_t\rho_n + \partial_x(\rho_n u_n) = 0\,, \\ &\partial_t u_n + \partial_x\Big(\frac{u_n^2}{2}\Big) + \partial_x\Big(\sum_{1\le\ell\le N}\rho_\ell\Big) = 0, \end{aligned}$$

with an energy,

$$\mathscr{E}(f) = \frac{1}{2}\int_{\mathbb{R}_x}\left(\sum_{1\le n\le N}\rho_n(t,x)|u_n(t,x)|^2 + \Big(\sum_{1\le n\le N}\rho_n(t,x)\Big)^2\right)dx, \tag{25}$$

which is not (as observed in details in [6]) always convex near a constant state.

For instance with $N = 2$, it is strictly convex near $(\rho_-, u_-, \rho_+, u_+) = (1, -a, 1, a)$ for $a^2 > 2$, and not convex otherwise. Therefore the Theorem 7 can be extended near constant states which ensure the convexity of $\mathscr{E}(U)$, while for perturbations near other initial states the Cauchy problem is ill-posed in any Sobolev and stability requires analyticity of the initial perturbation as in the Theorem 5.

5 Local-in-Time Stability of the One Bump Profile Solution

As observed in the Sect. 3.4, the evolution of a one bump profile can be described either with the velocity variables $v_\pm(t,x,\sigma)$ or with the fluid variables $U(t,x,\sigma) = (\rho(t,x,\sigma), u(t,x,\sigma))$ according to the equations,

$$\partial_t v_\pm + \partial_x\Big(\frac{v_\pm^2}{2}\Big) + \partial_x\int_0^1 (v_+(t,x,\sigma) - v_-(t,x,\sigma))d\sigma = 0\,, \tag{26}$$

or with

$$\begin{aligned}
&\rho(t,x,\sigma) = v_+(t,x,\sigma) - v_-(t,x,\sigma) \quad \text{and}\\
&u(t,x,\sigma) = (v_+(t,x,\sigma) + v_-(t,x,\sigma))/2,\\
&\partial_t\rho(t,x,\sigma) + \partial_x(\rho(t,x,\sigma)u(t,x,\sigma)) = 0,\\
&\partial_t u(t,x,\sigma) + \partial_x\left(\frac{1}{2}u^2(t,x,\sigma) + \frac{1}{8}\rho^2(t,x,\sigma)\right) + \partial_x\int_0^1 \rho(t,x,a)da = 0.
\end{aligned} \tag{27}$$

This system can be reformulated as

$$\partial_t U + \partial_x F(U) = 0, \tag{28}$$

with $U(\sigma) \mapsto F(U)(\sigma)$ a twice Gâteaux-differentiable operator in $\mathbb{L}^\infty(0,1)$. Moreover the energy of these solutions in the $v_\pm(t,x,\sigma)$ or in the $(\rho(t,x,\sigma), u(t,x,\sigma))$ representation is given by

$$
\begin{aligned}
\mathscr{E}(f)(t) &= \frac{1}{2}\left(\int_{\mathbb{R}\times\mathbb{R}} v^2 f(t,x,v)dxdv + \int_{\mathbb{R}} \left(\rho_f(t,x)\right)^2 dx\right)\\
&= \int_{\mathbb{R}_x}\left[\frac{1}{6}\int_0^1 (v_+^3(t,x,\sigma) - v_-^3(t,x,\sigma))d\sigma\right.\\
&\quad \left. + \frac{1}{2}\left(\int_0^1 (v_+(t,x,\sigma) - v_-(t,x,\sigma))d\sigma\right)^2\right] dx,\\
&= \frac{1}{2}\int_{\mathbb{R}}\int_0^1 \left(\rho(t,x,\sigma)u^2(t,x,\sigma) + \frac{1}{12}\rho^3(t,x,\sigma)\right) d\sigma dx\\
&\quad + \frac{1}{2}\int_{\mathbb{R}}\left(\int_0^1 \rho(t,x,\sigma)d\sigma\right)^2 dx.
\end{aligned}
\tag{29}
$$

Therefore, equipped with a convex entropy (29), the system (27) has for the Cauchy problem a unique solution according to the

Theorem 8. *For any set of initial data*

$$(\tilde{\rho}^0(x,\sigma), u^0(x,\sigma)) \in \mathbb{L}^\infty(0,1;H^3(\mathbb{R}_x))$$

there exists a finite time T such that the Cauchy problem

$$
\begin{aligned}
&\partial_t\rho(t,x,\sigma) + \partial_x(\rho(t,x,\sigma)u(t,x,\sigma)) = 0,\\
&\partial_t u(t,x,\sigma) + \partial_x\left(\frac{1}{2}u^2(t,x,\sigma) + \frac{1}{8}\rho^2(t,x,\sigma)\right) + \partial_x\int_0^1 \rho(t,x,a)da = 0,\\
&\rho(0,x,\sigma) = C + \tilde{\rho}^0(x,\sigma), \quad u(0,x,\sigma) = u^0(x,\sigma) \quad \text{with} \quad \rho(0,x,\sigma) \geq c > 0,
\end{aligned}
\tag{30}
$$

has a unique solution $(\rho(t,x,\sigma) = C + \tilde{\rho}(t,x,\sigma) \geq c > 0,\ u(t,x,\sigma))$ with

$$(\tilde{\rho}(t,x,\sigma), u(t,x,\sigma)) \in L^\infty(0,T;\mathbb{L}^\infty(0,1;H^3(\mathbb{R}_x))).$$

Corollary 2. *Let us consider an initial profile*

$$f^0(x,v) = \mathrm{Y}(v)f_+^0(x,v) + (1-\mathrm{Y}(v))f_-^0(x,v),$$

with $f_+^0(v)$ decreasing and $f_-^0(v)$ increasing such that the functions $(\rho^0(x,\sigma), u^0(x,\sigma))$ given by the formulas,

$$f_\pm(v_\pm(\sigma)) = \sigma, \quad \rho^0(x,\sigma) = v_+(x,\sigma) - v_-(x,\sigma),$$

and

$$u^0(x,\sigma) = (v_+(x,\sigma) + v_-(x,\sigma))/2,$$

satisfy the hypothesis of the Theorem 8. Then the distribution function,

$$f(t,x,v) = \int_0^1 (\Upsilon(v_+(t,x,\sigma) - v) - \Upsilon(v_-(t,x,\sigma) - v))d\sigma, \tag{31}$$

is, for $0 < t < T$ *(with* T *given by the Theorem 8), a solution of the problem*

$$\partial_t f + v\partial_x f - \partial_x \left(\int_{\mathbb{R}} f(t,x,v)dv \right) \partial_v f = 0,$$

with initial data $f(0,x,v) = f^0(x,v)$.

Proof. With $U(t,x,\sigma) = (\rho(t,x,\sigma), u(t,x,\sigma))$, the Cauchy problem (30) can be written according to the formulas

$$\partial_t U + \partial_x F(U) = 0,$$
$$F(U) = \begin{cases} \rho(t,x,\sigma)u(t,x,\sigma), \\ \frac{1}{2}u^2(t,x,\sigma) + \frac{1}{8}\rho^2(t,x,\sigma) + \int_0^1 \rho(t,x,a)da\,. \end{cases} \tag{32}$$

This system has the convex energy

$$\eta(U) = \frac{1}{2}\int_{\mathbb{R}}\int_0^1 (\rho(t,x,\sigma)u^2(t,x,\sigma) + \frac{1}{12}\rho^3(t,x,\sigma))d\sigma dx$$
$$+ \frac{1}{2}\int_{\mathbb{R}} \left(\int_0^1 \rho(t,x,\sigma)d\sigma \right)^2 dx.$$

Therefore (cf. Theorem 6) $D_U^2\eta(U)$ is a well defined positive symmetrizer and local-in-time estimates can be obtained by considering the expression

$$\partial_x^3(\partial_t U + D_U F(U)\partial_x U = 0),$$

on which we can apply the symmetrizer integral operator $D_U^2\eta(U)$ from the left, and proceeding as in the classical case (cf. [1, 9, 21]) to complete the proof of the Theorem 8.

Now, given the functions $v_\pm(t,x,\sigma)$, the fluid variables $(\rho(t,x,\sigma), u(t,x,\sigma))$ are recovered by the formulas

$$v_\pm(t,x,\sigma) = u(t,x,\sigma) \pm \frac{1}{2}\rho(t,x,\sigma).$$

Obviously one has $v_+(t,x,\sigma) \geq v_-(t,x,\sigma)$ and the equations

$$\partial_t(\partial_\sigma v_\pm) + v_\pm \partial_x(\partial_\sigma v_\pm) + (\partial_\sigma v_\pm)\partial_x v_x = 0,$$

imply that the monotonicity of the functions $\sigma \mapsto v_{\pm}(t,x,\sigma)$ are preserved by the dynamics [1, 4, 8]. Eventually one uses the formula (31) to reconstruct the solution of the V–D–B equation. □

Remark 5. Since the equivalent system (26) has also a convex entropy it can be also diagonalized and this leads to a formulation in term of generalized Riemann invariants giving also a stability result but with well adapted regularity of the initial data with respect to the variable σ. This was done in [4] by N. Besse following a method introduced by Teshukov [27].

Remark 6. With no surprise there is a good agreement between the results for the linearized problem and the nonlinear one. The above theorems concerning the "waterbag" and the "mono-kinetic" equations are the counterpart of the stability results near a one bump profile which is the object of the Theorem 2. In particular for the "mono-kinetic" equation this profile is a Dirac mass. Then direct computation shows that the dispersion relation has no complex value solution [1]. The same remarks is also valid for "multi-kinetic model". For instance with $N = 2$, the dispersion relation for the linearized model is

$$1 = \int_{\mathbb{R}} \frac{G'(v)}{v - \omega} dv = \frac{1}{(a-\omega)^2} + \frac{1}{(a+\omega)^2},$$

which has real solutions if $a^2 > 2$, and complex solutions, i.e. unstable modes, otherwise.

6 Wigner or Semi-Classical Limit of Solutions of the Nonlinear Schrödinger Equation

6.1 Formal Derivations

The connection of the Schrödinger equation with a self-consistent potential to the Vlasov equation, via the Wigner limit, is at present very well documented. For instance for the Schrödinger–Poisson equation, i.e. with a self-consistent defocusing Coulomb type potential of the form

$$\int_{\mathbb{R}^d} \frac{1}{|x-y|^{(d-2)}} |\psi(y)|^2 dy$$

not only (with well adapted initial data) the problem is uniformly well-posed but convergence of the Wigner transform, on an arbitrary large time is proven (cf. for instance [20] or [12]).

For the present discussion one starts with a family $\{\psi_\hbar(t,x,\sigma)\}_{\sigma\in M}$, solution of the following quadratic nonlinear Schrödinger equation

$$i\hbar\partial_t\psi_\hbar(t,x,\sigma) = -\frac{\hbar^2}{2}\Delta_x\psi_\hbar(t,x,\sigma) + \left(\int_M |\psi_\hbar(t,x,\sigma)|^2 d\sigma\right)\psi_\hbar(t,x,\sigma).$$

Given the potential

$$\mathscr{V}_\hbar(t,x) = \int_M |\psi_\hbar(t,x,\sigma)|^2 d\sigma,$$

the time-dependent equation

$$i\hbar\partial_t\theta_\hbar(t,x,\sigma) = -\frac{\hbar^2}{2}\Delta_x\theta_\hbar(t,x,\sigma) + \mathscr{V}_\hbar(t,x)\theta_\hbar(t,x,\sigma),$$

defines by the formula

$$\{\theta_\hbar(t,x,\sigma)\}_{\sigma\in M} = U_\hbar(t)\{\theta_0(t,x,\sigma)\}_{\sigma\in M},$$

a family of unitary operators $U_\hbar(t)$ acting in the space $L^\infty(M\,;L^2(0,T;L^2(\mathbb{R}^d_x)))$. Then one introduces the projection operator

$$K_\hbar(t,x,y) = \int_M \psi_\hbar(t,x,\sigma)\otimes\overline{\psi_\hbar(t,y,\sigma)}d\sigma,$$

with energy

$$\mathscr{E}_\hbar(K_\hbar) = \mathrm{Trace}\left(-\frac{\hbar^2}{2}\Delta_x K_\hbar + \mathscr{V}_\hbar K_\hbar\right) < \infty.$$

The operator $K_\hbar$ is a solution of the Von Neumann–Heisenberg equation,

$$\frac{d}{dt}K_\hbar = -\frac{1}{i\hbar}[H_\hbar, K_\hbar] = -\frac{1}{i\hbar}\left[\frac{\delta\mathscr{E}_\hbar}{\delta K_\hbar}(K_\hbar), K_\hbar\right],$$

with

$$H_\hbar = \left(-\frac{\hbar^2}{2}\Delta_x + \mathscr{V}_\hbar\right).$$

Eventually for the Wigner transform of the Von Neumann–Heisenberg equation, which involves the Weyl symbol $W_\hbar(t,x,v)$ defined by the Wigner transform of $K_\hbar$,

$$W_\hbar(t,x,v) = \frac{1}{(2\pi)^d}\int_{\mathbb{R}^d} e^{-iy\cdot v}K_\hbar\Big(t, x+\frac{\hbar}{2}y, x-\frac{\hbar}{2}y\Big)dy,$$

one has "at the formal level" (i.e. assuming all sufficient conditions to pass to the limit) the following convergences as $\hbar \to 0$ (Wigner or semi-classical limit):

$$W_\hbar(t,x,v) \longrightarrow W(t,x,v),$$

$$\mathcal{E}_\hbar(K_\hbar) \longrightarrow \frac{1}{2}\int_{\mathbb{R}^d}\left(\int_{\mathbb{R}^d}|v|^2 W(t,x,v)dv + \left(\int_{\mathbb{R}^d} W(t,x,v)dv\right)^2\right)dx,$$

$$\partial_t W + v\cdot\nabla_x W - \nabla_x\left(\int_{\mathbb{R}^d} W(t,x,w)dw\right)\cdot\nabla_v W = 0.$$

In order to consider "mixed states" as in [20] and connect with Zakharov–Grenier formula (10), we now assume that the functions $\psi_\hbar(t,x,\sigma)$ can be written as

$$\psi_\hbar(t,x,\sigma) = a_\hbar(t,x,\sigma)e^{i\frac{S_\hbar(t,x,\sigma)}{\hbar}},$$

with $a_\hbar$ and $S_\hbar$ "uniformly regular" with respect to $\hbar$, then for the Wigner transform one has:

$$\begin{aligned}
&\lim_{\hbar\to 0} W_\hbar(K_\hbar(t,x,y))\\
&\quad= \lim_{\hbar\to 0}\int_M d\sigma\frac{1}{(2\pi)^d}\int_{\mathbb{R}^d} e^{iv\cdot y} a_\hbar(t,x+\frac{\hbar}{2}y,\sigma)e^{i\frac{S_\hbar(t,x+\frac{\hbar}{2}y,\sigma)}{\hbar}}\\
&\quad\quad\times\overline{a_\hbar(t,x-\frac{\hbar}{2}y,\sigma)e^{-i\frac{S_\hbar(t,x-\frac{\hbar}{2}y,\sigma)}{\hbar}}}dy\\
&\quad= \int_M |a(t,x,\sigma)|^2\delta(v-\nabla_x S(t,x,\sigma))d\sigma = \int_M \rho(t,x,\sigma)\delta(v-u(t,x,\sigma))d\sigma.
\end{aligned}$$

Taking "formally" the limit $\hbar \to 0$, one obtains with $\rho = \lim_{\hbar\to 0} a_\hbar \overline{a}_\hbar$ and $u = \lim_{\hbar\to 0}\nabla_x S_\hbar$,

$$\begin{aligned}
&\partial_t\rho(t,x,\sigma) + \nabla_x\cdot(\rho(t,x,\sigma)u(t,x,\sigma)) = 0,\\
&\partial_t u(t,x,\sigma) + u(t,x,\sigma)\cdot\nabla_x u(t,x,\sigma) + \nabla_x\int_M \rho(t,x,a)da = 0,
\end{aligned}\tag{33}$$

which is the Benney or V–D–B equation in the Zakharov–Grenier representation. Moreover, on the other hand, with

$$\psi_\hbar(t,x,\sigma) = a_\hbar(t,x,\sigma)e^{i\frac{S_\hbar(t,x,\sigma)}{\hbar}}, \quad \text{and} \quad w_\hbar(t,x,\sigma) = \nabla_x S_\hbar(t,x,\sigma),$$

the equation

$$i\hbar\partial_t\psi_\hbar(t,x,\sigma) = -\frac{\hbar^2}{2}\Delta_x\psi_\hbar(t,x,\sigma) + \left(\int_M |\psi_\hbar(t,x,\sigma)|^2 d\sigma\right)\psi_\hbar(t,x,\sigma),$$

is equivalent to the system

$$\partial_t a_\hbar(t,x,\sigma)+w_\hbar(t,x,\sigma)\cdot\nabla_x a_\hbar(t,x,\sigma)+\frac{1}{2}a_\hbar(t,x,\sigma)\nabla_x\cdot w_\hbar(t,x,\sigma) = \frac{i\hbar}{2}\Delta_x a_\hbar(t,x,\sigma)\,,$$

$$\partial_t w_\hbar(t,x,\sigma)+w_\hbar(t,x,\sigma)\cdot\nabla_x w_\hbar(t,x,\sigma)+\nabla_x\int_M a_\hbar(t,x,\sigma)\bar{a}_\hbar(t,x,\sigma)d\sigma=0. \tag{34}$$

Remark 7. The above representation is a variant both of the Madelung transform (where the amplitude $a_\hbar$ is taken real) and of the WKB method which is a Taylor expansion. As a consequence $a_\hbar(t,x,\sigma)$ does not remain real for $t\neq 0$ and x real, while $w_\hbar(t,x,\sigma)=\nabla_x S_\hbar(t,x,\sigma)$ remains real for x real. This representation already appeared in [7] and [13] . It was used by Grenier [15, 16] for the validation of the semiclassical limit. Here we apply it both in the analytical and the Sobolev setting (cf. Theorem 9 and Theorem 10) to validate the formal convergence by proving convenient uniform a priori estimates. With no surprise these estimates are in full agreement with the well-posedness or ill-posedness results given above for the V–D–B equation.

6.2 Convergence Proof for Analytic Initial Data

With the notations introduced in the Sect. 3.6, the counterpart of the Theorem 5 turns out to be the

Theorem 9. *There exists $\lambda>0$, depending only on the dimension d, and the constant parameters $s_0>0$, $r>0$ and $\gamma\in(0,1)$ (and in particular independent of $\hbar$) such that for any*

$$(a_\hbar(0,x,\sigma),w_\hbar(0,x,\sigma)=\nabla_x S_\hbar(0,x,\sigma))\in\mathcal{H}^s, \tag{35}$$

with $||(a_\hbar(0),w_\hbar(0))||_{s_0}<r$, there exists on the time interval $(0,\frac{s_0}{\lambda})$ a solution

$$(a_\hbar(t,x,\sigma),w_\hbar(t,x,\sigma)=\nabla_x S_\hbar(t,x,\sigma))\in\mathcal{H}^{s_0-\lambda t},$$

with $\|(a_\hbar(t),w_\hbar(t)\|_{s_0-\lambda t}<r$, of the problem (34) with initial data (35). Moreover these solutions are uniformly bounded (with respect to $\hbar$) in $\mathcal{H}^{s_0-\lambda t}$, so that they converges, as $\hbar\to 0$, to the solutions of Zakharov–Benney equation (33) given by the Theorem 5.

Proof. First observe that any function $x\mapsto f(x)$ defined for $x\in\mathbb{R}^d$, and which is the restriction of an analytic function $f(x+iy)$, defined for $|y_i|<s$, $(i=1,\dots,d)$, can be represented (with the Payley–Wiener Theorem [24]) by the formula

$$f(x) = \int_{\mathbb{R}^d} e^{ix\cdot\xi}\hat{f}(\xi)d\xi,$$

with $\hat{f}(\xi)$ decaying exponentially for $|\xi| \to \infty$. Hence the complex conjugate

$$\overline{f(x)} = \int_{\mathbb{R}^d} e^{-ix\cdot\xi}\overline{\hat{f}(\xi)}d\xi,$$

is also the Fourier transform of a function with the same exponential decay and therefore can be extended as analytic function in the complex domain according to the formula:

$$f^*(x+iy) = \int_{\mathbb{R}^d} e^{-i(x+iy)\cdot\xi}\overline{\hat{f}(\xi)}d\xi. \tag{36}$$

Of course such extension does not coincide with the complex conjugate of $f(x+iy)$ for $y \neq 0$, but it belongs to the same class (in term of regularity) of analytical functions. With this remark in mind, one introduces the analytic extension $(a_\hbar(t, x+iy, \sigma), a^*_\hbar(t, x+iy, \sigma), w_\hbar(t, x+iy, \sigma))$ of $(a_\hbar(t, x, \sigma), \bar{a}_\hbar(t, x, \sigma), w_\hbar(t, x, \sigma))$ and write the system (34) in the equivalent form: for $z = x+iy \in \mathbb{C}^d$,

$$\begin{aligned}
&\partial_t w_\hbar(t,z,\sigma) + w_\hbar(t,z,\sigma)\cdot\nabla_z w_\hbar(t,z,\sigma) + \nabla_z \int_M a_\hbar(t,z,\sigma)a^*_\hbar(t,z,\sigma)d\sigma = 0,\\
&\partial_t a_\hbar(t,z,\sigma) + w_\hbar(t,z,\sigma)\cdot\nabla_z a_\hbar(t,z,\sigma) + \frac{1}{2}a_\hbar(t,z,\sigma)\nabla_z\cdot w_\hbar(t,z,\sigma)\\
&\quad = \frac{i\hbar}{2}\Delta_z a_\hbar(t,z,\sigma),\\
&\partial_t a^*_\hbar(t,z,\sigma) + w_\hbar(t,z,\sigma)\cdot\nabla_z a^*_\hbar(t,z,\sigma) + \frac{1}{2}a^*_\hbar(t,z,\sigma)\nabla_z\cdot w_\hbar(t,z,\sigma)\\
&\quad = \frac{-i\hbar}{2}\Delta_z a^*_\hbar(t,z,\sigma).
\end{aligned} \tag{37}$$

With the notations

$$U = \begin{pmatrix} w_\hbar(t,z,\sigma)\\ a_\hbar(t,z,\sigma)\\ a^*_\hbar(t,z,\sigma)\end{pmatrix}, \qquad L_\hbar = \begin{pmatrix} 0\\ \frac{i\hbar}{2}\Delta_z\\ -\frac{i\hbar}{2}\Delta_z\end{pmatrix},$$

and

$$F(U) = -\begin{pmatrix} w_\hbar(t,z,\sigma)\cdot\nabla_z w_\hbar(t,z,\sigma) + \nabla_z \int_M a_\hbar(t,z,\sigma)a^*_\hbar(t,z,\sigma)d\sigma\\ w_\hbar(t,z,\sigma)\cdot\nabla_z a_\hbar(t,z,\sigma) + \frac{1}{2}a_\hbar(t,z,\sigma)\nabla_z\cdot w_\hbar(t,z,\sigma)\\ w_\hbar(t,z,\sigma)\cdot\nabla_z a^*_\hbar(t,z,\sigma) + \frac{1}{2}a^*_\hbar(t,z,\sigma)\nabla_z\cdot w_\hbar(t,z,\sigma)\end{pmatrix}, \tag{38}$$

the system becomes

$$\partial_t U = F(U) + L_\hbar(U),$$

which, using a Duhamel's formula, implies

$$U(t) = e^{tL_\hbar}(U_0) + \int_0^t e^{(t-\tau)L_\hbar} F(U(\tau))d\tau = \Phi(U). \tag{39}$$

Since F is bilinear (and linear with respect to the first-order derivative) and since $e^{tL_\hbar}$ is, for any $\hbar$, a unitary operator in H^s, then for any $0 < s' < s < s_0$, one has

$$\|e^{(t-\tau)L_\hbar} F(U(\tau)) - e^{(t-\tau)L_\hbar} F(V(\tau))\|_{s'} \leq \frac{C}{s-s'} \|U(\tau)-V(\tau)\|_s (\|U(\tau)\|_s + \|V(\tau)\|_s),$$

with C depending only on the dimension and in particular not on $\hbar$. Next one uses the $\|\cdot\|_{s_0}^{\gamma}$-norm of the Banach space $X_{s_0}^{\gamma}$, i.e.

$$\|U\|_{s_0}^{\gamma} = \sup_{0 \leq s+\lambda t < s_0} (s_0 - s - \lambda t)^{\gamma} \|U(t)\|_s,$$

and following Safonov [26] shows that

$$\|\Phi(U) - \Phi(V)\|_{s_0}^{\gamma} \leq \frac{2^{\gamma+1} C(\|U(\tau)\|_s + \|V(\tau)\|_s)}{\gamma \lambda} \|U(\tau) - V(\tau)\|_{s_0}^{\gamma}.$$

Hence for λ chosen large enough (with respect to C and r), Φ preserves the ball $\mathscr{B}_{s_0}^0(r)$, and is a contraction in $X_{s_0}^{\gamma} \cap \mathscr{B}_{s_0}^0(r)$. The rest of the proof follows. □

Remark 8. The above proof is simpler than the forerunner result of Gérard [11], and it also provides an extension to mixed states as considered by Lions and Paul [20]. This is essentially due to the fact that [26] version of the Cauchy–Kowalewski Theorem is very well adapted to the problem in the representation proposed by Grenier [15, 16].

6.3 *Convergence Proof for Finite Time with Finite Sobolev Regularity*

With no surprise, the stability result for the limit equation should find their counterpart at the level of the convergence. In particular the local-in-time stability in Sobolev spaces has been proven for the mono-kinetic solutions. Mono-kinetic solutions correspond to the ($\hbar \to 0$)-limit of the Wigner transform of a "pure WKB-state", i.e.

$$W_\hbar\left(a_\hbar(t,x)e^{i\frac{S_\hbar(t,x)}{\hbar}}\otimes a_\hbar(t,y)^* e^{-i\frac{S_\hbar(t,y)}{\hbar}}\right)\underset{\hbar\to 0}{\longrightarrow}\rho(t,x)\delta(v-u(t,x)).$$

The validity of such convergence comes from standard uniform estimates due to Grenier [15, 16]. For comparison with the rest of the present contribution, this result is recalled below.

Theorem 10 (Grenier [16]). *Let $s > d/2+2$, let $S^0(x)\in H^s(\mathbb{R}^d)$ and $a^0(x,\hbar)$ be a sequence of functions uniformly bounded in $H^s(\mathbb{R}^d)$. Then there exist $T>0$, and solutions*

$$\psi_\hbar(t,x) = a_\hbar(t,x)e^{i\frac{S_\hbar(t,x)}{\hbar}},$$

to the Cauchy problem

$$i\hbar\partial_t\psi_\hbar = -\frac{\hbar^2}{2}\Delta_x\psi_\hbar + |\psi_\hbar|^2\psi_\hbar,\qquad \psi_\hbar(0,x) = a^0(x,\hbar)e^{i\frac{S^0_\hbar(x)}{\hbar}}.$$

Moreover, $a_\hbar(t,x)$ and $S_\hbar(t,x)$ are bounded in $L^\infty(0,T;H^s(\mathrm{IR}^d))$ uniformly in $\hbar$.

To prove this theorem, Grenier starts from the following system

$$\begin{aligned}
&\partial_t w_\hbar(t,x) + w_\hbar(t,x)\nabla_x w_\hbar(t,x) + \nabla_x(\alpha_\hbar^2(t,x)+\beta_\hbar^2(t,x)) = 0,\\
&\partial_t\alpha_\hbar(t,x) + w_\hbar(t,x)\cdot\nabla_x\alpha_\hbar(t,x) + \frac{1}{2}\alpha_\hbar(t,x)\nabla_x\cdot w_\hbar(t,x) = -\frac{\hbar}{2}\Delta_x\beta_\hbar(t,x),\\
&\partial_t\beta_\hbar(t,x) + w_\hbar(t,x)\cdot\nabla_x\beta_\hbar(t,x) + \frac{1}{2}\beta_\hbar(t,x)\nabla_x\cdot w_\hbar(t,x) = \frac{\hbar}{2}\Delta_x\alpha_\hbar(t,x),
\end{aligned}\tag{40}$$

which corresponds to the restriction to the real domain of (37) and where $\alpha_\hbar(t,x)$ and $\beta_\hbar(t,x)$ denote respectively the real and imaginary part of $a_\hbar(t,x)$. He observes that this system can be symmetrized by a strictly positive matrix S and this will lead to the standard a priori estimates of hyperbolic systems of conservation laws [9]. In fact the existence of such strictly positive symmetrizer is a consequence of the fact that the mass (i.e. with $\rho_\hbar(t,x) = \alpha_\hbar(t,x)^2+\beta_\hbar(t,x)^2)$ and the energy of the system,

$$\frac{1}{2}\int_{\mathbb{R}^d}\left(w_\hbar(t,x)^2+\rho_\hbar(t,x)\right)\rho_\hbar(t,x)dx,$$

are a strictly convex invariants.

Hence in $\mathbb{R}^d$ with $s > d/2+2$, there exists an $\hbar$-independent function $(a_0(0,\cdot),w_0(0,\cdot))$ such that

$$\|(a_\hbar(0,\cdot),w_\hbar(0,\cdot))\|_{H^s}\longrightarrow\|(a_0(0,\cdot),w_0(0,\cdot))\|_{H^s},\qquad \text{as }\hbar\to 0,$$

and the system (40) has for $t < T(\|(a_\hbar(0,\cdot), w_\hbar(0,\cdot))\|_{H^s})$ a unique solution satisfying the estimate

$$\|(a_\hbar(t,\cdot), w_\hbar(t,\cdot))\|_{H^s} \leq C(\|(a_\hbar(0,\cdot), w_\hbar(0,\cdot))\|_{H^s}).$$

7 Conclusion

The fact that in the V–D–B equation, the operator,

$$f \mapsto E_f = -\partial_x \int f(t,x,v)dv,$$

is local in x and of degree 1 has in the present contribution the following consequences. The well-posedness of the Cauchy problem depends drastically on the initial data and this is related to the convexity of the energy (in a convenient class of solutions).

With such locality the notion of invariants in the sense of Hamiltonian systems, and the notion of conserved quantities for conservation laws do coincide.

The V–D–B equation appears also as the semi-classical or Wigner limit (with $\hbar \to 0$) of solutions of the nonlinear self-consistent Schrödinger equation. Such limit can formally be described and proven in the general case, i.e. for mixed WKB-states initial data only when the initial data are analytic. Otherwise that would be in contradiction with the cases where the limit Cauchy problem is not well-posed. On the other hand it is only for pure WKB-states initial data that the limit (which will be a mono-kinetic solution) is proven with finite Sobolev type regularity.

The above observations remain true in a one-dimensional space variable where not only the nonlinear Schrödinger equation but also its generalization as system of coupled equations (for mixed states) are integrable (cf. Zakharov [28]). This confers to the V–D–B equation a status of quasi-integrable equation with an infinity of invariant quantities, limit of the corresponding invariants at the level of the Schrödinger equations. The above properties being in some sense algebraic, the proof of convergence with analyticity hypothesis seems well adapted to such considerations even if convergence proofs have been obtained (as in the d-dimensional case) for a genuine scalar equation (not a system) either as above by the theorem of Grenier, or in the spirit of scattering theory by Jin, Levermore and McLaughlin [18].

References

1. Bardos, C., Besse, N.: The Cauchy problem for the Vlasov–Dirac–Benney equation and related issues in fluid mechanics and semi-classical limits. Kin. Relat. Models **6**(4), 893–917 (2013)
2. Bardos, C., Nouri, A.: A Vlasov equation with Dirac potential used in fusion plasmas. J. Math. Phys. **53**(11), 115621–115637 (2012)

3. Benney, D.J.: Some properties of long nonlinear waves. Stud. Appl. Math. **52**, 45–50 (1973)
4. Besse, N.: On the waterbag continuum. Arch. Ration. Mech. Anal. **199**(2), 453–491 (2011)
5. Berk, H., Nielsen, C., Robert, K.: Phase space hydrodynamics of equivalent non linear systems: experimental and computational observations. Phys. Fluids **13**(4), 980–985 (1967)
6. Besse, N., Berthelin, F., Brenier, Y., Bertrand, P.: The multi-water-bag equations for collisionless kinetic modeling. Kin. Relat. Models **2**(1), 39–90 (2009)
7. Chazarain, J.: Spectre d'un hamiltonien quantique et mécanique classique. Commun. Partial Differ. Equ. **5**(6), 595–644 (1980)
8. Crandall, M.G., Tartar, L.: Some relations between nonexpansive and order preserving mappings. Proc. Am. Math. Soc. **78**, 385–390 (1980)
9. Dafermos, C.: Hyperbolic Conservation Laws in Continuum Physics. Springer, New York (2000)
10. Degond, P.: Spectral theory of the linearized Vlasov-Poisson equation. Trans. Am. Math. Soc. **294**(2), 435–453 (1986)
11. Gérard, P.: Remarques sur l'analyse semi-classique de l'équation de Schrödinger non linéaire. Séminaire sur les Equations aux Dérivées Partielles 1992–1993, Exp. No. XIII, 13 pp. Ecole Polytech., Palaiseau (1993)
12. Gérard, P., Markowich, P., Mauser, N., Poupaud, F.: Homogenization limits and Wigner transforms. Commun. Pure Appl. Math. **50**(4), 323–379 (1997)
13. Graffi, S., Martinez, A., Pulvirenti, M.: Mean-field approximation of quantum systems and classical limit. Math. Models Methods Appl. Sci. **13**(1), 59–73 (2003)
14. Grenier, E.: Oscillations in quasineutral plasmas. Commun. Partial Differ. Equ. **21**(3–4), 363–394 (1996)
15. Grenier, E.: Limite semi-classique de l'équation de Schrödinger non linéaire en temps petit. C. R. Acad. Sci. Paris Sér. I Math. **320**(6), 691–694 (1995)
16. Grenier, E.: Semiclassical limit of the nonlinear Schrödinger equation in small time. Proc. Am. Math. Soc. **126**(2), 523–530 (1998)
17. Jabin, P.E., Nouri, A.: Analytic solutions to a strongly nonlinear Vlasov. C. R. Math. Acad. Sci. Paris **349**(9–10), 541–546 (2011)
18. Jin, S., Levermore, C.D., McLaughlin, D.W.: The semiclassical limit of the defocusing NLS hierarchy. Commun. Pure Appl. Math. **52**(5), 613–654 (1999)
19. Kruskal, M.D.: Hydromagnetics and the theory of plasma in a strong magnetic field, and the energy principles for equilibrium and for stability. 1960 La théorie des gaz neutres et ionisés (Grenoble, 1959) pp. 251–274. Hermann, Paris; Wiley, New York
20. Lions, P.-L., Paul, T.: Sur les mesures de Wigner. Rev. Mat. Iberoamericana **9**(3), 553–618 (1993)
21. Majda, A.: Compressible fluid flow and systems of conservation laws in several space variables. Applied Mathematical Sciences, vol. 53. Springer, New York (1984)
22. Morrison, P.J.: Hamiltonian and action principle formulations of plasma physics. Phys. Plasmas **12**, 058102 (2005)
23. Morrison, P.J.: Hamiltonian description of the ideal fluid. Rev. Mod. Phys. **70**(2), 467–521 (1998)
24. Paley, R.C., Wiener, N.: Fourier transforms in the complex plane, vol. 19. AMS, USA (1934)
25. Penrose, O.: Electronic instabilities of a non uniform plasma. Phys. of Fluids **3**(2), 258–265 (1960)
26. Safonov, M.: The abstract Cauchy-Kovalevskaya theorem in a weighted Banach space. Commun. Pure Appl. Math. **48**(6), 629–637 (1995)
27. Teshukov, V.M.: On hyperbolicity of long-wave equations. Soviet Math. Dokl. **32**, 469–473 (1985)
28. Zakharov, V.E.: Benney equations and quasiclassical approximation in the inverse problem method. Funktsional. Anal. i Prilozhen. (Russian) **14**(2), 15–24 (1980)

Analysis of Enhanced Diffusion in Taylor Dispersion via a Model Problem

Margaret Beck, Osman Chaudhary, and C. Eugene Wayne

Dedicated to Walter Craig, with admiration and affection, on his 60th birthday.

Abstract We consider a simple model of the evolution of the concentration of a tracer, subject to a background shear flow by a fluid with viscosity $\nu \ll 1$ in an infinite channel. Taylor observed in the 1950s that, in such a setting, the tracer diffuses at a rate proportional to $1/\nu$, rather than the expected rate proportional to ν. We provide a mathematical explanation for this enhanced diffusion using a combination of Fourier analysis and center manifold theory. More precisely, we show that, while the high modes of the concentration decay exponentially, the low modes decay algebraically, but at an enhanced rate. Moreover, the behavior of the low modes is governed by finite-dimensional dynamics on an appropriate center manifold, which corresponds exactly to diffusion by a fluid with viscosity proportional to $1/\nu$.

1 Introduction

Taylor diffusion (or Taylor dispersion) describes an enhanced diffusion resulting from the shear in the background flow. First studied by Taylor in the 1950s [9, 10] many further authors have proposed refinements or extensions of this theory [1, 3, 7, 8]. In the present paper we show how in a simplified model of Taylor diffusion we can use center manifold theory to simply and rigorously predict the long-time behavior of the concentration of the tracer particle to any desired degree of accuracy. We note that in terms of prior work on this problem our approach is closest to that of Mercer and Roberts, [8], who also use a formal center manifold to approximate the Taylor dispersion problem. However, they construct their center-manifold in Fourier space, an approach which is difficult to make rigorous because there is no spectral gap between the center directions and the stable directions.

M. Beck • O. Chaudhary • C. Eugene Wayne (✉)
Department of Mathematics and Statistics, Boston University, Boston, MA, USA
e-mail: mabeck@math.bu.edu; oachaudh@bu.edu; cew@math.bu.edu

P. Guyenne et al. (eds.), *Hamiltonian Partial Differential Equations and Applications*, Fields Institute Communications 75,
DOI 10.1007/978-1-4939-2950-4_2

By introducing scaling variables we show that a spectral gap is created which allows us to rigorously apply existing center-manifold theorems to analyze the asymptotic behavior of the problem. We believe that a similar approach will also apply to the full Taylor diffusion problem and plan to consider that case in future work.

We consider the simplest situation in which Taylor dispersion is expected to occur, namely a channel with uniform cross-section and a shearing flow:

$$\begin{aligned} \partial_t u &= \nu \Delta u - A(1+\chi(y))\partial_x u\,, \quad -\infty < x < \infty\,, \; -\pi < y < \pi \\ u &= u(x,y,t), \qquad u_y(x,\pm 1,t) = 0. \end{aligned} \tag{1}$$

We assume that A is constant and that the background shear flow has been normalized so that $\int_{-\pi}^{\pi} \chi(y)dy = 0$. Thus, the mean velocity of the background flow is A and we can transform to a moving frame of reference $\tilde{x} = x - At$. In this new frame of reference (and dropping the tilde's to avoid cluttering the notation), we have

$$\partial_t u = \nu \Delta u - A\chi \partial_x u.$$

We begin with a formal calculation that will be justified in an appropriate sense in subsequent sections and that provides some intuition about the expected behavior of (1). Given the geometry of this situation it makes sense to expand u in terms of its y-Fourier series. Therefore, we write

$$u(x,y,t) = \sum_n \hat{u}_n(x,t)e^{iny}\,, \text{ and } \chi(y) = \sum_n \hat{\chi}_n e^{iny},$$

where

$$\hat{u}_n(x,t) = \frac{1}{2\pi}\int_{-\pi}^{\pi} e^{-iny}u(x,y,t)dy, \qquad \hat{\chi}_n = \frac{1}{2\pi}\int_{-\pi}^{\pi} e^{-iny}\chi(y)dy,$$

and we find (considering separately the case $n = 0$ and $n \neq 0$)

$$\begin{aligned} n = 0;\; \partial_t \hat{u}_0 &= \nu \partial_x^2 \hat{u}_0 - A\widehat{(\chi u_x)}_0 \\ n \neq 0;\; \partial_t \hat{u}_n &= \nu(\partial_x^2 - n^2)\hat{u}_n - A\widehat{(\chi u_x)}_n, \end{aligned} \tag{2}$$

where

$$\widehat{(\chi u_x)}_n = \sum_m \hat{\chi}_m (\hat{u}_{n-m})_x.$$

We now introduce scaling variables. These variables have often been used to analyze the asymptotic behavior of parabolic partial differential equations and they have the additional advantage that they frequently make it possible to apply invariant manifold theorems to these problems [5]. We expect that the modes with $n \neq 0$ will decay faster than those with $n = 0$, so we make a different scaling—of course we have to verify that the behavior of the solutions is consistent with this scaling. Let

$$\hat{u}_0(x,t) = \frac{1}{\sqrt{1+t}} w_0 \left(\frac{x}{\sqrt{1+t}}, \log(1+t) \right) \tag{3}$$

$$\hat{u}_n(x,t) = \frac{1}{(1+t)} w_n \left(\frac{x}{\sqrt{1+t}}, \log(1+t) \right) , \quad n \neq 0 . \tag{4}$$

In Sect. 3, below, we show that $\hat{u}_n(x,t)$ is basically an x derivative of a Gaussian, which generates the extra $t^{-1/2}$ decay. Proceeding, note that if we consider the advection term, the contribution to the $n = 0$ equation is of the form

$$\sum_{m \neq 0} \hat{\chi}_m (\hat{u}_{-m})_x ,$$

where we have no contribution from the term with $m = 0$ since $\hat{\chi}_m = 0$ because χ has zero average. The scaling in (3) and (4) was chosen so that the terms in this sum will have the same prefactor in t as all the remaining terms in the equation for w_0. More precisely, consider the various terms in the equation for $\hat{u}_0$. We find

$$\partial_t \hat{u}_0 = -\frac{1}{2} \frac{1}{(1+t)^{3/2}} w_0 - \frac{1}{2} \frac{1}{(1+t)^{3/2}} \xi \partial_\xi w_0 + \frac{1}{(1+t)^{3/2}} \partial_\tau w_0 ,$$

where we have introduced the new independent variables

$$\xi = \frac{x}{\sqrt{1+t}} , \quad \tau = \log(1+t) .$$

Likewise, we have

$$\partial_x^2 \hat{u}_0 = \frac{1}{(1+t)^{3/2}} \partial_\xi^2 w_0, \qquad \sum_{m \neq 0} \hat{\chi}_m (\hat{u}_{-m})_x = \frac{1}{(1+t)^{3/2}} \sum_{m \neq 0} \hat{\chi}_m (w_{-m})_\xi .$$

Thus, the equation for w_0 becomes

$$\partial_\tau w_0 = \mathscr{L} w_0 - A \sum_{m \neq 0} \hat{\chi}_m (w_{-m})_\xi , \tag{5}$$

where

$$\mathscr{L} w = \nu \partial_\xi^2 w + \frac{1}{2} \partial_\xi (\xi w) .$$

Remark 1. The spectrum of $\mathscr{L}$ can be explicitly computed. See Sect. 2 for more details.

Repeating the calculation above for the evolution of the terms $\hat{u}_n$ with $n \neq 0$, one finds that the terms are *not* any longer of the same order in t. Consider first the advective term which now has the form

$$\sum_m \hat{\chi}_m (\hat{u}_{n-m})_x = \frac{1}{1+t}\hat{\chi}_n \partial_\xi w_0 + \frac{1}{(1+t)^{3/2}} \sum_{m\neq n} \hat{\chi}_m \partial_\xi w_{n-m} \,.$$

Working out the form of the remaining terms in (2), one finds

$$\left(\nu n^2 w_n + A\hat{\chi}_n \partial_\xi w_0\right) = e^{-\tau}\left((\mathscr{L}+1/2)w_n - \partial_\tau w_n\right) - Ae^{-\tau/2}\left(\sum_{m\neq n} \hat{\chi}_m \partial_\xi w_{n-m}\right) . \tag{6}$$

The terms on the right hand side of this expression should go to zero exponentially fast so that in the limit $\tau \to \infty$, w_n satisfies the simple algebraic equation

$$\nu n^2 w_n + A\hat{\chi}_n \partial_\xi w_0 = 0 \,. \tag{7}$$

Remark 2. This is reminiscent of various geometric singular perturbation arguments and we will expand more upon this point in Sect. 2.

If we are interested in the long time behavior of the system, we can conclude from (7) that

$$w_n = -\frac{A\hat{\chi}_n}{\nu n^2}\partial_\xi w_0 \,.$$

If we now insert this expression into (5) we find that

$$\partial_\tau w_0 = \mathscr{L} w_0 + \frac{A^2}{\nu}\sum_{m\neq 0}\frac{1}{m^2}\hat{\chi}_m \hat{\chi}_{-m} \partial_\xi^2 w_0 \,.$$

Since χ is real, $\hat{\chi}_{-m} = \overline{\hat{\chi}_m}$, and so the last term in the preceding equation can be rewritten as

$$\left(\frac{A^2}{\nu}\sum_{m\neq 0}\frac{|\hat{\chi}_m|^2}{m^2}\right)\partial_\xi^2 w_0,$$

which implies that the equation for w_0 becomes

$$\partial_\tau w_0 = \left(\nu + \frac{D_T}{\nu}\right)\partial_\xi^2 w_0 + \frac{1}{2}\partial_\xi(\xi w_0) .$$

This is just the heat equation written in terms of scaling variables, *but* we see that the diffusion constant ν has been replaced by the new diffusion constant $\nu + D_T/\nu$ where the Taylor correction is

$$D_T = A^2 \sum_{m\neq 0} \frac{|\hat{\chi}_m|^2}{m^2} .$$

That is to say, if we "undo" the change of variables, (3), and rewrite this equation in terms of the original variable $\hat{u}_0(x,t)$, we find

$$\partial_t \hat{u}_0 = (\nu + D_T/\nu)\partial_x^2 \hat{u}_0 .$$

Thus, we see that $\hat{u}_0$ (which gives the average, cross-channel concentration of the tracer particle) evolves diffusively, but with a greatly enhanced diffusion coefficient.

Remark 3. One can also check that the Taylor correction D_T to the diffusion rate computed above is the same as that given by the more traditional approaches cited earlier.

In order to justify the above formal calculation, we need to analyze the system (5) and (6), which we rewrite here:

$$\partial_\tau w_0 = \mathscr{L} w_0 - A \sum_{m\neq 0} \hat{\chi}_m (w_{-m})_\xi$$

$$\partial_\tau w_n = (\mathscr{L} + 1/2) w_n - A e^{\tau/2}\left(\sum_{m\neq n} \hat{\chi}_m \partial_\xi w_{n-m}\right) - e^\tau \left(\nu n^2 w_n + A\hat{\chi}_n \partial_\xi w_0\right) .$$

In particular, we would need to show that there is a center manifold given approximately by $\{w_n = -(A\hat{\chi}_n/(\nu n^2))\partial_\xi w_0\}$.

Rather than studying this full model of Taylor diffusion, we focus in this paper on the simplified model

$$\begin{aligned}\partial_\tau w &= \mathscr{L} w - \partial_\xi v \\ \partial_\tau v &= (\mathscr{L} + 1/2)v - e^\tau(\nu v + \partial_\xi w),\end{aligned} \tag{8}$$

which corresponds to just the modes w_0 and w_1 (or more generally, to w_0 and w_n, where n is the first integer for which $\hat{\chi}_n \neq 0$). The reader should note that this is not meant to be a physical model, but instead it is meant to be an analysis problem which reflects the core mathematical difficulties of analyzing the full Taylor Dispersion

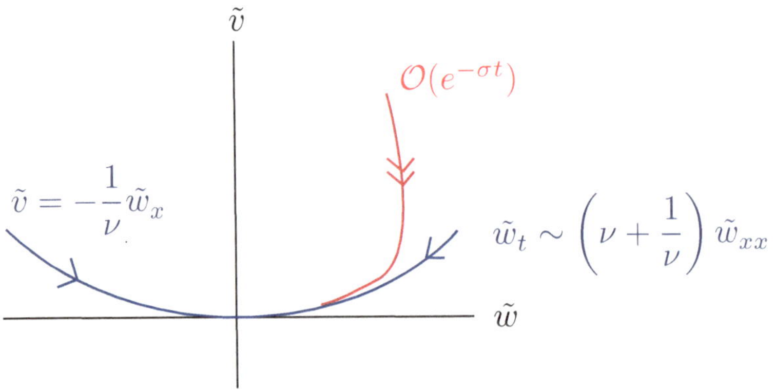

Fig. 1 Illustration of the invariant manifold one would expect for system (9), based upon the formal calculations

problem. Proceeding, note that the term proportional to $e^{\tau/2}$ has disappeared since, if only w_0 and w_1 are non-zero, that sum reduces to $\hat{\chi}_0 \partial_\xi w_1$, and $\hat{\chi}_0 = 0$. (We have also rescaled the variables so that the coefficient $A\hat{\chi}_1 = A\hat{\chi}_{-1} = 1$.) Also note that (8), written back in terms of the original variables, which we denote by $\tilde{w}(x,t)$ and $\tilde{v}(x,t)$, is given by

$$\begin{aligned} \tilde{w}_t &= \nu \tilde{w}_{xx} - \tilde{v}_x \\ \tilde{v}_t &= \nu \tilde{v}_{xx} - \nu \tilde{v} - \tilde{w}_x. \end{aligned} \tag{9}$$

The classical picture of a center manifold would imply (see Fig. 1) that solutions exponentially approach the invariant manifold (say at some rate σ), with the dynamics on the center manifold given by the heat equation with the new Taylor diffusion coefficient. However, our analysis of system (9), below, will show that the Taylor diffusion is really only affecting the lowest Fourier modes for $\tilde{w}$. Thus, the high Fourier modes still decay like $e^{-\nu|k_0|^2 t}$ for all $|k| \geq |k_0|$. Although this rate can be made uniform for $|k|$ sufficiently large, it does not reflect the large diffusion coefficient of order $1/\nu$. If there really was a center manifold as suggested by the formal calculations, with dynamics on the manifold given by $\tilde{w}_t = (\nu + 1/\nu)\tilde{w}_{xx}$, then on that manifold all Fourier modes would decay like $e^{-(1/\nu)|k|^2 t}$. Note that this does not contradict the fact that Taylor diffusion seems to be observable in numerical and physical experiments (see, for example, the original experiments by Taylor [9, 10]). The reason is that, for high modes with $|k| > |k_0|$, the uniform exponential decay in Fourier space implies exponential in time decay in physical space, whereas low modes exhibit only algebraic temporal decay. Since algebraic decay, even relative to a large diffusion coefficient, is slower than exponential decay, the fact that the high modes do not experience Taylor dispersion does not prevent the overall decay from being enhanced by this phenomenon.

While we believe that the picture sketched above applies to the full Taylor diffusion problem, as previously mentioned in this paper we analyze instead (8). For this coupled system of two partial differential equations we will show that

- The long-time behavior of solutions can be computed to any degree of accuracy by the solution on a (finite-dimensional) invariant manifold.
- To leading order, the long-time behavior on this invariant manifold agrees with that given by a diffusion equation with the enhanced Taylor diffusion constant.
- The expressions for the invariant manifolds can be computed quite explicitly, but we are not able to show that these expressions converge as the dimension of the manifold goes to infinity. Indeed, we believe on the basis of the argument above, that (8) probably does not have an infinite dimensional invariant manifold.

Our analysis will proceed as follows. First, in Sect. 2, we'll analyze the dynamics on the center manifold to see how the coupling between w and v affects the enhanced diffusion associated with the Taylor dispersion phenomenon. Intuitively, this is the key point of our result, and the latter sections can be thought of as justification of this calculation. Next, in Sect. 3, we will obtain some a priori estimates on the solutions $\tilde{w}$ and $\tilde{v}$ in Fourier space and show that, to leading order, the long-time dynamics are determined by the low Fourier modes. Finally, in Sect. 4, we show that the dynamics of these low Fourier modes are governed only by the dynamics on the center manifold analyzed in Sect. 2, and thus, to leading order, the solutions exhibit the above-described enhanced diffusion.

Our main result can be summarized in the following theorem, which is an abbreviation of Theorem 2.

Theorem 1. *Given any $M > 0$, there exist integers $m, N > 0$ such that, for initial data $(\tilde{w}_0, \tilde{v}_0) \in L^2(m)$ (see Definition 1), there exists a $(2N + 3)$ dimensional system of ordinary differential equations possessing an $(N + 2)$ dimensional center manifold, such that the long-time asymptotics of solutions of (9), up to terms of $\mathcal{O}(t^{-M})$, is given by the restriction of solutions of this system of ODEs to its center manifold. Moreover, the dynamics on this center manifold correspond to enhanced diffusion proportional to $\nu + 1/\nu$.*

Remark 4. In the more precise version of Theorem 1 stated in Sect. 4, there are ν−dependent constants that appear in the error term that make it clear that in order to actually see Taylor Dispersion in the system, one has to wait at least a time $t > \mathcal{O}(\frac{|\log \nu|}{\nu})$.

2 Dynamics on the Center Manifold

In this section we focus on a center-manifold analysis of the model equation (8). Our analysis justifies the formal lowest order approximation $\nu v + \partial_x w = 0$ and shows that to this order the solutions behave as if w was the solution of a diffusion equation with "enhanced" diffusion coefficient $\nu_T = (\nu + \frac{1}{\nu})$. Furthermore, the center-manifold machinery allows one to systematically (and rigorously) compute corrections to these leading order asymptotics to any order in time.

Remark 5. As noted in the introduction, we do not expect that the model equation (8) (or the full Taylor dispersion equation) has an exact, infinite dimensional center-manifold. What we will actually prove is that, up to any inverse power of time, $\mathcal{O}(t^{-M})$, there is a finite dimensional system of ordinary differential equations that approximates the solution of the PDE (8) up to corrections of $\mathcal{O}(t^{-M})$ and that this finite dimensional systems of ODE's has a center-manifold with the properties described above.

Because we expect $v \approx -\frac{1}{\nu}\partial_\xi w$ - i.e. because we expect v to behave at least asymptotically as a derivative, we define a new dependent variable u as

$$v = \partial_\xi u\,. \tag{10}$$

Inserting into the $\partial_\tau v$ equation in (8), we get

$$\begin{aligned}\partial_\tau(\partial_\xi u) &= \partial_\tau v = (\mathcal{L} + 1/2)\, v - e^\tau \left(\nu v + \partial_\xi w\right)\\ &= (\mathcal{L} + 1/2)\,\partial_\xi u - e^\tau \left(\nu \partial_\xi u + \partial_\xi w\right)\\ &= \partial_\xi \mathcal{L} u - e^\tau \left(\nu \partial_\xi u + \partial_\xi w\right)\end{aligned}$$

where we have used the fact that $\partial_\xi \mathcal{L} u = \mathcal{L}\partial_\xi u + \frac{1}{2}\partial_\xi u$. After antidifferentiating the last line with respect to ξ, we get a system in terms of w and u:

$$\begin{aligned}\partial_\tau w &= \mathcal{L} w - \partial_\xi{}^2 u\\ \partial_\tau u &= \mathcal{L} u - e^\tau\,(\nu u + w)\,.\end{aligned} \tag{11}$$

Remark 6. Note that, if $u \in L^2(m)$, the change of variables (10) implies that $\int_{-\infty}^{\infty} v(\xi,t)d\xi = 0$. We believe that, via minor modifications, our results can be extended to the case when $\int_{-\infty}^{\infty} v(\xi,t)d\xi \neq 0$. We plan to discuss such modifications in a future work, when we study the full model (5) and (6).

Studies of Taylor dispersion generally focus on localized tracer distributions. For that reason, and also because of the spectral properties of the operators $\mathcal{L}$ which we discuss further below, it is convenient to work in weighted Hilbert spaces.

Definition 1. The Hilbert space $L^2(m)$ is defined as

$$L^2(m) = \left\{ f \in L^2(\mathbb{R}) \mid \|f\|_m^2 = \int (1+\xi^2)^m |f(\xi)|^2 d\xi < \infty \right\}.$$

Note that we require the solutions of the equation to lie in these weighted Hilbert spaces when expressed in terms of the *scaling* variables. If we revert to the original variables then it is appropriate to study them in the time-dependent norms obtained from these as follows:

$$
\begin{aligned}
\|w(\xi,\tau)\|^2_{L^2(m)} &= \int (1+\xi^2)^m |w(\xi,\tau)|^2 d\xi \\
&= e^{\tau/2} \int (1+\xi^2)^m |\tilde{w}(e^{\tau/2}\xi, e^{\tau}-1)|^2 d\xi \\
&= \int (1+e^{-\tau}x^2)^m |\tilde{w}(x, e^{\tau}-1)|^2 dx \\
&= \sum_{\ell=0}^{m} \frac{C(m,\ell)}{(1+t)^{\ell}} \int x^{2\ell} |\tilde{w}(x,t)|^2 dx \,.
\end{aligned}
$$

Thus, when we study solutions of our model equations in the "original" variables, as opposed to the scaling variables, we will also consider the weighted L^2 norms of the functions, but the different powers of x will be weighted by a corresponding (inverse) power of t to account for the relationship between space and time encapsulated in the definition of the scaling variables. These norms are discussed further in Sect. 3.

Since we expect $\nu u + w \approx 0$, we further rewrite (11) by adding and subtracting $\frac{1}{\nu}\partial_\xi^2 w$ from the first equation and $\frac{1}{\nu}\partial_\xi^2 u$ from the second finally obtaining

$$
\begin{aligned}
\partial_\tau w &= \mathscr{L}_T w - \frac{1}{\nu}\left(\partial_\xi{}^2 w + \nu \partial_\xi{}^2 u\right) \\
\partial_\tau u &= \mathscr{L}_T u - \frac{1}{\nu}\partial_\xi{}^2 u - e^{\tau}(\nu u + w)\,,
\end{aligned} \tag{12}
$$

where

$$
\mathscr{L}_T \phi = \left(\nu + \frac{1}{\nu}\right)\partial_{\xi^2}\phi + \frac{1}{2}\partial_\xi(\xi\phi)\,.
$$

Thus, $\mathscr{L}_T$ is just the diffusion operator, written in terms of scaling variables, but with the enhanced, Taylor diffusion rate, $\nu_T = \nu + 1/\nu$.

The operators $\mathscr{L}_T$ have been analyzed in [6]. In particular, their spectrum can be computed in the weighted Hilbert spaces $L^2(m)$ and one finds

$$
\sigma(\mathscr{L}_T) = \left\{\lambda \in \mathbb{C} \mid \Re(\lambda) \le \frac{1}{4} - \frac{m}{2}\right\} \cup \left\{-\frac{k}{2} \mid k \in \mathbb{N}\right\}\,.
$$

Furthermore, the eigenfunctions corresponding to the isolated eigenvalues $\lambda_k = -k/2$ are given by the Hermite functions

$$\phi_0(\xi) = \frac{1}{\sqrt{4\pi \nu_T}} e^{-\xi^2/(4\nu_T)}\ , \quad \text{and } \phi_k(\xi) = \partial_\xi^k \phi_0(\xi)$$

and the corresponding spectral projections are given by the Hermite polynomials

$$H_k(\xi) = \frac{2^k (\nu_T)^k}{k!} e^{\xi^2/(4\nu_T)} \partial_\xi^k e^{-\xi^2/(4\nu_T)}\ .$$

Remark 7. The expressions in [6] for ϕ_k and H_k are derived in the case when the diffusion coefficient is 1. The expressions given here follow easily by the change of variables $\xi \to \xi/\sqrt{\nu_T}$. More explicitly, for the classical Hermite functions $\tilde{\phi}_0(y) = \frac{1}{\sqrt{4\pi}} e^{-y^2/4}$, $\tilde{\phi}_k(y) = \partial_y^k \phi_0$, and $\tilde{H}_k(y) = \frac{2^k}{k!} e^{y^2/4} \partial_y^k e^{-y^2/4}$, one has the orthonormality relations $\int \tilde{H}_k(y) \tilde{\phi}_\ell(y) dy = \delta_{k,\ell}$. Changing variables to $y = \xi/\sqrt{\nu_T}$ leads to the formulas for the eigenfunctions and spectral projections for $\mathscr{L}_T$. Note further that with this definition, the Hilbert space adjoint of $\mathscr{L}_T$ satisfies ${\mathscr{L}_T}^\dagger H_k = -\frac{k}{2} H_k$.

Given the spectrum of $\mathscr{L}_T$ discussed above, we expect that the leading order part of the solution as t tends to infinity will be associated with the eigenspace corresponding to eigenvalues closest to zero. With this in mind, fix an integer N and assume that $m > N + 1/2$. This insures that the spectrum of $\mathscr{L}_T$ has at least $N + 1$ isolated eigenvalues on the Hilbert space $L^2(m)$ and that the essential spectrum lies strictly to the left of the half-plane $\{\lambda \in \mathbb{C} \mid \Re(\lambda) < -N/2\}$. Now define P_N to be the spectral projection onto the first $N + 1$ eigenmodes

$$P_N w = \sum_{k=0}^{N} \alpha_k(\tau) \phi_k(\xi),$$

where

$$\alpha_k(\tau) = \langle H_k, w(\tau) \rangle_{L^2}.$$

We will write the solutions of (12) as

$$\begin{aligned} w &= P_N w + w_s \\ u &= P_N u + u_s. \end{aligned} \tag{13}$$

Based on the spectral picture and our discussion above, we expect that w^s and u^s will decay faster than $P_N w$ and $P_N u$ (a fact which we demonstrate in Sect. 4) and hence, since we are interested in the leading order terms in the long time behavior, we focus our attention on $P_N w$ and $P_N u$.

We will show that for any N the equations for $P_N w$ and $P_N u$ have an attractive center manifold and that the motion on this manifold reproduces and refines the expected Taylor diffusion.

If we apply the projection operator P_N to both of the equations in (12), we obtain

$$\sum_{k=0}^{N} \dot{\alpha_k}\phi_k = \sum_{k=1}^{N} -\frac{k}{2}\alpha_k\phi_k - \frac{1}{\nu}\sum_{k=0}^{N-2}(\alpha_k + \nu\beta_k)\phi_{k+2}$$
$$\sum_{k=0}^{N} \dot{\beta_k}\phi_k = \sum_{k=0}^{N} -\frac{k}{2}\beta_k\phi_k - \frac{1}{\nu}\sum_{k=0}^{N-2}\beta_k\phi_{k+2} - e^{\tau}\left(\sum_{k=0}^{N}(\nu\beta_k + \alpha_k)\,\phi_k\right).$$

Shifting indices and matching coefficients gives us the following system of ODEs for the coefficients α_k and β_k:

$$\begin{aligned}
\dot{\alpha_0} &= 0\\
\dot{\alpha_1} &= -\frac{1}{2}\alpha_1\\
\dot{\alpha_k} &= -\frac{k}{2}\alpha_k - \left(\frac{1}{\nu}\alpha_{k-2} + \beta_{k-2}\right) \text{ for } 2 \le k \le N\\
\dot{\beta_0} &= -e^{\tau}\,(\nu\beta_0 + \alpha_0)\\
\dot{\beta_1} &= -\frac{1}{2}\beta_1 - e^{\tau}\,(\nu\beta_1 + \alpha_1)\\
\dot{\beta_k} &= -\frac{k}{2}\beta_k - \frac{1}{\nu}\beta_{k-2} - e^{\tau}\,(\nu\beta_k + \alpha_k) \text{ for } 2 \le k \le N.
\end{aligned} \tag{14}$$

Note that these equations contain *no* contributions from the "stable" modes w^s and u^s. Note further that, because of the form of the equations, those with even indices k decouple from those with k odd. Thus, we can analyze these two cases separately. We'll provide the details for the case of k even below - the equations with k odd behave in a very similar fashion.

Remark 8. Note that if we multiply all of the equations in (14) by $e^{-\tau}$ and set $e^{-\tau} = \epsilon$ (since we are interested in large times) we get equations that are formally of classical singularly perturbed form. (However, the small parameter ϵ is time dependent here.) Invariant manifold theory has been a powerful tool in the rigorous analysis of singularly perturbed problems and that analogy will guide our use of the center-manifold theory in what follows.

In order to make the invariant manifold more apparent we rewrite the even index equations by rescaling the time variable as

$$\tau = \log(1 + t). \tag{15}$$

In analogy with the above remark about singularly perturbed systems, we are essentially switching to a "fast" version of our system by making this change of time variable. Continuing, we introduce a new dependent variable

$$\eta = e^{-\tau} = \frac{1}{1+t}. \tag{16}$$

Then, if we denote $\frac{d}{dt}$ by a prime $'$, we have

$$\begin{aligned}
\alpha_0' &= 0 \\
\alpha_k' &= -\eta\left(\frac{k}{2}\alpha_k + \frac{1}{\nu}\alpha_{k-2} + \beta_{k-2}\right) \\
\beta_0' &= -(\nu\beta_0 + \alpha_0) \\
\beta_k' &= -(\nu\beta_k + \alpha_k) - \eta\left(\frac{k}{2}\beta_k + \frac{1}{\nu}\beta_{k-2}\right) \\
\eta' &= -\eta^2,
\end{aligned} \tag{17}$$

where the values $2 \leq k \leq N$ are even. Notice the linearization of this system at the fixed point $\alpha_k = \beta_k = \eta = 0$ has eigenvalues $\lambda^c = 0$, with an $[N/2] + 2$ dimensional eigenspace and $\lambda^s = -\nu$, with an $[N/2] + 1$ dimensional eigenspace (here $[M]$ refers to the greatest integer less than or equal to M). We proceed by diagonalizing the linear part of the system via

$$\begin{aligned}
a_k &= \alpha_k \\
b_k &= \frac{1}{\nu}\alpha_k + \beta_k
\end{aligned} \tag{18}$$

which transforms (17) into

$$\begin{aligned}
a_0' &= 0 \\
a_k' &= -\eta\left(\frac{k}{2}a_k + b_{k-2}\right) \\
b_0' &= -\nu b_0 \\
b_k' &= -\nu b_k - \eta\left(\frac{k}{2}b_k - \frac{1}{\nu^2}a_{k-2} + \frac{2}{\nu}b_{k-2}\right) \\
\eta' &= -\eta^2,
\end{aligned} \tag{19}$$

where again $2 \leq k \leq N$ are even.

The remainder of this section is devoted to the analysis of these equations and we prove two main results:

- We first show that, for any N, (19) has a center-manifold of the type described in the introduction, and we derive explicit expressions for the functions whose graphs give the manifold. (See Propositions 1 and 2.)

- We derive the asymptotic (in τ) behavior of solutions of these equations. (See Propositions 3 and 4, and Corollary 1.)

We begin by noting that the linearization of (19) at the fixed point $a_k = b_k = \eta = 0$ has eigenvalues $\lambda^c = 0$, with an $[N/2]+2$ dimensional eigenspace and $\lambda^s = -\nu$, with an $[N/2]+1$ dimensional eigenspace. Thus, from the classical center-manifold theorem (say, for example, the center-manifold theorem proven in [4]), we know that (at least in a neighborhood of this point), there will be an invariant $[N/2]+2$ dimensional center manifold. We also know that, in a neighborhood of the origin, the center-manifold can be written as the graph of a function with components

$$b_k = h_k(a_N, \ldots, a_0, \eta). \tag{20}$$

In addition, because of the "lower triangular" form of the equations (i.e. the fact that the equations for a_k' and b_k' depend only on a_ℓ and b_ℓ with $\ell \le k$), we find that we can express the manifold as

$$b_k = h_k(a_k, a_{k-2}, \ldots, a_0, \eta) .$$

We now show that we can find explicit expressions for the functions h_k successively, starting with h_0 and then progressing through h_2, h_4, etc. What's more, these expressions hold for all $a_k, a_{k-2}, \ldots, a_0, \eta$, i.e. without the restriction to a small neighborhood that is inherent in general center-manifold theorems like that of [4].

We start with the equations for a_0 and b_0 which are just

$$\begin{aligned} a_0' &= 0 \\ b_0' &= -\nu b_0 . \end{aligned}$$

From this we see immediately that we can choose the invariant manifold to be the graph of $h_0 \equiv 0$. However, note that this example also reminds us that the center manifold is not unique, since we could also choose the center manifold to be given by the graph of $\tilde{h}_0(a_0, \eta) = K_0 e^{-\nu/\eta} a_0$. This is consistent with the theorems on the existence of center manifolds, since both of these manifolds have the same Taylor expansion to any finite order about $a_0 = \eta = 0$. For simplicity, in what follows we will always use the first function - i.e. we will take $h_0 \equiv 0$.

Now consider the center manifold for a_2 and b_2. Since the equations for a_0, a_2, b_0, b_2, η decouple from all other a_k and b_k, we expect the center manifold to be given by the graph of a function $b_2 = h_2(a_2, a_0, \eta)$. In fact, as we show below, it has no dependence on a_2 - i.e. we can take $b_2 = h_2(a_0, \eta)$. In this case the equation for the invariance of the graph of this function takes the form

$$(D_{a_0} h_2) a_0' + (D_\eta h_2) \eta' = -\eta h_2 - \nu h_2 - \frac{2\eta}{\nu} h_0 + \frac{1}{\nu^2} \eta a_0 .$$

Inserting the equations for a_0' and η' and using the fact that $h_0 \equiv 0$, we find

$$-\eta^2(D_\eta h_2) = -\eta h_2 - \nu h_2 + \frac{1}{\nu^2}\eta a_0 \, .$$

We now show that h_2 is linear in a_0, so we write

$$h_2(a_0, \eta) = \phi_{2,0}(\eta) a_0 \, ,$$

and find

$$-\eta^2 \phi_{2,0}' = -(\eta + \nu)\phi_{2,0} + \frac{1}{\nu^2}\eta \, .$$

This equation is hard to solve in general due to the singular point at $\eta = 0$, but remarkably,

$$\phi_{2,0}(\eta) = \frac{\eta}{\nu^3}$$

is an exact solution (which goes to zero as $\eta \to 0$), so

$$h_2(a_0, \eta) = \frac{\eta a_0}{\nu^3}$$

is a function whose graph (together with that of $h_0 \equiv 0$) gives us the center manifold for the equations for a_0, a_2, b_0, b_2, η. Due to the singular point at $\eta = 0$, this may not be the only solution (just as in the case for h_0), but we are free to choose this special solution for h_2.

Next we consider the case of $h_4(a_4, a_2, a_0, \eta)$. Building on the examples above we show that

- h_4 is independent of a_4;
- h_4 is linear in a_2 and a_0.

If this is the case we can write

$$h_4(a_2, a_0, \eta) = \phi_{4,2}(\eta) a_2 + \phi_{4,0}(\eta) a_0 \, .$$

Inserting this form of the solution into the equation for the center-manifold, we find

$$\begin{aligned}\phi_{4,2}(\eta) a_2' + \phi_{4,0}(\eta) a_0' + (\phi_{4,2}'(\eta) a_2 + \phi_{4,0}'(\eta) a_0)\eta' \\ = -(\nu + 2\eta)(\phi_{4,2}(\eta) a_2 + \phi_{4,0}(\eta) a_0) + \frac{\eta a_2}{\nu^2} - \frac{2\eta^2 a_0}{\nu^4}\end{aligned}$$

where in the last term we have plugged in the expression for h_2. Inserting the equations for a_2' and η' and grouping the terms proportional to a_2 and a_0 we find two ODE's for the $\phi's$, namely

$$-\eta^2 \phi_{4,2}'(\eta) = -(\nu + \eta)\phi_{4,2}(\eta) + \frac{\eta}{\nu^2}$$

$$-\eta^2 \phi_{4,0}'(\eta) = -(\nu + 2\eta)\phi_{4,0}(\eta) - \frac{2\eta^2}{\nu^4}.$$

The first of these equations is the same as the equation for $\phi_{2,0}$ above so we have

$$\phi_{4,2}(\eta) = \frac{\eta}{\nu^3}.$$

The second equation is very similar and we find that it again has a simple, exact solution, namely

$$\phi_{4,0}(\eta) = -\frac{2\eta^2}{\nu^5}.$$

Thus, we also have an exact expression for the center-manifold in this case:

$$h_4(a_2, a_0, \eta) = \frac{\eta a_2}{\nu^3} - \frac{2\eta^2 a_0}{\nu^5}.$$

One can continue this procedure. For instance, for the function h_6, one obtains the formula

$$h_6(a_4, a_2, a_0, \eta) = \frac{\eta a_4}{\nu^3} - \frac{2\eta^2 a_2}{\nu^5} + \frac{5\eta^3 a_0}{\nu^7}.$$

This leads to the following

Proposition 1. *For any* $k = 0, 2, 4, \ldots$, *there exist constants* $\{\hat{H}(k, k-2\ell)\}$ *such that the graph of the function*

$$h_k(a_{k-2}, a_{k-4}, \ldots, a_0, \eta) = \sum_{\ell=1}^{k/2} \hat{H}(k, k-2\ell)\eta^\ell a_{k-2\ell} \tag{21}$$

gives the invariant manifold for b_k. *Furthermore, for any fixed* k, *the coefficients* $\{\hat{H}(k, k-2\ell)\}$ *can be explicitly determined, and the coefficients* $\hat{H}(k, p) \sim \mathscr{O}(\nu^{-(k-p)-1})$.

Proof. The proof proceeds inductively. Note that we have already verified the inductive hypothesis for $k = 0, 2, 4$. (We take the empty sum that occurs on the RHS of (21) when $k = 0$ to correspond to $h_0 \equiv 0$.) Assume that it holds for all even integers less than or equal to $k-2$. We now show that it holds for h_k.

Inserting our inductive hypothesis into the invariance equation we find

$$\sum_{\ell=1}^{k/2} \hat{H}(k, k-2\ell)\eta^{\ell} a'_{k-2\ell} + \sum_{\ell=1}^{k/2} \ell \hat{H}(k, k-2\ell)\eta^{\ell-1} a_{k-2\ell}\eta'$$
$$= -\frac{k}{2}\eta h_k - \nu h_k - \frac{2}{\nu}\eta h_{k-2} + \frac{1}{\nu^2}\eta a_{k-2}$$
$$= -\sum_{\ell=1}^{k/2} \frac{k}{2}\hat{H}(k, k-2\ell)\eta^{\ell+1} a_{k-2\ell} - \sum_{\ell=1}^{k/2} \nu \hat{H}(k, k-2\ell)\eta^{\ell} a_{k-2\ell}$$
$$-\frac{2}{\nu}\sum_{\ell=1}^{k/2-1} \hat{H}(k-2, k-2-2\ell)\eta^{\ell+1} a_{k-2-2\ell} + \frac{1}{\nu^2}\eta a_{k-2}. \tag{22}$$

Inserting the equations for $a'_{k-2\ell}$ and η' into the first line of (22), one finds

$$\sum_{\ell=1}^{k/2} \hat{H}(k, k-2\ell)\eta^{\ell} \left(-\left(\frac{k-2\ell}{2}\right) \eta a_{k-2\ell} - \eta h_{k-2\ell-2} \right)$$
$$-\sum_{\ell=1}^{k/2} \ell \hat{H}(k, k-2\ell)\eta^{\ell+1} a_{k-2\ell}$$
$$= -\sum_{\ell=1}^{k/2} \frac{k}{2}\hat{H}(k, k-2\ell)\eta^{\ell+1} a_{k-2\ell} - \sum_{\ell=1}^{k/2} \hat{H}(k, k-2\ell)\eta^{\ell+1} h_{k-2\ell-2}. \tag{23}$$

Note that the first sum in the last line of (23) cancels the first sum on the RHS of (22). Thus, we can rewrite (22) and (23) as

$$\sum_{\ell=1}^{k/2} \nu \hat{H}(k, k-2\ell)\eta^{\ell} a_{k-2\ell} = \frac{1}{\nu^2}\eta a_{k-2} - \frac{2}{\nu}\sum_{\ell=1}^{k/2-1} \hat{H}(k-2, k-2\ell-2)\eta^{\ell+1} a_{k-2-2\ell}$$
$$+\sum_{\ell=1}^{k/2} \hat{H}(k, k-2\ell)\eta^{\ell+1} h_{k-2\ell-2}. \tag{24}$$

We now rewrite the last sum in this expression by using the inductive form of $h_{k-2\ell-2}$,

$$h_{k-2\ell-2} = \sum_{m=1}^{k/2-(\ell+1)} \hat{H}(k-2(\ell+1), k-2(\ell+m+1))\eta^{m} a_{k-2(\ell+m+1)} \,.$$

Thus,

$$\sum_{\ell=1}^{k/2} \hat{H}(k, k-2\ell)\eta^{\ell+1} h_{k-2\ell-2}$$

$$= \sum_{\ell=1}^{k/2} \sum_{m=1}^{k/2-(\ell+1)} \hat{H}(k, k-2\ell)\hat{H}(k-2(\ell+1), k-2(\ell+m+1))$$

$$\times \eta^{\ell+m+1} a_{k-2(\ell+m+1)}$$

$$= \sum_{p=3}^{k/2} \sum_{\ell=1}^{p-2} \hat{H}(k, k-2\ell)\hat{H}(k-2(\ell+1), k-2p)\eta^p a_{k-2p},$$

where in the last term we set $p = \ell+m+1$ and interchanged the order of summation. If in the last sum in the first line of (24) we also change the summation variable to $p = \ell + 1$ we find that (24) can finally be rewritten as

$$\sum_{\ell=1}^{k/2} \nu \hat{H}(k, k-2\ell)\eta^{\ell} a_{k-2\ell} = \frac{1}{\nu^2}\eta a_{k-2} - \frac{2}{\nu}\sum_{p=2}^{k/2} \hat{H}(k-2, k-2p))\eta^p a_{k-2p}$$

$$+ \sum_{p=3}^{k/2} \sum_{\ell=1}^{p-2} \hat{H}(k, k-2\ell)\hat{H}(k-2(\ell+1), k-2p)\eta^p a_{k-2p}. \tag{25}$$

We solve (25) for $\hat{H}(k, k-2\ell)$, beginning with $\hat{H}(k, k-2)$. Since the only term on the RHS of (25) proportional to a_{k-2} is the first term, and we obtain $\hat{H}(k, k-2) = \frac{1}{\nu^3}$, consistent with the inductive hypothesis. Next consider $\hat{H}(k, k-4)$. In this case, we consider all terms in (25) proportional to a_{k-4}. The only one comes from the second term on the RHS of the equation and we have $\hat{H}(k, k-4) = -\frac{2}{\nu^2}\hat{H}(k-2, k-4)$. The inductive hypothesis implies that $\hat{H}(k-2, k-4) \sim \mathcal{O}(\nu^{-3})$, so we find $\hat{H}(k, k-4) \sim \mathcal{O}(\nu^{-5})$ as required by the inductive hypothesis. We now continue to solve for the coefficients $\hat{H}(k, k-2\ell)$, $\ell = 3, 4, \ldots$, noting that in each case, the terms on the RHS of the equation proportional to $a_{k-2\ell}$ have coefficients that have already been determined at prior stages of the inductive process and that they are all $\mathcal{O}(\nu^{-2\ell-1}) = \mathcal{O}(\nu^{-(k-p)-1})$. □

We now describe the entirely analogous results for the modes α_k and β_k with k odd. If we introduce new variables t and η as in (15) and (16), and diagonalize the linear part of the resulting equations using the change of variables (18), we find:

$$a_1' = -\frac{1}{2}\eta a_1$$

$$a_k' = -\eta\left(\frac{k}{2}a_k + b_{k-2}\right)$$

$$\begin{aligned} b_1' &= -\left(\nu + \frac{1}{2}\eta\right) b_1 \\ b_k' &= -\nu b_k - \eta \left(\frac{k}{2} b_k - \frac{1}{\nu^2} a_{k-2} + \frac{2}{\nu} b_{k-2} \right) \\ \eta' &= -\eta^2, \end{aligned} \tag{26}$$

where the values $3 \leq k \leq N$ are odd this time.

Proceeding as before, consider first the equations for a_1, b_1, and η which decouple from all the rest of the equations. Then by inspection we see that, just as for b_0, the graph of the function $h_1(a_1, \eta) \equiv 0$ is an invariant center manifold for these equations. We now include the equations for a_3 and b_3 and, building on the experience from the even case, look for an invariant manifold of the form

$$b_3 = h_3(a_1, \eta) = \phi_{3,1}(\eta) a_1 \ .$$

Inserting this into the equations, we see that in order for this graph to be invariant, $\phi_{3,1}$ must satisfy

$$\phi_{3,1} a_1' + a_1 \phi_{3,1}' \eta' = -(\nu + \frac{3}{2}\eta)\phi_{3,1} a_1 + \frac{\eta}{\nu^2} a_1 - \frac{2\eta}{\nu} h_1 \ .$$

From the fact that $h_1 \equiv 0$ and the equation for a_1', we see that this reduces to the ODE for $\phi_{3,1}$

$$-\eta^2 \phi_{3,1}' = -(\nu + \eta)\phi_{3,1} + \frac{\eta}{\nu^2} \ .$$

This is the same equation satisfied by $\phi_{2,0}$ and thus we find

$$h_3(a_1, \eta) = \frac{\eta a_1}{\nu^3} \ .$$

Proceeding now as in the even case, we establish the following proposition by induction.

Proposition 2. *For any $k = 1, 3, 5, \ldots$, there exist constants $\{\hat{H}^{odd}(k, k-2\ell)\}$ such that the graph of the function*

$$h_k(a_{k-2}, a_{k-4}, \ldots, a_1, \eta) = \sum_{\ell=1}^{\frac{k-1}{2}} \hat{H}^{odd}(k, k-2\ell)\eta^{\ell} a_{k-2\ell}$$

gives the equation for the invariant manifold for b_k. Furthermore, for any fixed k, the coefficients $\{\hat{H}^{odd}(k, k-2\ell)\}$ can be explicitly determined, and the coefficients $\hat{H}^{odd}(k, p) \sim \mathcal{O}(\nu^{-(k-p)-1})$.

We conclude this section by using our expressions for the center-manifold to derive the asymptotic behavior of the coefficient functions a_k and b_k (or equivalently α_k and β_k.)

Begin by noting that from the general theory of center-manifolds, any solution with initial conditions in a neighborhood of the invariant manifold will approach the manifold at a rate $\sim \mathcal{O}(e^{-\nu t}) = \mathcal{O}(e^{-\nu(e^\tau - 1)})$. Thus, we can determine the long time asymptotics of all solutions in this neighborhood by focusing on the behavior of solutions on the invariant manifold. Note that this means, for solutions with sufficiently small initial conditions, that after a time τ such that $\nu e^\tau \gg 1$, we will be very close to the center-manifold and the behavior of solutions on this manifold will determine the asymptotic behavior of solutions after this time. Reverting from our rescaled time τ to the original time t in the problem this means that solutions on the center-manifold will determine the behavior of solutions for times $t > \mathcal{O}(\frac{1}{\nu})$, which is the expected timescale for Taylor Dispersion to occur. At the moment, it appears our results only hold for solutions with small initial conditions. However, it turns out our formulas for the center manifolds (which are defined globally) are also *globally attracting* on the timescale $t > \mathcal{O}(\frac{|\log \nu|}{\nu})$. We provide details in the Appendix.

We proceed with our calculation of the asymptotics of the quantities a_k and b_k. As in the case of the calculation of the manifold we focus separately on the coefficients with even and odd indices. Starting with the coefficients with k even, note that we obviously have $\alpha_0 =$ constant, so we begin with $k = 2$.

Given

$$a_2' = -\eta(a_2 + b_0),$$

we can simplify this by noting that $b_0 = h_0 \equiv 0$ on the center-manifold. Finally, it's simpler to solve this differential equation by reverting from the t variables to $\tau = \log(1 + t)$; keeping Remark 8 about singularly perturbed systems in mind, notice we are essentially switching to the "slow" version of the system (which gives the dynamics on the center manifold). The equation then reduces to

$$\dot{a}_2 = -a_2 ,$$

from which we can immediately conclude that

$$a_2(\tau) \sim \mathcal{O}(e^{-\tau}) .$$

Next consider a_4, for which we have (again, rewriting things in terms of the temporal variable τ)

$$\dot{a}_4 = -2a_4 - b_2 = -2a_4 - \frac{e^{-\tau} a_0}{\nu^3} ,$$

where the last equality used the fact that $b_2 = h_2(a_0, \eta) = \frac{\eta a_0}{\nu^2}$ on the center-manifold. Solving this equation using the method of variation of constants, we find that

$$a_4(\tau) \sim \mathcal{O}(\frac{e^{-\tau}}{\nu^3}) \,.$$

As a last explicit example, consider the case of a_6 where we have

$$\dot{a}_6 = -3a_6 - b_4 = -3a_6 - \frac{e^{-\tau} a_2}{\nu^3} + \frac{2e^{-2\tau} a_0}{\nu^5} \,.$$

Finally, since a_0 is constant and $a_2(\tau) \sim \mathcal{O}(e^{-\tau})$, we see that the asymptotic behavior of a_6 is

$$a_6(\tau) \sim \mathcal{O}(\frac{e^{-2\tau}}{\nu^5}) \,.$$

We can generalize these results in the following

Proposition 3. *Suppose $k = 4, 6, \dots$ is an even, positive integer. On the center manifold of the system of equations* (19)*, the variables a_k have the following asymptotic behavior:*

$$|a_k(\tau)| \leq \begin{cases} \frac{C(N,k) e^{-\frac{k}{4}\tau}}{\nu^{k-1}} & : k = 0 \; mod \; 4 \\ \frac{C(N,k) e^{-\frac{k+2}{4}\tau}}{\nu^{k-1}} & : k = 2 \; mod \; 4 . \end{cases}$$

Note that once we have these formulas, the expressions for the center-manifold immediately imply the following.

Corollary 1. *Suppose $k = 4, 6, \dots$ is an even, positive integer. On the center manifold of the system of equations* (19)*, the variables b_k have the following asymptotic behavior:*

$$|b_k(\tau)| \leq \begin{cases} \frac{C(N,k) e^{-\frac{k+4}{4}\tau}}{\nu^{k+1}} & : k = 0 \; mod \; 4 \\ \frac{C(N,k) e^{-\frac{k+2}{4}\tau}}{\nu^{k+1}} & : k = 2 \; mod \; 4 . \end{cases}$$

Proof. The proof of Proposition 3 is a straightforward induction argument. Suppose that we have demonstrated that the estimates hold for $k = 4, 6, \dots, k_0$. We then show that it holds for $k_0 + 2$. The equation of motion for a_{k_0+2} is

$$\dot{a}_{k_0+2} = -\frac{k_0 + 2}{2} a_{k_0+2} - h_{k_0}(a_{k_0-2}, a_{k_0-4}, \dots, a_0, e^{-\tau}).$$

Inserting the formula for h_{k_0} from Proposition 1 and solving using Duhamel's formula, we obtain the bound

$$|a_{k_0+2}| \le \frac{C(N)}{\nu} \sum_{\ell=1}^{k_0/2} a_{k_0-2\ell} \frac{\eta^\ell}{\nu^{2\ell}}. \tag{27}$$

Consider the case $k_0 = 0 \bmod 4$. Then

$$k_0 - 2\ell = \begin{cases} 2 \bmod 4 & \text{if } \ell \text{ is odd} \\ 0 \bmod 4 & \text{if } \ell \text{ is even} \end{cases}$$

and correspondingly from the induction hypothesis,

$$|a_{k_0-2\ell}| \le \begin{cases} \frac{C(N)e^{-\frac{-k_0-2\ell+2}{4}\tau}}{\nu^{k-1}} & \text{if } \ell \text{ is odd} \\ \frac{C(N)e^{-\frac{-k_0-2\ell}{4}\tau}}{\nu^{k-1}} & \text{if } \ell \text{ is even.} \end{cases}$$

Inserting into (27), using the fact that $\eta = e^{-\tau}$, and splitting the sum into even and odd ℓ, we obtain

$$|a_{k_0+2}| \le \frac{C(N)}{\nu} \left\{ \sum_{\ell=1,\ell odd}^{k_0/2-1} \frac{e^{-\frac{(k_0-2\ell+2)\tau}{4}} e^{-\ell\tau}}{\nu^{k_0-2\ell-1}\nu^{2\ell}} + \sum_{\ell=2,\ell even}^{k_0/2-2} \frac{e^{-\frac{(k_0-2\ell)\tau}{4}} e^{-\ell\tau}}{\nu^{k_0-2\ell-1}\nu^{2\ell}} + \frac{a_0 e^{-\frac{k_0\tau}{2}}}{\nu^{k_0}} \right\}. \tag{28}$$

Notice we have to separate out the $\ell = k_0/2$ term because this corresponds to a_0, which is actually constant. We are interested in locating the slowest decaying terms. These terms will have, in the exponent, the least negative coefficients on τ. For $\ell \ge 1$ odd, the coefficients in the exponent are

$$-\frac{k_0 - 2\ell + 2}{4} - \ell = -\frac{k_0}{4} - \frac{1}{2} - \frac{\ell}{2} \tag{29}$$

which are least negative when $\ell = 1$. The corresponding coefficient in the exponent is $-\frac{k_0+4}{4}$, and so the slowest decaying term from the ℓ odd sum is $\mathcal{O}(e^{-\frac{k_0+4}{4}\tau})$. We determine the slowest decaying term in the ℓ even sum. For $\ell \ge 2$ even, the coefficients in the exponent are

$$-\frac{k_0 - 2\ell}{4} - \ell = -\frac{k_0}{4} - \frac{\ell}{2} \tag{30}$$

which are least negative when $\ell = 2$. The corresponding coefficient in the exponent is again $-\frac{k_0+4}{4}$, and so the slowest decaying term from the ℓ odd sum is again $\mathcal{O}(e^{-\frac{k_0+4}{4}\tau})$. Lastly, we determine the ν dependence of the constant. The largest power of ν in the denominator comes from $\ell = k_0/2$ and is $\frac{1}{\nu^{k_0+1}}$. Therefore we have

$$|a_{k_0+2}| \le \frac{C(N)}{\nu^{k_0+1}} e^{-\frac{k_0+4}{4}\tau}.$$

Recalling that we are in the case $k_0 = 0 \bmod 4$ (so that $k_0 + 2 = 2 \bmod 4$), we have verified the claim in this case. The case $k_0 = 2 \bmod 4$ follows similarly. Once Proposition 3 is established, a nearly identical calculation establishes Corollary 1. □

The coefficients a_k and b_k, with k odd, can be estimated in an entirely analogous fashion to obtain the following proposition.

Proposition 4. *Suppose $k = 1, 3, \dots$ is an odd, positive integer. On the center manifold of the system of equations (26), the variables a_k have the following asymptotic behavior:*

$$|a_k(\tau)| \le \begin{cases} \frac{C(N,k)e^{-\frac{k+1}{4}\tau}}{\nu^{k-1}} & : k = 1 \; mod \; 4 \\ \frac{C(N,k)e^{-\frac{k+3}{4}\tau}}{\nu^{k-1}} & : k = 3 \; mod \; 4. \end{cases} \tag{31}$$

If $k = 3, 5, \dots$ (recall that $b_1 \equiv 0$ on the center manifold), the corresponding coefficients b_k satisfy the estimates

$$|b_k(\tau)| \le \begin{cases} \frac{C(N,k)e^{-\frac{5+k}{4}\tau}}{\nu^{k+1}} & : k = 1 \; mod \; 4 \\ \frac{C(N,k)e^{-\frac{3+k}{4}\tau}}{\nu^{k+1}} & : k = 3 \; mod \; 4. \end{cases} \tag{32}$$

3 A Priori Estimates via the Fourier Transform

In order to show that the center manifold, discussed in the previous section, really does describe the leading order large-time behavior of solutions of (9), we need to make our discussion before Theorem 1 in the introduction more precise (which basically says Taylor Dispersion only happens for low wavenumbers). We'll have to undo the scaling variables, and switch to the Fourier side; this way we can precisely cut-off wavenumbers larger than, say $|k_0| \approx \frac{\nu}{2}$ and quantify how fast these "high" wavenumber terms decay. To do this in a way that is consistent with the analysis in Sect. 2, we need to introduce a new norm $|||\cdot|||$, which, when applied to functions on the Fourier side, is equivalent to the $L^2(m)$ norm applied to their real-space scaling variables counterparts.

The main result (see Theorem 2 in Sect. 4) depends on estimates of the solution in $L^2(m)$. With this in mind, we note that

$$\|w(\tau)\|_{L^2(m)} \le C(m)(t+1)^{1/4} \sqrt{\sum_{j=0}^{m} \left\| \frac{1}{(1+t)^{j/2}} \partial_k^j \hat{w}(\cdot, t) \right\|_{L^2}^2} =: |||\tilde{w}(t)|||,$$

and below we will bound each partial derivative of $\hat{w}(k,t)$. Note that $\|\cdot\|_{L^2(m)}$ and $|||\cdot|||$ are indeed equivalent norms, which follows from the fact that

$$\|\partial_k^j \hat{w}(\cdot,t)\|_{L^2}^2 \leq C\int (1+x^j)^2 |\tilde{w}(x,t)|^2 dx \leq C(t+1)^{j-1/2}\|w(\tau)\|_{L^2(j)}^2,$$

which in turn implies that $|||\tilde{w}(t)||| \leq C(m)\|w(\tau)\|_{L^2(m)}$.

Consider Eq. (9). Let $\hat{w} = \mathscr{F}\tilde{w}$ and $\hat{v} = \mathscr{F}\tilde{v}$, where $\mathscr{F}$ sends a function to its Fourier transform. We obtain

$$\frac{d}{dt}\begin{pmatrix}\hat{w}\\ \hat{v}\end{pmatrix} = A(k)\begin{pmatrix}\hat{w}\\ \hat{v}\end{pmatrix}, \qquad A(k) = \begin{pmatrix}-\nu k^2 & -ik\\ -ik & -\nu(k^2+1)\end{pmatrix}.$$

The solution to this equation is

$$\begin{pmatrix}\hat{w}(k,t)\\ \hat{v}(k,t)\end{pmatrix} = e^{A(k)t}\begin{pmatrix}\hat{w}_0(k)\\ \hat{v}_0(k)\end{pmatrix} \quad\Rightarrow\quad \begin{pmatrix}\tilde{w}(x,t)\\ \tilde{v}(x,t)\end{pmatrix} = \mathscr{F}^{-1}[e^{A(k)t}] * \begin{pmatrix}\tilde{w}_0(x)\\ \tilde{v}_0(x)\end{pmatrix}.$$

To understand these solutions, we must understand $e^{A(k)t}$, which we'll do by diagonalizing $A(k)$. The eigenvalues of A are given by

$$\lambda_\pm(k,\nu) = -\nu k^2 - \frac{\nu}{2} \pm \frac{1}{2}\sqrt{\nu^2 - 4k^2},$$

and the corresponding eigenvectors are

$$v_\pm(\lambda,k) = \begin{pmatrix} ik\\ -\nu k^2 - \lambda_\pm(k,\nu)\end{pmatrix} = \begin{pmatrix} ik\\ \frac{\nu}{2} \mp \frac{1}{2}\sqrt{\nu^2-4k^2}\end{pmatrix}.$$

We put these into the columns of a matrix $S = [v_+, v_-]$ and obtain

$$S = \begin{pmatrix} ik & ik\\ \frac{1}{2}[\nu - \sqrt{\nu^2-4k^2}] & \frac{1}{2}[\nu + \sqrt{\nu^2-4k^2}]\end{pmatrix}$$

$$S^{-1} = \frac{1}{ik\sqrt{\nu^2-4k^2}}\begin{pmatrix} \frac{1}{2}[\nu + \sqrt{\nu^2-4k^2}] & -ik\\ \frac{1}{2}[-\nu + \sqrt{\nu^2-4k^2}] & ik\end{pmatrix}.$$

We then have $A = S\Lambda S^{-1}$, where $\Lambda = \mathrm{diag}(\lambda_+, \lambda_-)$.

Remark 9. Note that S becomes singular when $k = \pm\nu/2$, because for that value of k there is a double eigenvalue, and a slightly different decomposition of A, reflecting the resultant Jordan block structure, is necessary. This will be dealt with in the proof of Proposition 5. We do not highlight this issue in the below formulas for the solution, as we wish to focus on the intuition for how to decompose solutions, which does not depend on this singularity.

Hence,

$$e^{A(k,\nu)t} = S(k,\nu)\begin{pmatrix} e^{\lambda_+(k,\nu)t} & 0 \\ 0 & e^{\lambda_-(k,\nu)t}\end{pmatrix} S^{-1}(k,\nu),$$

or explicitly

$$\begin{aligned}
\hat{w}(k,t) &= \frac{ik(e^{\lambda_- t} - e^{\lambda_+ t})}{\sqrt{\nu^2 - 4k^2}}\hat{v}_0 \\
&\quad + \frac{1}{2}\left(\frac{(-\nu + \sqrt{\nu^2 - 4k^2})e^{\lambda_- t} + (\nu + \sqrt{\nu^2 - 4k^2})e^{\lambda_+ t}}{\sqrt{\nu^2 - 4k^2}}\right)\hat{w}_0 \\
\hat{v}(k,t) &= \frac{1}{2}\left(\frac{(-\nu + \sqrt{\nu^2 - 4k^2})e^{\lambda_+ t} + (\nu + \sqrt{\nu^2 - 4k^2})e^{\lambda_- t}}{\sqrt{\nu^2 - 4k^2}}\right)\hat{v}_0 \\
&\quad - \frac{ik(e^{\lambda_+ t} - e^{\lambda_- t})}{\sqrt{\nu^2 - 4k^2}}\hat{w}_0,
\end{aligned} \tag{33}$$

which we'll abbreviate as

$$\hat{w}(k,t) = (f_1(k)\hat{w}_0(k) + f_2(k)\hat{v}_0(k))\, e^{\lambda_+(k)t} + g(k)e^{\lambda_-(k)t} \tag{34}$$

and similarly for $\hat{v}$. The motivation for separating the solution in this way is the fact that $\mathrm{Re}(\lambda_-(k)) \leq -\nu/2$, and so any component of the solution that includes a factor of $e^{\lambda_-(k)t}$ will decay exponentially in time, even for k near zero. Hence, it is primarily the first term, above, involving $e^{\lambda_+(k)t}$ that we must focus our attention on. We'll proceed with the analysis only for $\hat{w}$; all of the results for $\hat{v}$ are analogous.

Remark 10. In order to justify the difference of $(t+1)^{-1/2}$ in the scaling variables for $\tilde{w}$ and $\tilde{v}$, corresponding to (3), we need to show that $\tilde{v}$ decays faster than $\tilde{w}$ by this amount. This can be seen from the above expression for solutions. In particular, for k near zero, say $|k| < \nu/2$, we have

$$e^{A(k,\nu)t} \sim \frac{e^{-\nu k^2 t}}{ik\nu}\begin{pmatrix} 1 & \frac{k}{\nu} \\ \frac{k}{\nu} & -\frac{k^2}{\nu^2}\end{pmatrix}.$$

An extra factor of k corresponds to an x-derivative, and so the v component does decay faster by a factor of $t^{-1/2}$.

We will split the analysis into "high" and "low" frequencies using a cutoff function and Taylor expansion about $k = 0$. Define

$$\Omega^> = \left\{|k| > \frac{\sqrt{15}\nu}{8}\right\}, \quad \Omega^< = \left\{|k| \leq \frac{\sqrt{15}\nu}{8}\right\},$$

and let $\psi(k)$ be a smooth cutoff function equalling 1 on $\Omega^<$ and zero for $|k| \geq \frac{\sqrt{15}\nu}{8} + \nu^2$. We then write $\hat{w}$ as

$$\begin{aligned}\hat{w}(k,t) &= \psi(k)\hat{w}(k,t) + (1-\psi(k))\,\hat{w}(k,t)\\ &=: \psi(k)\left(f_1(k)\hat{w}_0(k) + f_2(k)\hat{v}_0(k)\right)e^{\lambda_+(k)t} + \psi(k)g(k)e^{\lambda_-(k)t}\\ &\quad + \hat{w}_{high}(k,t).\end{aligned}$$

Again, the motivation is to focus on the part of the solution that does not decay exponentially in time. This does not necessarily occur if k is small, which is exactly where $\psi(k) \neq 0$.

Notice that, on $\Omega^<$, we can write $\lambda_+(k) = -\left(\nu + \frac{1}{\nu}\right)k^2 + \Lambda(k)$, where

$$\Lambda(k) = \frac{\nu}{2}\sum_{n=2}^{\infty}\binom{1/2}{n}(-1)^n\left(\frac{4k^2}{\nu^2}\right)^n. \tag{35}$$

In $\Omega^<$, $4k^2/\nu^2 < 15/16 < 1$, and so the above series is convergent. It will also be important that it starts with four powers of k. More precisely,

$$\Lambda(k) = \frac{8k^4}{\nu^3}\sum_{n=0}^{\infty}\binom{1/2}{n+2}(-1)^n\left(\frac{4k^2}{\nu^2}\right)^n.$$

This representation for $\Lambda(k)$ holds by similar reasoning whenever $\psi(k) \neq 0$. We now write

$$\begin{aligned}\hat{w}(k,t) &= \psi(k)e^{-\nu_T k^2 t}e^{\Lambda(k)t}\left(f_1(k)\hat{w}_0(k) + f_2(k)\hat{v}_0(k)\right) + \psi(k)g(k)e^{\lambda_-(k)t}\\ &\quad + \hat{w}_{high}(k,t)\\ &=: \psi(k)e^{-\nu_T k^2 t}\bar{w}(k,t) + \psi(k)g(k)e^{\lambda_-(k)t} + \hat{w}_{high}(k,t),\end{aligned}$$

where $\nu_T = \nu + \frac{1}{\nu}$ and

$$\bar{w}(k,t) = e^{\Lambda(k)t}\left(f_1(k)\hat{w}_0(k) + f_2(k)\hat{v}_0(k)\right). \tag{36}$$

The purpose of this last part of our decomposition of solutions is to emphasize that, to leading order, the decay of the low modes will be determined by the term $e^{-\nu_T k^2 t}$. Therefore, the Taylor dispersion phenomenon is also apparent in Fourier space.

Finally, we Taylor expand the quantity $\bar{w}$ into a polynomial of degree N, plus a remainder term:

$$\bar{w}(k,t) = \sum_{j=0}^{N}\frac{\partial_k^j\bar{w}(0,t)}{j!}k^j + \left[\bar{w}(k,t) - \sum_{j=0}^{N}\frac{\partial_k^j\bar{w}(0,t)}{j!}k^j\right] =: \bar{w}_{low}^N + \bar{w}_{low}^{res}.$$

Thus, we have (suppressing some of the k and t dependence for notational convenience)

$$\hat{w}(k,t) = \psi e^{-\nu_T k^2 t}\left(\bar{w}^N_{low} + \bar{w}^{res}_{low}\right) + \psi g e^{\lambda_-(k)t} + \hat{w}_{high}. \tag{37}$$

The main results of this section are

Proposition 5. *There exists a constant C, independent of* ν *and the initial data, such that*

$$\|\partial^j_k \hat{w}_{high}\|_{L^2} + \|\partial^j_k(\psi g e^{\lambda_- t})\|_{L^2} \leq C\nu^{-2-j} e^{-\frac{\nu}{8}t}(\|\hat{w}_0\|_{C^j} + \|\hat{v}_0\|_{C^j}).$$

Proposition 6. *There exists a constant C such that*

$$\left\| \frac{1}{(1+t)^{\frac{j}{2}}} \partial^j_k\left(\psi e^{-\nu_T k^2 t}\bar{w}^{res}_{low}\right)\right\|_{L^2} \leq \frac{C}{\nu^{\frac{N}{4}+\frac{j}{2}} t^{\frac{N}{4}+\frac{1}{2}}}(\|\hat{w}_0\|_{C^{N+j}} + \|\hat{v}_0\|_{C^{N+j}}).$$

The constant C depends on N, but it is independent of ν.

Here $\|f\|_{C^j} = \sum_{s=0}^{j} \sup_{k\in\mathbb{R}} |\partial^s_k f(k)|$. These results imply that $\hat{w}_{high}$ and $g\psi e^{\lambda_- t}$ decay exponentially in t, and are thus higher-order, while $\hat{w}^{res}_{low}$ decays algebraically, at a rate that can be made large by choosing N (which will correspond to the dimension of the center manifold from Sect. 2) large. In the next section, Sect. 4, it will be shown that the behavior of the remaining term, $\bar{w}^N_{low}$, is governed by the dynamics on the center manifold, in which one can directly observe the Taylor dispersion phenomenon.

Proof of Proposition 5. Notice that, for $k \in \Omega^>$ (the support of $\hat{w}_{high}$), the eigenvalues $\lambda_\pm(k)$ both lie in a sector with vertex at $(\mathrm{Re}\lambda, \mathrm{Im}\lambda) = (-\nu k^2 - \nu/4, 0)$. Therefore, to obtain the desired bound, we need to determine the effect of the derivatives ∂^j_k. Such a derivative could potentially be problematic, due to the factors of $\sqrt{\nu^2 - 4k^2}$, which can be zero in $\Omega^>$. (This is exactly due to the Jordan block structure at $k = \pm\nu/2$.) To work around this, we use the fact that we can equivalently write

$$\begin{pmatrix} \hat{w}(k,t) \\ \hat{v}(k,t) \end{pmatrix} = e^{A(k)t}\begin{pmatrix} \hat{w}_0(k) \\ \hat{v}_0(k) \end{pmatrix}$$

and bound derivatives of this expression for $k \in \Omega^>$. Such derivatives either fall on the initial conditions, which leads to the dependence of the constant on the C^j norms of $\hat{v}_0$ and $\hat{w}_0$, or the derivatives can fall on the exponential. In the latter case, using the fact that

$$A'(k) = \begin{pmatrix} -2\nu k & -i \\ -i & -2\nu k \end{pmatrix},$$

which behaves no worse that $\mathcal{O}(k)$, we obtain terms of the form (writing $\hat{U} = (\hat{w}_0, \hat{v}_0)$ for convenience)

$$\|(kt)^p e^{A(k)t} \partial_k^q \hat{U}_0\|_{L^2}^2 \leq C \|\hat{U}_0\|_{C^q}^2 \int |kt|^{2p} \|e^{A(k)t}\|^2 dk.$$

Next, note that $\|e^{A(k)t}\| \leq C\nu^{-2} e^{-\nu(k^2+1/4)t}$. This follows essentially from the above-mentioned bound on the real part of $\lambda_\pm$ in $\Omega^>$. One needs to be a bit careful when $k = \pm\nu/2$, as there $\lambda_+ = \lambda_-$. This changes the bound from $\sim e^{\lambda_+(k)t}$ to $\sim \nu t e^{\lambda_+(k)t}$, but this power of t can be absorbed into the exponential since $\text{Re}(\lambda_+) < -\nu k^2 - \nu/4 - \delta\nu$ for some $\delta > 0$ that is independent of ν. The factor of ν^{-2} that appears is related to the fact that $\|S^{-1}\| = \mathcal{O}(\nu^{-2})$ for $k \in \Omega^>, k \neq \pm\nu/2$. Thus, we have

$$\begin{aligned}
\|(kt)^p e^{A(k)t} \partial_k^q \hat{U}_0\|_{L^2}^2 &\leq C\nu^{-4} \|\hat{U}_0\|_{C^q}^2 \int |kt|^{2p} e^{-2(\nu k^2+\nu/4)t} dk \\
&\leq C\nu^{-4-p-1/2} \|\hat{U}_0\|_{C^q}^2 t^{p-1/2} e^{-\nu t/2} \\
&\leq C\nu^{-4-2p} e^{-\nu t/4} \|\hat{U}_0\|_{C^q}^2,
\end{aligned}$$

which proves the result for $\hat{w}_{high}$. A similar proof works for the $\|\partial_k^j(\psi g e^{\lambda_- t})\|_{L^2}$ term. □

Proof of Proposition 6. We now derive bounds on the residual term $\psi e^{-\nu_T k^2 t} \bar{w}_{low}^{res}$. Recall the integral formula for the Taylor Remainder:

$$\bar{w}_{low}^{res}(k,t) = \int_0^k \int_0^{k_1} \cdots \int_0^{k_N} \partial_{k_{N+1}}^{N+1} \bar{w}(k_{N+1}, t) d_{k_{N+1}} d_{k_N} \ldots d_{k_1}. \tag{38}$$

With this formula in mind, we want to derive bounds on the derivatives of $\bar{w}(k,t)$, but we need only deal with $k \in \Omega^<$, since we are ultimately estimating the size of $\psi e^{-\nu_T k^2 t} \bar{w}_{low}^{res}$.

Recall that

$$\bar{w} = e^{\Lambda(k)t} (f_1 \hat{w}_0 + f_2 \hat{v}_0).$$

The functions f_1 and f_2 are smooth in $\Omega^<$, so our estimate will depend on derivatives of the initial data and derivatives of $e^{\Lambda(k)t}$. However, the reader should note that f_1 and f_2 are also dependent on ν, but their derivatives give us inverse powers of ν no worse than any others appearing in this section, so we choose not to explicitly keep track of these powers. The following lemma will be used in estimating these derivatives:

Lemma 1. *Let $\Phi(k,t) = k^d e^{-\nu_T k^2 t}$. Then*

$$\|\Phi(\cdot,t)\|_{L^2} \le C(d)(\nu_T t)^{-\frac{2d+1}{4}}.$$

Proof. Use the fact that $\int_{\mathbb{R}} e^{-x^2/4}dx = 2\sqrt{\pi}$ and change variables. □

With this lemma in mind, we need to keep track of the powers of k and t that appear in $\partial_k^j e^{\Lambda(k)t}$. To see why one would expect the powers of ν and t appearing in Proposition 6, consider the following formal calculation. Recall from the Taylor expansion of $\Lambda(k)$, we have $\bar{w} \approx e^{-\frac{k^4}{\nu^3}t}$.

We are essentially estimating

$$\|\partial_k^j e^{-\nu_T k^2 t} \bar{w}_{low}^{res}\|_{L^2},$$

with the aid of the estimate

$$\|k^d e^{-\nu_T k^2 t}\|_{L^2} \le C(d)\, (\nu_T t)^{-\frac{d}{2}-\frac{1}{4}}$$

and the Taylor Remainder formula

$$\bar{w}_{low}^{res}(k,t) = \int_0^k \int_0^{k_1} \cdots \int_0^{k_N} \partial_{k_{N+1}}^{N+1} \bar{w}(k_{N+1},t) d_{k_{N+1}} d_{k_N} \ldots d_{k_1}. \tag{39}$$

We'll proceed by finding bounds on $\partial_k^J e^{-\frac{k^4}{\nu^3}t}$, and plug into (39) with $J = N+1$. We'll make the following changes of variable: we set

$$\begin{aligned} T &= \frac{t}{\nu^3} \\ x &= T^{1/4} k \end{aligned} \tag{40}$$

so that

$$\bar{w} = e^{-x^4}$$

and

$$\partial_k^J \bar{w} = T^{J/4} \partial_x^J \bar{w}. \tag{41}$$

Let's proceed by computing x–derivatives of $\bar{w}$, only taking into account what powers of x appear at each stage. In the following, a prime means ∂_x. We compute

$$\begin{aligned} \bar{w}' &\sim x^3 e^{-x^4} \\ \bar{w}'' &\sim \left(x^2 + x^6\right) e^{-x^4} \\ \bar{w}''' &\sim \left(x + x^5 + x^9\right) e^{-x^4}. \end{aligned}$$

In particular, notice that the powers of x that appear in the J–th derivative can be obtained from the powers of x that appear in the $J-1$st derivative by subtracting one from each power appearing (where only nonnegative powers are permitted), and also adding three to each power appearing:

$$\bar{w}^{(4)} \sim \left(x^0 + x^4 + x^8 + x^{12}\right) e^{-x^4}$$
$$\bar{w}^{(5)} \sim \left(x^3 + x^7 + x^{11} + x^{15}\right) e^{-x^4}$$
$$\bar{w}^{(6)} \sim \left(x^2 + x^6 + x^{10} + x^{14} + x^{18}\right) e^{-x^4}.$$

In general, we have

$$\partial_x^J \bar{w} \sim \sum_{l=0}^{J-2} x^{R+4l} e^{-x^4}$$

where $R = (-J) \bmod 4$. In the original variables, we have, using (41) and (40),

$$\partial_k^J \bar{w} \sim \left(\frac{t}{\nu^3}\right)^{J/4} \sum_{l=0}^{J-2} \left(\left(\frac{t}{\nu^3}\right)^{1/4} k\right)^{R+4l} e^{-\frac{k^4}{\nu^3}t},$$

or more precisely,

$$|\partial_k^J \bar{w}| \leq C(J) \left(\frac{t}{\nu^3}\right)^{J/4} \sum_{l=0}^{J-2} \left(\left(\frac{t}{\nu^3}\right)^{1/4} |k|\right)^{R+4l} e^{-\frac{k^4}{\nu^3}t}.$$

Combining with the Taylor Remainder formula and setting $J = N + 1$, we have

$$\|\psi e^{-\nu_T k^2 t} \bar{w}_{low}^{res}\|_{L^2} \leq C(N) \sum_{l=0}^{N-1} \frac{t^{(1/4)(R+N+1)+l}}{\nu^{(3/4)(R+N+1)+3l}} \|k^{N+1+R+4l} e^{-\nu_T k^2 t}\|_{L^2}.$$

Using the estimate (1), we get

$$\begin{aligned}
\|\psi e^{-\nu_T k^2 t} \bar{w}_{low}^{res}\|_{L^2} &\leq C(N) \sum_{l=0}^{N-1} \frac{t^{-1/4R-1/4(N+1)-l-1/4}}{\nu^{1/4R+1/4(N+1)+l-1/4}} \\
&= C(N)\, (\nu t)^{-1/4R-1/4(N+1)} \frac{t^{-1/4}}{\nu^{-1/4}} \sum_{l=0}^{N-1} \left(\nu^{-1} t^{-1}\right)^l \\
&= C(N)\, (\nu t)^{-1/4R-1/4(N+1)} \frac{t^{-1/4}}{\nu^{-1/4}} \left(\frac{1-\left(\nu^{-1}t^{-1}\right)^N}{1-\left(\nu^{-1}t^{-1}\right)}\right).
\end{aligned}$$

Therefore if $t > \frac{2}{\nu}$, we have

$$\|\psi e^{-\nu_T k^2 t}\bar{w}^{res}_{low}\|_{L^2} \le 2C(N)\,(\nu t)^{-1/4R-1/4(N+1)}\,\frac{t^{-1/4}}{\nu^{-1/4}},$$

which implies that

$$\|\psi e^{-\nu_T k^2 t}\bar{w}^{res}_{low}\|_{L^2} \le \frac{C(N)}{\nu^{\frac{N}{4}} t^{\frac{N}{4}+\frac{1}{2}}}$$

as reflected in Proposition 6. This concludes the formal calculation. We proceed with deriving the precise estimate.

Because k is small in $\Omega^<$, powers of k are helpful, so we only need to record the smallest power of k relative to the largest power of t. We obtain additional powers of t when a derivative falls on the exponential (as opposed to any factors in front of it), which creates not only powers of t but powers of $(\Lambda'(k)t)$. When derivatives fall on factors of $\Lambda'(k)$ in front of the exponential, we obtain fewer powers of k but no additional powers of t. Using (35), we see that $\Lambda'(k) \sim k^3/\nu^3$, and so $\partial_k^j e^{\Lambda(k)t}$ will lead to terms of the form

$$\left(\frac{k^3 t}{\nu^3}\right)^q \left(\frac{k^2 t}{\nu^3}\right)^{l_1} \left(\frac{kt}{\nu^3}\right)^{l_2} \left(\frac{t}{\nu^3}\right)^{l_3} e^{\Lambda(k)t}, \qquad q + 2l_1 + 3l_2 + 4l_3 = j.$$

This implies that

$$|\partial_k^j \bar{w}(k,t)| \le C(\|\hat{w}_0\|_{C^j} + \|\hat{v}_0\|_{C^j}) \left| \left(\frac{k^3 t}{\nu^3}\right)^q \left(\frac{k^2 t}{\nu^3}\right)^{l_1} \left(\frac{kt}{\nu^3}\right)^{l_2} \left(\frac{t}{\nu^3}\right)^{l_3} e^{\Lambda(k)t} \right|$$

for any $q + 2l_1 + 3l_2 + 4l_3 = j$. Using the fact that, on $\Omega^<$, $|e^{\Lambda(k)t}| \le 1$, as well as (38), we find

$$\begin{aligned} &\|\psi e^{-\nu_T k^2 t}\bar{w}^{res}_{low}\|_{L^2} \\ &\quad \le C(\|\hat{w}_0\|_{C^{N+1}} + \|\hat{v}_0\|_{C^{N+1}}) \left\| \psi e^{-\nu_T k^2 t} \left| \left(\frac{t}{\nu^3}\right)^{q+l_1+l_2+l_3} k^{3q+2l_1+l_2+N+1} \right| \right\|_{L^2}. \end{aligned}$$

Note the extra $N+1$ powers of k come from the $N+1$ antiderivatives in the Taylor Remainder formula. We need to estimate

$$\left\| \psi e^{-\nu_T k^2 t} \left(\frac{t}{\nu^3}\right)^{q+l_1+l_2+l_3} k^{3q+2l_1+l_2+N+1} \right\|_{L^2},$$

where

$$q + 2l_1 + 3l_2 + 4l_3 = N + 1 \quad \Rightarrow \quad \frac{N+1}{4} = \frac{q}{4} + \frac{l_1}{2} + \frac{3l_2}{4} + l_3. \tag{42}$$

We being by noting that, since $k \in \Omega^<$,

$$\begin{aligned}\left|\left(\frac{t}{\nu^3}\right)^{q+l_1+l_2+l_3} k^{3q+2l_1+l_2+N+1}\right| &= \left| t^{q+l_1+l_2+l_3} \frac{k^{\frac{5}{2}q+3l_1+\frac{7}{2}l_2+4l_3}}{\nu^{\frac{3}{2}q+2l_1+\frac{5}{2}l_2+3l_3}} \left(\frac{k}{\nu}\right)^{\frac{3}{2}q+l_1+\frac{l_2}{2}} \right| \\ &\le C \left| t^{q+l_1+l_2+l_3} \frac{k^{\frac{5}{2}q+3l_1+\frac{7}{2}l_2+4l_3}}{\nu^{\frac{3}{2}q+2l_1+\frac{5}{2}l_2+3l_3}} \right|,\end{aligned}$$

where C is independent of ν. Therefore, since $\nu_T \sim \nu^{-1}$,

$$\begin{aligned}&\left\| \psi e^{-\nu_T k^2 t} \left(\frac{t}{\nu^3}\right)^{q+l_1+l_2+l_3} k^{3q+2l_1+l_2+N+1} \right\|_{L^2} \\ &\le C \left\| \psi e^{-\nu_T k^2 t} t^{q+l_1+l_2+l_3} \frac{k^{\frac{5}{2}q+3l_1+\frac{7}{2}l_2+4l_3}}{\nu^{\frac{3}{2}q+2l_1+\frac{5}{2}l_2+3l_3}} \right\|_{L^2} \\ &\le C \frac{t^{q+l_1+l_2+l_3}}{\nu^{\frac{3}{2}q+2l_1+\frac{5}{2}l_2+3l_3}} (\nu_T t)^{-\frac{1}{4}-\frac{1}{2}\left(\frac{5}{2}q+3l_1+\frac{7}{2}l_2+4l_3\right)} \\ &\le C \frac{t^{q+l_1+l_2+l_3-\frac{1}{4}-\frac{1}{2}\left(\frac{5}{2}q+3l_1+\frac{7}{2}l_2+4l_3\right)}}{\nu^{\frac{3}{2}q+2l_1+\frac{5}{2}l_2+3l_3-\frac{1}{4}-\frac{1}{2}\left(\frac{5}{2}q+3l_1+\frac{7}{2}l_2+4l_3\right)}} \\ &= C \frac{t^{-\frac{N+1}{4}-\frac{1}{4}}}{\nu^{\frac{N}{4}}},\end{aligned}$$

where we used (42) in the last equality.

Using a similar calculation, we can bound the L^2 norm of each jth derivative of this remainder term. One can show that for each integer triple $l+s+r=j$, we have

$$\| \partial_k^l \psi \partial_k^s e^{-\nu_T k^2 t} \partial_k^r \bar{w}_{low}^{res} \| \le C(\|\hat{w}_0\|_{C^j} + \|\hat{v}_0\|_{C^j}) \frac{t^{-\frac{N}{4}-\frac{1}{2}}}{\nu^{N/4}} \left(\frac{t}{\nu}\right)^{\frac{s+r}{2}}.$$

The proposition follows from the fact that $s+r \le j$. □

Remark 11. The key point is that we can analyze the asymptotic behavior of $\hat{w}$ and $\hat{v}$ to any given order of accuracy $\mathcal{O}(t^{-M})$ (when $t > \mathcal{O}(\frac{1}{\nu})$) by choosing N (and hence m) sufficiently large and studying only the behavior of $e^{-\nu_T k^2 t} \bar{w}_{low}^N$ and $e^{-\nu_T k^2 t} \bar{v}_{low}^N$.

4 Decomposition of Solutions and Proof of the Main Result

In this final section, we state and prove our main result.

Theorem 2. *Given any* $M > 0$*, let* $N \geq 4M$*, and let* $m > N + 1/2$*. If the initial values* $\tilde{w}_0, \tilde{v}_0$ *of* (9) *lie in the space* $L^2(m)$*, then there exists a constant* $C = C(m, N, \tilde{w}_0, \tilde{v}_0)$ *and approximate solutions* w_{app}*,* v_{app}*, computable in terms of the* $2N + 3$ *dimensional system of ODEs* (14)*, such that*

$$\|w(\xi,\tau) - w_{app}(\xi,\tau)\|_{L^2(m)} + \|v(\xi,\tau) - v_{app}(\xi,\tau)\|_{L^2(m)} \leq \frac{C}{\nu^{\frac{N}{4}+\frac{m}{2}}} e^{-M\tau}$$

for all τ *sufficiently large. The approximate solutions* w_{app} *and* v_{app} *satisfy equations* (49) *and* (50) *respectively. The functions* $\phi_j(\xi)$ *are the eigenfunctions of the operator* $\mathscr{L}_T$ *(corresponding to diffusion with constant* $\nu_T = \nu + \frac{1}{\nu}$ *in scaling variables) in the space* $L^2(m)$*. The quantities* $\alpha_k(\tau)$ *and* $\beta_k(\tau)$ *solve system* (14) *and have the following asymptotics, obtainable via a reduction to an* $N + 2$*-dimensional center manifold:*

$$|\alpha_k(\tau)| \leq \begin{cases} \frac{C(N,k)e^{-\frac{k}{4}\tau}}{\nu^{k-1}} & : k = 0 \bmod 4 \\ \frac{C(N,k)e^{-\frac{k+1}{4}\tau}}{\nu^{k-1}} & : k = 1 \bmod 4 \\ \frac{C(N,k)e^{-\frac{k+2}{4}\tau}}{\nu^{k-1}} & : k = 2 \bmod 4 \\ \frac{C(N,k)e^{-\frac{k+3}{4}\tau}}{\nu^{k-1}} & : k = 3 \bmod 4. \end{cases} \tag{43}$$

$$|\beta_k(\tau)| \leq \begin{cases} \frac{C(N,k)e^{-\frac{k}{4}\tau}}{\nu^{k+1}} & : k = 0 \bmod 4 \\ \frac{C(N,k)e^{-\frac{k+1}{4}\tau}}{\nu^{k+1}} & : k = 1 \bmod 4 \\ \frac{C(N,k)e^{-\frac{k+2}{4}\tau}}{\nu^{k+1}} & : k = 2 \bmod 4 \\ \frac{C(N,k)e^{-\frac{k+3}{4}\tau}}{\nu^{k+1}} & : k = 3 \bmod 4. \end{cases} \tag{44}$$

Remark 12. As we will see in the course of the proof of the theorem, $\tau > \mathscr{O}(\log(\frac{|\log \nu|}{\nu}))$ (or equivalently $t > \mathscr{O}(\frac{|\log \nu|}{\nu})$) will suffice for these estimates to hold.

Proof of Theorem 2. We first concentrate on defining w_{app} and v_{app} and establishing the error estimates in Theorem 2; this process will mainly use results from Sect. 3. Recall the decomposition of $\hat{w}$ from Sect. 3:

$$\hat{w}(k,t) = \psi e^{-\nu_T k^2 t}\left(\bar{w}_{low}^N + \bar{w}_{low}^{res}\right) + \psi g e^{\lambda_-(k)t} + \hat{w}_{high}. \tag{45}$$

The main results of Sect. 3 essentially said $\hat{w} \approx \psi e^{-\nu_T k^2 t}\bar{w}_{low}^N$, with errors (measured in the $||| \cdot |||$ norm introduced in that section) either algebraically or exponentially decaying. More precisely, using Propositions 5 and 6, we obtain

$$|||\hat{w} - \psi e^{-\nu_T k^2 t} \bar{w}^N_{low}||| \leq C \left(\frac{1}{\nu^{\frac{N}{4}+\frac{m}{2}} t^{\frac{N}{4}+\frac{1}{2}}} + \frac{1}{\nu^{m+2}} e^{-\frac{\nu}{8}t} \right),$$

where C is independent of ν. For some t sufficiently large, we can "absorb" the exponentially decaying term into the algebraically decaying term; i.e.

$$\frac{1}{\nu^{m+2}} e^{-\frac{\nu}{8}t} < \frac{1}{\nu^{\frac{N}{4}+\frac{m}{2}} t^{\frac{N}{4}+\frac{1}{2}}}.$$

We want to quantify how large t must be for the above inequality to hold. However, there are several other places in this section where terms of the form $\nu^{-p} e^{-\frac{\nu}{A}t}$ appear, which we wish to absorb into algebraically decaying errors. For this reason, we state and prove the following lemma:

Lemma 2. *Let $A, M, \ell, p > 0$ with $\nu > 0$ as before. Then there exists a constant $C = C(M, A) > 0$ such that for all $t > \frac{C}{\nu} \log(\nu^{\ell-p-M})$, we have the inequality*

$$\frac{1}{\nu^p} e^{-\frac{\nu}{A}t} < \frac{1}{\nu^\ell t^M}.$$

Remark 13. Note in particular that since $\tau = \log(1+t)$, the inequality $t > \frac{C}{\nu} \log(\nu^{\ell-p-M})$ essentially translates to $\tau > \mathcal{O}(\log(\frac{|\log \nu|}{\nu}))$.

Proof of Lemma 2. We introduce a few new quantities to simplify the notation: we let $d = \nu^{p-\ell}$ and we let $a = \nu/A$. Then the target estimate in the lemma reads

$$t^M e^{-at} < d.$$

Now set $f_\lambda(t) = t^M e^{-a\lambda t}$ where $0 < \lambda < 1$ is fixed. Now, the target estimate in the lemma reads

$$f_\lambda(t) e^{-a(1-\lambda)t} < d.$$

Using basic calculus, we find that the maximum value of f_λ lies at $t = \frac{M}{a\lambda}$, and for $t > \frac{M}{a\lambda}$, we have $f_\lambda(t) < \left(\frac{M}{a\lambda e}\right)^M$. Therefore, if

$$\left(\frac{M}{a\lambda e}\right)^M e^{-a(1-\lambda)t} < d,$$

we have the target estimate. The above inequality holds for

$$t > \frac{-1}{a(1-\lambda)} \log\left(d \left(\frac{a\lambda e}{M} \right)^M \right),$$

or, substituting $a = \nu/A$ and $d = \nu^{p-\ell}$, we have

$$t > \left(\frac{A}{1-\lambda}\right)\frac{1}{\nu}\left(\log(\nu^{\ell-p-M}) + M\left(\log(M) + \log(\frac{A}{\lambda e})\right)\right).$$

The time estimate in the lemma is just a less precise version of this inequality. This concludes the proof of lemma 2. □

Next, we apply the lemma. Using the definition of $\tilde{w}$ and inverting the Fourier Transform, we obtain, for t sufficiently large,

$$|||\tilde{w}(x,t) - \mathscr{F}^{-1}[\psi(k)e^{-\nu_T k^2 t}\bar{w}_{low}^N](x,t)||| \leq \frac{C}{\nu^{\frac{N}{4}+\frac{m}{2}}}(1+t)^{-N/4}. \tag{46}$$

Proceeding, notice we can "drop" the cutoff function ψ in the above estimate with only an exponentially decaying penalty: due to the fact that

$$|\psi(k)e^{-\nu_T k^2 t}\bar{w}_{low}^N - e^{-\nu_T k^2 t}\bar{w}_{low}^N| = |(\psi(k)-1)e^{-\nu_T k^2 t}\bar{w}_{low}^N| = 0$$

for $|k| \leq \frac{\sqrt{15\nu}}{8}$, which implies that

$$\|\partial_k^j\left((\psi(k)-1)e^{-\nu_T k^2 t}\bar{w}_{low}^N\right)\|_{L^2} \leq \frac{C}{\nu^{2j}}e^{-\frac{\nu}{8}t}.$$

From here on out, we will sometimes suppress the ν-dependence of the constants for notational convenience. Proceeding, we define our approximate solution in x and t variables:

$$\mathscr{F}^{-1}[e^{-\nu_T k^2 t}\bar{w}_{low}^N](x,t) \equiv \tilde{w}_{app}(x,t),$$

which gives us the estimate

$$|||\tilde{w}(x,t) - \tilde{w}_{app}(x,t)||| \leq C(1+t)^{-N/4}.$$

This is just estimate (46) without the cutoff function; it holds for t sufficiently large as in Lemma 2. Therefore, using scaling variables and defining

$$\tilde{w}_{app}(x,t) \equiv \frac{1}{\sqrt{1+t}}w_{app}(\xi,\tau),$$

we have the estimate, which holds for $\tau > \mathscr{O}(\log(\frac{|\log\nu|}{\nu}))$,

$$\|w(\xi,\tau) - w_{app}(\xi,\tau)\|_{L^2(m)} \leq \frac{C}{\nu^{\frac{N}{4}+\frac{m}{2}}}e^{-\frac{N}{4}\tau}.$$

(This holds since the $|||\cdot|||$ and $||\cdot||_{L^2(m)}$ norms are equivalent in the way made precise at the beginning of Sect. 3.) Using similar calculations, we have functions $\tilde{v}_{app}(x,t)$ and $v_{app}(\xi,\tau)$ satisfying

$$|||\tilde{v}(x,t)-\tilde{v}_{app}(x,t)|||\leq C(1+t)^{-N/4}$$

and

$$\|v(\xi,\tau)-v_{app}(\xi,\tau)\|_{L^2(m)}\leq \frac{C}{\nu^{\frac{N}{4}+\frac{m}{2}}}e^{-\frac{N}{4}\tau}.$$

This establishes the error estimates in Theorem 2; the remainder of the section is devoted to making more explicit the relationship between our approximate solutions w_{app}, v_{app}, and our center manifold calculations in Sect. 2.

Observe that

$$\begin{aligned}\tilde{w}_{app}(x,t) &= \sum_{j=0}^{N}\frac{\partial_k^j\bar{w}(0,t)}{j!}\mathscr{F}^{-1}[k^j e^{-\nu_T k^2 t}](x,t)\\ &= \sum_{j=0}^{N}\frac{\partial_k^j\bar{w}(0,t)}{j!(i)^j}\partial_x^j\mathscr{F}^{-1}[e^{-\nu_T k^2 t}](x,t)\\ &= \sum_{j=0}^{N}\frac{\partial_k^j\bar{w}(0,t)}{j!(i)^j}\partial_x^j\left(\frac{1}{\sqrt{4\pi\nu_T t}}e^{-\frac{x^2}{4\nu_T t}}\right).\end{aligned}$$

Defining new scaling variables

$$\tilde{\xi}:=\frac{x}{\sqrt{t}},\qquad \tilde{\tau}:=\log(t),$$

and defining

$$\tilde{w}_{app}(x,t):=\frac{1}{\sqrt{t}}w_{app}(\tilde{\xi},\tilde{\tau}),$$

gives us

$$\begin{aligned}w_{app}(\tilde{\xi},\tilde{\tau}) &= \sum_{j=0}^{N}\frac{\partial_k^j\bar{w}(0,e^{\tilde{\tau}})}{j!(i)^j}e^{-\frac{j}{2}\tilde{\tau}}\partial_{\tilde{\xi}}^j\left(\phi_0(\tilde{\xi})\right)\\ &= \sum_{j=0}^{N}\frac{\partial_k^j\bar{w}(0,e^{\tilde{\tau}})}{j!(i)^j}e^{-\frac{j}{2}\tilde{\tau}}\phi_j(\tilde{\xi}),\end{aligned}\tag{47}$$

where the $\phi_j(\tilde{\xi})$ above are again the eigenfunctions of the operator $\mathscr{L}_T$ on the space $L^2(m)$.

We now show that the coefficients in (47) can be expressed in terms of the functions $\{\alpha_k(\tau), \beta_k(\tau)\}$ from Sect. 2, demonstrating that the leading order asymptotic behavior of the solution is determined by the center-manifold.

First, recall from Sect. 3, formulas (34) and (36) that

$$\bar{w}(k,t) = e^{\nu_T k^2 t}\hat{w}(k,t) + g(k)e^{\lambda_- t}.$$

Differentiating, we have

$$\begin{aligned}\partial_k^j \bar{w}(k,t) &= \sum_{l=0}^{j} \binom{j}{l} \partial_k^{j-l}(e^{\nu_T k^2 t})\partial_k^l \hat{w}(k,t) + \partial_k^j(g(k)e^{\lambda_- t}) \\ &= \sum_{l=0}^{j} \binom{j}{l} (\nu_T t)^{\frac{j-l}{2}} P^{j-l}(\sqrt{\nu_T t}k)e^{\nu_T k^2 t}\partial_k^l \hat{w}(k,t) + \partial_k^j(g(k)e^{\lambda_- t}),\end{aligned}$$

where P^{j-l} is a polynomial of degree $j-l$. Setting $k=0$, and substituting $t = e^{\tilde{\tau}}$, we have

$$\partial_k^j \bar{w}(0, e^{\tilde{\tau}}) = \sum_{l=0}^{j} C_{j,l}^{\nu} e^{(\frac{j-l}{2})\tilde{\tau}} \partial_k^l \hat{w}(0, e^{\tilde{\tau}}) + \mathscr{O}(e^{-\frac{\nu}{2}e^{\tilde{\tau}}}), \tag{48}$$

where $C_{j,l}^{\nu} = \binom{j}{l}\nu_T^{\frac{j-l}{2}} P^{j-l}(0)$, and $\partial_k^j(g(k)e^{\lambda_- t})|_{k=0}$ is $\mathscr{O}(e^{-\frac{\nu}{2}e^{\tilde{\tau}}})$ since $\lambda_-(0) = -\nu$. We will proceed by computing the derivatives $\partial_k^j \hat{w}(0,t)$ in terms of the α_j from Sect. 2.

Recall from Sect. 2, formula (13), that we have the decomposition (using the original scaling variables ξ and τ)

$$w(\xi,\tau) = w_c(\xi,\tau) + w_s(\xi,\tau), \qquad w_c(\xi,\tau) = \sum_{j=0}^{N} \alpha_j(\tau)\varphi_j(\xi), \qquad w_s = (w - w_c),$$

$$v(\xi,\tau) = v_c(\xi,\tau) + v_s(\xi,\tau), \qquad v_c(\xi,\tau) = \sum_{j=0}^{N} \beta_j(\tau)\varphi_j(\xi), \qquad v_s = (v - v_c),$$

and note the following.

Lemma 3. *$\int \xi^k w_s(\xi,\tau)d\xi = \int \xi^k v_s(\xi,\tau)d\xi = 0$ for all $k \le N$.*

Proof. We will prove the result for w_s only, as the proof for v_s is analogous. Note that

$$w_s = w - P_n w = w - \sum_{j=0}^{N} \langle H_j, w\rangle \phi_j,$$

and so $\langle H_k, w_s \rangle = 0$ for all $k \leq N$. We'll proceed by induction on k. The $k = 0$ case follows because $\xi^0 = 1 = H_0(\xi)$. Next,

$$0 = \langle (H_{k+1} - H_k), w_s \rangle = c_{k+1} \int \xi^{k+1} w_s(\xi) d\xi + \sum_{j=0}^{k} c_k \int \xi^k w_s(\xi) d\xi$$
$$= c_{k+1} \int \xi^{k+1} w_s(\xi) d\xi$$

by the inductive assumption. Since $c_{k+1} \neq 0$, the result follows. □

Using this lemma, we can compute

$$\partial_k^l \hat{w}(0,t) = \left[\partial_k^l \int e^{ikx} \tilde{w}(x,t) dx \right] |_{k=0}$$
$$= \left[\partial_k^l \int e^{ik\sqrt{t+1}\xi} w(\xi,\tau) d\xi \right] |_{k=0}$$
$$= (i\sqrt{t+1})^l \int \xi^l w(\xi,\tau) d\xi$$
$$= (i\sqrt{t+1})^l \int \xi^l w_c(\xi,\tau) d\xi$$

for all $l \leq N$, and similarly for $\hat{v}$. As a result, we have a relationship between $\partial_k^l \hat{w}(0,t)$ and the quantities α_r: from Sect. 2

$$\partial_k^l \hat{w}(0,t) = \frac{1}{\sqrt{1+t}} \sum_{r=0}^{N} \alpha_r(\log(1+t)) \int x^l \phi_r(\frac{x}{\sqrt{1+t}}) dx,$$

or equivalently

$$\partial_k^l \hat{w}(0, e^{\tilde{\tau}}) = \sum_{r=0}^{N} \alpha_r(\log(1+e^{\tilde{\tau}}))(1+e^{\tilde{\tau}})^{\frac{l}{2}} \int \xi^l \phi_r(\xi) d\xi.$$

Inserting into (48), we obtain

$$\partial_k^j \bar{w}(0, e^{\tilde{\tau}}) = \sum_{l=0}^{j} C_{j,l}^{\nu} e^{(\frac{j-l}{2})\tilde{\tau}} \sum_{r=0}^{N} \alpha_r(\log(1+e^{\tilde{\tau}}))(1+e^{\tilde{\tau}})^{\frac{l}{2}} \int \xi^l \phi_r(\xi) d\xi + \mathcal{O}(e^{-\frac{\nu}{2} e^{\tilde{\tau}}})$$
$$= e^{\frac{j}{2}\tilde{\tau}} \sum_{r=0}^{N} \alpha_r(\log(1+e^{\tilde{\tau}})) \sum_{l=0}^{N} (1+e^{-\tilde{\tau}})^{\frac{l}{2}} C_{j,l}^{\nu} \int \xi^l \phi_r(\xi) d\xi + \mathcal{O}(e^{-\frac{\nu}{2} e^{\tilde{\tau}}}).$$

Therefore we can replace the coefficients in (47) and write

$$w_{app}(\tilde{\xi},\tilde{\tau}) = \sum_{j=0}^{N}\left(\frac{1}{j!(i)^j}\sum_{r=0}^{N}\alpha_r(\log(1+e^{\tilde{\tau}}))\sum_{l=0}^{j}(1+e^{-\tilde{\tau}})^{\frac{l}{2}}C^{\nu}_{j,l}\int \xi^l\phi_r(\xi)d\xi\right)\phi_j(\tilde{\xi}), \quad (49)$$

where $C^{\nu}_{j,l} \sim \nu^{-j}$ (see the line after (48)), and where we also have omitted an error term of $\mathcal{O}(e^{-\frac{\nu}{2}e^{\tilde{\tau}}})$ (This can be absorbed into the estimate in the original definition of w_{app} by applying Lemma 2 with $\tilde{\tau} = \log(t)$.) Analogous calculations give us a similar result for v:

$$v_{app}(\tilde{\xi},\tilde{\tau}) = \sum_{j=0}^{N}\left(\frac{1}{j!(i)^j}\sum_{r=0}^{N}\beta_r(\log(1+e^{\tilde{\tau}}))\sum_{l=0}^{j}(1+e^{-\tilde{\tau}})^{\frac{l}{2}}D^{\nu}_{j,l}\int \xi^l\phi_r(\xi)d\xi\right)\phi_j(\tilde{\xi}). \quad (50)$$

This completes the proof of Theorem 2.

Remark 14. Note that Theorem 2 is stated in terms of the original scaling variables ξ and τ. Since $\tau = \log(1+e^{\tilde{\tau}})$, errors in $\tilde{\tau}$, for large $\tilde{\tau}$, are equivalent to errors in τ, for large τ.

Acknowledgements The work of OC and CEW was supported in part by the NSF through grant DMS-1311553. The work of MB was supported in part by a Sloan Fellowship and NSF grant DMS-1411460. MB and CEW thank Tasso Kaper and Edgar Knobloch for pointing out a possible connection between their prior work in [2] and the phenomenon of Taylor dispersion, and we all gratefully acknowledge the many insightful and extremely helpful comments of the anonymous referee.

Appendix: Convergence to the Center Manifold

The purpose of this appendix is to show that the center manifold constructed in Sect. 2 attracts all solutions. We know that we can construct the invariant manifold for the Eq. (21) (for a'_k and b'_k) globally since we have explicit formulas which hold for all values of a_k and η. In this note we show that any trajectory will converge toward the center manifold on a time scale of $\mathcal{O}(1/\nu)$.

Write $b_k = B_k + h_k(a_{k-2},\ldots,a_0,\eta)$. We'll prove that for any choice of initial conditions B_k goes to zero like $\sim e^{-\nu t}$.

First note that

$$b'_k = B'_k + \sum_{\ell=0,even}^{k-2}(\partial_{a_\ell}h_k)a'_\ell + (\partial_\eta h_k)\eta'$$

$$= B_k' - \sum_{\ell=2,even}^{k-2} (\partial_{a_\ell} h_k)\eta\frac{\ell}{2}a_\ell - \sum_{\ell=2,even}^{k-2} (\partial_{a_\ell} h_k)\eta b_{\ell-2} - (\partial_\eta h_k)\eta^2$$

$$= B_k' - \eta \sum_{\ell=2,even}^{k-2} (\partial_{a_\ell} h_k)B_{\ell-2} + \sum_{\ell=2,even}^{k-2} (\partial_{a_\ell} h_k)(-\eta\frac{\ell}{2}a_\ell - \eta h_{\ell-2}) - (\partial_\eta h_k)\eta^2.$$

Thus, using the equation for b_k' in (21) we have

$$B_k' - \eta \sum_{\ell=2,even}^{k-2} (\partial_{a_\ell} h_k)B_{\ell-2} + (\nu + \eta\frac{k}{2})B_k + \frac{2\eta}{\nu}B_{k-2}$$
$$= -(\nu + \eta\frac{k}{2})h_k + \frac{\eta}{\nu^2}a_{k-2}$$
$$- \sum_{\ell=2,even}^{k-2} (\partial_{a_\ell} h_k)(-\eta\frac{\ell}{2}a_\ell - \eta h_{\ell-2}) - (\partial_\eta h_k)\eta^2 - \frac{2\eta}{\nu}h_{k-2}.$$

The key observation is that the terms on the right hand side precisely represent the invariance equation that defines h_k (and hence they all cancel), leaving the equation

$$B_k' = -(\nu + \eta\frac{k}{2})B_k - \frac{2\eta}{\nu}B_{k-2} + \eta \sum_{\ell=2,even}^{k-2} (\partial_{a_\ell} h_k)B_{\ell-2}. \tag{51}$$

We now see that the system of equations for B_k is homogeneous, linear, upper-triangular (but non-autonomous), and hence can be analyzed inductively. We'll show that

$$|B_k(t)| < \frac{C(N)}{\nu^{k/2}}e^{-\nu t}$$

for $t > \frac{1}{\nu}$. We'll only prove this for the even-indexed subsystem; the proof for the odd-subsystem is analogous. Notice that the base case, $k = 0$, holds since (51) for $k = 0$ reads $B_0' = -\nu B_0$, which implies $B_0 \sim e^{-\nu t}$. Now let's proceed with the induction argument: assume for $j = 0, 2, \ldots k$, that

$$|B_j(t)| < \frac{C(N)}{\nu^{j/2}}e^{-\nu t}.$$

Next, we write the equation for B_{k+2} from (51) with a key difference in the way the last term is written:

$$B_{k+2}' = -\left(\nu B_{k+2} + \eta\frac{k+2}{2}B_{k+2}\right) - \frac{2\eta}{\nu}B_k + \frac{\eta}{\nu}\sum_{\ell=1}^{\frac{k}{2}} C_{k+2-2\ell}^{k+2}\frac{\eta^\ell}{\nu^{2\ell}}B_{k-2\ell}. \tag{52}$$

This last sum is obtained from reindexing ℓ and using the fact that the formula for h_k in Proposition 1 implies that $(\partial_{a_\ell} h_k)$ consists of a single term. Let's proceed by noting that the equation

$$y' = -(\nu y + a\eta y)$$

has the exact solution

$$y = e^{-\nu(t-t_0)}(1+t)^{-a}(1+t_0)^a.$$

To derive this solution, it may help to recall that $\eta = \frac{1}{1+t}$. Applying this to (52) with $a = \frac{k+2}{2}$ and using Duhamel's formula, we obtain

$$\begin{aligned} B_{k+2}(t) &= e^{-\nu(t-t_0)}(1+t)^{-\frac{k+2}{2}}(1+t_0)^{\frac{k+2}{2}} B_{k+2}(0) + D_k^{k+2}(t) \\ &+ \sum_{\ell=1}^{\frac{k+2}{2}-1} D_{k-2\ell}^{k+2}(t) \end{aligned} \tag{53}$$

where the Duhamel terms $D_{k-2\ell}^{k+2}$ satisfy

$$\begin{aligned} &D_{k-2\ell}^{k+2}(t) \\ &\sim \frac{C}{\nu^{2\ell+1}} \int_{t_0}^{t} e^{-\nu(t-s)}(1+t)^{-\frac{k+2}{2}}(1+s)^{\frac{k+2}{2}}(1+s)^{-\ell-1}\frac{1}{\nu^{\frac{k-2\ell}{2}}}e^{-\nu s}ds. \end{aligned} \tag{54}$$

Notice in the above Duhamel term, we have substituted, using the induction hypothesis $|B_{k-2\ell}(t)| \leq \frac{C}{\nu^{\frac{k-2\ell}{2}}}e^{-\nu t}$ (we also assume $t > t_0 > \frac{1}{\nu}$). These Duhamel terms are the most slowly decaying terms in the solution formula (53). Proceeding, we simplify (54) and obtain (for all ℓ),

$$\begin{aligned} D_{k-2\ell}^{k+2}(t) &\sim \frac{C}{\nu^{\frac{k}{2}+\ell+1}} e^{-\nu t}(1+t)^{-\left(\frac{k+2}{2}\right)}\left((1+t)^{\frac{k}{2}-\ell+1} - (1+t_0)^{\frac{k}{2}-\ell+1}\right) \\ &= \frac{C}{\nu^{\frac{k}{2}+1}} e^{-\nu t}\left(\frac{1}{\nu^\ell(1+t)^\ell} - \frac{1}{\nu^\ell(1+t_0)^\ell}\frac{(1+t_0)^{\frac{k}{2}+1}}{(1+t)^{\frac{k}{2}+1}}\right). \end{aligned}$$

Now since $t > t_0 > \frac{1}{\nu}$, we obtain

$$|D_{k-2\ell}^{k+2}(t)| \leq \frac{C}{\nu^{\frac{k+2}{2}}} e^{-\nu t},$$

and subsequently we obtain, for $t > t_0 > \frac{1}{\nu}$,

$$|B_{k+2}(t)| \leq \frac{C}{\nu^{\frac{k+2}{2}}} e^{-\nu t}$$

as desired. □

References

1. Aris, R.: On the dispersion of a solute in a fluid flowing through a tube. Proc. Roy. Soc. Lond. Ser. A **235**(1200), 67–77 (1956)
2. Beck, M., Eugene Wayne, C.: Metastability and rapid convergence to quasi-stationary bar states for the two-dimensional Navier-Stokes equations. Proc. Roy. Soc. Edinb. Sect. A Math. **143**, 905–927, 10 (2013)
3. Chatwin, P.C., C.M. Allen, C.M.: Mathematical models of dispersion in rivers and estuaries. Ann. Rev. Fluid Mech. **17**, 119–149 (1985)
4. Chen, X.-Y., Hale, J. K., Tan, B.: Invariant foliations for C^1 semigroups in Banach spaces. J. Differ. Equ. **139**(2), 283–318 (1997)
5. Eugene Wayne, C.: Invariant manifolds for parabolic partial differential equations on unbounded domains. Arch. Ration. Mech. Anal. **138**(3), 279–306 (1997)
6. Gallay, T., Eugene Wayne, C.: Invariant manifolds and the long-time asymptotics of the Navier-Stokes and vorticity equations on $\mathbb{R}^2$. Arch. Ration. Mech. Anal. **163**(3), 209–258 (2002)
7. Latini, M., Bernoff, A.J.: Transient anomalous diffusion in Poiseuille flow. J. Fluid Mech. **441**(8), 399–411, (2001)
8. Mercer, G.N., Roberts, A.J.: A centre manifold description of contaminant dispersion in channels with varying flow properties. SIAM J. Appl. Math. **50**(6), 1547–1565 (1990)
9. Taylor, G.: Dispersion of soluble matter in solvent flowing slowly through a tube. Proc. Roy. Soc. Lond. Ser. A **219**(1137), 186–203 (1953)
10. Taylor, G.: Dispersion of matter in turbulent flow through a tube. Proc. Roy. Soc. Lond. Ser. A **223**(1155), 446–468 (1954)

Normal Form Transformations for Capillary-Gravity Water Waves

Walter Craig and Catherine Sulem

Abstract This paper addresses the equations of capillary-gravity waves in a two-dimensional channel of finite or infinite depth. These equations are considered in the framework of Hamiltonian systems, for which the Hamiltonian energy has a convergent Taylor expansion in canonical variables near the equilibrium solution. We give an analysis of the Birkhoff normal form transformation that eliminates third-order non-resonant terms of the Hamiltonian. We also provide an analysis of the dynamics of remaining resonant triads in certain cases, related to Wilton ripples.

1 Introduction

The system of equations for free surface water waves is known to have a formulation as a Hamiltonian partial differential equation [3, 14]. In this article, we consider the case of capillary-gravity waves in a two-dimensional channel with periodic lateral boundary conditions, with either finite of infinite depth. The Hamiltonian is analytic in natural canonical conjugate variables, and the nth term of its Taylor expansion about equilibrium is associated with n-wave interactions. We perform a Birkhoff normal form transformation to eliminate all non-resonant cubic terms from the Hamiltonian. We show that this transformation is a well-defined and continuous canonical change of variables in a neighbourhood of zero in a fixed Sobolev space, and moreover it is C^1 in the sense that its Jacobian is a bounded map on a slightly larger Sobolev space.

This work is motivated by a number of open questions in the theory of water waves. The first is the question of long-time existence of solutions for small initial data. In the case of infinite horizontal extent, there have been a number of recent

W. Craig
The Fields Institute, 222 College Street, Toronto, ON, Canada
Department of Mathematics, McMaster University, Hamilton, ON, Canada
e-mail: craig@math.mcmaster.ca

C. Sulem (✉)
Department of Mathematics, University of Toronto, Toronto, ON, Canada
e-mail: sulem@math.toronto.edu

P. Guyenne et al. (eds.), *Hamiltonian Partial Differential Equations and Applications*, Fields Institute Communications 75,
DOI 10.1007/978-1-4939-2950-4_3

results, including [8, 13] who give exponentially long existence times and [1, 10] who extended this to global solutions. These are results for infinite depth and zero surface tension. The question in the periodic case is open, and more difficult because of the lack of dispersive decay estimates. It is important because of the effort to derive a rigorous justification of the modulational approximation offered by the NLS equation, in the more natural setting in which the solutions are not dispersing to zero. The NLS approximation has been given a rigorous justification in the case of an infinite horizontal domain by Totz and Wu [12] and by Düll, et al. [6], following the initial analysis of Craig, et al. [4].

Another interesting aspect of a rigorous normal form transformation is that it exhibits certain classes of special solutions. In the setting of this paper, the remaining resonant terms after the third order Birkhoff normal form transformation take the form of coupled resonant triads, related to the classical Wilton ripples. In the presence of a single resonant triad, the dynamics is that of the integrable three-wave system. In cases of higher numbers of coupled triads, the dynamics are more complicated [2, 9]. The normal forms transformation in the present paper gives a rigorous justification of the behaviour of resonant triads models, over long time intervals, for the initial value problem.

A useful aspect of normal forms given through canonical transformation is that, first of all, they preserve the Hamiltonian character of the underlying equations of motion and the principle of conservation of energy. Secondly, these transformations can in principle be repeated, resulting in a normal form for higher order terms in the Hamiltonian, and eliminating non-resonant higher order nonlinearities. On a formal level, normal forms are described up to fourth order in [7].

Finally, normal form transformations play a central role in Zakharov's theory of wave turbulence, In this, nonlinear wave interactions are reduced to resonant submanifolds under canonical changes of variables [11, 15, 16]. Any effort to make a rigorous analysis of this picture of wave turbulence will need to understand the analytic properties of such transformations.

In the present paper, Sect. 2 describes the Hamiltonian for capillary-gravity water waves, and transforms it to complex symplectic coordinates. In Sect. 3, we describe the Birkhoff normal form and solve the cohomological equation. Section 4 gives the key result of the paper, namely that the time-one solution map of the Hamiltonian vector field is well-defined and continuous in an appropriately defined scale of energy spaces. This result is based on energy estimates for solutions. Moreover, the solution map of the vector field is smooth on this scale of spaces in the sense that the Jacobian of the solution and subsequent higher derivatives are bounded in slightly larger spaces of the scale. Section 5 is a study of the normal form Hamiltonian itself, truncated at fourth order, in certain specific cases of triad and multiple triad interaction. In the case of a single triad, the system reduces to two decoupled subsystems of one degree of freedom apiece, corresponding to two independent copies of the three-wave resonant system. We also consider a particular case of two coupled resonant triads giving rise to a Hamiltonian system with two degrees of freedom for which we find the stationary points and analyze their respective stability. The normal forms transformation gives a rigorous justification of the relevance of these finite dimensional dynamical systems to model the dynamics of the full water wave system, at least over long periods of time.

2 Water Waves Equations

2.1 *The Classical Equations and Their Hamiltonian Formulation*

The classical water problem refers to the movement of an ideal incompressible fluid in the presence of gravity and surface tension. Making the usual oceanographers assumption that the fluid is irrotational, it is described by a potential flow $u = \nabla\varphi$ satisfying

$$\Delta\varphi = 0 \,, \tag{1}$$

in the fluid domain $S(t;\eta) = \{(x,y) : x \in \mathbb{R}, -h < y < \eta(x,t)\}$, where η is the surface elevation. The boundary condition on the fixed bottom $\{y = -h\}$ of the fluid is

$$-\partial_y\varphi(x,-h) = 0 \,. \tag{2}$$

On the interface $\{y = \eta(x,t)\}$, two boundary conditions are imposed, namely

$$\begin{aligned} &\partial_t\eta = \partial_y\varphi - \partial_x\eta\,\partial_x\varphi \,, \\ &\partial_t\varphi + \frac{1}{2}|\nabla\varphi|^2 + g\eta - \sigma\partial_x\Big(\frac{\partial_x\eta}{(1+|\partial_x\eta|^2)^{1/2}}\Big) = 0 \,, \end{aligned}$$

where g is the acceleration of gravity and σ the coefficient of surface tension. We assume periodic boundary condition in horizontal direction, $\eta(x+2\pi) = \eta(x), \xi(x+2\pi) = \xi(x)$.

It is well-known that this system has a Hamiltonian formulation [14] with canonical variables $\eta(x,t)$ and $\xi(x,t) = \varphi(x,\eta(x,t),t)$, in the form

$$\partial_t\begin{pmatrix}\eta\\ \xi\end{pmatrix} = \begin{pmatrix}0 & I\\ -I & 0\end{pmatrix}\begin{pmatrix}\delta_\eta H\\ \delta_\xi H\end{pmatrix} = J\,\delta H \tag{3}$$

with the Hamiltonian being given by the expression of the total energy

$$\begin{aligned} H(\eta,\xi) &= \frac{1}{2}\int\int_{-h}^{\eta(x)} |\nabla\varphi|^2\,dydx + \int(\frac{g}{2}\eta^2 + \sigma\sqrt{1+|\partial_x\eta|^2})\,dx \\ &= \int(\frac{1}{2}\xi(x)G(\eta)\xi(x) + \frac{g}{2}\eta^2(x) + \sigma\sqrt{1+|\partial_x\eta|^2})\,dx \,. \end{aligned} \tag{4}$$

Here $G(\eta)$ is the Dirichlet-Neumann operator which associates to the Dirichlet data ξ on the curve $y = \eta(x)$ the normal derivative of the harmonic function φ, with a normalized factor, namely, $\partial_n\varphi\sqrt{1+|\partial_x\eta|^2}$. The other conserved quantities are the mass $M(\eta,\xi) = \int \eta(x)dx$ and the horizontal momentum $I(\eta,\xi) = \int \xi(x)\partial_x\eta(x)dx$. Defining the Poisson bracket as

$$\{F,G\} = \int (\partial_\eta F \partial_\xi G - \partial_\xi F \partial_\eta G)dx,$$

one can check that the conserved quantities M and I Poisson-commute with the Hamiltonian H,

$$\{H,M\} = 0, \qquad \{H,I\} = 0.$$

2.2 Complex Symplectic Coordinates

The Dirichlet-Neumann operator is analytic in η, given in Taylor series by

$$G(\eta)\xi = \sum_{m=0}^{\infty} G^{(m)}(\eta),$$

with the property that each term in the Taylor expansion $G^{(m)}$ is homogeneous of degree m, $G^{(m)}(\lambda\eta) = \lambda^m G^{(m)}(\eta)$. In particular, the two first terms in the expansion are $G^{(0)} = D\tanh(hD)$, $G^{(1)} = D\eta D - G^{(0)}\eta G^{(0)}$ where $D = (1/i)\partial_x$. In turn, the Hamiltonian has an expansion in the form

$$H(\eta,\xi) = H^{(2)} + H^{(3)} + \ldots. + H^{(m)} + R^{(m+1)} \tag{5}$$

where

$$H^{(2)} = \frac{1}{2}\int_0^{2\pi} \left(\xi G^{(0)}\xi + g\eta^2 + \sigma|\partial_x\eta|^2\right)dx,$$

$$H^{(3)} = \frac{1}{2}\int_0^{2\pi} \xi(D\eta D - G^{(0)}\eta G^{(0)})\xi dx,$$

with similar expressions for higher order $H^{(m)}$, and where $R^{(m+1)}$ is the Taylor remainder.

We consider a periodic setting, i.e. $\eta(x+2\pi k,t) = \eta(x,t)$ and $\xi(x+2\pi k,t) = \xi(x,t)$, writing η and ξ as Fourier series

$$\eta(x) = \frac{1}{\sqrt{2\pi}}\sum_k \eta_k e^{ikx}, \quad \xi(x) = \frac{1}{\sqrt{2\pi}}\sum_k \xi_k e^{ikx}.$$

Since mass is conserved, we can assume, without loss of generality, that the *zero*th Fourier coefficient η_0 vanishes. Then,

$$H^{(2)} = \frac{1}{2}\sum_k k\tanh(hk)|\xi_k|^2 + (g+\sigma k^2)|\eta_k|^2),$$

$$H^{(3)} = \frac{1}{2\sqrt{2\pi}} \sum_{k_1+k_2+k_3=0} (-k_1k_3 - G^{(0)}_{k_1}G^{(0)}_{k_3})\, \xi_{k_1}\eta_{k_2}\xi_{k_3},$$

where $G^{(0)}_k = k\tanh(hk)$. Also note that the *zero*th Fourier coefficient ξ_0 of ξ does not appear in the Hamiltonian. It is convenient to introduce the complex symplectic coordinates

$$z_k = \frac{1}{\sqrt{2}}(a_k\eta_k + ia_k^{-1}\xi_k), \tag{6}$$

or equivalently,

$$\eta_k = \frac{1}{\sqrt{2}}a_k^{-1}(z_k + \bar{z}_{-k})\,,\quad \xi_k = \frac{1}{\sqrt{2}i}a_k(z_k - \bar{z}_{-k}), \tag{7}$$

with the coefficients a_k defined by

$$a_k^2 = \left(\frac{g+\sigma k^2}{k\tanh(hk)}\right)^{1/2}.$$

The dispersion relation

$$\omega_k^2 = (g+\sigma k^2)k\tanh(hk),$$

expresses the temporal frequencies of the normal modes of the linearized system, as given by the quadratic Hamiltonian $H^{(2)}$. A key distinction between the case of pure gravity waves and gravity—capillary waves is that the dispersion relation grows as a $3/2$ power in wavenumber k in the latter case, as compared with a $1/2$ power in the case of pure gravity waves. Using the dispersion relation, we have the identities $a_k^2\omega_k = g+\sigma k^2$ and $\omega_k/a_k^2 = k\tanh(hk) = G_k^{(0)}$.

In terms of complex symplectic coordinates, the system (3) becomes

$$\partial_t z = \frac{1}{i}\partial_{\bar{z}}H. \tag{8}$$

The quadratic part $H^{(2)}$ of the Hamiltonian takes the simple form

$$H^{(2)} = \sum_k \omega_k|z_k|^2 \tag{9}$$

while the cubic order term $H^{(3)}$ is

$$H^{(3)} = \frac{1}{8\sqrt{\pi}} \sum_{k_1+k_2+k_3=0} (k_1k_3 + G_1G_3)\frac{a_1a_3}{a_2}(z_1 - \bar{z}_{-1})(z_2 + \bar{z}_{-2})(z_3 - \bar{z}_{-3}) \quad (10)$$

where, for simplicity, we have dropped the k indices and denoted $z_j = z_{k_j}$, $a_j = a_{k_j}$, and $G_k = G_k^{(0)}$. We will use this notation when there is no possible confusion.

3 Birkhoff Normal Forms

A Birkhoff normal form is a canonical change of variables up to a given order m, so that the Taylor expansion of the transformed Hamiltonian up to order m contains only resonant terms. A term in the Hamiltonian $H(z)$ is resonant at order m when

$$\sum_{j=1}^{l} \omega_{k_j} - \sum_{j=l+1}^{m} \omega_{k_j} = 0$$

and $k_1 + \ldots k_l + k_{l+1} + \ldots k_m = 0$. We do not include $k = 0$ in the sums because we have assumed that the zero modes of η and ξ vanish. In particular, a resonant triad takes the form

$$\omega_{k_1} - \omega_{k_2} - \omega_{k_3} = 0, \quad k_1 + k_2 + k_3 = 0, \quad k_j \neq 0. \quad (11)$$

In the presence of surface tension and gravity, there are possible non trivial resonant triads. The resulting gravity-capillary waves are known as Wilton ripples, at least this applies to the standing wave solutions. In the case of a periodic domain, generically these resonant triads do not appear, but for certain choices of parameters (g, h, σ) there can be a finite number of such triads. The maximum wave number k_j involved in a resonant triad is bounded by a constant $C = C(g, h, \sigma)$ that depends locally uniformly upon these parameters.

3.1 *Canonical Transformations*

We perform a canonical change of variables

$$\tau : v = (\eta, \xi) \to w = (\eta', \xi') \quad (12)$$

on the Hamiltonian

$$\tilde{H}(w) = H(v) = H \circ \tau^{-1}(w)$$

by the Lie method of giving τ as the time-one flow associated to a Hamiltonian K:

$$\frac{d}{ds}\psi_s = X^K(\psi_s), \text{ with } \psi_s(w)_{|s=0} = w, \ \tilde{H}(w) = H(\psi_s(w))_{|s=-1}. \tag{13}$$

This is a canonical transformation preserving the Hamiltonian character of the system. A Taylor series expansion near $s = 0$ of the new Hamiltonian $\tilde{H}$ gives

$$\tilde{H}(w) = H(\psi_s(w))_{|s=0} - \frac{d}{ds}H(\psi_s(w))_{|s=0} + \frac{1}{2}\frac{d^2}{ds^2}H(\psi_s(w))_{|s=0} - \dots \tag{14}$$

As a formal expression at least in the above equation, we have

$$\begin{aligned} H(\psi_s(w))_{|s=0} &= H(w) \\ \frac{d}{ds}H(\psi_s(w))_{|s=0} &= \int (\partial_\eta H\frac{d\eta}{ds} + \partial_\xi H\frac{d\xi}{ds})\, dx = \int (\partial_\eta H \partial_\xi K - \partial_\xi H \partial_\eta K) dx \\ &\equiv \{H, K\}\,, \end{aligned}$$

with similar formulas for higher orders of s-derivatives, thus giving the expression

$$\tilde{H}(w) = H(w) - \{K, H\}(w) + \frac{1}{2}\{K, \{K, H\}\}(w) + \dots. \tag{15}$$

3.2 Third-Order Cohomological Equation

The expression (15) represents an ordering of H and K in terms of powers of homogeneity with respect to the variables $z, \bar{z}$. Returning to the expansion (5) of H in terms of (η, ξ), we apply the canonical transformation associated to Hamiltonian K on each term: The transformed Hamiltonian has the form [5]

$$\begin{aligned} \tilde{H}(w) = &H^{(2)}(w) + H^{(3)}(w) + \dots \\ &-\{K, H^{(2)}\}(w) - \{K, H^{(3)}\}(w) - \dots. \\ &+\frac{1}{2}\{K, \{K, H^{(2)}\}\}(w) + \frac{1}{2}\{K, \{K, H^{(3)}\}\}(w) + \dots \end{aligned} \tag{16}$$

If K is homogeneous of degree m, its Poisson bracket with $H^{(n)}$ (homogeneous of degree n) will be of degree $m + n - 2$. Thus if we can find $K = K^{(3)}$ homogeneous of degree 3 satisfying the relation

$$\{H^{(2)}, K^{(3)}\} + H^{(3)} = 0, \tag{17}$$

we will have eliminated the cubic terms in the transformed Hamiltonian $\tilde{H}$.

The proposition below states that it is indeed possible to solve the cohomological equation (17) explicitly, removing all cubic terms except the resonant terms of $H^{(3)}$.

Proposition 1. *The solution of the cohomological equation (17) is given by*

$$
\begin{aligned}
K^{(3)} =& \frac{1}{\sqrt{\pi}} \sum_{k_1+k_2+k_3=0} (k_1k_3 + G_1G_3)\frac{a_1a_3}{a_2}\frac{z_1z_2z_3 - \bar{z}_{-1}\bar{z}_{-2}\bar{z}_{-3}}{\omega_1 + \omega_2 + \omega_3} \\
&- \sum_{k_1+k_2+k_3=0} (k_1k_3 + G_1G_3)\frac{a_1a_3}{a_2}2\frac{z_1z_2\bar{z}_{-3} - \bar{z}_{-1}\bar{z}_{-2}z_3}{\omega_1 + \omega_2 - \omega_3} \\
&+ \sum_{k_1+k_2+k_3=0} (k_1k_3 + G_1G_3)\frac{a_1a_3}{a_2}\frac{z_1\bar{z}_{-2}z_3 - \bar{z}_{-1}z_2\bar{z}_{-3}}{\omega_1 - \omega_2 + \omega_3} + P,
\end{aligned} \tag{18}
$$

where the three sums are performed for triads (k_1, k_2, k_3), with $k_1 + k_2 + k_3 = 0$ excluding the resonant terms for which the corresponding denominator vanishes. The term P consists of the finite sum of exceptional terms. That is, it consists of the non resonant terms of $K^{(3)}$ for which (k_1, k_2, k_3) possesses a resonant triad. Generically, $P = 0$.

Proof. This equation can be solved easily in complex symplectic coordinates which diagonalize the linear operation of taking Poisson bracket with $H^{(2)}$ (the $\mathrm{ad}_{H^{(2)}}$ action). Indeed, the Poisson bracket of $H^{(2)}$ acting on monomials of the form $z_{k_1}z_{k_2}\bar{z}_{-k_3}$ is simply a multiplicative factor:

$$
\{H^{(2)}, z_{k_1}z_{k_2}\bar{z}_{-k_3}\} = \frac{1}{i}(\omega_{k_1} + \omega_{k_2} - \omega_{k_3})z_{k_1}z_{k_2}\bar{z}_{-k_3}. \tag{19}
$$

We thus look for $K^{(3)}$ in the form of a linear combination of all possible monomials of degree 3 of the form similar to that above and we identify the coefficients, which is possible as long as the corresponding mutliplicative factor $(\omega_1 \pm \omega_2 \pm \omega_3)$ does not vanish. This leads to $K^{(3)}$ given in the form of (18) where, for simplicity, we have denoted $z_j = z_{k_j}$.

It is useful for the analysis to rewrite $K^{(3)}$ in terms of the (η, ξ) variables. After some algebraic manipulations, one finds

$$
\begin{aligned}
K^{(3)} =& \frac{1}{\sqrt{2\pi}} \sum_{k_1+k_2+k_3=0} \frac{k_1k_3 + G_1G_3}{d(\omega_1, \omega_2, \omega_3)}\Big[a_1^2\omega_1(\omega_1^2 - \omega_2^2 - \omega_3^2)\eta_{k_1}\eta_{k_2}\xi_{k_3} \\
&+ \frac{a_1^2a_3^2}{a_2^2}\omega_1\omega_2\omega_3\eta_{k_1}\xi_{k_2}\eta_{k_3} + \frac{1}{2a_2^2}\omega_2(\omega_1^2 - \omega_2^2 + \omega_3^2)\xi_{k_1}\xi_{k_2}\xi_{k_3}\Big] + P
\end{aligned} \tag{20}
$$

where

$$
d(\omega_1, \omega_2, \omega_3) = (\omega_1 + \omega_2 + \omega_3)(\omega_1 + \omega_2 - \omega_3)(\omega_1 - \omega_2 + \omega_3)(\omega_1 - \omega_2 - \omega_3), \tag{21}
$$

and the summation is performed over all triads (k_1, k_2, k_3) such that $k_1 + k_2 + k_3 = 0$ and $d(\omega_1, \omega_2, \omega_3) \neq 0$. Finally, the term P in the RHS of (20) contains finite sums of the alternate triplets to resonant triads. Namely, it contains the terms in the first and last sums in (18) corresponding to triads for which the denominator $\omega_1 + \omega_2 - \omega_k$ of the second sum vanishes, and the terms of the first and second sums in (18) corresponding to triads for which the denominator $\omega_1 - \omega_2 - \omega_k$ of the third sum vanishes. □

Remark 1. This calculation has been performed in the case of finite depth $0 < h < \infty$. In the case $h = \infty$, the same expression holds, with the substitution $G^{0)} = |D|$.

3.3 Transformation to Third-Order Normal Form

The new coordinates $(\tilde{\eta}, \tilde{\xi})$ are obtained as the solutions at $s = -1$ of the system of equations

$$\frac{d}{ds}\begin{pmatrix}\eta\\ \xi\end{pmatrix} = \begin{pmatrix}0 & I\\ -I & 0\end{pmatrix}\begin{pmatrix}\partial_\eta K^{(3)}\\ \partial_\xi K^{(3)}\end{pmatrix} := X^{K^{(3)}} \tag{22}$$

with the (initial) condition at $s = 0$ being the original variables $(\eta, \xi)(t)$. Equivalently, in Fourier space,

$$\begin{aligned}\frac{d}{ds}\eta_{-k} &= \partial_{\xi_k} K^{(3)}\\ \frac{d}{ds}\xi_{-k} &= -\,\partial_{\eta_k} K^{(3)}\end{aligned} \tag{23}$$

with the RHS given by

$$\begin{aligned}\sqrt{2\pi}\,\partial_{\xi_k} K^{(3)} = &\Sigma'_{k_1+k_2+k=0}\Big[\frac{k_1 k + G_1 G_k}{d_{k_1 k_2 k}}(g+\sigma k_1^2)(\omega_1^2 - \omega_2^2 - \omega_k^2)\eta_{k_1}\eta_{k_2}\\ &+ \frac{k_1 k_2 + G_1 G_2}{d_{k_1 k_2 k}}(g+\sigma k_1^2)(g+\sigma k_2^2)k\tanh(hk)\eta_{k_1}\eta_{k_2}\Big]\\ &+ \Big[\frac{k_1 k + G_1 G_k}{d_{k_1 k_2 k}} k_2\tanh(hk_2)(\omega_1^2 - \omega_2^2 + \omega_k^2)\xi_{k_1}\xi_{k_2}\\ &+ \frac{k_1 k_2 + G_1 G_2}{2 d_{k_1 k_2 k}} k\tanh(hk)(\omega_1^2 - \omega_k^2 + \omega_2^2)\xi_{k_1}\xi_{k_2}\Big],\end{aligned} \tag{24}$$

and

$$\sqrt{2\pi}\,\partial_{\eta_k} K^{(3)} = \Sigma'_{k_1+k_2+k=0}\Big[\frac{kk_2 + G_k G_2}{d_{kk_1k_2}}(g+\sigma k^2)(\omega_k^2 - \omega_1^2 - \omega_2^2)\eta_{k_1}\xi_{k_2}$$
$$+ \frac{k_1k_2 + G_1G_2}{d_{k_1kk_2}}(g+\sigma k_1^2)(\omega_1^2 - \omega_k^2 - \omega_2^2)\eta_{k_1}\xi_{k_2}$$
$$+ 2\frac{kk_1 + G_kG_1}{d_{kk_2k_1}}(g+\sigma k^2)(g+\sigma k_1^2)k_2\tanh(hk_2)\eta_{k_1}\xi_{k_2}\Big], \tag{25}$$

where $d_{k_1k_2k_3} = d(\omega_1, \omega_2, \omega_3)$, and the notation Σ' indicates that the summation is performed over all triads (k_1, k_2, k_3) satisfying $k_1 + k_2 + k_3 = 0$ and $d_{k_1k_2k_3} \neq 0$. In case of the presence of resonances, there is a finite number of exceptional terms. For convenience of estimates, we assume in this and the following section that we are in the generic case and there are no resonant triads.

4 Analysis of the Normal Form Transformation

Let H^r denote the Sobolev space of order r, equipped with the norm $\|f\|^2_{H^r} = \sum_k \langle k\rangle^{2r}|f_k|^2$, where $\langle\, k\,\rangle = (1+|k|^2)^{1/2}$. Define the energy norm

$$\|(\eta,\xi)\|^2_{E^r} = \frac{1}{2}(\langle \eta, \sigma|D|^2\eta\rangle_r + \langle \xi, G_0\xi\rangle_r) \tag{26}$$

and the energy space $E^r \simeq H^{r+1} \times H^{r+1/2}$. We denote B_R be the ball centered at the origin, of radius R of the energy space. Define a transformation $w = \tau(v) := \psi_s(v)|_{s=-1}$ given by the time-one solution map of the Hamiltonian vector field $X^{K^{(3)}}$ (22) with the auxiliary Hamiltonian $K^{(3)}$. This map is well defined and continuous in a neighbourhood of the origin $B_R \subseteq E^r$ because of the following result.

Theorem 1. *There exists $R_0 > 0$ such that, for all $R < R_0$, the canonical transformation $\tau : v \to w$ defined in* (13) *is continuous on $B_R \subseteq E^r$, with continuous inverse τ^{-1}, and it satisfies $\tau : B_{R/2} \to B_R$ and $\tau^{-1} : B_{R/2} \to B_R$. This transformation removes all non resonant cubic terms from the Hamiltonian. The Jacobian $\partial_{(\eta,\xi)}\tau$ of the transformation is bounded on the energy space $E^{r-1/2} \to E^{r-1/2}$.*

The proof of the existence of the mapping τ is based on an energy estimate for the vector field (23), and the result for the Jacobian follows from a similar energy estimate for the variational equation of (23). From this, the existence of the solution to (22) for $s \in [-1, 1]$ is obtained by a fixed point argument for a sequence of approximations, under the condition that the ball $B_R \subseteq E^r$ in which one takes the initial data is of sufficiently small radius.

Theorem 2. *Given $(\eta, \xi) \in E^r$, the vector field $X^{K^{(3)}}$ satisfies the following energy inequality*

$$|\langle(\eta, \xi), X^{K^{(3)}}(\eta, \xi)\rangle_{E^r}| \leq C\|(\eta, \xi)\|_{E^r}^3. \tag{27}$$

The proof of Theorem 2 is given in Sects. 4.3 and 4.4, using some technical results presented in Sects. 4.1 and 4.2.

Theorem 3. *The flow $\psi_s(\eta, \xi)$ of the vector field $X^{K^{(3)}}$ satisfies the following estimates on the scale of energy spaces E^p, $0 \leq p \leq r$. For $\|(\eta, \xi)\|_{E^r} \leq R$, and $|s| < s_R$, the Jacobian satisfies*

$$\|\big(\partial_{(\eta,\xi)}\psi_s(\eta, \xi) - I\big)\,(\tilde{\eta}, \tilde{\xi})^{\mathrm{T}}\|_{E^{r-1/2}} \;\leq\; C_1\, R\,\|(\tilde{\eta}, \tilde{\xi})\|_{E^{r-1/2}}. \tag{28}$$

Higher derivatives of the flow satisfy

$$\|\partial^q_{(\eta,\xi)}\psi_s(\eta, \xi)\|_{E^{r-q/2}} \;\leq\; C_{r,q}. \tag{29}$$

It follows from this result that the transformation $\tau(\eta, \xi) = \psi_{s=-1}(\eta, \xi)$ is smooth on the scale of spaces E^p, $0 \leq p \leq r$, in the sense that for $(\eta, \xi) \in B_R \subseteq E^r$, the derivatives $\partial^q_{(\eta,\xi)}\tau \;:\; E^{r-q/2} \to E^{r-q/2}$ are continuous. For this, we require $R \leq R_0$, so that the guaranteed existence time for the flow satisfies $s_R > 1$. The proof of Theorem 3 is given in Sect. 4.5.

Proof of Theorem 1. The canonical transformation that we seek is designed as the time $s = -1$ image of the solution map for Eq. (22). The question is whether the solution exists and its regularity with respect to its initial data $(\eta, \xi)|_{s=0} \in E^r$. We will show that the solution map exists and is continuous on each ball $B_R \subset E^r$ for some interval $-s_R < s < s_R$, and that for sufficiently small R the bound $s_R > 1$. The desired transformation is $\tau = \psi_{s=-1}$ while $\tau^{-1} = \psi_{s=1}$. The vector field is not Lipschitz continuous in any reasonable Banach space, not even locally. We proceed with a strategy for the existence which is well-known from the theory of symmetric hyperbolic systems. Namely, one solves an approximate equation which has smooth solutions and takes the limit. In the case at hand,

$$\frac{d}{ds}\begin{pmatrix}\eta\\ \xi\end{pmatrix} = X^{K^{(3)}}(\eta, \xi) + \alpha\Delta\begin{pmatrix}\eta\\ \xi\end{pmatrix} \tag{30}$$

with parameter $0 < \alpha \ll 1$. Using the energy estimates of $X^{K^{(3)}}$ of Theorem 2,

$$\begin{aligned}\frac{d}{ds}\|(\eta^{(\alpha)}, \xi^{(\alpha)})\|_{E^r}^2 &\leq C\|(\eta^{(\alpha)}, \xi^{(\alpha)})\|_{E^r}^3\\ &\quad - \alpha\|(\eta^{(\alpha)}, \xi^{(\alpha)})\|_{E^{r+1}}^2\,.\end{aligned} \tag{31}$$

With initial data $(\eta^{(\alpha)}, \xi^{(\alpha)}) = (\eta, \xi) \in B_R$, solutions $(\eta^{(\alpha)}(s), \xi^{(\alpha)}(s))$ exist over a time interval $(-s_R, s_R)$, uniformly in α. This gives rise to a family of continuous curves $(\eta^{(\alpha)}(s), \xi^{(\alpha)}(s)) \in C((-s_R, s_R); E^r)$ which is bounded and equicontinuous in $C((-s_R, s_R); E^0)$ whose limit points as $\alpha \to 0$ are solutions of (22). The limit is unique in E^0 and hence in all E^r as well. When one takes $R < 1/C_0$, the time of existence satisfies $s_R > 1$ and the image of B_R, $\psi_{\pm 1}(B_R) \subset B_{R/2}$. □

4.1 *Lower Bound for the Denominator* $d(\omega_1, \omega_2, \omega_3)$

In this section, we examine the denominator $d(\omega_1, \omega_2, \omega_3)$ defined in (21) that appears in the formulas (24) and (25) of the vector field.

Lemma 1. *The set of resonant triads (k_1, k_2, k_3), i.e. such that $k_1 + k_2 + k_3 = 0$ and $d(\omega_1, \omega_2, \omega_3) = 0$ exists only on a compact set W of $\mathbb{R}^3$.*

Proof. This is due to the change of concavity of the curve ω versus k. Assume without loss of generality that $|k_1| > |k_2|$, $k_1 > 0$ and $k_2 < 0$. For $\omega_1 - \omega_2 - \omega_3$ to vanish, one needs that the curve given by the graph of $\omega(k)$, denoted $\mathscr{C}$, starting from the origin point of coordinates O intersects the curve $\mathscr{C}'$ that starts for the point $O' = (k_2, \omega_2)$. For this to happen, the origin O' needs to be in the region where the curve $\mathscr{C}$ is concave. There are thus only a finite number of triads that are resonant. Because of the periodic boundary conditions, for generic values of the parameters g, h, σ, there are no resonances. However, there are cases in which triads resonances and multiple coupled triad resonances occur. □

In order to proceed with the estimates, we divide the plane (k_1, k_2) in 4 sectors represented in Fig. 1 and defined as follows, using that $k_1 + k_2 + k_3 = 0$,

Sector (i) = $\{(k_1, k_2),\ |k_1| < \frac{1}{5}|k_2|, \text{ and } |k_3| \sim |k_2|\}$,
Sector (ii) = $\{(k_1, k_2),\ |k_2| < \frac{1}{5}|k_1|, \text{ and } |k_3| \sim |k_1|\}$,
Sector (iii) = $\{(k_1, k_2),\ |k_3| < \frac{1}{5}|k_1|, \text{ and } |k_2| \sim |k_1|\}$,
Sector (iv) = $\{(k_1, k_2),\ \text{all } |k_3|, |k_1|, |k_2| \text{ are comparable size}\}$.

Lemma 2. *The denominator $d(\omega_1, \omega_2, \omega_3)$ is bounded from below as follows:*

In region (iv), $d(\omega_1, \omega_2, \omega_3) \geq C\big(\langle k_1\rangle^{3/2} + \langle k_2\rangle^{3/2} + \langle k_3\rangle^{3/2}\big)^4$.
In regions (i), (ii) and (iii), $d(\omega_1, \omega_2, \omega_3) \geq C\langle k_1\rangle^2\langle k_2\rangle^2\langle k_3\rangle^2$.

Proof. First of all, $\omega(k) \sim \langle k\rangle^{3/2}$ for large $|k|$. Also, the expression $d(\omega_1, \omega_2, \omega_3)$ is even in k_1, k_2, k_3 and invariant after permutations of the arguments. It is thus sufficient to find for example a lower bound in sectors (iv) and (ii).

In sector (iv), $|k_3|, |k_1|, |k_2|$ are comparable size. Thus, each term in the product $d(\omega_1, \omega_2, \omega_3)$ is bounded from below by $C(\langle k_1\rangle^{3/2} + \langle k_2\rangle^{3/2} + \langle k_3\rangle^{3/2})$ and the estimate given in Lemma 2 is straightforward.

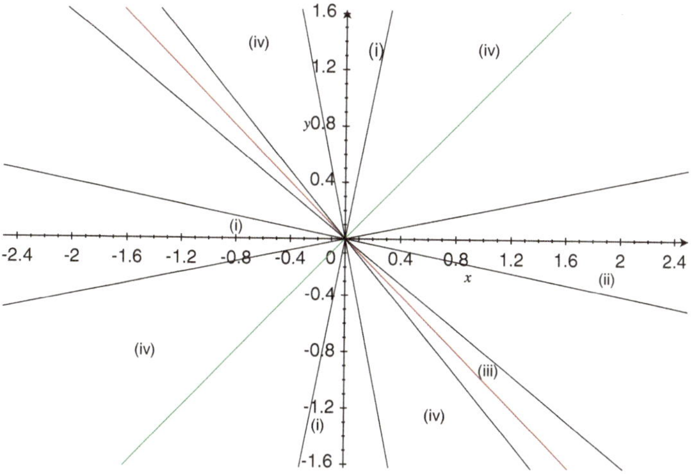

Fig. 1 Division of the plane (k_1, k_2)

In sector (ii) where $|k_2| < \frac{1}{5}|k_1|$, two of the factors appearing in the denominator are easily bounded :

$$\omega_1 + \omega_2 + \omega_3 \geq \langle k_1 \rangle^{3/2} + \langle k_2 \rangle^{3/2} + \langle k_3 \rangle^{3/2}$$

and

$$\omega_1 - \omega_2 + \omega_3 \geq \omega_3 \geq \langle k_3 \rangle^{3/2} .$$

In this sector, $\langle k_3 \rangle \sim \langle k_1 \rangle \geq 5\langle k_2 \rangle$. Thus

$$\omega_1 - \omega_2 + \omega_3 \geq C(\langle k_1 \rangle^{3/2} + \langle k_2 \rangle^{3/2} + \langle k_3 \rangle^{3/2}) .$$

It remains to bound the two other factors. Consider the factor $\omega_1 - \omega_2 - \omega_3$. In the region of (ii) where $k_1 > 0,\ k_2 > 0$, we have that $k_3 < 0$ and $|k_3| > |k_1|$. Thus, for some s_0 , $|k_1| < s_0 < |k_3|$,

$$\omega_3 - \omega_1 + \omega_2 = (|k_3| - |k_1|)\omega'(s_0) + \omega_2 \geq \langle k_2 \rangle\ C\langle k_3 \rangle^{1/2}. \tag{32}$$

However, in the region of (ii) where $k_1 > 0,\ k_2 < 0$, we have that $k_3 < 0$ and $|k_3| < |k_1|$. Thus,

$$\begin{aligned} \omega_1 - \omega_3 - \omega_2 &= \int_{|k_3|}^{|k_1|} \omega'(s)ds - \int_0^{|k_2|} \omega'(s)ds \\ &= |k_2|\ (\omega'(s_1) - \omega'(s_2)) , \end{aligned} \tag{33}$$

for some $|k_3| < s_1 < |k_1|$ and $0 < s_2 < |k_2|$. Since $|k_2| \leq |k_1|/5$, we have also $|k_2| \leq |k_3|/4$, and $|\sup_{s_1} \omega'(s_1) - \inf_{s_2} \omega'(s_2)| \geq C\langle k_3\rangle^{1/2}$ for k_3 sufficiently large. The region of (ii) where $k_1 < 0$ is treated similarly as well as the last factor in the denominator $\omega_1 + \omega_2 - \omega_3$. This concludes the proof of the lower bound for the denominator $d_{k_1 k_2 k_3}$. □

4.2 Auxiliary Estimates

Lemma 3. *Let $r_0 > 1/2$ and $A(p,q) = \sum_{k+k_1+k_2=0} A_{kk_1k_2} p_{k_1} q_{k_2}$. Suppose that for $k + k_1 + k_2 = 0$, the kernel $A_{kk_1k_2}$ satisfies*

$$|A_{kk_1k_2}| \leq C_0 \min\{\langle k_1\rangle^{-r_0}, \langle k_2\rangle^{-r_0}\}.$$

Then,

$$\|A(p,q)\|_{\ell^2} \leq C\|p\|_{\ell^2} \, \|q\|_{\ell^2} . \tag{34}$$

Proof. We divide the bilinear form into two sums, the sum over k_1 such that $|k| \leq 2|k_1|$ and the sum over k_1 such that $2|k_1| \leq |k|$.

In the first sum, the argument satisfies $|k| \leq 2|k_1$, therefore $|k_2| \leq 3|k_1|$. We first bound the ℓ^2 -norm by the norm of the operator with kernel $A_{kk_1k_2}p_{k_1}$,

$$\begin{aligned} C_1 := \sup_k \sum_{k_2} |A_{kk_1k_2}p_{k_1}| &\leq \sup_k \sum_{k_2} \frac{C_0}{\langle k_1\rangle^{r_0}} |p_{k_1}| \\ &\leq \sup_k \Big(\sum_{k_2} \frac{C_0^2}{\langle k+k_2\rangle^{2r_0}}\Big)^{1/2} \sup_k \Big(\sum_{k_1} |p_{k_1}|^2\Big)^{1/2} \\ &\leq C_0 \|p\|_{\ell^2}. \end{aligned} \tag{35}$$

Also,

$$\begin{aligned} C_2 := \sup_{k_2} \sum_{k} |A_{kk_1k_2}p_{k_1}| &\leq \sup_{k_2} \sum_{k} \frac{C_0}{\langle k_1\rangle^{r_0}} |p_{k_1}| \\ &\leq \sup_{k_2} \sum_{k} \frac{C_0}{\langle k+k_2\rangle^{r_0}} |p_{k+k_2}| \\ &\leq \sup_{k_2} \Big(\sum_{k} \frac{C_0^2}{\langle k+k_2\rangle^{2r_0}}\Big)^{1/2} \sup_k \Big(\sum_{k} |p_{k+k_2}|^2\Big)^{1/2} \\ &\leq C_0 \|p\|_{\ell^2}. \end{aligned} \tag{36}$$

The ℓ^2-norm of the first sum $A_1(p,q)$ is thus bounded by

$$\|A_1(p,q)\| \leq \sqrt{C_1 C_2}\|q\|_{\ell^2} \leq C_0\|p\|_{\ell^2}\|q\|_{\ell^2} . \tag{37}$$

The second summation A_2 follows similarly, given that the conditions

$$k_1 + k_2 + k = 0 , \quad \text{and} \quad 2|k_1| \leq |k|$$

imply

$$2(|k| - |k_2|) \leq 2|k + k_2| = 2|k_1| \leq |k|,$$

therefore, $|k| \leq 2|k_2|$. exchanging indices k_1 and k_2, and the role of p and q, the estimate for A_2 follows. □

This lemma also holds in the case of $k \in \mathbb{R}$ rather than $k \in \mathbb{Z}$. That is to say, for the problem posed on all of $\mathbb{R}$ rather than the periodic case.

The estimate obtained in the above lemma is however not strong enough to control all the terms appearing in estimates of Theorem 2 because the hypothesis on $A_{kk_1k_2}$ is too symmetric with respect to k_1, k_2. We sometimes need to examine the different regions separately and establish estimates accordingly.

Lemma 4. *Suppose the bilinear form $A(p,q)$ with kernel $A_{kk_1k_2}$ satisfies*

$$A(p,q) = \sum_{k+k_1+k_2=0} A_{kk_1k_2} p_{k_1} q_{k_2}$$

with

$$|A_{kk_1k_2}| \leq C_0 \langle k_1 \rangle^{-r_0} , \quad r_0 > 1 .$$

Then

$$\|A(p,q)\|_{\ell^2} \leq C|p|_{\ell^\infty} \|q\|_{\ell^2} \leq C\|p\|_{\ell^2} \|q\|_{\ell^2} . \tag{38}$$

The roles of p and q in this lemma can be inverted, therefore one achieves the same conclusion (38) if $|A_{kk_1k_2}| \leq C_0\langle k_2 \rangle^{-r_0}$ instead.

Proof. From the hypothesis, we have that

$$|A_{kk_1k_2} p_{k_1}| \leq a(k_1) = C_0 \langle k_1 \rangle^{-r_0} |p|_{\ell^\infty} . \tag{39}$$

For $r_0 > 1$, this is an ℓ^1-sequence. This majorant provides estimates of the norm of the linear operator with kernel $A_{kk_1k_2}p_{k_1}|_{k+k_1+k_2=0}$:

$$\sup_k \sum_{k_2} |A_{kk_1k_2}p_{k_1}| \le \sum_{k_2} C_0 |p_{k_1}|_{\ell^\infty} \frac{1}{\langle k+k_2\rangle^{r_0}}$$
$$\le C_0' |p_{k_1}|_{\ell^\infty} \text{ when } r_0 > 1. \tag{40}$$

Similarly,

$$\sup_{k_2} \sum_{k} |A_{kk_1k_2}p_{k_1}| \le C_0' |p_{k_1}|_{\ell^\infty} . \tag{41}$$

The bilinear ℓ^2 estimate follows from the simple fact that $|p|_{\ell^\infty} \le \|p\|_{\ell^2}$ on sequence spaces. □

The analysis also encounters the Dirichlet-Neumann operator in various terms and sectors. For example, in sector (i) (where $|k_1| \ll |k_2|, |k|$) the terms that stem from the near-commutator nature of the operator is exhibited by the fact that k and k_2 have opposite signs. Therefore

$$|kk_2 + G_kG_2| \le C\, e^{-h(|k|+|k_2|)} . \tag{42}$$

As a consequence, we have the following proposition.

Proposition 2. *Assume that the bilinear form $B(p,q) = \sum_{k+k_1+k_2=0} B_{kk_1k_2}p_{k_1}q_{k_2}$ has a kernel $b(k,k_1,k_2)$ satisfying one of conditions below :*

(a) its support is included in sector (i), and the near-commutator term is given by $kk_2 + G_kG_2$.
(b) its support is in sector (ii) and the near-commutator term is given by $kk_1 + G_kG_1$.

Then

$$\|B(p,q)\|_{\ell^2} \le C \|p\|_{\ell^2} \, \|q\|_{\ell^2} . \tag{43}$$

Proposition 3. *The bilinear form $B(p,q) = \sum_{k+k_1+k_2=0} B_{kk_1k_2}p_{k_1}q_{k_2}$ with kernel $B_{kk_1k_2}$ such that*

(a) it is supported in region (iii), and
(b) for $s - \beta = r_0 > 1$ has growth bounds of the form:

$$|B_{kk_1k_2}| = b(k,k_1,k_2)\frac{\langle k\rangle^s}{\langle k_1\rangle^s\langle k_2\rangle^s} , \text{ with } |b(k,k_1,k_2)| \le \langle k\rangle^\beta + \langle k_1\rangle^\beta + \langle k_2\rangle^\beta$$

gives rise to the estimate

$$\|B(p,q)\|_{\ell^2} \le C \|p\|_{\ell^2} \, \|q\|_{\ell^2} . \tag{44}$$

Proof. In region (iii), we have $|k| \ll |k_1|$ and $|k_2|$. Therefore

$$|B_{kk_1k_2}| = |b(k,k_1,k_2)|\frac{1}{\langle k_1\rangle^{s/2}\langle k_2\rangle^{s/2}}\,\frac{\langle k\rangle^s}{\langle k_1\rangle^{s/2}\langle k_2\rangle^{s/2}}$$
$$\leq \frac{\langle k\rangle^\beta + \langle k_1\rangle^\beta + \langle k_2\rangle^\beta}{\langle k_1\rangle^{s/2}\langle k_2\rangle^{s/2}} \leq \frac{C}{\langle k_1\rangle^{s-\beta}} \tag{45}$$

and because $|k_1|$ and $|k_2|$ are of the same order, this satisfies the hypothesis of Lemma 3. □

Proposition 4. *The bilinear form* $B(p,q) = \sum_{k+k_1+k_2=0} B_{kk_1k_2} p_{k_1} q_{k_2}$ *with kernel* $B_{kk_1k_2}$ *such that*

(a) it is supported in region (i), and
(b) for $s - \beta \geq r_0 > 1$ *has growth bounds of the form:*

$$|B_{kk_1k_2}| = b(k,k_1,k_2)\frac{\langle k\rangle^s}{\langle k_1\rangle^s\langle k_2\rangle^s}\,, \text{ with } |b(k,k_1,k_2)| \leq \langle k_1\rangle^\beta \tag{46}$$

gives rise to the estimate

$$\|B(p,q)\|_{\ell^2} \leq C\|p\|_{\ell^2}\,\|q\|_{\ell^2}\,. \tag{47}$$

Proof. The region (i) is defined as the sector in the (k_1,k_2) plane where $|k_1| \ll |k_2|, |k|$. Therefore,

$$|B_{kk_1k_2}| = |b(k,k_1,k_2)|\,\frac{\langle k\rangle^s}{\langle k_1\rangle^s\langle k_2\rangle^s} \leq \langle k_1\rangle^{\beta-s}. \tag{48}$$

For $s - \beta = r_0 > 1$, the kernel satisfies the hypothesis of Lemma 4. □

Note: The same conclusion holds under the hypothesis that (a) the kernel $B_{kk_1k_2}$ is supported in region (ii) and (b) the estimate (46) is true with the role of k_1 and k_2 exchanged.

4.3 Energy Estimates for Vector Field $X^{K^{(3)}}$

In this section, we give energy estimates for the vector field (22), whose solutions taken at time $s = -1$ is the desired canonical transformation to the Birkhoff normal form at third order.

The evolution of energy norms are expressed as

$$\frac{d}{ds}\|(\eta,\xi)\|^2_{E^r} = 2\mathrm{Re}\Big(\langle \eta, \sigma|D|^2\partial_s\eta\rangle_r + \langle \xi, G_0\partial_s\xi\rangle_r\Big)$$
$$= 2\mathrm{Re}\Big(\langle \sigma|D|^2\eta, \partial_\xi K^{(3)}\rangle_r - \langle G_0\xi, \partial_\eta K^{(3)}\rangle_r\Big). \tag{49}$$

The RHS of (49) contains cancellations that are quite subtle, leading to the following estimate.

Lemma 5. *Fix $r_0 > 2$. For all $r \geq r_0$, there is a bound in the form*

$$|\langle \sigma |D|^2 \eta, \partial_\xi K^{(3)} \rangle_r - \langle G_0 \xi, \partial_\eta K^{(3)} \rangle_r| \leq C_{r_0} \|(\eta, \xi)\|^3_{E^r}. \tag{50}$$

Proof. The proof is somewhat computational, indeed even the expression of the LHS of (50) is rather long. For convenience, we will assume that there are no resonant triplets, namely that $[H^{(3)}] = 0$, otherwise stated as $P = 0$. The difference in (50) if it were not zero would only lead to a compact perturbation of the RHS. In this setting,

$$\begin{aligned}
&\langle \sigma |D|^2 \eta, \partial_\xi K^{(3)} \rangle_r - \langle G_0 \xi, \partial_\eta K^{(3)} \rangle_r \\
&= \Sigma'_{k_1+k_2+k_3=0} \Big[\frac{k_1 k_3 + G_1 G_3}{d_{123}} (g + \sigma k_1^2)(\omega_1^2 - \omega_2^2 - \omega_3^2) \\
&\quad + \frac{k_1 k_2 + G_1 G_2}{d_{123}} (g + \sigma k_1^2)(g + \sigma k_2^2) G_3 \Big] \sigma |k_3|^2 \, \eta_1 \eta_2 \eta_3 \langle k_3 \rangle^{2r} \\
&\quad + \Sigma'_{k_1+k_2+k_3=0} \Big[\frac{k_1 k_3 + G_1 G_3}{d_{123}} G_2 (\omega_1^2 - \omega_2^2 + \omega_3^2) \\
&\quad + \frac{k_2 k_3 + G_2 G_3}{2 d_{123}} G_1 (\omega_3^2 - \omega_1^2 + \omega_2^2) \Big] \sigma |k_1|^2 \, \eta_1 \xi_2 \xi_3 \, \langle k_1 \rangle^{2r} \\
&\quad - \Sigma'_{k_1+k_2+k_3=0} \Big[\frac{k_2 k_3 + G_2 G_3}{d_{123}} (g + \sigma k_3^2)(\omega_3^2 - \omega_1^2 - \omega_2^2) \\
&\qquad \frac{k_1 k_2 + G_1 G_2}{d_{123}} (g + \sigma k_1^2)(\omega_1^2 - \omega_2^2 - \omega_3^2) \\
&\qquad 2 \frac{k_1 k_3 + G_1 G_3}{d_{123}} (g + \sigma k_1^2)(g + \sigma k_3^2) G_2 \Big] \, G_3 \eta_1 \xi_2 \xi_3 \, \langle k_3 \rangle^{2r}.
\end{aligned} \tag{51}$$

The indices have been relabeled to exhibit cancellations. The notation $\Sigma'_{k_1+k_2+k_3=0}$ is as above, summation over all $k_j \in \mathbb{Z} \backslash \{0\}$, and $d_{123} = d_{k_1 k_2 k_3}$.

Estimates for the sums involving $\eta_1 \eta_2 \eta_3$ an $\eta_1 \xi_2 \xi_3$ are treated independently. We will start with the first quantity in the RHS of (51) involving $\eta_1 \eta_2 \eta_3$. Energy estimates for η in terms of E^r-norms are tantamount to ℓ^2 estimates for $p = \sigma^{1/2} |D| \langle D \rangle^r \eta$, leading to the sum

$$\begin{aligned}
&\Sigma'_{k_1+k_2+k_3=0} \Big[\frac{k_1 k_3 + G_1 G_3}{d_{123}} (g + \sigma k_1^2)(\omega_1^2 - \omega_2^2 - \omega_3^2) \\
&\quad + \frac{k_1 k_2 + G_1 G_2}{d_{123}} (g + \sigma k_1^2)(g + \sigma k_2^2) G_3 \Big] \frac{\sigma^{1/2} |k_3| \, \langle k_3 \rangle^r}{\sigma |k_1| \langle k_1 \rangle^r |k_2| \langle k_2 \rangle^r} \, p_1 p_2 p_3.
\end{aligned} \tag{52}$$

The expression (52) has the form $\sum_{k_1+k_2+k_3} b_{k_1k_2k_3}p_1p_2p_3$ for which we write

$$|\sum_{k_1+k_2+k_3=0} b_{k_1k_2k_3}p_1p_2p_3| \leq \| \sum_{k_1+k_2+k_3=0} b_{k_1k_2k_3}p_1p_2\|_{\ell^2}\|p\|_{\ell^2}. \tag{53}$$

In order to bound the above sum, we separate the contribution of each sector (i)–(iv) of the lattice (k_1, k_2). Most of the work consists in finding appropriate bounds for the kernel $b_{k_1k_2k_3}$ in order to apply Lemma 4 and Propositions 2–4. This will be the procedure to estimate each term of (51). For the term (52), the kernel $b_{k_1k_2k_3}$ identifies to

$$\begin{aligned} b_{k_1k_2k_3} &= \frac{k_1k_3 + G_1G_3}{d_{123}}(g+\sigma k_1^2)(\omega_1^2 - \omega_2^2 - \omega_3^2)\frac{\sigma^{1/2}|k_3|\,\langle k_3\rangle^r}{\sigma|k_1|\langle k_1\rangle^r|k_2|\langle k_2\rangle^r} \\ &+ \frac{k_1k_2 + G_1G_2}{d_{123}}(g+\sigma k_1^2)(g+\sigma k_2^2)G_3\,\frac{\sigma^{1/2}|k_3|\,\langle k_3\rangle^r}{\sigma|k_1|\langle k_1\rangle^r|k_2|\langle k_2\rangle^r} \\ &=: \frac{N_a}{D_a} + \frac{N_b}{D_b} \end{aligned} \tag{54}$$

where $\frac{N_a}{D_a}$ and $\frac{N_b}{D_b}$ identify respectively to the first and second term of the RHS of (54). Next, we bound each of these terms in the different sectors.

Sector (i): where $|k_1| \ll |k_2|, |k_3|$.

The numerator N_a is bounded above by

$$|N_a| \leq \langle k_1\rangle^3\left(\langle k_1\rangle^3 + \langle k_2\rangle^3 + \langle k_3\rangle^3\right)\langle k_3\rangle^{r+1},$$

while the denominator D_a is bounded below by

$$|D_a| \geq \langle k_1\rangle^{3+r}\langle k_2\rangle^{3+r}\langle k_3\rangle^2.$$

In this sector, $\langle k_2\rangle \sim \langle k_3\rangle$, therefore

$$\frac{|N_a|}{|D_a|} \leq C_0\langle k_1\rangle^{-r}. \tag{55}$$

Taking $r \geq r_0 > 1$, the result of Lemma 4 applied to this term in the sector (i), implies the bound $\|p\|_{\ell^2}^3$.

Considering the term N_b/D_b in this sector, the analogous estimate is that

$$|N_b| \leq \langle k_1\rangle^3\langle k_2\rangle^3\langle k_3\rangle^{2+r},$$

while

$$|D_b| \geq \langle k_1\rangle^{3+r}\langle k_2\rangle^{3+r}\langle k_3\rangle^2,$$

with the conclusion that again

$$\frac{|N_b|}{|D_b|} \le C_0 \langle k_1 \rangle^{-r}. \tag{56}$$

Sector (ii): where $|k_2| \ll |k_1|, |k_3|$.

The numerator N_a is bounded above by the quantity

$$|N_a| \le |k_1 k_3 + G_1^{(0)} G_3^{(0)}| \langle k_1 \rangle^2 (\langle k_1 \rangle^3 + \langle k_2 \rangle^3 + \langle k_3 \rangle^3) \langle k_3 \rangle^{1+r}.$$

In this sector, both $|k_1|$ and $|k_3|$ are large and comparable, of opposite sign, and the commutator estimate (43) holds, which compensates any polynomial growth. Therefore

$$\frac{|N_a|}{|D_a|} \le C_0 e^{-h(|k_1|+|k_3|)/2}. \tag{57}$$

Term N_b/D_b: The numerator behaves as

$$|N_b| \le \langle k_1 \rangle^3 \langle k_2 \rangle^3 \langle k_3 \rangle^{2+r},$$

while the denominator is as above, hence

$$\frac{|N_b|}{|D_b|} \le C_0 \langle k_2 \rangle^{-r}, \tag{58}$$

again giving rise the ℓ^2-estimates in p.

In *Sector (iii)* : where $|k_3| \ll |k_1|, |k_2|$, the weights in the denominator are dominant, so that

$$|N_a| \le \langle k_1 \rangle^3 (\langle k_1 \rangle^3 + \langle k_2 \rangle^3 + \langle k_3 \rangle^3) \langle k_3 \rangle^{3+r}$$

$$|N_b| \le \langle k_1 \rangle^3 \langle k_2 \rangle^3 \langle k_3 \rangle^3,$$

while

$$|D_a|, |D_b| \ge \langle k_1 \rangle^{3+r} \langle k_2 \rangle^{3+r} \langle k_3 \rangle^2,$$

so that

$$\frac{|N_b|}{|D_b|}, \frac{|N_b|}{|D_b|} \le C_0 \langle k_1 \rangle^{-r/2} \langle k_2 \rangle^{-r/2}. \tag{59}$$

Finally consider sector (iv) in which all $|k_1|, |k_2|$ and $|k_3|$ are comparable. In this region,

$$|D_a|,\ |D_b| \geq \langle k_1\rangle^{3+r}\langle k_2\rangle^{3+r}\langle k_3\rangle^2,$$

which easily dominate the numerators

$$|N_a|,\ |N_b| \leq \langle k_1\rangle^3 \left(\langle k_1\rangle^3 + \langle k_2\rangle^3 + \langle k_3\rangle^3\right) \langle k_3\rangle^{3+r},$$

giving an estimate in terms of $\|p\|^3_{\ell^2}$ as long as $r \geq r_0 > 1/2$.

We now turn to terms in the RHS of (51) that involve $\eta_1\xi_2\xi_3$. As before, we introduce the weighted quantities $p = \sigma^{1/2}|D|\langle D\rangle^r\eta$, $q = G^{(0)}\langle D\rangle^r\xi$. We rewrite the last five terms of the RHS of (51) in terms of p_1, q_2 and q_3 and symmetrize with respect to the indices 2 and 3. This leads to nine terms that we label (a_1) to (a_9) and we list below. Each term will be examined in the previously defined four sectors of the lattice (k_1, k_2). In some cases, the estimate will be similar to what we have already seen and the bound is straightforward. In other cases, the previous analysis will not be sufficient. However, by combining terms, we will be able to control them thanks to cancellations. The three first terms are

$$\begin{aligned}
&\Sigma'_{k_1+k_2+k_3=0}\frac{1}{2d_{123}}p_1q_2q_3\Big[\\
&(a_1)\quad (k_1k_3+G_1G_3)G_2(\omega_1^2-\omega_2^2+\omega_3^2)\sigma|k_1|^2\\
&(a_2)\quad (k_1k_2+G_1G_2)G_3(\omega_1^2-\omega_2^2+\omega_3^2)\sigma|k_1|^2\\
&(a_3)\quad (k_2k_3+G_2G_3)G_2(\omega_1^2-\omega_3^2+\omega_2^2)\sigma|k_1|^2\Big]\frac{\langle k_1\rangle^r}{\sigma^{1/2}|k_1|\langle k_1\rangle^r G_2\langle k_2\rangle^r G_3\langle k_3\rangle^r}.
\end{aligned}$$

The next 4 terms are

$$\begin{aligned}
&-\Sigma'_{k_1+k_2+k_3=0}\frac{1}{2d_{123}}p_1q_2q_3\Big[\\
&(a_4)\quad (k_2k_3+G_2G_3)(g+\sigma k_3^2)(\omega_3^2-\omega_1^2-\omega_2^2)G_3\frac{\langle k_3\rangle^r}{\sigma^{1/2}|k_1|\langle k_1\rangle^r G_2\langle k_2\rangle^r G_3\langle k_3\rangle^r}\\
&(a_5)\quad +(k_2k_3+G_2G_3)(g+\sigma k_2^2)(\omega_2^2-\omega_1^2-\omega_3^2)G_2\frac{\langle k_2\rangle^r}{\sigma^{1/2}|k_1|\langle k_1\rangle^r G_2\langle k_2\rangle^r G_3\langle k_3\rangle^r}\\
&(a_6)\quad +(k_1k_2+G_1G_2)(g+\sigma k_1^2)(\omega_1^2-\omega_2^2-\omega_3^2)G_3\frac{\langle k_3\rangle^r}{\sigma^{1/2}|k_1|\langle k_1\rangle^r G_2\langle k_2\rangle^r G_3\langle k_3\rangle^r}\\
&(a_7)\quad +(k_1k_3+G_1G_3)(g+\sigma k_1^2)(\omega_1^2-\omega_3^2-\omega_2^2)G_2\frac{\langle k_2\rangle^r}{\sigma^{1/2}|k_1|\langle k_1\rangle^r G_2)\langle k_2\rangle^r G_3\langle k_3\rangle^r}\Big]
\end{aligned}$$

and finally the last two terms are

$$- \Sigma'_{k_1+k_2+k_3=0} \frac{1}{2d_{123}} p_1 q_2 q_3 \Big[$$

$$(a_8) \quad 2(k_1k_3 + G_1G_3)(g + \sigma k_1^2)(g + \sigma k_3^2)G_2G_3 \frac{\langle k_3\rangle^r}{\sigma^{1/2}|k_1|\langle k_1\rangle^r G_2 \langle k_2\rangle^r G_3 \langle k_3\rangle^r}$$

$$(a_9) \quad + 2(k_1k_2 + G_1G_2)(g + \sigma k_1^2)(g + \sigma k_2^2)G_3G_2 \frac{\langle k_2\rangle^r}{\sigma^{1/2}|k_1|\langle k_1\rangle^r G_2 \langle k_2\rangle^r G_3 \langle k_3\rangle^r} \Big].$$

One has to count the factors of k_1, k_2, k_3 for each term and each sector. The general principle is that we need at least three factors of $|k_1|$ in the numerator for sector (i) and we check that this is the case for all nine terms. In sector (ii) [resp. (iii)] , we need at least 2.5 factors of $|k_2|$ in the numerator [resp. 2.5 factors of $|k_3|$].

In Sector (ii), the terms that present a difficulty are those labelled $(a_2), (a_3), (a_4), (a_6)$. Note that all the weight can be shifted to $\langle k_1\rangle^r$ or $\langle k_3\rangle^r$ (or $\langle k_1\rangle^{r/2}\langle k_3\rangle^{r/2}$, modulo a factor of $\langle k_1\rangle - \langle k_3\rangle \sim |k_2|$). We write only the relevant terms of the sum of these four contributions, the numerator of which is

$$\begin{aligned}&\sigma(\omega_1^2 - \omega_3^2)\langle k_1\rangle^{r/2}\langle k_3\rangle^{r/2} \\ &\Big[(k_1k_2 + G_1G_2)G_3|k_1|^2 - (k_2k_3 + G_2G_3)G_1|k_1|^2 \\ &\qquad + (k_2k_3 + G_2G_3)G_3|k_3|^2 - (k_1k_2 + G_1G_2)G_3|k_1|^2\Big].\end{aligned}$$

The first and last term cancel exactly and the expression reduces to

$$\begin{aligned}&\sigma(\omega_1^2 - \omega_3^2)\langle k_1\rangle^{r/2}\langle k_3\rangle^{r/2}\Big[- (k_2k_3 + G_2G_3)(G_1|k_1|^2 - G_3|k_3|^2)\Big] \\ &\sim O(|k_2|^3)\langle k_1\rangle^{r/2}\langle k_3\rangle^{r/2}(\langle k_1\rangle^5 + \langle k_3\rangle^5).\end{aligned} \tag{60}$$

The corresponding denominator is

$$\begin{aligned}D &= d_{123}\sigma^{1/2}|k_1|\, G_2^{1/2}\langle k_2\rangle^r\, G_3^{1/2}\langle k_3\rangle^r \\ &\ge \langle k_1\rangle^3\langle k_2\rangle^{r+5/2}\langle k_3\rangle^{r+5/2}.\end{aligned} \tag{61}$$

Counting the powers in the numerator and denominator, we have

$$\frac{N}{D} \le \langle k_2\rangle^{1/2-r}\langle k_1\rangle^{-1/2}, \tag{62}$$

which allows us to conclude the result in this case.

In Sector (iii), the terms that are critical are $(a_1), (a_3), (a_5).(a_7)$. Proceeding as above for Sector (ii), we write the relevant terms for the sum of these four contributions. In the numerator, we have

$$(\omega_1^2 - \omega_2^2)\sigma \langle k_1\rangle^{r/2}\langle k_2\rangle^{r/2}$$
$$\Big[(k_1k_3 + G_1G_3)G_2|k_1|^2 - (k_2k_3 + G_2G_3)G_1|k_1|^2$$
$$+ (k_2k_3 + G_2G_3)G_2|k_2|^2 - (k_1k_3 + G_1G_3)G_2|k_1|^2\Big]. \tag{63}$$

The first and last term in the brackets cancel and the expression becomes

$$(\omega_1^2 - \omega_2^2)\sigma \langle k_1\rangle^{r/2}\langle k_2\rangle^{r/2}(k_2k_3 + G_2G_3)(-G_1|k_1|^2 + G_2|k_2|^2) + O(|k_3|^4)$$
$$= O(|k_3|^3).$$

The numerator N and denominator D of terms $(a_1) + (a_3) + (a_5) + (a_7)$ in sector (iii) thus satisfy

$$N \le \langle k_1\rangle^{r/2}\langle k_2\rangle^{r/2}\langle k_3\rangle^3(\langle k_1\rangle^5 + \langle k_2\rangle^5) \quad ; \quad D \le \langle k_1\rangle^3\langle k_2\rangle^{r+5/2}\langle k_3\rangle^{r+5/2},$$

leading to

$$\frac{N}{D} \le \langle k_3\rangle^{1/2-r}\langle k_1\rangle^{1/2}$$

and thus the application of In sector (iv), $|k_1| \sim |k_2| \sim |k_3|$ and the bounds are straightforward. This concludes the proof of Lemma 5. □

4.4 Energy Estimates for the Variational Equation

The vector field in question is $X^{K^{(3)}}(\eta, \xi) = (\partial_\xi K^{(3)}, -\partial_\eta K^{(3)})$ where $(z = \eta, \xi) \in E^r$. Denote the variations of the orbit by $\tilde{z} = (\tilde{\eta}, \tilde{\xi}) = (\delta\eta, \delta\xi)$. These satisfy

$$\frac{d}{ds}\begin{pmatrix}\tilde{\eta}\\ \tilde{\xi}\end{pmatrix} = \partial_z X^{K^{(3)}}(\eta, \xi)\begin{pmatrix}\tilde{\eta}\\ \tilde{\xi}\end{pmatrix}$$
$$= \begin{pmatrix}\partial_\eta\partial_\xi K^{(3)} & \partial_\xi^2 K^{(3)}\\ -\partial_\eta^2 K^{(3)} & -\partial_\eta\partial_\xi K^{(3)}\end{pmatrix}\begin{pmatrix}\tilde{\eta}\\ \tilde{\xi}\end{pmatrix}. \tag{64}$$

The goal is to prove an energy estimate for solutions of (64) of the form:

Lemma 6.

$$\frac{d}{ds}\|\tilde{z}\|_{E^r}^2 \le C_0\|z\|_{E^{r'}}\|\tilde{z}\|_{E^r}^2 \tag{65}$$

where $r' \ge r + 1/2$.

Proof. The first task is to calculate an expression in Fourier coordinates for the RHS of (64).

$$\sqrt{2\pi}\,\partial_\eta\partial_\xi K^{(3)}\tilde{\eta}_2 = \sum_{k_1+k_2+k=0} \frac{1}{d_{12k}}\Big[(k_2k + G_2G_k)(g+\sigma k_2^2)(\omega_2^2 - \omega_1^2 - \omega_k^2)$$
$$(k_1k + G_1G_k)(g+\sigma k_1^2)(\omega_1^2 - \omega_2^2 - \omega_k^2)$$
$$2(k_1k_2 + G_1G_2)(g+\sigma k_1^2)(g+\sigma k_2^2)G_k\Big]\eta_1\tilde{\eta}_2$$

$$\sqrt{2\pi}\,\partial_\xi^2 K^{(3)}\tilde{\xi}_2 = \sum_{k_1+k_2+k=0} \frac{1}{d_{12k}}\Big[(k_2k + G_2G_k)G_1(\omega_2^2 - \omega_1^2 + \omega_k^2)$$
$$(k_1k + G_1G_k)G_2(\omega_1^2 - \omega_2^2 + \omega_k^2)$$
$$(k_1k_2 + G_1G_2)G_k(\omega_1^2 - \omega_k^2 + \omega_2^2)\Big]\xi_1\tilde{\xi}_2$$

$$\sqrt{2\pi}\,\partial_\eta^2 K^{(3)}\tilde{\eta}_2 = \sum_{k_1+k_2+k=0} \frac{1}{d_{12k}}\Big[2(k_2k + G_2G_k)(g+\sigma k_2^2)(g+\sigma k^2)G_1$$
$$(k_1k + G_1G_k)(g+\sigma k^2)(\omega_k^2 - \omega_2^2 - \omega_k^2)$$
$$(k_1k_2 + G_1G_2)(g+\sigma k_2^2)(\omega_2^2 - \omega_k^2 - \omega_1^2)\Big]\xi_1\tilde{\eta}_2$$

$$\frac{d}{ds}\|\tilde{z}\|_{E^r}^2 = 2\mathrm{Re}\Big[\langle\tilde{\eta},\ \sigma|D|^2(\partial_\eta\partial_\xi K^{(3)}\tilde{\eta} + \partial_\xi^2 K^{(3)}\tilde{\xi})\rangle_r$$
$$- \langle\tilde{\xi},\ G(\partial_\eta^2 K^{(3)}\tilde{\eta} + \partial_\xi\partial_\eta K^{(3)}\tilde{\xi})\rangle_r\Big]. \tag{66}$$

This expression is the sum of 12 terms that we list below and denote (a) to (ℓ).

$$(a) = \sum \frac{1}{d_{123}}\Big[2(k_1k_2 + G_1G_2)(g+\sigma k_1^2)(g+\sigma k_2^2)G_3$$
$$(b) \qquad + (k_1k_3 + G_1G_3)(g+\sigma k_1^2)(\omega_1^2 - \omega_2^2 - \omega_3^2)$$
$$(c) \qquad + (k_2k_3 + G_2G_3)(g+\sigma k_2^2)(\omega_2^2 - \omega_1^2 - \omega_3^2)\Big]\sigma k_3^2\ \eta_1\tilde{\eta}_2\tilde{\eta}_3\langle k_3\rangle^{2r}$$

$$(d) \quad + \sum \frac{1}{d_{123}}\Big[(k_1k_2 + G_1G_2)G_3(\omega_1^2 - \omega_3^2 + \omega_2^2)$$
$$(e) \qquad + (k_1k_3 + G_1G_3)G_2(\omega_1^2 - \omega_2^2 + \omega_3^2)$$

$$
\begin{aligned}
&(f) \qquad + (k_2k_3 + G_2G_3)G_1(\omega_2^2 - \omega_1^2 + \omega_3^2)\Big]\sigma k_3^2\, \xi_1\tilde{\xi}_2\tilde{\eta}_3\langle k_3\rangle^{2r} \\
&(g) \quad -\sum \frac{1}{d_{123}}\Big[2(k_1k_3 + G_1G_3)(g + \sigma k_1^2)(g + \sigma k_3^2)G_2 \\
&(h) \qquad + (k_2k_3 + G_2G_3)(g + \sigma k_3^2)(\omega_3^2 - \omega_1^2 - \omega_2^2) \\
&(i) \qquad + (k_1k_2 + G_1G_2)(g + \sigma k_1^2)(\omega_1^2 - \omega_2^2 - \omega_3^2)\Big]G_3\, \eta_1\tilde{\xi}_2\tilde{\xi}_3\langle k_3\rangle^{2r} \\
&(j) \quad -\sum \frac{1}{d_{123}}\Big[2(k_2k_3 + G_2G_3)(g + \sigma k_2^2)(g + \sigma k_3^2)G_1 \\
&(k) \qquad + (k_1k_3 + G_1G_3)(g + \sigma k_3^2)(\omega_3^2 - \omega_1^2 - \omega_2^2) \\
&(\ell) \qquad + (k_1k_2 + G_1G_2)(g + \sigma k_2^2)(\omega_2^2 - \omega_3^2 - \omega_1^2)\Big]G_3\, \xi_1\tilde{\eta}_2\tilde{\xi}_3\langle k_3\rangle^{2r}.
\end{aligned}
\tag{67}
$$

Terms $(a), (b), (c)$ of the form $\eta_1\tilde{\eta}_2\tilde{\eta}_3$ will be estimated in terms of the r'-order energy norm of $z = (\eta, \xi)$ and r-order energy norm of $\tilde{z} = (\tilde{\eta}, \tilde{\xi})$. Similarly to the energy estimates of Sect. 4.3, we introduce $p = \sigma^{1/2}|D|\langle D\rangle^{r'}\eta$ and $\tilde{p} = \sigma^{1/2}|D|\langle D\rangle^{r}\tilde{\eta}$, and write $\langle k_3\rangle^{2r}\eta_1\tilde{\eta}_2\tilde{\eta}_3 = p_1\tilde{p}_2\tilde{p}_3\frac{\langle k_3\rangle^{r}}{\sigma^{3/2}|k_1|\langle k_1\rangle^{r'}|k_2|\langle k_2\rangle^{r}|k_3|}$. We seek estimates in terms of the ℓ^2 norm of p and $\tilde{p}$., and we need to examine the corresponding kernels in order to satisfy the hypotheses of Propositions 1–3. We consider the expressions of the kernels in sectors:

In sector (i) where $|k_1| \ll |k_2| \sim |k_3|$, we need only to analyze $(a) + (b)$. Their numerators and denominators satisfy

$$|N_a| \le \langle k_1\rangle^3\langle k_2\rangle^3\langle k_3\rangle^{3+r}$$

$$|D_a| \ge \langle k_1\rangle^{3+r'}\langle k_2\rangle^{3+r}\langle k_3\rangle^3$$

$$|N_b| \le \langle k_1\rangle^3(\langle k_2\rangle^3 + \langle k_3\rangle^3)\langle k_3\rangle^{3+r}$$

$$|D_b| \ge \langle k_1\rangle^{3+r'}\langle k_2\rangle^{3+r}\langle k_3\rangle^3,$$

thus

$$\frac{N_a}{D_a}, \frac{N_b}{D_b} \le C_0\langle k_1\rangle^{-r'},$$

where we assume $r' > 1$.

In sector (ii) where $|k_2| \ll |k_1| \sim |k_3|$, terms $(a) + (c)$ are relevant.

$$|N_a| \le \langle k_1\rangle^3\langle k_2\rangle^3\langle k_3\rangle^{3+r}$$

$$|D_a| \ge \langle k_1\rangle^{3+r'}\langle k_2\rangle^{3+r}\langle k_3\rangle^3$$

thus

$$\frac{N_a}{D_a} \le C_0 \langle k_2 \rangle^{-r}. \tag{68}$$

Here we need $r' \ge r$. Also,

$$|N_c| \le \langle k_2 \rangle^3 \langle k_3 \rangle^{3+r} (\langle k_1 \rangle^3 + \langle k_3 \rangle^3)$$

$$|D_c| \ge \langle k_1 \rangle^{3+r'} \langle k_2 \rangle^{3+r} \langle k_3 \rangle^3$$

$$\frac{N_c}{D_c} \le C_0 \langle k_2 \rangle^{-r} \tag{69}$$

since $r' \ge r$. Estimates in sectors (iii) and (iv) are straightforward, using that $r' > 1$.

We turn to terms $(g), (h), (i)$ which are of the form $\eta_1 \tilde{\xi}_2 \tilde{\xi}_3$. Introducing $\tilde{q} = G^{1/2} \langle D \rangle^r \tilde{\xi}$, we write $\langle k_3 \rangle^{2r} \eta_1 \tilde{\xi}_2 \tilde{\xi}_3 = p_1 \tilde{q}_2 \tilde{q}_3 \frac{\langle k_3 \rangle^r}{\sigma^{1/2} |k_1| \langle k_1 \rangle^{r'} |G_2|^{1/2} \langle k_2 \rangle^r G_3^{1/2}}$.

In sector (i) where $|k_1| \ll |k_2| \sim |k_3|$, only $(g) + (i)$ count:

$$|N_g| \le \langle k_1 \rangle^3 \langle k_2 \rangle \langle k_3 \rangle^{4+r}$$

$$|D_g| \ge \langle k_1 \rangle^{3+r'} \langle k_2 \rangle^{2.5+r} \langle k_3 \rangle^{2.5}$$

$$|N_i| \le \langle k_1 \rangle^3 \langle k_2 \rangle^2 \langle k_3 \rangle^{3+r}$$

$$|D_i| \ge \langle k_1 \rangle^{3+r'} \langle k_2 \rangle^{2.5+r} \langle k_3 \rangle^{2.5},$$

thus in sector (i)

$$\frac{N_g}{D_g}, \frac{N_i}{D_i} \le C_0 \langle k_1 \rangle^{-r'}. \tag{70}$$

In sector (ii) where $|k_2| \ll |k_1| \sim |k_3|$, $(h) + (i)$ are relevant.

$$|N_h| \le \langle k_2 \rangle^2 \langle k_3 \rangle^{4+r} (\langle k_1 \rangle^2 + \langle k_3 \rangle^2)$$

$$|D_h| \ge \langle k_1 \rangle^{3+r'} \langle k_2 \rangle^{2.5+r} \langle k_3 \rangle^{2.5}$$

$$|N_i| \le \langle k_1 \rangle^3 \langle k_2 \rangle^2 \langle k_3 \rangle^{1+r} (\langle k_1 \rangle^2 + \langle k_3 \rangle^2)$$

$$|D_i| \ge \langle k_1 \rangle^{3+r'} \langle k_2 \rangle^{2.5+r} \langle k_3 \rangle^{2.5}.$$

We have used that $|\omega_1 - \omega_3| \le |k_2| \langle k_1 \rangle^2$. Note that this is where we need the hypothesis that $r' \ge r + 1/2$. Estimates in sectors (iii) and (iv) are straightforward.

We now consider the terms involving $\xi_1\tilde{\eta}_2\tilde{\xi}_3$, that is $((d)+(e)+(f)+(j)+(k)+(\ell)$. We need to relabel the indices of $(d)+(e)+(f)$ in order to take advantage of cancellations. We are considering the following sum:

$$\begin{aligned}
&(d) \quad \sum \frac{1}{d_{123}}\Big[(k_1k_3+G_1G_3)G_2(\omega_1^2-\omega_2^2+\omega_3^2)\\
&(e) \qquad + (k_1k_2+G_1G_2)G_3(\omega_1^2-\omega_3^2+\omega_2^2)\\
&(f) \qquad + (k_2k_3+G_2G_3)G_1(\omega_2^2-\omega_1^2+\omega_3^2)\Big]\sigma k_2^2\langle k_2\rangle^{2r}\\
&(j) \quad -\sum \frac{1}{d_{123}}\Big[2(k_2k_3+G_2G_3)(g+\sigma k_2^2)(g+\sigma k_3^2)G_1\\
&(k) \qquad + (k_1k_3+G_1G_3)(g+\sigma k_3^2)(\omega_3^2-\omega_1^2-\omega_2^2)\\
&(\ell) \qquad + (k_1k_2+G_1G_2)(g+\sigma k_2^2)(\omega_2^2-\omega_3^2-\omega_1^2)\Big]G_3\langle k_3\rangle^{2r}\\
&\qquad \times \frac{1}{G_1^{1/2}\langle k_1\rangle^{r'}\sigma^{1/2}|k_2|\langle k_2\rangle^r G_3^{1/2}\langle k_3\rangle^r}\, p_1\tilde{q}_2\tilde{p}_3
\end{aligned} \tag{71}$$

where we are using the notation :

$$p_1 = G_1^{1/2}\langle k_1\rangle^{r'}\xi_1, \quad \tilde{q}_2 = \sigma^{1/2}|k_2|\langle k_2\rangle^r\tilde{\eta}_2, \quad \tilde{p}_3 = G_3^{1/2}\langle k_3\rangle^r\tilde{\xi}_3,$$

and we seek estimates in terms of the ℓ^2-norm of $p, \tilde{q}, \tilde{p}$.

The critical quantities that one must analyze depend upon sectors. In sector (iv) where $|k_1| \sim |k_2| \sim |k_3|$, all terms are tame and estimates follow, under the condition $r' > 1$. After inspection, we see that the critical terms are :

$(d)+(e)+(k)+(\ell)$ in sector (i)
$(j)+(e)$ in sector (ii)
$(d)+(f)$ in sector (iii).

Writing the principal terms of $(d)+(e)+(k)+(\ell)$ in sector (i) , we get

$$\begin{aligned}
&\frac{\sigma}{d_{123}}(\omega_2^2-\omega_3^2)\Big[(k_1k_3+G_1G_3)(G_3k_3^2\langle k_3\rangle^{2r}-G_2k_2^2\langle k_2\rangle^{2r})\\
&\qquad + (k_1k_2+G_1G_2)G_3k_2^2(\langle k_2\rangle^{2r}-\langle k_3\rangle^{2r})\Big]\\
&\qquad \times \frac{1}{G_1^{1/2}\langle k_1\rangle^{r'}\sigma^{1/2}|k_2|\langle k_2\rangle^r G_3^{1/2}\langle k_3\rangle^r}\, p_1\tilde{q}_2\tilde{p}_3.
\end{aligned}$$

The numerator and denominator N and D of this expression are bounded as follows:

$$|N| \le \langle k_1\rangle^3(\langle k_2\rangle + \langle k_3\rangle)^{5+2r}$$
$$|D| \ge \langle k_1\rangle^{2.5+r'}\langle k_2\rangle^{2.5+r}\langle k_3\rangle^{2.5+r}$$

leading to

$$\frac{N}{D} \leq C \langle k_1 \rangle^{-r'} .$$

We can then apply Proposition 3 with $r' > 1$.

We now turn to the principal terms of $(j) + (\ell)$ in sector (ii). Proceeding as above, we get

$$|N_j| \leq \langle k_2 \rangle^3 \langle k_3 \rangle^{4+r} \langle k_1 \rangle$$
$$|N_\ell| \leq \langle k_2 \rangle^3 \langle k_1 \rangle \langle k_3 \rangle^{1+2r} (\langle k_1 \rangle^3 + \langle k_3 \rangle^3)$$
$$|D_j|, \ |D_\ell| \geq \langle k_1 \rangle^{2.5+r'} \langle k_2 \rangle^{3+r} \langle k_3 \rangle^{2.5+r},$$

leading to the desired estimate if $r \geq r$, and $r' > 1$, Finally terms $(d) + (f)$ are bounded by:

$$|N_d| \leq \langle k_3 \rangle^2 \langle k_1 \rangle (\langle k_1 \rangle^2 + \langle k_2 \rangle^2) \langle k_2 \rangle^{3+2r}$$
$$|N_f| \leq \langle k_3 \rangle^2 \langle k_2 \rangle^{3+2r} \langle k_1 \rangle (\langle k_1 \rangle^2 + \langle k_2 \rangle^2)$$
$$|D_d|, \ |D_f| \geq \langle k_1 \rangle^{2.5+r'} \langle k_2 \rangle^{3+r} \langle k_3 \rangle^{2.5+r}.$$

Again here, one can apply Proposition 3 if $r' \geq r + 1/2$. This concludes the proof of Theorem 2. □

4.5 Smoothness Estimates for the Transformation

Proof. This section presents the proof of Theorem 3. The variational equation about a solution $\psi_s(\eta, \xi)$ is given by (64), and by Lemma 6, the linearized equation has solutions which satisfy the energy estimate with a loss of $1/2$ derivative:

$$|\langle (\tilde{\eta}, \tilde{\xi}), \ \partial_{(\eta,\xi)} X^{K^{(3)}}(\eta, \xi) \, (\tilde{\eta} \ \tilde{\xi})^T \rangle_{E^{r-1/2}}| \ \leq \ \|(\tilde{\eta}, \tilde{\xi})\|^2_{E^{r-1/2}} \tag{72}$$

for $\psi_s(\eta, \xi) \in B_R \subseteq E^r$. Therefore by Gronwall's lemma, for $s < s_R$

$$\|\partial_{(\eta,\xi)} \psi_s - I\|_{E^{r-1/2}} \leq C R. \tag{73}$$

Estimates of higher derivatives of the flow $\psi_s(\eta, \xi)$, $s < s_R$, follow with a similar argument. Namely, setting $z = (\eta, \xi)$, higher derivatives of the flow satisfy the inhomogeneous equations

$$\frac{\partial}{ds}(\partial_z^p \psi_s) = \partial_z X^{K^{(3)}}(\partial_z^p \psi_s) + \sum_{\substack{p_1+p_2=p \\ 1\le p_1,p_2<p}} C_{p_1 p_2} \langle \partial_z^2 X^{K^{(3)}}(\partial_z^{p_1}\psi_s), (\partial_z^{p_2}\psi_s) \rangle . \tag{74}$$

Since $X^{K^{(3)}}$ is quadratic in its arguments, no other terms appear. Furthermore, since $p \geq 2$, $(\partial_z^p \psi_s)|_{s=0} = 0$, therefore by induction,

$$\begin{aligned} \|\partial_z^p \psi_s\|_{E^{r-p/2}} &\leq C_{rp} \sup_{|s|<s_R} \sum_{\substack{p_1+p_2=p \\ 1\le p_1,p_2<p}} \|\partial_z^{p_1}\psi_s\|_{E^{r-p/2+1/2}} \|\partial_z^{p_2}\psi_s\|_{E^{r-p/2+1/2}} \\ &\leq C_{rp} \end{aligned}$$

using the induction hypothesis that $\|\partial_z^{p_1}\psi_s\|_{E^{r-p/2+1/2}} \leq C_{r,p-1}$ for all $1 \leq p_1 \leq p-1$. □

5 Resonant Triads

The change of variables that we have introduced eliminates non resonant cubic terms in the Hamiltonian. In the new variables, the only third order terms that remain correspond to resonant triads. The dynamics of the resonant subsystems therefore will dominate the behavior of the full system of water waves for long periods of time. There have been a number of formal studies of the behavior of resonant triads for the system of water wave equations with surface tension, including [2, 9]. The normal forms analysis of our work gives a rigorous justification of these studies. Several special configurations of coupled resonant triads are considered in these papers, including of course the simplest triad of three interaction resonant modes which is isolated (in Fourier space) from the dynamics of the other modes, resonant or not. In this section we describe this simple configuration in the framework of a Hamiltonian system, and examine the stability of its periodic orbits. We also consider the case of two coupled resonant triads, in a different setting of coupled resonances than those of Hammack and Henderson [9]. We do not give an exhaustive analysis of all of the possible resonant cases.

5.1 Single Resonant Triad

Assume that the triads $(\pm k_1^0, \pm k_2^0, \pm k_3^0)$ are resonant; that is, $k_1^0 + k_2^0 + k_3^0 = 0$ and $\omega_1 - \omega_2 - \omega_3 = 0$, which is the standard resonant triad (modulo a possible reindexing of wave numbers). As before we use the notation that $\omega_i = \omega_{k_i^0}$. For fixed spatial period such resonances are nongeneric, but they will exist for certain values of the physical parameters. From the choice of signs we have assumed that

$|k_1^0| > |k_2^0|$, and $|k_1^0| > |k_3^0|$, and therefore we may take $k_1^0 > 0 > k_2^0, k_3^0$. Returning to the expression of the transformed Hamiltonian, and retaining only the resonant modes, the quadratic and cubic terms, the truncated Hamiltonian is given by

$$H_+ = H_+^{(2)} + H_+^{(3)} = \sum_{j=\pm 1}^{\pm 3} \omega_j z_j \bar{z}_j$$
$$+ C_+^{(3)}(z_1\bar{z}_{-2}\bar{z}_{-3} + \bar{z}_{-1}z_2 z_3 + z_{-1}\bar{z}_2\bar{z}_3 + \bar{z}_1 z_{-2} z_{-3}). \quad (75)$$

This is a system of 6 degrees of freedom, with the three modes $(z_1, \bar{z}_2, \bar{z}_3)$ decoupled from $(\bar{z}_{-1}, z_2, z_3)$. Written in terms of symplectic polar coordinates (phase and square-root of the amplitude), $z_j = \sqrt{R_j}e^{i\theta_j}$, the Hamiltonian H_+ takes the form

$$H_+ = \sum_1^3 (\omega_j R_j + \omega_{-j} R_{-j}) + 2C_+^{(3)}\Big(\sqrt{R_1 R_{-2} R_{-3}}\cos(\theta_1 - \theta_{-2} - \theta_{-3})$$
$$+ \sqrt{R_{-1}R_2R_3}\cos(\theta_{-1} - \theta_2 - \theta_3)\Big). \quad (76)$$

Only two angles appear in the expression of H_+, one for each system of three degrees of freedom. Thus there are four conserved quantities, and the system reduces to two decoupled systems each with one degree of freedom and therefore integrable, and indeed it consists of two independent copies of the well-known three-wave resonant system. Specifically, perform a change of variable in the form of a simultaneous rotation,

$$\Phi = \begin{pmatrix} \phi_1 \\ \phi_{-2} \\ \phi_{-3} \end{pmatrix} = A \begin{pmatrix} \theta_1 \\ \theta_{-2} \\ \theta_{-3} \end{pmatrix}, \quad I = \begin{pmatrix} I_1 \\ I_{-2} \\ I_{-3} \end{pmatrix} = A \begin{pmatrix} R_1 \\ R_{-2} \\ R_{-3} \end{pmatrix} \quad (77)$$

where $A = (a)_{ij}$ is a 3×3 rotation, making the natural choice to set, $\phi_1 = \frac{1}{\sqrt{3}}(\theta_1 - \theta_{-2} - \theta_{-3})$, and $\omega_1 R_1 + \omega_2 R_{-2} + \omega_3 R_{-3} = \Omega_3 I_{-3}$, where $\Omega_3 = \sqrt{\omega_1^2 + \omega_2^2 + \omega_3^2}$. The matrix A has the form

$$A = \begin{pmatrix} 1/\sqrt{3} & -1/\sqrt{3} & -1/\sqrt{3} \\ (\omega_3 - \omega_2)/N_2 & (\omega_3 - \omega_1)/N_2 & (-\omega_2 - \omega_1)/N_2 \\ \omega_1/\Omega_3 & \omega_2/\Omega_3 & \omega_3/\Omega_3 \end{pmatrix}$$

with $N_2^2 = (\omega_3 - \omega_2)^2 + (\omega_3 - \omega_1)^2 + (\omega_2 + \omega_1)^2$. In the new variables, the first term of H_+ is written as $\Omega_3 I_3 + \Omega_3 I_{-3}$. Expressed in these action-angle variables, the Hamiltonian for the system involving $(R_1, R_{-2}, R_{-3}, \theta_1, \theta_{-2}, \theta_{-3})$ is

$$H_+ = \Omega_3 I_{-3} + 2C^{(3)}\sqrt{R_1 R_{-2} R_{-3}}\cos(\sqrt{3}\phi_1),$$

where $R_j = R_j(I_1)$ depend linearly upon (I_1, I_{-2}, I_{-3}). The Hamiltonian involving $(R_{-1}, R_2, R_3, \theta_{-1}, \theta_2, \theta_3)$ is similar. The equations of motion are

$$\begin{aligned} \dot{I}_1 &= -\partial_{\phi_1} H_+ \\ \dot{\phi}_1 &= \partial_{I_1} H_+ \end{aligned} \tag{78}$$

along with the two cyclic variables (ϕ_{-2}, ϕ_{-3}) and the canonically conjugate conserved quantities (I_{-2}, I_3) which are considered as parameters.

$$\begin{aligned} \dot{I}_{-2} &= -\partial_{\phi_{-2}} H_+ = 0\,, \quad \dot{I}_{-3} = -\partial_{\phi_{-3}} H_+ = 0 \\ \dot{\phi}_{-2} &= \partial_{I_{-2}} H_+\,, \quad \dot{\phi}_{-3} = \partial_{I_{-3}} H_+\,. \end{aligned} \tag{79}$$

The range of I_1 such that $R_1(I), R_{-2}(I), R_{-3}$ are all positive, which is an interval with endpoints I_1^+ and I_1^- depending parametrically on I_{-2} and I_{-3}. The endpoint I_1^- is characterized by the vanishing of R_1, while I_1^+ is defined by a zero of R_{-2} or R_{-3}, whichever vanishes first. There is one exceptional case, determined by a choice of parameters (I_{-2}, I_{-3}) such that R_{-2} and R_{-3} vanish simultaneously in I_1. In this case the factor $\sqrt{R_1 R_{-2} R_{-3}}$ vanishes linearly in I_1 at I_1^+. The system for (I_1, ϕ_1) (and for (I_{-1}, ϕ_{-1}) respectively) can be analysed through its phase plane.

Firstly, the phase plane $\{(I_1, \phi_1) : I_1^- \leq I_1 \leq I_1^+, 0 \leq \phi_1 < 2\pi/\sqrt{3}\}$ is identified as a sphere $\mathbb{S}^2$, with polar coordinate singularities at the endpoints $I_1 = I_1^-$ and $I_1 = I_1^+$ defining the poles. Typical orbits are time periodic, meaning that a typical orbit for the full system (78), (79) will be quasiperiodic with three basic frequencies. Lower dimensional tori are found through the stationary points of the system (78), which occur when

$$\sqrt{3}\phi_1 = 0, \pi\,, \quad \partial_{I_1}(R_1 R_{-2} R_{-3}) = (R_{-2}R_{-3} - R_1 R_{-3} - R_1 R_{-2}) / \sqrt{3 R_1 R_{-2} R_{-3}} = 0\,.$$

By inspection, except for one particular case, there are only two such stationary points per sphere, whose locations are at points $\phi_1^0 = 0, \pi/\sqrt{3}$ and $I_1 = I_1^0(I_{-2}, I_{-3})$ where $\sqrt{R_1 R_{-2} R_{-3}}$ achieves its maximum. Both are stable periodic orbits, as can be seen from the variational equation of the vector field at the stationary point in question, namely

$$\begin{aligned} J\partial^2_{I_1,\phi_1} H_+ &= \begin{pmatrix} 0 & 2C^{(3)}\partial^2_{I_1}\sqrt{R_1 R_{-2} R_{-3}}\cos(\sqrt{3}\phi_1) \\ 2C^{(3)}\sqrt{R_1 R_{-2} R_{-3}}\cos(\sqrt{3}\phi_1) & 0 \end{pmatrix} \\ &= \begin{pmatrix} 0 & A \\ B & 0 \end{pmatrix} \end{aligned}$$

where A and B have opposite signs because of the character of a maximum.

The exception occurs in the case in which the parameters (I_{-2}, I_{-3}) have been adjusted such that R_{-2} and R_{-3} vanish simultaneously in I_1. This corresponds to a stationary point of the vector field itself, rather than a polar coordinate singularity, and for the full system (78), (79) this stationary point corresponds to a basic periodic orbit of Lyapunov type that is guaranteed by the Lyapunov center theorem, associated with the highest frequency ω_1 of the resonant triad system. This periodic orbit is unstable, as the variational equation of the vector field at this point is

$$J\partial^2_{I_1,\phi_1}H_+ = \begin{pmatrix} -2C^{(3)}\partial_{I_1}\sqrt{R_1R_{-2}R_{-3}}\sin(\sqrt{3}\phi_1) & 2C^{(3)}\partial^2_{I_1}\sqrt{R_1R_{-2}R_{-3}}\cos(\sqrt{3}\phi_1) \\ 0 & 2C^{(3)}\partial_{I_1}\sqrt{R_1R_{-2}R_{-3}}\sin(\sqrt{3}\phi_1) \end{pmatrix}$$
$$= \begin{pmatrix} A & C \\ 0 & -A \end{pmatrix}$$

for $A = -4C^{(3)}\sqrt{R_1(I_1^+)}\sin(\sqrt{3}\phi_1)$ real valued, which gives the Lyapunov exponent in explicit terms. There is a family of orbits homoclinic to this Lyapunov-type periodic orbit, consisting of $I_1(t)$ running down along the coordinate axis $\{\phi_1 = \pi/2\}$, through the south pole of the sphere at $I_1 = I_1^-$ and then back up the coordinate axis $\{\phi_1 = 3\pi/2\}$.

5.2 *Mutliple Resonant Triads*

There are numerous possibilities for multiple coupled resonant triads in this case of a non-zero coefficient of surface tension, number of which are discussed in references [2, 9]. In the present paper we will analyse one case that does not appear in these articles, of a system of two coupled resonant triads satisfying the resonance relations

$$k_1 + k_2 + k_3 = 0\,, \quad -k_1 + k_2 + k_4 = 0\,,$$
$$k_2\,,\ k_3 < 0 < k_1 < k_4$$
$$\omega_1 - \omega_2 - \omega_3 = 0\,, \quad \omega_4 - \omega_1 - \omega_2 = 0\,.$$

This is among the simplest situations that is possible with multiple coupled triads. For further simplicity we consider only standing wave solutions, namely we impose Neumann boundary conditions on two vertical walls of the fluid domain at $\{x = 0\}$ and $\{x = \pi\}$, which has the effect that in our complex symplectic coordinates, $z_k = z_{-k}$. The water waves Hamiltonian truncated at third order becomes

$$H_+ = H_+^{(2)} + H_+^{(3)} = \sum_{j=1}^{4} \omega_j |z_j|^2 + \left[c_1^{(3)} z_1\bar{z}_2\bar{z}_3 + c_2^{(3)}\bar{z}_1\bar{z}_2 z_4 + c.c.\right]. \qquad (80)$$

Introducing symplectic polar coordinates $z_j = \sqrt{R_j}e^{i\theta_j}$, we have

$$H_+^{(2)} = \sum_{j=1}^{4} \omega_j R_j \ ,$$

$$H_+^{(3)} = 2c_1^{(3)} \sqrt{R_1 R_2 R_3} \cos(\theta_1 - \theta_2 - \theta_3) + 2c_2^{(3)} \sqrt{R_1 R_2 R_4} \cos(\theta_4 - \theta_1 - \theta_2) \ .$$

Perform a symplectic change of coordinates $I = AR$ and $\Phi = A\Theta$, where

$$A = \begin{pmatrix} \frac{1}{\sqrt{3}} & -\frac{1}{\sqrt{3}} & -\frac{1}{\sqrt{3}} & 0 \\ -\frac{1}{\sqrt{3}} & -\frac{1}{\sqrt{3}} & 0 & \frac{1}{\sqrt{3}} \\ b_1 & b_2 & b_3 & b_4 \\ \frac{\omega_1}{\Omega_4} & \frac{\omega_2}{\Omega_4} & \frac{\omega_3}{\Omega_4} & \frac{\omega_4}{\Omega_4} \end{pmatrix}$$

where $\Omega_4^2 = \omega_1^2 + \omega_2^2 + \omega_3^2 + \omega_4^2$, and where b_j are chosen appropriately so that the matrix A is orthogonal. The Hamiltonian (80) becomes

$$H_+ = \Omega_4 I_4 + \left[2c_1^{(3)} \sqrt{R_1 R_2 R_3} \cos(\sqrt{3}\phi_1) + 2c_2^{(3)} \sqrt{R_1 R_2 R_4} \cos(\sqrt{3}\phi_2)\right] , \quad (81)$$

with $R_j = R_j(I_1, I_2; I_3, I_4)$ which are affine linear in the action variables (I_1, I_2). The two angles (ϕ_3, ϕ_4) do not appear in the Hamiltonian $H_=$ in (81), they are cyclic variables, and their canonically conjugate variables (I_3, I_4) are integrals of motion. This is a Hamiltonian system with two degrees of freedom, given by the Hamiltonian $H_+^{(3)}(I_1, I_2, \phi_1, \phi_2)$ described in (81), and posed on the manifold $M := \{(I_1, I_2, \phi_1, \phi_2) \ : \ R_j(I) \geq 0 \, , (\phi_1, \phi_2) \in \mathbb{T}^2\}$. Topologically, taking into account the polar coordinate singularities at the poles $R_j(I) = 0$, the manifold M is either a $\mathbb{S}^2 \times \mathbb{S}^2$ or a $\mathbb{S}^3 \times \mathbb{S}^1$, depending upon the values of the parameters (I_3, I_4).

The dynamics of a system with two degrees of freedom can be quite complex in general. We will restrict ourselves to considering special structures in this phase space, namely periodic orbits and lower dimensional tori (quasi-periodic motion with two independent frequencies), all of which can be analysed through an inspection of the vector field $X^{H_+^{(3)}}$ on the phase space M. From this point on, denote the relevant action—angle variables by $\mathrm{t}(I, \Phi) := (I_1, I_2, \phi_1, \phi_2)$; the phase space M is coordinatized by (I, Φ) except for the poles of the spheres corresponding to one or several of the polar coordinate singularities $R_j(I) = 0$. The equations of motion restricted to M are given by

$$\dot{\Phi} = \partial_I H_+^{(3)} \ , \qquad \dot{I} = -\partial_\Phi H_+^{(3)} \ . \quad (82)$$

Stationary points of (82) give rise to periodic, or more generally quasi-periodic orbits for the full system governed by the Hamiltonian (80). We first treat stationary points of (82) interior to the coordinate chart $M^0 := \{R_j(I) > 0 \ : \ \forall j\}$. Such stationary points satisfy

$$\partial_{I_j} H_+^{(3)} = \left[2c_1^{(3)} \partial_{I_j}(\sqrt{R_1R_2R_3})\cos(\sqrt{3}\phi_1) + 2c_2^{(3)}\partial_{I_j}(\sqrt{R_1R_2R_4})\cos(\sqrt{3}\phi_2)\right]$$
$$-\partial_{\phi_1} H_+^{(3)} = 2\sqrt{3}c_1^{(3)}\sqrt{R_1R_2R_3}\sin(\sqrt{3}\phi_1)$$
$$-\partial_{\phi_1} H_+^{(3)} = 2c_2^{(3)}\sqrt{R_1R_2R_4}\sin(\sqrt{3}\phi_2).$$

Since $\sqrt{R_1R_2R_3}, \sqrt{R_1R_2R_4} > 0$ in M^0, stationary points may only occur where $\phi_1 = 0, \pi/\sqrt{3}$ and $\phi_2 = 0, \pi/\sqrt{3}$. The case where both $\phi_1, \phi_2 = 0$ is called the *in-phase* solutions, and when $\phi_1 = 0$ but $\phi_2 = \pi/\sqrt{3}$ these are *out-of-phase* solutions. All other choices of ϕ_j reduce to these two cases, using possibly a time reversal. Thus such critical points are characterized by the condition that

$$0 = \partial_I\left(\sqrt{R_1R_2}(c_1^{(3)}\sqrt{R_3} \pm c_2^{(3)}\sqrt{R_4})\right)$$

where the plus sign corresponds to the in-phase case and the minus sign is for the out-of-phase solutions.

Proposition 5. *For fixed parameter values $I_3 = a_3$, $I_4 = a_4$, the in-phase case has two stationary points on the manifold M , which are both stable. The resulting solutions of the full system* (80) *are generically quasi-periodic with two independent frequencies, and are geometrically distinct but related one to the other by a time reversal.*

Proof. The Hamiltonian $H_+^{(3)}$, when evaluated on the hypersurface $(\phi_1, \phi_2) = 0$ as indicated as being necessary in the paragraph above, is positive and can only vanish when either $R_1(I) = 0$ or $R_2(I) = 0$, a subset of the boundaries of the coordinate chart M^0. On the boundary sets defined by $R_3(I) = 0$ or $R_4(I) = 0$, the Hamiltonian $H_+^{(3)} > 0$ while its outward unit normal derivative is negative. Hence any maximum I^* is an interior point. Since both functions $R_1R_2R_3$ and $R_1R_2R_4$ are cubic polynomials when considered as functions of (I_1, I_2),and furthermore R_3 is independent of I_4 while R_4 is independent of I_3, there can be at most one interior maximum, and there are no other critical points. This critical point is identified as the in-phase quasi-periodic solution of the system (80).

The statement of stability of this stationary point follows from an analysis of the first variation of the vector field $J\partial_{(\Phi,I)}H_+^{(3)}$ at the critical point (Φ^*, I^*). Namely

$$\begin{pmatrix} \partial_\Phi\partial_I H_+^{(3)} & \partial_I^2 H_+^{(3)} \\ -\partial_\Phi^2 H_+^{(3)} & -\partial_\Phi\partial_I H_+^{(3)} \end{pmatrix} = \begin{pmatrix} 0 & \partial_I^2 H_+^{(3)} \\ -\partial_\Phi^2 H_+^{(3)} & 0 \end{pmatrix} \tag{83}$$

where

$$-\partial_\Phi^2 H_+^{(3)} = \begin{pmatrix} 6c_1^{(3)}\sqrt{R_1R_2R_3} & 0 \\ 0 & 6c_2^{(3)}\sqrt{R_1R_2R_4} \end{pmatrix}.$$

The eigenvalues μ of (83) are given by

$$\mu^2 = 3\big(c_1^{(3)}\sqrt{R_1R_2R_3}h_{11} + c_2^{(3)}\sqrt{R_1R_2R_4}h_{22}\big) \qquad (84)$$
$$\pm 3\sqrt{\big((c_1^{(3)}\sqrt{R_1R_2R_3}h_{11} + c_2^{(3)}\sqrt{R_1R_2R_4}h_{22})^2 - 4c_1^{(3)}c_2^{(3)}R_1R_2\sqrt{R_3R_4}\det(h)\big)}$$

where $h = (h_{j\ell}) = \partial^2_{I_{j\ell}} H_+^{(3)}(\Phi^*, I^*)$, and all other expressions are also evaluated at $(\Phi, I) = (\Phi^*, I^*)$. The constants $c_j^{(3)}$ are both positive. Because I^* is a nondegenerate maximum, $h_{11} = \partial^2_{I_1} H_+^{(3)}(\Phi^*, I^*) < 0$ and $h_{22} = \partial^2_{I_2} H_+^{(3)}(\Phi^*, I^*) < 0$, while $\det(h) > 0$. In addition the radicand of (84) also satisfies

$$(c_1^{(3)}\sqrt{R_1R_2R_3}h_{11} + c_2^{(3)}\sqrt{R_1R_2R_4}h_{22})^2 - 4c_1^{(3)}c_2^{(3)}R_1R_2\sqrt{R_3R_4}\det(h)$$
$$= (c_1^{(3)}\sqrt{R_1R_2R_3}h_{11} - c_2^{(3)}\sqrt{R_1R_2R_4}h_{22})^2 + 4c_1^{(3)}c_2^{(3)}R_1R_2\sqrt{R_3R_4}h_{12}^2 ,$$

which is nonnegative. The radicand also satisfies

$$(c_1^{(3)}\sqrt{R_1R_2R_3}h_{11} + c_2^{(3)}\sqrt{R_1R_2R_4}h_{22})^2 - 4c_1^{(3)}c_2^{(3)}R_1R_2\sqrt{R_3R_4}\det(h)$$
$$< (c_1^{(3)}\sqrt{R_1R_2R_3}h_{11} + c_2^{(3)}\sqrt{R_1R_2R_4}h_{22})^2 .$$

Therefore both roots μ^2 of (84) are negative, and the eigenvalues μ of (83) arise in pure imaginary complex conjugate pairs. □

Proposition 6. *For $(I_3, I_4) = (a_3, a_4)$ fixed, the out-of-phase solutions on M are either two or four in number, and are geometrically distinct but are interchanged pairwise by time reversal of the system. The resulting solutions of the full system* (80) *are quasi-periodic with generically two independent frequencies, and they are all unstable.*

Proof. The Hamiltonian $H_+^{(3)}$, when evaluated on the hypersurface $\Phi^* := (\phi_1 = 0, \phi_2 = \pi/\sqrt{3})$ is either everywhere positive, or else changes sign, and in the latter case there is only one component of each sign. It is

$$H_+^{(3)}|_{\Phi=\Phi^*} = 2c_1^{(3)}\sqrt{R_1R_2R_3} - 2c_2^{(3)}\sqrt{R_1R_2R_4} . \qquad (85)$$

The Hamiltonian $H_+^{(3)}$ vanishes on the boundaries of the region $M = \{(I_1, I_2) : R_j(I) > 0\}$ for which either $R_1(I) = 0$ or $R_2(I) = 0$. Since $R_3(I)$ is decreasing in I_1 and independent of I_2, while $R_4(I)$ is independent of I_1 and increasing in I_2, the boundary component on which $R_4(I) = 0$ is always nonempty, and on it $H_+^{(3)}|_{\Phi=\Phi^*} > 0$. The boundary component defined by $R_3(I) = 0$ may be empty, in which case $H_+^{(3)}|_{\Phi=\Phi^*} > 0$ throughout the region. If it is not empty, $H_+^{(3)}|_{\Phi=\Phi^*} < 0$ on this set, while its outward normal derivative is positive, giving rise to a region in which $H_+^{(3)}|_{\Phi=\Phi^*}$ is negative. Because of the monotonicity properties of $R_3(I)$ and

$R_4(I)$ there can be at most one component of each sign, and because of the affine linear character of the $R_j(I)$, the maximum and minimum critical points are unique and nondegenerate.

As mentioned above, one deduces from the monotonicity of the $R_j(I)$ that the boundary components of M^0 corresponding to $R_1(I) = 0$ and $R_4(I) = 0$ are always nonempty. On the other hand, under different choices of parameters $(I_3, I_4) = (a_3, a_4)$ it could be that there are nonempty boundary components for both $R_2(I)$ and $R_3(I)$, or it could be that one of the two is empty. In the former case, the manifold $M \simeq \mathbb{S}^2 \times \mathbb{S}^2$, while if one boundary component is empty, then $M \simeq \mathbb{S}^3 \times \mathbb{S}^1$.

The statement of instability of these orbits comes again from an inspection of the spectrum of the variational equation (83) at the stationary points (Φ^*, I^*), give in this case by

$$\begin{pmatrix} \partial_\Phi \partial_I H_+^{(3)} & \partial_I^2 H_+^{(3)} \\ -\partial_\Phi^2 H_+^{(3)} & -\partial_\Phi \partial_I H_+^{(3)} \end{pmatrix} = \begin{pmatrix} 0 & \partial_I^2 H_+^{(3)} \\ -\partial_\Phi^2 H_+^{(3)} & 0 \end{pmatrix} \tag{86}$$

where as before,

$$-\partial_\Phi^2 H_+^{(3)} = \begin{pmatrix} 6c_1^{(3)} \sqrt{R_1 R_2 R_3} & 0 \\ 0 & -6c_2^{(3)} \sqrt{R_1 R_2 R_4} \end{pmatrix} .$$

The eigenvalues of (86) are expressed by

$$\begin{aligned} \mu^2 &= 3\big(c_1^{(3)} \sqrt{R_1 R_2 R_3} h_{11} - c_2^{(3)} \sqrt{R_1 R_2 R_4} h_{22}\big) \\ &\quad \pm 3\sqrt{\big((c_1^{(3)} \sqrt{R_1 R_2 R_3} h_{11} - c_2^{(3)} \sqrt{R_1 R_2 R_4} h_{22})^2 + 4c_1^{(3)} c_2^{(3)} R_1 R_2 \sqrt{R_3 R_4} \det(h)\big)} \end{aligned} \tag{87}$$

where $h = \partial_I^2 H_+^{(3)}(\Phi^*, I^*)$ as above. For I^* a maximum critical point, $h_{11}, h_{22} < 0$ and $\det(h) > 0$, and we observe that the radicand is positive and that the radical also dominates the first term $3(c_1^{(3)} \sqrt{R_1 R_2 R_3} h_{11} - c_2^{(3)} \sqrt{R_1 R_2 R_4} h_{22})$ in absolute value. That is to say, of the roots μ^2 of (87), one is positive and one is negative, leading to the eigenvalues μ of (86) consisting of a pure imaginary eigenvalue pair and a pair of real eigenvalues, one of which is positive; this does not give rise to a fully whiskered torus for the full system, but one whose normal environment consists of two center tangent vectors as well as a one dimensional stable and a one dimensional unstable component of the tangent space of M.

In cases in which there is a nontrivial negative minimum critical point, giving rise to a second pair of stationary points (Φ^*, I^*), we have $\det(h) > 0$ while $h_{11}, h_{22} > 0$. The radicand of (87) is again positive, the radical dominates the first term, and therefore there is again a pair of real eigenvalues and a pair of pure imaginary and complex conjugate eigenvalues of the variational equation (86).

The remaining phase space orbit of interest is the stationary point at which $R_1(I)$ and $R_2(I)$ vanish simultaneously, which occurs on the boundary of the coordinate chart M^0. □

Proposition 7. *In the situation in which $R_1(I)$ and $R_2(I)$ vanish simultaneously on the boundary of the coordinate chart M^0, there is an additional quasi-periodic orbit of system* (80). *This orbit can be either stable or unstable, depending upon the values of the integrals of motion (I_3, I_4). But even in the stable case its variational equation has double pure imaginary roots, which have opposite Krein signature and are therefore unstable under generic perturbations.*

Proof. For this case one must work in another, and local coordinate system. Both $R_1(I) = 0 = R_2(I)$ when

$$R_1 = \frac{1}{\sqrt{3}}(I_1 - I_2) + K_1 = 0\,, \quad R_2 = -\frac{1}{\sqrt{3}}(I_1 + I_2) + K_2 = 0\,,$$

where $K_1 := b_1 I_3 + \omega_1/\omega_4 I_4$ and $K_2 := b_2 I_3 + \omega_2/\omega_4 I_4$, constants set by the values of the two integrals of motion. This simultaneous zero exists whenever the subset of the boundary of M^0 defined by $\{R_2(I) = 0\}$ is nonempty, and it occurs when $I_1 = I_1^0 := -\sqrt{3}/2(K_1 - K_2)$ and $I_2 = I_2^0 := \sqrt{3}/2(K_1 + K_2)$. Using than $R_j(I) = |z_j|^2, j = 1, 2$, then the Hamiltonian $H_+^{(3)}$ can be written in terms of z_1, z_2 as

$$\begin{aligned} H_+^{(3)} &= 3c_1^{(3)}\sqrt{R_3}|z_1||z_2|\cos(\theta_1 - \theta_2 - \theta_3) \pm 3c_2^{(3)}\sqrt{R_4}|z_1||z_2|\cos(\theta_4 - \theta_1 - \theta_2) \\ &= 3c_1^{(3)}(z_1\bar{z}_2\bar{z}_3 + \bar{z}_1 z_2 z_3) \pm 3c_2^{(3)}(\bar{z}_1\bar{z}_2 z_4 + z_1 z_2 \bar{z}_4) \end{aligned}$$

where to lowest order in $(I_1 - I_1^0, I_2 - I_2^0)$, the variables (z_3, z_4) are constants (Z_3, Z_4), set by the value of the conserved quantities (I_3, I_4). The resulting variational equation in the variables (z_1, z_2) is as follows:

$$\begin{aligned} \dot{z}_1 &= i\big(3c_1^{(3)} Z_3 z_2 \pm 3c_2^{(3)} Z_4 \bar{z}_2\big) \\ \dot{z}_2 &= i\big(3c_1^{(3)} \bar{Z}_3 z_1 \pm 3c_2^{(3)} Z_4 \bar{z}_1\big)\,, \end{aligned} \tag{88}$$

which is written in the form

$$\begin{pmatrix} \dot{z}_1 \\ \dot{\bar{z}}_1 \\ \dot{z}_2 \\ \dot{\bar{z}}_2 \end{pmatrix} = i \begin{pmatrix} 0 & 0 & a & b \\ 0 & 0 & -\bar{b} & -\bar{a} \\ \bar{a} & b & 0 & 0 \\ -\bar{b} & -a & 0 & 0 \end{pmatrix} \begin{pmatrix} z_1 \\ \bar{z}_1 \\ z_2 \\ \bar{z}_2 \end{pmatrix},$$

where $a = 3c_1^{(3)} Z_3$ and $b = \pm 3c_2^{(3)} Z_4$. The eigenvalues of this matrix are $\mu = \pm i\sqrt{|a|^2 - |b|^2}$ which are all double roots; they are of course pure imaginary complex conjugates when $|a|^2 > |b|^2$, and real when otherwise, dictating the stability of the system (88). However even in the stable case the two pairs of double eigenvalues have opposite Krein signature, as can be seen by the non-positive definiteness of the Hamiltonian giving (88). Thus even the pure imaginary case is libel to instability under perturbation. □

Acknowledgements WC is partially supported by the Canada Research Chairs Program and NSERC through grant number 238452–11. CS is partially supported by NSERC through grant number 46179–13 and Simons Foundation Fellowship 265059. CS would like to extend her warmest wishes to Walter on the occasion of his 60th birthday.

References

1. Alazard, T., Delort, J.-M.: Global solutions and asymptotic behavior for two dimensional gravity water waves. (2013). arXiv:1305.4090. Preprint
2. Chow, C., Henderson, D., Segur, H.: A generalized stability criterion for resonant triad interactions. J. Fluid Mech. **319**, 67–76 (1996)
3. Craig, W., Sulem, C.: Numerical simulation of gravity waves. J. Comput. Phys. **108**, 73–83 (1993)
4. Craig, W., Sulem, C., Sulem, P.-L.: Nonlinear modulation of gravity waves: a rigorous approach. Nonlinearity **5**, 497–522 (1992)
5. Craig, W., Worfolk, P.: An integrable normal form for water waves in infinite depth. Physica D **84**, 515–531 (1994)
6. Düll, W.P., Schneider, G., Wayne, C.E.: Justification of the non- linear Schrödinger equation for the evolution of gravity driven 2D surface water waves in a canal of finite depth (2013). Preprint
7. Dyachenko, A.I., Zakharov, V.E.: A dynamic equation for water waves in one horizontal dimension, Eur. J. Mech. B/Fluids **32**, 17–21 (2012)
8. Germain, P., Masmoudi, N., Shatah, J.: Global solutions for the gravity water waves equation in dimension 3. Ann. Math. **175**, 691–754 (2012)
9. Hammack, J., Henderson, D.: Resonant interactions among surface water waves. Annu. Rev. Fluid Mech. **25**, 55–97 (1993). Annual Reviews, Palo Alto, CA
10. Ionescu, A., Pusateri, F.: Global solutions for the gravity water waves system in 2d. (2013). arXiv:1303.5357v2. Preprint
11. Nazarenko, S.: Wave Turbulence. Lecture Notes in Physics, vol. 825. Springer, Heidelberg (2011). xvi+279 pp. ISBN: 978-3-642-15941-1
12. Totz, D., Wu, S.: A rigorous justification of the modulation approximation to the 2D full water wave problem. Commun. Math. Phys. **310**, 817–883 (2012)
13. Wu, S.: Almost global wellposedness of the 2-D full water wave problem. Invent. Math. **177**, 45–135 (2009)
14. Zakharov, V.E.: Stability of periodic waves of finite amplitude on the surface of a deep fluid. J. Appl. Mech. Tech. Phys. **9**, 1990–1994 (1968)
15. Zakharov, V.E.: Statistical theory of gravity and capillary waves on the surface of a finite-depth fluid, Eur. J. Mech. B/Fluids **18**, 327–344 (1999)
16. Zakharov, V.E., L'vov, V.S., Falkovich, G.: Komogorov Spectra of Turbulence. Springer, Berlin (1992)

On a Fluid-Particle Interaction Model: Global in Time Weak Solutions Within a Moving Domain in $\mathbb{R}^3$

Stefan Doboszczak and Konstantina Trivisa

Dedicated to Walter Craig, mentor and friend, in honor of his 60th birthday.

Abstract A fluid-particle interaction model is presented for the evolution of particles dispersed in a fluid. The fluid flow is governed by the Navier-Stokes equations for a compressible fluid while the evolution of the particle densities is given by the Smoluchowski equation. The coupling between the dispersed and dense phases is obtained through the drag forces that the fluid and particles exert mutually. In the present context, the flow occupies a physical domain Ω_t with boundary Γ_t both of which vary in time. Global-in-time weak solutions are obtained using an approach based on penalization of the boundary behavior and viscosity in the weak formulation.

1 Introduction

Particles in fluids (liquids or gases) appear in many practical applications in chemical engineering, atmospheric sciences, fluid mechanics, geology and biology [1, 2, 5, 6]. In this framework, *sprays* can be thought as gases in which droplets constitute a dispersed phase, and are typically governed by an equation of fluid dynamics and a kinetic equation (cf. Williams [23], Caflisch and Papanicolaou [7]). In this modeling, the gas is characterized by macroscopic variables depending on time t and the position x, namely, the fluid density $\rho = \rho(t,x)$ and the velocity field $\mathbf{u} = \mathbf{u}(t,x)$.

In order to describe the dispersed phase (the droplets), their distribution function in the phase space is employed, defined as $f_\varepsilon = f_\varepsilon(t,x,\xi) \geq 0$, and denoting the density of droplets which at time t and position x have velocity ξ. More precisely, the distribution function $f_\varepsilon(t,x,\xi)$ is the solution to the dimensionless Vlasov-Fokker-

S. Doboszczak • K. Trivisa (✉)
Department of Mathematics, University of Maryland, College Park, MD, USA
e-mail: doboss27@math.umd.edu; trivisa@math.umd.edu

P. Guyenne et al. (eds.), *Hamiltonian Partial Differential Equations and Applications*, Fields Institute Communications 75,
DOI 10.1007/978-1-4939-2950-4_4

Planck equation, see [8],

$$\partial_t f_\varepsilon + \frac{1}{\sqrt{\varepsilon}}\Big(\xi \cdot \nabla_x f_\varepsilon - \nabla_x \Phi \cdot \nabla_\xi f_\varepsilon\Big) = \frac{1}{\varepsilon}\,\mathrm{div}_\xi\Big(\big(\xi - \sqrt{\varepsilon}\mathbf{u}_\varepsilon\big)f_\epsilon + \nabla_\xi f_\varepsilon\Big).$$

Here, $\epsilon > 0$ is a dimensionless parameter, and Φ is an external potential (i.e. gravity). The drag force is independent of the fluid density ρ_ε, but follows Stokes' law and is therefore proportional to the relative velocity of the fluid and the particles given by

$$F(t,x,\xi) = 6\pi\mu a(\xi - \mathbf{u}_\varepsilon(t,x)),$$

that is the drag force is in this context proportional to the fluctuations of the microscopic velocity $\xi \in \mathbb{R}^3$ around the fluid velocity field $\mathbf{u}_\varepsilon$.

The coupling between the kinetic and the fluid equations is obtained through the friction forces that the fluid and the particles exert mutually,

$$\begin{aligned}
&\partial_t f_\varepsilon + \frac{1}{\sqrt{\varepsilon}}\Big(\xi \cdot \nabla_x f_\varepsilon - \nabla_x \Phi \cdot \nabla_\xi f_\varepsilon\Big) = \frac{1}{\varepsilon}\mathrm{div}_\xi\Big(\big(\xi - \sqrt{\varepsilon}\mathbf{u}_\varepsilon\big)f_\epsilon + \nabla_\xi f_\varepsilon\Big),\\
&\partial_t \rho_\varepsilon + \mathrm{div}_x(\rho_\varepsilon \mathbf{u}_\varepsilon) = 0,\\
&\partial_t(\rho_\varepsilon \mathbf{u}_\varepsilon) + \mathrm{div}_x(\rho_\varepsilon \mathbf{u}_\varepsilon \otimes \mathbf{u}_\varepsilon) + \nabla_x p(\rho_\varepsilon) - \mu\Delta_x \mathbf{u}_\epsilon + \beta\rho_\varepsilon \nabla_x \Phi = F_\varepsilon.
\end{aligned} \tag{1}$$

The right hand-side of the momentum equation in the Navier-Stokes system takes into account the action of the cloud of particles on the fluid through the forcing term

$$F_\varepsilon = \int_{\mathbb{R}^3}\Big(\frac{\xi}{\sqrt{\varepsilon}} - \mathbf{u}_\varepsilon(t,x)\Big) f_\epsilon(t,x,\xi)\,\mathrm{d}\xi.$$

The density of the particles $\eta_\varepsilon(t,x)$ is related to the probability distribution function $f_\varepsilon(t,x,\xi)$ through the relation

$$\eta_\varepsilon(t,x) = \int_{\mathbb{R}^3} f_\varepsilon(t,x,\xi)\,\mathrm{d}\xi.$$

The resulting formal macroscopic fluid-particle system obtained via the standard Hilbert-expansion procedure, as $\varepsilon \to 0$, through the scaling limit in (1) is governed by the primitive conservation equations for fluid-particle flows in the *bubbling regime* [9]. In this regime the particles are supposed to have negligible density with respect to the fluid, and thus, due to buoyancy effects, will typically move upwards in a system under gravity. From this phenomenon the regime bears the name of *bubbling*. We remark that different macroscopic equations can be obtained by various scaling limits [8].

The resulting system describes the evolution of particles dispersed in a viscous compressible fluid and is expressed by the conservation of fluid mass, the balance of momentum and the balance of particle density often referred to as the Smolukowski equation. The state of such flows is, in general, characterized by the macroscopic variables: the total fluid mass density $\rho = \rho(t, x)$, the velocity field $\mathbf{u} = \mathbf{u}(t, x)$ as well as the density of particles in the mixture $\eta = \eta(t, x)$, depending on the time $t \in (0, T)$ and the Eulerian spatial coordinate x in the moving domain $\Omega_t \subset \mathbb{R}^N$, $N = 3$. This system is presented below:

$$\partial_t \rho + \operatorname{div}_x(\rho \mathbf{u}) = 0, \tag{2a}$$

$$\partial_t(\rho \mathbf{u}) + \operatorname{div}_x(\rho \mathbf{u} \otimes \mathbf{u}) + \nabla_x(p(\rho) + \eta) = \operatorname{div}_x \mathbb{S} - (\eta + \beta\rho)\nabla_x \Phi, \tag{2b}$$

$$\partial_t \eta + \operatorname{div}_x(\eta(\mathbf{u} - \nabla_x \Phi)) - \Delta_x \eta = 0. \tag{2c}$$

The physical properties of the mixture are given through the following constitutive relations.

- The pressure p is taken to be

$$p(\rho) := a\rho^{\gamma}, \text{ with } a > 0 \text{ and } \gamma > \frac{3}{2}. \tag{3}$$

  The total pressure $P = P(\rho, \eta)$ in the mixture depends on the density of the particles and the density of the fluid and is given by

$$P(\rho, \eta) = p(\rho) + \eta.$$

- The viscous stress tensor $\mathbb{S} = \mathbb{S}(\nabla_x \mathbf{u})$ is assumed to satisfy *Newton's Law for Viscosity* which requires that

$$\mathbb{S} = \mu(\nabla_x \mathbf{u} + \nabla_x \mathbf{u}^T) + \lambda \operatorname{div}_x \mathbf{u}\, \mathbb{I}, \tag{4}$$

  where μ and λ are constant viscosity coefficients satisfying

$$\mu > 0, \ \lambda + \frac{2}{3}\mu \geq 0,$$

  and so

$$\operatorname{div}_x \mathbb{S}(\nabla_x \mathbf{u}) = \mu \Delta_x \mathbf{u} + \lambda \nabla_x \operatorname{div}_x \mathbf{u}.$$

The boundary of the domain Ω_t occupied by the fluid and the particles is described by means of a given velocity field $\mathbf{V}(t, \mathbf{x})$, where $t \geq 0$ and $\mathbf{x} \in \mathbb{R}^3$. More precisely, assuming $\mathbf{V}$ is sufficiently regular, we solve the associated system of differential equations

$$\begin{cases} \dfrac{d}{dt}\mathbf{X}(t, \mathbf{x}) = \mathbf{V}(t, \mathbf{X}(t, \mathbf{x})), \ t > 0, \\ \mathbf{X}(0, \mathbf{x}) = \mathbf{x}, \end{cases}$$

and set

$$\Omega_t = \mathbf{X}(t, \Omega_0), \text{ where } \Omega_0 \subset \mathbb{R}^3 \text{ is a given initial domain,}$$
$$\Gamma_t = \partial\Omega_t, \text{ and } Q^f = \{(t, x) \mid t \in (0, T), x \in \Omega_t\}.$$

We also define here the "solid" part of the domain,

$$Q^s = ((0, T) \times D) \backslash \overline{Q^f}.$$

The *no-slip boundary conditions* on the solid wall are imposed, meaning

$$\mathbf{u}(t, \cdot)\big|_{\Gamma_t} = \mathbf{V}(t, \cdot)\big|_{\Gamma_t}, \text{ for any } t \geq 0. \tag{5}$$

In addition, a no-flux condition for particle density holds,

$$(\nabla_x \eta + \eta \nabla_x \Phi) \cdot \nu = 0 \text{ on } (0, T) \times \Gamma_t, \tag{6}$$

with $\nu(t, x)$ denoting the outer normal vector to the boundary Γ_t.

The external potential

$$\Phi : \mathbb{R}^3 \to \mathbb{R}^+$$

typically represents the effects of gravity and buoyancy and β in (2b) is assumed to be non-zero for bounded domains and positive for unbounded domains.

Several results presented in the literature are obtained under certain assumptions concerning the geometry of Ω and the external potential Φ under the generic name of *confinement conditions*. Let us remark that the external potential Φ is always defined up to a constant. Therefore, for bounded from below external potentials Φ, we can always assume without loss of generality, by adding a suitable constant, that

$$\inf_{x \in \Omega} \Phi(x) = 0. \tag{7}$$

The following definition will be relevant in the subsequent sections, see [10].

Definition 1. Given a domain $\Omega \in C^{2,\nu}$, $\nu > 0$, $\Omega \subset \mathbb{R}^3$, and given a bounded from below external potential $\Phi : \Omega \longrightarrow \mathbb{R}_0^+$ satisfying (7), we will say that (Ω, Φ) verifies the confinement hypotheses **(HC)** for the two-phase flow system (2a)–(2c) coupled with no-flux boundary conditions (5)–(6) whenever:

(HC-Bounded) If Ω is bounded, $\Phi \in W^{1,\infty}(\Omega)$ and the sub-level sets $\{\Phi < k\}$ are connected in Ω for any $k > 0$.
(HC-Unbounded) If Ω is unbounded, we assume that $\Phi \in W^{1,\infty}_{loc}(\Omega)$, $\beta > 0$, the sub-level sets $\{\Phi < k\}$ are connected in Ω for any $k > 0$,

$$e^{-\Phi/2} \in L^1(\Omega),$$

and

$$|\Delta\Phi(x)| \leq c_1|\nabla_x\Phi(x)| \leq c_2\Phi(x), \ |x| > R, \tag{8}$$

for some large $R > 0$.

Remark 1. The confinement assumption (**HC**) has physical relevance in our setting as it is verified for several domains Ω with Φ being the gravitational potential. For instance,

1. when $\Omega = \{x \in \mathbb{R}^3 \mid (x_1, x_2) \in [a,b]^2, \, x_3 \in [0,H]\}$ and $\Phi(x) = gx_3$, where $\beta = 1 - \frac{\rho_F}{\rho_P}$.
2. when $\Omega = \{x \in \mathbb{R}^3 \mid (x_1, x_2) \in [a,b]^2, \, x_3 > 0\}$ and $\Phi(x) = gx_3$, where $\beta = 1 - \frac{\rho_F}{\rho_P}$ and $\rho_F < \rho_P$.
3. when $\Omega = \mathbb{R}^3 \setminus \overline{B(0,R)}$ and $\Phi(x) = g|x|$, where $B(0,R)$ is the ball centered at the origin with radius R and $\beta > 0$.

Here, ρ_F and ρ_P are the typical mass density of fluid and particles, respectively. Note that 1 corresponds to the standard bubbling case (see [8]) in which particles move upwards due to buoyancy.

Remark 2. For our situation involving a time dependent domain Ω_t, it suffices that if $\Omega_t \subset D$, where D is a fixed reference domain, then (D, Φ) satisfies the confinement hypotheses.

Our problem is supplemented with initial data $\{\eta_0, \rho_0, \mathbf{u}_0\}$ such that

$$\begin{aligned} \eta(0,x) &= \eta_0 \in (L^2 \cap L^1_+)(\mathbb{R}^3), \\ \rho(0,x) &= \rho_0 \in (L^\gamma \cap L^1_+)(\mathbb{R}^3), \\ (\rho\mathbf{u})(0,x) &= \mathbf{m}_0 \in (L^{\frac{6}{5}} \cap L^1)(\mathbb{R}^3). \end{aligned} \tag{9}$$

Our aim is to study system (2a)–(2c) in a spatial domain with boundary varying in time, along with the previous assumptions.

1.1 Strategy

The main ingredients of our approach can be formulated as follows:

- For the construction of a suitable approximating scheme *penalizing* the boundary behavior, extra diffusion and viscosity terms are introduced in the weak formulation. The central component of this approach is the addition of a singular term

$$\int_0^\tau \int_{\Omega_t} \frac{\chi(\mathbf{u} - \mathbf{V})}{\varepsilon} \cdot \boldsymbol{\varphi}\, dxdt, \quad \varepsilon > 0 \text{ small}, \tag{10}$$

  in the momentum equation. This extra term models solid obstacles as porous media, with porosity and viscous permeability approaching zero, see Angot et al.[3]. Effectively, the problem is reformulated over a fixed domain such that the fluid is allowed to "flow" through solid obstacles.
- In addition to (10), we introduce *variable* shear viscosity coefficients $\mu = \mu_\omega$ and $\lambda = \lambda_\omega$, vanishing outside the fluid domain and remaining positive within the fluid domain, to take care of extra stress terms that appear in the "solid" domain.
- In constructing the approximating problem we employ a number of ingredients: time discretization h, a parameter δ which enables us to introduce an *artificial pressure* essential for the establishment of suitable pressure estimates and parameters ε and ω for the penalization of the boundary behavior and viscosity. Keeping $h, \delta, \epsilon, \omega$ fixed, we solve the modified problem in a (bounded) reference domain $D \subset \mathbb{R}^3$ chosen in such a way that

$$\Omega_t \subset \overline{\Omega_t} \subset D \text{ for any } t \geq 0.$$

  Letting $h \to 0$ and $\delta \to 0$ in the spirit of the analysis in [10] we obtain the solution $(\rho, \mathbf{u}, \eta)_{\omega,\varepsilon}$ within the fixed reference domain.
- We take the initial densities (ρ_0, η_0) vanishing outside Ω_0, and letting the penalization $\varepsilon \to 0$ we obtain a "two-phase" model consisting of the *fluid region* and the *solid region* separated by impermeable boundary. We show that the densities vanish in the "solid" part of the reference domain, specifically on $((0, T) \times D) \setminus \overline{Q^f}$.
- The penalization ϵ is taken to vanish and then we perform the limit $\omega \to 0$.

The main contribution of the present article to the existing theory can be characterized as follows:

- The present work investigates the dynamics of a mixture of particles dispersed in a viscous, compressible fluid within a moving domain $\Omega_t \subset \mathbb{R}^3$. The global existence of weak solutions within a moving domain in $\mathbb{R}^3$ is obtained by establishing the convergence of a Brinkman type penalization.

- The framework presented here relies on physically grounded principles and provides a description of the dynamics of fluid-particle interaction. The investigation of fluid-particle interaction flows have potential applications in atmospheric sciences, biological sciences and medicine (tumor growth, asthma, etc.).

For related works on the mathematical analysis of models of compressible fluids within moving domains we refer the reader to Feireisl et. al. [16, 17]. In the context of tumor growth models in the general three dimensional Euclidean setting we refer the reader to Donatelli and Trivisa [12–14] and to Friedman [18] in the context of radially symmetric solutions. There is a substantial literature on the mathematical analysis of fluid-particle interaction models. We refer the reader to Ballew and Trivisa [4], Carrillo, et al. [8, 9], Carrillo, et al. [10], Mellet and Vasseur [20, 21] and the references therein. The existence theory for the barotropic Navier-Stokes system on *fixed* spatial domains in the framework of weak solutions was developed in the seminal work of Lions [19].

1.2 Outline

The paper is organized as follows: Sect. 1 presents the motivation, modeling and introduces the necessary preliminary material and the overall strategy. Section 2 provides the weak formulations and states the main result. In addition, the penalized problem is introduced along with a suitable approximation scheme. The central component of the approximating procedure is the addition of a singular forcing term

$$\int_0^\tau \int_{\Omega_t} \frac{\chi(\mathbf{u} - \mathbf{V})}{\varepsilon} \cdot \varphi \, \mathrm{d}x\mathrm{d}t, \quad \varepsilon > 0 \text{ small},$$

penalizing the velocity on the boundary of the fluid-particle domain in the variational formulation of the momentum equation. In order to treat the moving boundary, an additional penalization on the viscosity is required. In Sect. 3 we collect all the uniform bounds satisfied by the solution of the penalization scheme. In Sect. 3.2 through Sect. 3.5 the singular limits for $\varepsilon \to 0, \omega \to 0$ are performed successively. A key part in the penalization limit is to get rid of the terms supported in the solid part $((0, T) \times D) \backslash \overline{Q^f}$. This part of the analysis is presented in Sect. 3.4 The main issue is to describe the evolution of the interface Γ_t. To that effect we employ elements from the so-called *level set method* (cf. Osher and Fedwik [22]). Finally, an Appendix of earlier results [10] on global existence and asymptotic analysis of solutions for two different types of domains.

2 Free Energy Solutions: Global in Time Existence Within a Moving Domain $\Omega_t \in \mathbb{R}^3$

2.1 Weak Formulation and Main Results

We begin with the notion of free energy solutions of the original problem. In order to state the following definition concisely, we make use of the notation $L^p(0,T;L^q(\Omega_t))$, and variants thereof. For such a function $F = F(t,x)$, this is to be interpreted in the sense that the maps

$$t \mapsto \|F(t,\cdot)\|_{L^q(\Omega_t)}$$

belong to $L^p(0,T)$, and similarly for other spaces. Since we make no use of the structure of such "evolving spaces,"aside from ensuring the weak formulations are well-defined, this suffices for our purposes.

Definition 2 (Free Energy Solutions). Let us assume that (Ω_t, Φ) satisfy the confinement hypotheses (**HC**). We say that $\{\rho, \mathbf{u}, \eta\}$ is a free-energy solution of problem (2a)–(2c) with initial and boundary data satisfying (9)–(6) respectively provided that the following hold:

- $\rho \geq 0$ represents a renormalized solution of Eq. (2a) on a time-space cylinder $(0,\infty) \times \Omega_t$. More precisely, for any test function $\varphi \in \mathscr{D}([0,T) \times \overline{\Omega_t})$, any $T > 0$, and any b such that

$$b \in L^\infty \cap C[0,\infty),\ B(\rho) = \rho B(1) + \rho \int_1^\rho \frac{b(z)}{z^2}\,\mathrm{d}z,$$

  the following integral identity holds:

$$\int_0^\infty \int_{\Omega_t} \Big(B(\rho)\partial_t\varphi + B(\rho)\mathbf{u}\cdot\nabla_x\varphi - b(\rho)\mathrm{div}_x\mathbf{u}\varphi\Big)\,\mathrm{d}x\mathrm{d}t = -\int_{\Omega_0} B(\rho_0)\varphi(0,\cdot)\,\mathrm{d}x. \tag{11}$$

- The balance of momentum holds in distributional sense, namely

$$\begin{aligned}
&\int_0^\infty \int_{\Omega_t} \Big(\rho\mathbf{u}\cdot\partial_t\varphi + \rho\mathbf{u}\otimes\mathbf{u} : \nabla_x\varphi + (p(\rho)+\eta)\,\mathrm{div}_x\varphi\Big)\,\mathrm{d}x\mathrm{d}t = \\
&\int_0^\infty \int_{\Omega_t} (\mu\nabla_x\mathbf{u} + \lambda\mathrm{div}_x\mathbf{u}\mathbb{I}) : \nabla_x\varphi - (\eta + \beta\rho)\nabla_x\Phi\cdot\varphi\,\mathrm{d}x\mathrm{d}t \\
&- \int_{\Omega_0} \rho_0\mathbf{u}_0\cdot\varphi(0,\cdot)\,\mathrm{d}x
\end{aligned} \tag{12}$$

for any test function $\varphi \in \mathscr{D}([0,T);\mathscr{D}(\overline{\Omega_t};\mathbb{R}^3))$ and any $T > 0$ satisfying $\varphi|_{\partial\Omega_t} = 0$.

All quantities appearing in (12) are supposed to be at least integrable. In particular, the velocity field $\mathbf{u}$ belongs to the space $L^2(0,T;W^{1,2}(\Omega_t;\mathbb{R}^3))$, therefore it is legitimate to require $\mathbf{u}$ to satisfy the boundary conditions (5) in the sense of traces.

- The integral identity

$$\int_0^\infty \int_{\Omega_t} \eta\partial_t\varphi + \eta\mathbf{u}\cdot\nabla_x\varphi - \eta\nabla_x\Phi\cdot\nabla_x\varphi - \nabla_x\eta\cdot\nabla_x\varphi \,\mathrm{d}x\,\mathrm{d}t = -\int_{\Omega_0} \eta_0\varphi(0,\cdot)\,\mathrm{d}x \quad (13)$$

is satisfied for test functions $\varphi \in \mathscr{D}([0,T)\times\overline{\Omega_t})$ and any $T > 0$.

All quantities appearing in (13) must be at least integrable on $(0,T)\times D$. In particular, η belongs to $L^2(0,T;L^3(\Omega_t)) \cap L^1(0,T;W^{1,\frac{3}{2}}(\Omega_t))$.

- Given the total free-energy of the system by

$$E(\rho,\mathbf{u},\eta)(t) := \int_{\Omega_t} \left(\frac{1}{2}\rho|\mathbf{u}|^2 + \frac{a}{\gamma-1}\rho^\gamma + \eta\log\eta + (\beta\rho+\eta)\Phi\right)\,\mathrm{d}x,$$

$E(\rho,\mathbf{u},\eta)(t)$ is finite and bounded by the initial energy of the system, i.e., $E(\rho,\mathbf{u},\eta)(t) \le E(\rho_0,\mathbf{u}_0,\eta_0)$ for a.e. $t > 0$. Moreover, the following free energy-dissipation inequality holds

$$\int_0^\infty \int_{\Omega_t} (\mu|\nabla_x\mathbf{u}|^2 + \lambda|\mathrm{div}_x\mathbf{u}|^2 + |2\nabla_x\sqrt{\eta} + \sqrt{\eta}\nabla_x\Phi|^2)\,\mathrm{d}x\mathrm{d}t$$
$$\le E(\rho_0,\mathbf{u}_0,\eta_0) + C(1 + ||\mathbf{V}||_{L^\infty(0,T;\Omega_t)}). \quad (14)$$

2.2 *Penalization*

We introduce the penalization as follows. Following the strategy of Angot et. al. [3], see also [16] we fix a reference spatial domain $D \subset \mathbb{R}^3$ containing Ω_0 and such that

$$\mathbf{V}|_{\partial D} = 0. \quad (15)$$

System (2a)–(2c) is replaced by a penalized problem

$$\partial_t\rho + \mathrm{div}_x(\rho\mathbf{u}) = 0, \quad (16a)$$

$$\partial_t(\rho\mathbf{u}) + \mathrm{div}_x(\rho\mathbf{u}\otimes\mathbf{u}) + \nabla_x(p(\rho)+\eta)$$
$$= \mathrm{div}_x\mathbb{S}_\omega - (\eta+\beta\rho)\nabla_x\Phi - \frac{1}{\varepsilon}\chi(\mathbf{u}-\mathbf{V}) \quad (16b)$$

$$\partial_t\eta + \mathrm{div}_x(\eta(\mathbf{u}-\nabla_x\Phi)) - \Delta_x\eta = 0, \quad (16c)$$

considered in the cylinder $(0,T) \times D$. The function

$$\chi(t,x) = \begin{cases} 0 & \text{if } t \in (0,T), x \in \Omega_t \\ 1 & \text{otherwise} \end{cases} \tag{17}$$

is used to separate the fluid and "solid" domains and represents a weak solution to the transport equation

$$\begin{cases} \partial_t \chi + \mathbf{V} \cdot \nabla_x \chi = 0 \\ \chi(0,\cdot) = \mathbb{1}_D - \mathbb{1}_{\Omega_0}. \end{cases} \tag{18}$$

Problem (16) is supplemented with the boundary conditions

$$\mathbf{u}\big|_{\partial D} = \mathbf{V}\big|_{\partial D}, \tag{19}$$

$$(\nabla_x \eta + \eta \nabla_x \Phi) \cdot \nu\big|_{\partial D} = 0 \tag{20}$$

with ν denoting the outer normal vector to the boundary ∂D, and initial conditions

$$\rho(0,\cdot) = \rho_{0,\varepsilon} \geq 0, \quad (\rho\mathbf{u})(0,\cdot) = (\rho\mathbf{u})_{0,\varepsilon}, \quad \eta(0,\cdot) = \eta_{0,\varepsilon} \geq 0, \tag{21}$$

to be specified in Theorem 2.

In order to eliminate extra stresses that appear we introduce a variable shear viscosity coefficient $\mu = \mu_\omega(t,\mathbf{x})$ where, $\mu = \mu_\omega$ remains strictly positive in Q^f but vanishes in the complement as $\omega \to 0$, namely μ_ω is taken such that

$$\mu_\omega \in C_c^\infty\left([0,T] \times \mathbb{R}^3\right), \quad 0 < \underline{\mu}_\omega \leq \mu_\omega(t,\mathbf{x}) \leq \mu \text{ in } [0,T] \times D,$$

$$\mu_\omega = \begin{cases} \mu = const > 0 & \text{in } Q^f \\ \mu_\omega \to 0 & \text{a.e. in } ((0,T) \times D) \backslash Q^f. \end{cases}$$

We penalize the coefficient $\lambda = \lambda_\omega(t,x)$ exactly the same way. Finally we modify the initial data

$$(\rho\mathbf{u})_{0,\varepsilon,\omega} = \frac{|(\rho\mathbf{u})_{0,\varepsilon,\omega}|^2}{\rho_{0,\varepsilon,\omega}} = 0, \quad \text{whenever } \rho_{0,\varepsilon,\omega} = 0.$$

The weak formulation of the penalized problem reads.

Definition 3 (Free Energy Solutions of the Penalized Problem). Let us assume that (D,Φ) satisfies the confinement hypotheses **(HC)**. We say that

$\{\rho_{\varepsilon,\omega}, \mathbf{u}_{\varepsilon,\omega}, \eta_{\varepsilon,\omega}\}$ is a free-energy solution of problem (16) with initial and boundary data satisfying (19)–(21) respectively provided that the following hold:

- $\rho_{\epsilon,\omega} \geq 0$ represents a renormalized solution of equation (16c) on a time-space cylinder $(0,\infty) \times D$, that is, for any test function $\varphi \in \mathscr{D}([0,T) \times \overline{D})$, any $T > 0$, and any b such that

$$b \in L^\infty \cap C[0,\infty),\; B(\rho_{\varepsilon,\omega}) = \rho_{\varepsilon,\omega} B(1) + \rho_{\varepsilon,\omega} \int_1^{\rho_{\varepsilon,\omega}} \frac{b(z)}{z^2}\,\mathrm{d}z,$$

  the following integral identity holds:

$$\int_0^\infty \int_D \Big(B(\rho_{\varepsilon,\omega})\partial_t \varphi + B(\rho_{\varepsilon,\omega})\mathbf{u}_{\varepsilon,\omega} \cdot \nabla_x \varphi - b(\rho_{\varepsilon,\omega})\mathrm{div}_x \mathbf{u}_{\varepsilon,\omega}\varphi \Big)\,\mathrm{d}x\mathrm{d}t$$
$$= - \int_D B(\rho_{0,\varepsilon,\omega})\varphi(0,\cdot)\,\mathrm{d}x. \tag{22}$$

- The balance of momentum holds in distributional sense, namely

$$\int_0^\infty \int_D \Big(\rho_{\varepsilon,\omega}\mathbf{u}_{\varepsilon,\omega} \cdot \partial_t \varphi + \rho_{\varepsilon,\omega}\mathbf{u}_{\varepsilon,\omega} \otimes \mathbf{u}_{\varepsilon,\omega} : \nabla_x \varphi + (p(\rho_{\varepsilon,\omega}) + \eta_{\varepsilon,\omega})\,\mathrm{div}_x \varphi \Big)\,\mathrm{d}x\mathrm{d}t \tag{23}$$

$$= \int_0^\infty \int_D (\mu_\omega \nabla_x \mathbf{u}_{\varepsilon,\omega} + \lambda \mathrm{div}_x \mathbf{u}_{\varepsilon,\omega}\mathbb{I}) : \nabla_x \varphi - (\eta_{\varepsilon,\omega} + \beta\rho_{\varepsilon,\omega})\nabla_x \Phi \cdot \varphi\,\mathrm{d}x\mathrm{d}t$$
$$+ \int_0^\infty \int_D \frac{\chi(\mathbf{u}_{\varepsilon,\omega} - \mathbf{V})}{\varepsilon} \cdot \varphi\,\mathrm{d}x\mathrm{d}t - \int_D (\rho\mathbf{u})_{0,\varepsilon,\omega} \cdot \varphi(0,\cdot)\,\mathrm{d}x \tag{24}$$

  for any test function $\varphi \in \mathscr{D}([0,T);\mathscr{D}(\overline{D};\mathbb{R}^3))$ and any $T > 0$ satisfying $\varphi|_{\partial D} = 0$.

  All quantities appearing in (24) are supposed to be at least integrable. In particular, the velocity field $\mathbf{u}_{\epsilon,\omega}$ belongs to the space $L^2(0,T;W^{1,2}(D;\mathbb{R}^3))$, therefore it is legitimate to require $\mathbf{u}_{\epsilon,\omega}$ to satisfy the boundary conditions (19) in the sense of traces.
- The integral identity

$$\int_0^\infty \int_D \eta_{\varepsilon,\omega}\partial_t \varphi + \eta_{\varepsilon,\omega}\mathbf{u}_{\varepsilon,\omega} \cdot \nabla_x \varphi - \eta_{\varepsilon,\omega}\nabla_x \Phi \cdot \nabla_x \varphi - \nabla_x \eta_{\varepsilon,\omega} \cdot \nabla_x \varphi\,\mathrm{d}x\mathrm{d}t$$
$$= - \int_D \eta_{0,\varepsilon,\omega}\varphi(0,\cdot)\,\mathrm{d}x \tag{25}$$

is satisfied for test functions $\varphi \in \mathscr{D}([0,T) \times \overline{D})$ and any $T > 0$. All quantities appearing in (25) must be at least integrable on $(0,T) \times D$. In particular, $\eta_{\epsilon,\omega}$ belongs to $L^2(0,T;L^3(D)) \cap L^1(0,T;W^{1,\frac{3}{2}}(D))$.

- Given the total free-energy of the system by

$$E(\rho_{\varepsilon,\omega}, \mathbf{u}_{\varepsilon,\omega}, \eta_{\varepsilon,\omega})(t)$$
$$:= \int_D \left(\frac{1}{2}\rho_{\varepsilon,\omega}|\mathbf{u}_{\varepsilon,\omega}|^2 + \frac{a}{\gamma-1}\rho_{\varepsilon,\omega}^{\gamma} + \eta_{\varepsilon,\omega}\log\eta_{\varepsilon,\omega} + (\beta\rho_{\varepsilon,\omega} + \eta_{\varepsilon,\omega})\Phi \right) \mathrm{d}x,$$

$E(\rho_{\varepsilon,\omega}, \mathbf{u}_{\varepsilon,\omega}, \eta_{\varepsilon,\omega})(t)$ is finite and the following free energy dissipation inequality holds

$$E(\rho_{\varepsilon,\omega}, \mathbf{u}_{\varepsilon,\omega}, \eta_{\varepsilon,\omega})(\tau) + \int_0^\tau \int_D (\mu_\omega|\nabla_x \mathbf{u}_{\epsilon,\omega}|^2 + \lambda_\omega|\mathrm{div}_x \mathbf{u}_{\varepsilon,\omega}|^2$$
$$+|2\nabla_x\sqrt{\eta_{\varepsilon,\omega}} + \sqrt{\eta_{\varepsilon,\omega}}\nabla_x\Phi|^2)\,\mathrm{d}x\mathrm{d}t$$
$$\leq E(\rho_{0,\varepsilon,\omega}, \mathbf{u}_{0,\varepsilon,\omega}, \eta_{0,\varepsilon,\omega}) - \int_0^\tau \int_D \frac{\chi}{\epsilon}(\mathbf{u}_{\epsilon,\omega} - \mathbf{V}) \cdot \mathbf{u}_{\epsilon,\omega}\,\mathrm{d}x\mathrm{d}t. \quad (26)$$

2.3 Main Result

We are now ready to state the main result of this article.

Theorem 1. *Let $\Omega_0 \subset \overline{\Omega_0} \subset D \subset \mathbb{R}^3$ be a bounded domain with boundary of class $C^{2+\nu}$, $\nu > 0$. Assume that the pressure p is given by* (3)*, with $\gamma > 3/2$, and that Φ satisfies the confinement hypothesis* **(HC)***. Let $\mathbf{V}$ be a given vector field belonging to $C^{2+\nu}([0,T] \times \overline{D}; \mathbb{R}^3)$,*

$$\mathbf{V}|_{\partial D} = 0.$$

Finally, we suppose that the initial data satisfy (9) *and*

$$\rho_{0,\varepsilon} \to \rho_0 \;\; in \;\; L^\gamma(D), \;\; \rho_0|_{\Omega_0} > 0, \;\; \rho_0|_{D\setminus\Omega_0} = 0, \quad (27)$$

$$(\rho\mathbf{u})_{0,\varepsilon} \to (\rho\mathbf{u})_0 \;\; in \;\; L^1(D;\mathbb{R}^3), \;\; (\rho\mathbf{u})_0|_{D\setminus\Omega_0} = 0, \quad (28)$$

$$\int_D \frac{|(\rho\mathbf{u})_{0,\varepsilon}|^2}{\rho_{0,\varepsilon}}\,\mathrm{d}x < c, \quad (29)$$

$$\eta_{0,\varepsilon} \to \eta_0 \;\; in \;\; L^2(D), \;\; \eta_0|_{\Omega_0} > 0, \;\; \eta_0|_{D\setminus\Omega_0} = 0, \quad (30)$$

where c is independent of $\varepsilon \to 0$. Then any sequence $\{\rho_\varepsilon, \mathbf{u}_\varepsilon, \eta_\varepsilon\}_{\varepsilon>0}$ of free energy solutions to problem (16)*, in the sense of Definition 3, contains a subsequence such that*

$$\rho_\varepsilon \to \rho \ \textit{in} \ C_{weak}([0,T]; L^\gamma(D)) \cap L^\gamma(Q^f), \tag{31}$$

$$\mathbf{u}_\varepsilon \rightharpoonup \mathbf{u} \ \textit{in} \ L^2(0,T; W_0^{1,2}(D;\mathbb{R}^3)), \ \ \mathbf{u} = \mathbf{V} \ \textit{a.e. in} \ Q^s, \tag{32}$$

$$\eta_\varepsilon \rightharpoonup \eta \ \textit{in} \ L^2(0,T; L^{\frac{3}{2}}(D)) \cap L^p(0,T; W^{1,q}(D)) \ \textit{with} \ p,q > 1, \tag{33}$$

with ρ, η vanishing in Q^s. Finally, $(\rho, \mathbf{u}, \eta)$ represents a free energy solution of problem (2a)–(2c) *in the sense of Definition 2.*

2.4 Construction of Approximate Solutions Within D

The construction of the approximate solutions

$$(\rho_{\varepsilon,\omega}, \mathbf{u}_{\varepsilon,\omega}, \eta_{\varepsilon,\omega})$$

within the fixed reference domain D relies

- on the time-discretization of the system (16) with the aid of a parameter h transforming the system into an elliptic-parabolic system, and
- on the *artificial pressure* with replacing the pressure term $p(\rho)$ with the term $p_\delta(\rho_\delta) = p(\rho_\delta) + \delta\rho^6$ with the aid of a parameter δ which enables us to establish suitable pressure estimates.

Keeping ϵ, ω fixed, we solve the modified problem in a (bounded) reference domain $D \subset \mathbb{R}^3$ chosen in such way that

$$\overline{\Omega_t} \subset D \text{ for any } t \geq 0.$$

Letting $h \to 0$ and $\delta \to 0$ in the spirit of the analysis in (cf. [10]) we obtain the existence of a weak solution $(\rho, \mathbf{u}, \eta)_{\varepsilon,\omega}$ within the fixed reference domain D in the sense of Definition 3. We refer the reader to Appendix where relevant results are presented.

3 Energy Estimates

3.1 Modified Energy Inequality

Choosing as a test function $\varphi = \psi_n(t)\mathbf{V}, \psi_n \in C_c^\infty[0,T), \psi_n \to \mathbb{1}_{[0,\tau)}$ in (24) and adding to the inequality (26), we find that

$$\begin{aligned}
&\int_D \left(\frac{1}{2}\rho_{\varepsilon,\omega}|\mathbf{u}_{\varepsilon,\omega}|^2 + \frac{a}{\gamma-1}\rho_{\varepsilon,\omega}^\gamma + \eta_{\varepsilon,\omega}\log\eta_{\varepsilon,\omega} + (\beta\rho_{\varepsilon,\omega} + \eta_{\varepsilon,\omega})\Phi \right)(\tau,\cdot)\,\mathrm{d}x \\
&+ \int_0^\tau\int_D \left(\mu_\omega|\nabla_x\mathbf{u}_{\varepsilon,\omega}|^2 + \lambda|\mathrm{div}_x\mathbf{u}_{\varepsilon,\omega}|^2 + |2\nabla_x\sqrt{\eta_{\varepsilon,\omega}} + \sqrt{\eta_{\varepsilon,\omega}}\nabla_x\Phi|^2 \right)\mathrm{d}x\mathrm{d}t \\
&+ \frac{1}{\varepsilon}\int_0^\tau\int_D \chi|\mathbf{u}_{\varepsilon,\omega} - \mathbf{V}|^2\,\mathrm{d}x\mathrm{d}t \\
\le &\int_D \left(\frac{1}{2}\frac{|(\rho\mathbf{u})_{0,\varepsilon,\omega}|^2}{\rho_{0,\varepsilon,\omega}} + \frac{a}{\gamma-1}\rho_{0,\varepsilon,\omega}^\gamma + \eta_{0,\varepsilon,\omega}\log\eta_{0,\varepsilon,\omega} + (\beta\rho_{0,\varepsilon,\omega} + \eta_{0,\varepsilon,\omega})\Phi \right)\mathrm{d}x \\
&+ \int_D (\rho_{\epsilon,\omega}\mathbf{u}_{\epsilon,\omega}\cdot\mathbf{V})(\tau,\cdot) - (\rho\mathbf{u})_{0,\varepsilon,\omega}\cdot\mathbf{V}(0,\cdot)\,\mathrm{d}x \\
&+ \int_0^\tau\int_D \mathbb{S}_{\varepsilon,\omega} : \nabla_x\mathbf{V} - \rho_{\epsilon,\omega}\mathbf{u}_{\epsilon,\omega}\cdot\partial_t\mathbf{V} - \rho_{\epsilon,\omega}\mathbf{u}_{\epsilon,\omega}\otimes\mathbf{u}_{\epsilon,\omega} : \nabla_x\mathbf{V} \\
&- (\eta_{\epsilon,\omega} + \beta\rho_{\epsilon,\omega})\nabla_x\Phi\cdot\mathbf{V} - \left(\frac{a}{\gamma-1}\rho_{\epsilon,\omega}^\gamma + \eta_{\epsilon,\omega} \right)\mathrm{div}_x\mathbf{V}\,\mathrm{d}x\mathrm{d}t
\end{aligned} \tag{34}$$

for a.a. $\tau \in (0,T)$. This yields uniform bounds on $(\rho_{\epsilon,\omega}, \mathbf{u}_{\epsilon,\omega}, \eta_{\epsilon,\omega})$ independent of $\varepsilon \to 0$ provided $\mathbf{V}$ is sufficiently smooth.

In accordance with the boundary conditions (19) and (20), the total fluid and particle mass

$$M_{\rho,\varepsilon,\omega} = \int_D \rho_{\varepsilon,\omega}(t,\cdot)\,\mathrm{d}x = \int_D \rho_{0,\varepsilon,\omega}\,\mathrm{d}x \tag{35}$$

$$M_{\eta,\varepsilon,\omega} = \int_D \eta_{\varepsilon,\omega}(t,\cdot)\,\mathrm{d}x = \int_D \eta_{0,\varepsilon,\omega}\,\mathrm{d}x \tag{36}$$

are constants of motion (see [10], Lemma 3.13). The following bounds, uniform in ϵ, ω, are evident from a quick inspection of (34):

$$\{\sqrt{\rho_{\varepsilon,\omega}}\mathbf{u}_{\varepsilon,\omega}\}_{\{\varepsilon,\omega>0\}} \text{ bounded in } L^\infty(0,T;L^2(D;\mathbb{R}^3)) \tag{37}$$

$$\{\rho_{\varepsilon,\omega}\}_{\{\varepsilon,\omega>0\}} \text{ bounded in } L^\infty(0,T;L^\gamma(D)) \tag{38}$$

$$\{\nabla_x \mathbf{u}_{\varepsilon,\omega}\}_{\{\varepsilon,\omega>0\}} \text{ bounded in } L^2(0,T;L^2(D;\mathbb{R}^3 \times \mathbb{R}^3)) \tag{39}$$

$$\{\mathrm{div}_x \mathbf{u}_{\varepsilon,\omega}\}_{\{\varepsilon,\omega>0\}} \text{ bounded in } L^2(0,T;L^2(D)) \tag{40}$$

$$\{\nabla_x \sqrt{\eta_{\varepsilon,\omega}}\}_{\{\varepsilon,\omega>0\}} \text{ bounded in } L^2(0,T;L^2(D;\mathbb{R}^3)). \tag{41}$$

In addition,

$$\int_0^\tau \int_D \chi |\mathbf{u}_{\varepsilon,\omega} - \mathbf{V}|^2 \mathrm{d}x\,\mathrm{d}t = \int_{Q^s} |\mathbf{u}_{\varepsilon,\omega} - \mathbf{V}|^2\, \mathrm{d}x\mathrm{d}t \leq \varepsilon c, \tag{42}$$

for a.a. $\tau \in (0,T)$ with c independent of ε, ω, where we used the definition of $\chi(t,x)$.

Using the embedding of $W^{1,2}(D)$ in $L^6(D)$ (since $D \subset \mathbb{R}^3$) on the last bound listed above, it is clear that $\{\eta_{\varepsilon,\omega}\}_{\{\varepsilon,\omega>0\}} \in_b L^1(0,T;L^3(D))$. This, and mass conservation implies

$$\{\eta_{\varepsilon,\omega}\}_{\{\varepsilon,\omega>0\}} \in_b L^1(0,T;L^3(D)) \cap L^\infty(0,T;L^1(D)). \tag{43}$$

Using this result, and that

$$2\nabla_x\sqrt{\eta} = \frac{\nabla_x \eta}{\sqrt{\eta}},$$

it is also clear that

$$\{\eta_{\varepsilon,\omega}\}_{\{\varepsilon,\omega>0\}} \in_b L^1(0,T;W^{1,\frac{3}{2}}(D)) \cap L^2(0,T;W^{1,1}(D)). \tag{44}$$

By Poincaré's inequality and (39), we get that

$$\{\mathbf{u}_{\varepsilon,\omega}\}_{\{\varepsilon,\omega>0\}} \text{ bounded in } L^2(0,T;W_0^{1,2}(D;\mathbb{R}^3)). \tag{45}$$

3.2 *Pressure Estimates and Pointwise Convergence of the Fluid Density*

The detailed analysis in [16] yields the estimates needed to deal with the nonlinear pressure, $p(\rho) = a\rho^\gamma$, obtain pointwise convergence of the fluid density ρ, and pass to the limit in (22), (23). In particular,

$$\int_K p(\rho_{\varepsilon,\omega})\rho_{\varepsilon,\omega}^{\nu}\,dxdt \leq c(K) \text{ for any compact } K \subset Q^f, \tag{46}$$

and these estimates can be extended up to the boundary, and

$$\rho_{\varepsilon,\omega} \to \rho_\omega \text{ in } L^q((0,T)\times D) \text{ for any } 1 \leq q < \gamma.$$

3.3 *Singular Limits*

3.3.1 The Limit $\varepsilon \to 0$

Combining (38), (45) with equation (22) we may infer that

$$\rho_{\varepsilon,\omega} \to \rho_\omega \quad \text{in } C_{weak}([0,T];L^\gamma(D)), \tag{47}$$

$$\mathbf{u}_{\varepsilon,\omega} \rightharpoonup \mathbf{u}_\omega \quad \text{in } L^2(0,T;W_0^{1,2}(D;\mathbb{R}^3)), \tag{48}$$

passing to subsequences if necessary. Moreover as a consequence of (42),

$$\mathbf{u}_\omega = \mathbf{V} \quad \text{a.e. in } Q^s, \tag{49}$$

again after passing to a subsequence. From (43) and interpolation we get that

$$\eta_{\varepsilon,\omega} \to \eta_\omega \quad \text{in } L^2(0,T;L^{\frac{3}{2}}(D)). \tag{50}$$

To deal with the $\nabla_x\eta_{\varepsilon,\omega}$ term in (25), we can interpolate in (44) and conclude that

$$\nabla_x\eta_{\varepsilon,\omega} \rightharpoonup \nabla_x\eta_\omega \text{ in } L^p(0,T;L^q(D)), \tag{51}$$

for some $p,q > 1$.

3.4 *Convergence in the Set Q^s*

The convergence of the densities in the “solid” part of the domain play a crucial role in the analysis. That

$$\rho(t,x) = 0 \quad \text{for a.a. } (t,x) \in Q^s$$

holds has been worked out in [16]. The proof relies on regularizing the equation of continuity (2a) and employing the commutator lemma of DiPerna and Lions [11]. It remains to show that

$$\eta(t,x) = 0 \quad \text{for a.a. } (t,x) \in Q^s.$$

Before proving the following lemmas, first we set some notation. Recall that the cutoff function $\chi(t,x)$ satisfies the transport equation (18). In anticipation of using a suitable (smooth) test function, consider instead the unique function $\overline{\chi} \in C^\infty(\mathbb{R}^3)$ solving

$$\partial_t \overline{\chi} + \mathbf{V} \cdot \nabla_x \overline{\chi} = 0 \quad t > 0, x \in \mathbb{R}^3,$$

with the initial data satisfying

$$C^\infty(\mathbb{R}^3) \ni \overline{\chi}(0,\cdot) = \begin{cases} > 0 & x \in D \backslash \Omega_0 \\ < 0 & x \in \Omega_0 \cup (\mathbb{R}^3 \backslash \bar{D}) \end{cases}, \quad \nabla_x \overline{\chi}_0 \neq 0 \quad \text{on } \partial\Omega_0.$$

We define the level-set test function,

$$\varphi_\xi = \begin{cases} 1 & \overline{\chi} \geq \xi \\ \dfrac{\overline{\chi}}{\xi} & 0 \leq \overline{\chi} < \xi \\ 0 & \overline{\chi} < 0 \end{cases} = \min\left\{\frac{\overline{\chi}}{\xi}, 1\right\}^+, \tag{52}$$

supported on $D \backslash \Omega_\tau$, see [17, 22].

Lemma 1. *Let* $\eta_{\varepsilon,\omega} \in L^2(0,T;L^3(D)) \cap L^1(0,T;W^{1,\frac{3}{2}}(D))$, $\eta_{\varepsilon,\omega} \geq 0$, $\mathbf{u}_{\varepsilon,\omega} \in L^2(0,T;W^{1,2}(D;\mathbb{R}^3))$ *be a weak solution of* (25), *that is,*

$$\int_0^\infty \int_D \eta_{\varepsilon,\omega} \partial_t \varphi + \eta_{\varepsilon,\omega} \mathbf{u}_{\varepsilon,\omega} \cdot \nabla_x \varphi - \eta_{\varepsilon,\omega} \nabla_x \Phi \cdot \nabla_x \varphi - \nabla_x \eta_{\varepsilon,\omega} \cdot \nabla_x \varphi \,\mathrm{d}x\mathrm{d}t$$
$$= -\int_D \eta_{0,\varepsilon,\omega} \varphi(0,\cdot)\,\mathrm{d}x, \tag{53}$$

holds for all $\varphi \in \mathscr{D}([0,T) \times \overline{D})$ *and any* $T > 0$. *Let the initial data satisfy*

$$\eta_0 \in L^2(D) \cap L^1_+(D), \quad \eta_0\big|_{D\backslash\Omega_0} = 0.$$

Then for $\xi > 0$ *and* $\overline{\chi}$ *defined as above, it holds that*

$$\lim_{\xi \to 0} \frac{1}{\xi} \int_0^\tau \int_{\{0 \leq \bar{\chi} < \xi\}} (\eta \nabla_x \Phi + \nabla_x \eta) \cdot \nabla_x \bar{\chi} \,\mathrm{d}x\mathrm{d}t = 0, \tag{54}$$

for any $\tau > 0$.

Proof. Plugging (52) into (53) and rearranging we get that

$$\frac{1}{\xi}\int_0^\tau \int_{\{0\le\overline{\chi}<\xi\}} (\eta_{\varepsilon,\omega}\nabla_x\Phi + \nabla_x\eta_{\varepsilon,\omega})\cdot\nabla_x\overline{\chi}\,\mathrm{d}x\mathrm{d}t$$
$$= \frac{1}{\xi}\int_0^\tau \int_{\{0\le\overline{\chi}<\xi\}} \eta_{\varepsilon,\omega}(\mathbf{u}_{\varepsilon,\omega}-\mathbf{V})\cdot\nabla_x\overline{\chi}\,\mathrm{d}x\mathrm{d}t + \int_D \eta_{0,\varepsilon,\omega}\varphi_\xi(0,\cdot)\,\mathrm{d}x. \tag{55}$$

Since we can pass $\varepsilon,\omega\to 0$ on the left side in (55), it suffices to show that right side vanishes as we take $\varepsilon,\omega\to 0$ and $\xi\to 0$ successively. First,

$$\lim_{\varepsilon,\omega\to 0}\int_D \eta_{0,\varepsilon,\omega}\varphi_\xi(0,\cdot)\mathrm{d}x = \int_{\Omega_0}\eta_0\varphi_\xi(0,\cdot)\,\mathrm{d}x = 0,$$

since on Ω_0, we have $\overline{\chi}(0,\cdot)<0$ and so $\varphi_\xi(0,\cdot)=0$. Now,

$$\lim_{\varepsilon,\omega\to 0}\frac{1}{\xi}\int_0^\tau \int_{\{0\le\overline{\chi}<\xi\}} \eta_{\varepsilon,\omega}(\mathbf{u}_{\varepsilon,\omega}-\mathbf{V})\cdot\nabla_x\overline{\chi}\,\mathrm{d}x\mathrm{d}t$$
$$= \frac{1}{\xi}\int_0^\tau \int_{\{0\le\overline{\chi}<\xi\}} \eta(\mathbf{u}-\mathbf{V})\cdot\nabla_x\overline{\chi}\,\mathrm{d}x\mathrm{d}t = 0,$$

since $\mathbf{u}=\mathbf{V}$ a.e. in $D\backslash\Omega_0$, i.e. where $\overline{\chi}\ge 0$, using (49). Letting $\xi\to 0$ concludes the proof of the lemma. □

Lemma 2. *Under the same conditions as Lemma 1, the following holds,*

$$\eta(\tau,\cdot)|_{D\backslash\Omega_\tau} = 0 \quad \textit{for a.a. } \tau\in[0,T].$$

Proof. First note that by choosing a test function having the form

$$\varphi_n = \psi_n(t)\varphi(t,x),\ \varphi\in C_c^\infty([0,T)\times\bar{D}),\ \psi_n\to \mathbb{1}_{[0,\tau)} \text{ as } n\to\infty,$$

and $\psi_n\in C^\infty[0,T)$, we can rewrite the weak form (53) as

$$\int_D \eta_{\varepsilon,\omega}(\tau,\cdot)\varphi(\tau,\cdot) - \eta_{0,\varepsilon,\omega}\varphi(0,\cdot)\,\mathrm{d}x = \int_0^\tau\int_D \eta_{\varepsilon,\omega}(\partial_t\varphi + \mathbf{u}_{\varepsilon,\omega}\cdot\nabla_x\varphi)$$
$$- (\eta_{\varepsilon,\omega}\nabla_x\Phi + \nabla_x\eta_{\varepsilon,\omega})\cdot\nabla_x\varphi\,\mathrm{d}x\mathrm{d}t, \tag{56}$$

for any $\varphi\in C_c^\infty([0,T)\times\bar{D})$. It suffices to establish that

$$\int_{D\backslash\Omega_\tau}\eta(\tau,\cdot)\,\mathrm{d}x = 0, \quad a.a\ \tau\in(0,T).$$

Inserting φ_ξ into (56), using the initial conditions, and letting $\varepsilon, \omega \to 0$ yields,

$$\int_D \eta(\tau,\cdot)\varphi_\xi(\tau,\cdot)\,\mathrm{d}x = \frac{1}{\xi}\int_0^\tau \int_{\{0\le\overline{\chi}<\xi\}} \eta(\mathbf{u}-\mathbf{V})\cdot\nabla_x\overline{\chi} - (\eta\nabla_x\Phi + \nabla_x\eta)\cdot\nabla_x\overline{\chi}\,\mathrm{d}x\,\mathrm{d}t. \quad (57)$$

Since $\varphi_\xi(\tau,\cdot) \to \mathbb{1}_{D\setminus\Omega_\tau}$ as $\xi \to 0$ in any $L^p(D), p < \infty$, and $\eta \in L^2(0,T;L^3(D))$, the left-hand side of (57) converges to

$$\int_{D\setminus\Omega_\tau} \eta(\tau,\cdot)\,\mathrm{d}x,$$

as $\xi \to 0$. Finally, using Lemma 1 and that $\mathbf{u} = \mathbf{V}$ for any $\xi > 0$, it is clear the right hand side of (57) vanishes as $\xi \to 0$. □

3.5 The Limit $\omega \to 0$

Performing the limit $\varepsilon \to 0$, we arrive at the weak formulation of the momentum satisfied, except for the following term

$$\int_0^\infty \int_D (\mu_\omega \nabla_x \mathbf{u}_\omega + \lambda_\omega \mathrm{div}_x \mathbf{u}_\omega \mathbb{I}) : \nabla_x \varphi\,\mathrm{d}x\mathrm{d}t. \quad (58)$$

Using the viscosity penalization (22) (similarly for λ_ω), vanishing in $((0,T)\times D)\setminus Q^f$ and using that $\mathbf{u}_\omega = \mathbf{V}$ here, we conclude that

$$\int_0^T \int_{D\setminus\Omega_t} (\mu_\omega \nabla_x \mathbf{u}_\omega + \lambda_\omega \mathrm{div}_x \mathbf{u}_\omega \mathbb{I}) : \nabla_x \varphi\,\mathrm{d}x\mathrm{d}t \to 0 \text{ as } \omega \to 0.$$

We can now pass all terms in the weak formulation as $\omega \to 0$, using the same estimates in the previous sections. In order to obtain the limiting energy inequality, we first state the following lemma. See Corollary 2.2 in [15] for the proof.

Lemma 3. *Let $O \subset \mathbb{R}^m$ be a bounded measurable set, and $\{\mathbf{v}_n\}_{n=1}^\infty$ a sequence of functions such that*

$$\mathbf{v}_n \to \mathbf{v} \ \textit{weakly in} \ L^1(O;\mathbb{R}^n).$$

Let $\Phi : \mathbb{R}^n \to (\infty,\infty]$ be a convex lower semi-continuous function. Then $\Phi(\mathbf{v}) : O \to \mathbb{R}$ is integrable, and

$$\int_O \Phi(\mathbf{v})\,\mathrm{d}\mathbf{y} \le \liminf_{n\to\infty} \int_O \Phi(\mathbf{v}_n)\,\mathrm{d}\mathbf{y}.$$

Using this lemma, the previously derived estimates, and Lemma 2, it is now easy to pass $\epsilon, \omega \to 0$ in (26) to derive the energy inequality (14).

Appendix

Free Energy Solutions: Global in Time Existence Within Bounded and Unbounded Domains $\Omega \subset \mathbb{R}^3$

In this section, we present for completeness the main results on global existence of free energy solutions to (2a)–(2c) for both bounded and unbounded fixed domains. We impose the no-slip boundary condition for the velocity and no-flux for the particle density, which in the present context has the form

$$\mathbf{u}\big|_{\partial\Omega} = (\nabla_x \eta + \eta \nabla_x \Phi) \cdot \nu\big|_{\partial\Omega} = 0 \quad \text{for a.a. } t \in (0, T), \tag{59}$$

with ν denoting the outer normal vector to the boundary $\partial\Omega$. The usual pressure and stress tensor, (3), (4), are imposed.

Our problem is supplemented with the initial data $\{\eta_0, \rho_0, \mathbf{u}_0\}$ such that

$$\begin{aligned} \eta(0,x) &= \eta_0 \in L^2(\Omega) \cap L^1_+(\Omega), \\ \rho(0,x) &= \rho_0 \in L^\gamma(\Omega) \cap L^1_+(\Omega), \\ (\rho\mathbf{u})(0,x) &= \mathbf{m}_0 \in L^{\frac{6}{5}}(\Omega) \cap L^1(\Omega). \end{aligned} \tag{60}$$

The total energy of the mixture is given by

$$E(\eta, \rho, \mathbf{u})(t) := \int_\Omega \frac{1}{2}\rho(t)|\mathbf{u}(t)|^2 + \frac{a}{\gamma - 1}\rho^\gamma(t) + (\eta \log \eta)(t) + (\beta\rho + \eta)(t)\Phi \, \mathrm{d}x. \tag{61}$$

At the formal level, the total energy can be viewed as a Lyapunov function satisfying the *energy inequality*

$$\frac{dE}{dt} + \int_\Omega \mu|\nabla_x \mathbf{u}|^2 + \lambda|\mathrm{div}_x \mathbf{u}|^2 + |2\nabla_x\sqrt{\eta} + \sqrt{\eta}\nabla_x\Phi|^2 \, \mathrm{d}x \le 0. \tag{62}$$

Definition 4. Let us assume that (Ω, Φ) satisfy the confinement hypotheses **(HC)**. We say that $\{\rho, \mathbf{u}, \eta\}$ is a free-energy solution of problem (2a)–(2c) with boundary conditions (59) and initial data satisfying (60) provided that the following hold:

- The fluid density $\rho \ge 0$ represents a renormalized solution of Eq. (2a) on a time-space cylinder $(0, \infty) \times \Omega$, that is, for any test function $\varphi \in \mathcal{D}([0,T) \times \overline{\Omega})$, any $T > 0$, and any b such that

$$b \in L^\infty \cap C[0, \infty), \; B(\rho) = \rho B(1) + \rho \int_1^\rho \frac{b(z)}{z^2} \, \mathrm{d}z,$$

the following integral identity holds:

$$\int_0^{\infty}\int_{\Omega}\Big(B(\rho)\partial_t\varphi+B(\rho)\mathbf{u}\cdot\nabla_x\varphi-b(\rho)\mathrm{div}_x\mathbf{u}\varphi\Big)\,\mathrm{d}x\,\mathrm{d}t=-\int_{\Omega}B(\rho_0)\varphi(0,\cdot)\,\mathrm{d}x. \quad (63)$$

- The balance of momentum holds in a distributional sense, namely

$$\int_0^{\infty}\int_{\Omega}\Big(\rho\mathbf{u}\cdot\partial_t\varphi+\rho\mathbf{u}\otimes\mathbf{u}:\nabla_x\varphi+(p(\rho)+\eta)\mathrm{div}_x\varphi\Big)\,\mathrm{d}x\,\mathrm{d}t$$
$$=\int_0^{\infty}\int_{\Omega}-(\mu\nabla_x\mathbf{u}\ +\lambda\mathrm{div}_x\mathbf{u}\mathbb{I}):\nabla_x\varphi+(\eta+\beta\rho)\nabla_x\Phi\cdot\varphi\,\mathrm{d}x\,\mathrm{d}t$$
$$-\int_{\Omega}(\rho\mathbf{u})_0\cdot\varphi(0,\cdot)\,\mathrm{d}x \quad (64)$$

for any test function $\varphi\in\mathscr{D}([0,T);\mathscr{D}(\overline{\Omega};\mathbb{R}^3))$ and any $T>0$ satisfying $\varphi|_{\partial\Omega}=0$.

All quantities appearing in (64) are supposed to be at least integrable. In particular, the velocity field $\mathbf{u}$ belongs to the space $L^2(0,T;W^{1,2}(\Omega;\mathbb{R}^3))$, therefore it is legitimate to require $\mathbf{u}$ to satisfy the boundary conditions (59) in the sense of traces.

- The particle density $\eta\geq 0$ is a weak solution of (2c). In particular, the integral identity

$$\int_0^{\infty}\int_{\Omega}\eta\partial_t\varphi+\eta\mathbf{u}\cdot\nabla_x\varphi-\eta\nabla_x\Phi\cdot\nabla_x\varphi-\nabla_x\eta\cdot\nabla_x\varphi\,\mathrm{d}x\mathrm{d}t=-\int_{\Omega}\eta_0\varphi(0,\cdot)\,\mathrm{d}x \quad (65)$$

is satisfied for test functions $\varphi\in\mathscr{D}([0,T)\times\overline{\Omega})$ and any $T>0$.

All quantities appearing in (65) must be at least integrable on $(0,T)\times\Omega$. In particular, η belongs $L^2(0,T;L^3(\Omega))\cap L^1(0,T;W^{1,\frac{3}{2}}(\Omega))$.

- Given the total free-energy of the system by

$$E(\rho,\mathbf{u},\eta)(t):=\int_{\Omega}\left(\frac{1}{2}\rho|\mathbf{u}|^2+\frac{a}{\gamma-1}\rho^{\gamma}+\eta\log\eta+(\beta\rho+\eta)\Phi\right)\,\mathrm{d}x,$$

then $E(\rho,\mathbf{u},\eta)(t)$ is finite and bounded by the initial energy of the system, i.e., $E(\rho,\mathbf{u},\eta)(t)\leq E(\rho_0,\mathbf{u}_0,\eta_0)$ a.e. $t>0$. Moreover, the following free energy-dissipation inequality holds

$$\int_0^{\infty}\int(\mu|\nabla_x\mathbf{u}|^2+\lambda|\mathrm{div}_x\mathbf{u}|^2+|2\nabla_x\sqrt{\eta}+\sqrt{\eta}\nabla_x\Phi|^2)\,\mathrm{d}x\mathrm{d}t\leq E(\rho_0,\mathbf{u}_0,\eta_0). \quad (66)$$

We can now state the main result on global existence. For the details of the proof we refer the reader to Carrillo, et al. [10].

Theorem 2. *Global Existence Let us assume that* (Ω, Φ) *satisfy the confinement hypotheses* ***(HC)****. Then, the problem (2a)–(2c) supplemented with boundary conditions (59) and initial data satisfying (60) admits a weak solution* $\{\rho, \mathbf{u}, \eta\}$ *on* $(0, \infty) \times \Omega$ *in the sense of Definition 4. In addition,*

i) the total fluid mass and particle mass given by

$$M_\rho(t) = \int_\Omega \rho(t, \cdot)\, dx \qquad \text{and} \qquad M_\eta(t) = \int_\Omega \eta(t, \cdot)\, dx,$$

respectively, are constants of motion.

ii) the density satisfies the higher integrability result

$$\rho \in L^{\gamma+\Theta}((0, T) \times \Omega),$$

for any $T > 0$, *where* $\Theta = \min\{\frac{2}{3}\gamma - 1, \frac{1}{4}\}$.

We can also completely characterize the large time behavior of free-energy solutions.

Theorem 3. *Large-time Asymptotics: Let us assume that* (Ω, Φ) *satisfy the confinement hypotheses* ***(HC)****. Then, for any free-energy solution* $(\rho, \mathbf{u}, \eta)$ *of the problem (2a)–(2c), in the sense of Definition 4, there exist universal stationary states* $\rho_s = \rho_s(x)$, $\eta_s = \eta_s(x)$, *such that*

$$\begin{cases} \rho(t) \to \rho_s \text{ strongly in } L^\gamma(\Omega), \\ \operatorname*{ess\,sup}_{\tau > t} \displaystyle\int_\Omega \rho(\tau)|\mathbf{u}(\tau)|^2\, dx \to 0, \\ \eta(t) \to \eta_s \text{ strongly in } L^{p_2}(\Omega) \text{ for } p_2 > 1. \end{cases}$$

Acknowledgements The work of S.D. was supported in part by the National Science Foundation under the grant DMS-1211519 and by the *John Osborn Memorial Summer Fellowship*. K.T. gratefully acknowledges the support in part by the National Science Foundation under the grant DMS-1211519 and by the Simons Foundation under the Simons Fellows in Mathematics Award 267399.

References

1. Amsden, A.A.: Kiva-3V Release 2, Improvements to Kiva-3V. Technical Report, Los Alamos National Laboratory (1999)
2. Amsden, A.A., O'Rourke, P.J., Butler, T.D.: Kiva-2, a computer program for chemical reactive flows with sprays. Technical Report. Los Alamos National Laboratory (1989)
3. Angot, P., Bruneau, Ch.-H., Fabrie, P.: A penalization method to take into account obstacles in incompressible viscous flows. Numer. Math. **81** (4), 497–520 (1999)

4. Ballew, J., Trivisa, K.: Weakly dissipative solutions and weak-strong uniqueness for the Navier-Stokes-Smoluchowski system. Nonlinear Anal. **91**, 1–19 (2013)
5. Baranger, C.: Modélisation, étude mathématique et simulation des collisions dans les fluides complexes. Théses ENS Cachan, Juin (2004)
6. Berres, S., Bürger, R., Karlsen, K.H., Rory, E.M.: Strongly degenerate parabolic-hyperbolic systems modeling polydisperse sedimentation with compression. SIAM J. Appl. Math. **64**, 41–80 (2003)
7. Caflisch, R., Papanicolaou, G.: Dynamic theory of suspensions with brownian effects. SIAM J. Appl. Math. **43**, 885–906 (1983)
8. Carrillo, J.A., Goudon, T.: Stability and asymptotic analysis of a fluid-particle interaction model. Commun. Partial Differ. Equ. **31**, 1349–1379 (2006)
9. Carrillo, J.A., Goudon, T., Lafitte, P.: Simulation of fluid and particles flows: asymptotic preserving schemes for bubbling and flowing regimes. J. Comput. Phys. **227**, 7929–7951 (2008)
10. Carrillo, J.A., Karper, T., Trivisa, K.: On the dynamics of a fluid-particle interaction model: The bubbling regime. Nonlinear Anal. **74**, 2778–2801 (2011)
11. DiPerna, R.J., Lions, P.-L.: Ordinary differential equations, transport theory and Sobolev spaces. Invent. Math. **98**, 511–547 (1989)
12. Donatelli, D., Trivisa, K.: On a nonlinear model for tumor growth: Global in time weak solutions. J. Math. Fluid Mech. **16**(4), 787–803 (2014)
13. Donatelli, D., Trivisa, K.: On a nonlinear model for tumor growth with drug application. To appear in Nonlinearity (2015)
14. Donatelli, D., Trivisa, K.: On a nonlinear model for tumor growth in a cellular medium. Preprint (2015)
15. Feireisl, E.: Dynamics of Viscous Compressible Fluids. Oxford University Press, Oxford (2003)
16. Feiresl, E., Neustupa, J., Stebel, J.: Convergence of a Brinkman-type penalization for compressible fluid flows. J. Differ. Equ. **250**(1), 596–606 (2011)
17. Feireisl, E., Kreml, O., Necasova, S., Neustupa, J., Stebel, J.: Weak solutions to the barotropic Navier-Stokes system with slip boundary conditions in time dependent domains. J. Differ. Equ. **254**, 125–140 (2013)
18. Friedman, A.: A hierarchy of cancer models and their mathematical challenges. Discrete Contin. Dyn. Syst. **4**(1), 147–159 (2004)
19. Lions, P.L.: Mathematical Topics in Fluid Dynamics Compressible Models, vol. 2. Oxford Science Publication, Oxford (1998)
20. Mellet, A., Vasseur, A.: Asymptotic analysis for a Vlasov-Fokker-Planck compressible Navier-Stokes system of equations. Commun. Math. Phys. **281**, 573–596 (2008)
21. Mellet, A., Vasseur, A.: Global weak solutions for a Vlasov-Fokker-Planck Navier-Stokes system of equations. Math. Models Methods Appl. Sci. **17**, 1039–1063 (2007)
22. Osher, S., Fedwik, R.: Level Set Methods and Dynamic Implicit Surfaces. Applied Mathematical Science, vol. 153. Springer, New York (2003)
23. Williams, F.A.: Combustion Theory. Benjamin/Cummings, Menlo Park, CA (1985)

Envelope Equations for Three-Dimensional Gravity and Flexural-Gravity Waves Based on a Hamiltonian Approach

Philippe Guyenne

Dedicated to Walter Craig on the occasion of his 60th birthday, with admiration and gratitude.

Abstract A Hamiltonian formulation for three-dimensional nonlinear flexural-gravity waves propagating at the surface of an ideal fluid covered by a thin ice sheet is presented. This is accomplished by introducing the Dirichlet–Neumann operator which reduces the original Laplace problem to a lower-dimensional system involving quantities evaluated at the fluid-ice interface alone. The ice-sheet model is based on the special Cosserat theory for hyperelastic shells, which yields a conservative and nonlinear expression for the bending force. By applying a Hamiltonian perturbation approach suitable for such a formulation, weakly nonlinear envelope equations for small-amplitude waves are derived. The various steps of this formal derivation are discussed including the modulational Ansatz, canonical transformations and expansions of the Hamiltonian. In particular, the contributions from higher harmonics are examined. Both cases of finite and infinite depth are considered, and comparison with direct numerical simulations is shown.

1 Introduction

Modulation theory is a well-established method in applied mathematics to study the long-time evolution and stability of oscillatory solutions to partial differential equations. Typical equations to which the theory is applied are nonlinear dispersive evolution equations describing wave phenomena that arise in physical applications. Examples include ocean waves as well as waves in optics and plasmas. The usual modulational Ansatz is to anticipate a weakly nonlinear monochromatic form for solutions, and to derive equations describing the evolution of their envelope. In two space dimensions (i.e. one-dimensional wave propagation), the first nontrivial terms typically yield the nonlinear Schrödinger (NLS) equation [38], while the

P. Guyenne (✉)
Department of Mathematical Sciences, University of Delaware, Newark, DE 19716, USA
e-mail: guyenne@udel.edu

P. Guyenne et al. (eds.), *Hamiltonian Partial Differential Equations and Applications*, Fields Institute Communications 75,
DOI 10.1007/978-1-4939-2950-4_5

Benney–Roskes–Davey–Stewartson (DS) system arises in three dimensions. The rigorous justification of these models is a challenging mathematical problem [13, 15, 27, 35] and recent breakthroughs have been made in the two-dimensional case [16, 40].

Of particular interest here are hydroelasticity problems dealing with the interaction between moving fluids and deformable bodies. Such problems not only entail considerable mathematical challenges but also have many engineering applications [26]. An important area of application is that devoted to hydroelastic (or flexural-gravity) waves in polar regions where water is frozen in winter and the resulting ice cover is transformed e.g. into roads and aircraft runways, and where air-cushioned vehicles are used to break the ice. A major difficulty in this problem has to do with modeling the ice deformations subject to water wave motions [36]. Theories based on potential flow and on the assumption that the ice cover may be viewed as a thin elastic sheet have been widely used [37]. In this context, most studies have considered linear approximations of the problem, which are valid only for small-amplitude water waves and ice deflections.

Intense waves-in-ice events, however, have also been reported and their analysis indicates that linear theories are not adequate for describing large-amplitude ice deflections [28]. In the last few decades, a number of numerical and theoretical investigations have used nonlinear models based on Kirchhoff–Love plate theory to analyze two-dimensional hydroelastic waves in ice sheets. For example, Forbes [17] computed finite-amplitude periodic waves by using a Fourier series expansion technique. Părău and Dias [32] derived a forced NLS equation for the envelope of ice-sheet deflections due to a moving load, and showed that solitary waves of elevation and depression exist for certain ranges of water depth. Bonnefoy et al. [4] examined numerically the same nonlinear problem of moving load on ice, through a high-order spectral approach, and found a good agreement with theoretical predictions of Părău and Dias [32]. Hegarty and Squire [25] simulated the interaction of large-amplitude water waves with a compliant floating raft such as a sea-ice floe, by expanding the solution as a series and evaluating it with a boundary integral method. Vanden-Broeck and Părău [42] computed periodic waves and generalized solitary waves on deep water by using a series truncation method. Milewski et al. [29] derived a defocusing NLS equation which indicates that small-amplitude solitary wavepackets do not exist on deep water. Their direct numerical simulations, based on conformal mapping, reveal however stable large-amplitude solitary waves of depression. Another nonlinear formulation is Plotnikov and Toland's adaptation of the special Cosserat theory for hyperelastic shells [34], which explicitly conserves elastic potential energy unlike Kirchhoff–Love theory. Guyenne and Părău [21–23] took advantage of this conservative property to write a Hamiltonian form of the flexural-gravity wave problem in arbitrary depth. Their asymptotic and numerical results were found to be consistent with those of Părău and Dias [32] and Milewski et al. [29]. In the long-wave regime, Xia and Shen [43] established a 5th-order Korteweg–de Vries equation for the nonlinear interaction of ice cover with shallow-water waves. However, a linear Euler–Bernoulli model was adopted for the ice cover.

There have been fewer studies of the three-dimensional nonlinear problem which has drawn serious attention only recently. In a weakly nonlinear setting similar to Xia and Shen's [43], Hărăguş-Courcelle and Ilichev [24] derived a Kadomtsev–Petviashvili equation for weakly three-dimensional flexural-gravity waves on shallow water, while Alam [3] obtained a DS system that admits fully localized dromion solutions. Părău and Vanden-Broeck [33] computed solitary lumps due to a steadily moving pressure, by solving the full nonlinear equations for the fluid combined with a linear Euler–Bernoulli ice sheet. More recently, Milewski and Wang [30] proposed a DS model based on the nonlinear formulation of Plotnikov and Toland [34]. These previous authors [3, 24, 30] used the standard method of multiple scales to derive their models.

In the present paper, we extend the theoretical results of Guyenne and Părău [21–23] to the three-dimensional case. After establishing the Hamiltonian formulation of the problem, we apply the perturbation approach of Craig et al. [8, 10] to deriving envelope equations for weakly nonlinear flexural-gravity waves in the modulational regime. This is accomplished by introducing and expanding the Dirichlet–Neumann operator (DNO) which allows us to reduce the original Laplace problem to a lower-dimensional system involving quantities evaluated at the fluid-ice interface alone. A new aspect of our contribution to this approach is the inclusion of higher harmonics in the modulational Ansatz and the associated canonical transformations. Both cases of finite and infinite depth are considered. The resulting NLS and DS equations resemble existing ones in their general forms, but details such as their numerical coefficients and the relation of their dependent variables to the original physical variables are different. An analysis of these models in the two-dimensional case is performed and their predictions are compared with direct numerical simulations of the full equations. We also explore the possibility of including the exact linear dispersion relation in these approximations to improve their dispersive properties.

The remainder of the paper is organized as follows. Section 2 presents the mathematical formulation of the three-dimensional hydroelastic problem in arbitrary depth. The DNO is introduced and the Hamiltonian equations of motion are established. From this Hamiltonian formulation, weakly nonlinear envelope equations for two- and three-dimensional small-amplitude waves are derived at a formal level in Sect. 3. The various steps of the perturbation method are discussed including the modulational Ansatz, canonical transformation and expansions of the Hamiltonian. Furthermore, comparison with two-dimensional direct numerical simulations is shown and models incorporating the exact linear dispersion relation are also examined. Finally, concluding remarks are given in Sect. 4.

2 Mathematical Formulation

2.1 *Equations of Motion*

We consider a three-dimensional fluid (e.g. water) of uniform depth h beneath a continuous thin elastic plate (e.g. a floating ice sheet). The fluid is assumed to be incompressible and inviscid, and the flow to be irrotational. The ice sheet is modeled by using the special Cosserat theory of hyperelastic shells in Cartesian coordinates (x, y, z) [34], with the horizontal (x, y)-plane being the bottom of the ice sheet at rest and the z-axis directed vertically upwards (see Fig. 1). The vertical deformation of the ice is denoted by $z = \eta(x, y, t)$. The fluid velocity potential $\Phi(x, y, z, t)$ satisfies the Laplace equation

$$\nabla^2 \Phi = 0 , \qquad \text{for } \mathbf{x} = (x, y)^\top \in \mathbb{R}^2 , \quad -h < z < \eta(x, y, t) . \tag{1}$$

The nonlinear boundary conditions at $z = \eta(x, y, t)$ are the kinematic condition

$$\eta_t + \Phi_x \eta_x + \Phi_y \eta_y = \Phi_z , \tag{2}$$

and the dynamic condition

$$\Phi_t + \frac{1}{2} |\nabla \Phi|^2 + g\eta + \frac{\mathscr{D}}{\rho} \mathscr{F} = 0 , \tag{3}$$

where

$$\mathscr{F} = \frac{2}{\sqrt{\mathscr{A}}} \left[\partial_x \left(\frac{1 + \eta_y^2}{\sqrt{\mathscr{A}}} \partial_x \mathscr{H} \right) - \partial_x \left(\frac{\eta_x \eta_y}{\sqrt{\mathscr{A}}} \partial_y \mathscr{H} \right) - \partial_y \left(\frac{\eta_x \eta_y}{\sqrt{\mathscr{A}}} \partial_x \mathscr{H} \right) \right.$$
$$\left. + \, \partial_y \left(\frac{1 + \eta_x^2}{\sqrt{\mathscr{A}}} \partial_y \mathscr{H} \right) \right] + 4\mathscr{H}^3 - 4\mathscr{K}\mathscr{H} ,$$

with

$$\mathscr{A} = 1 + \eta_x^2 + \eta_y^2 , \qquad \mathscr{K} = \frac{1}{\mathscr{A}^2} (\eta_{xx}\eta_{yy} - \eta_{xy}^2) ,$$

Fig. 1 Sketch of the hydroelastic problem

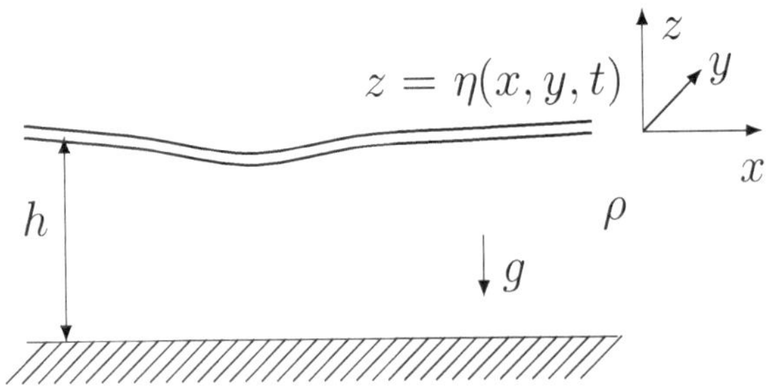

$$\mathscr{H} = \frac{1}{2\mathscr{A}^{3/2}}\Big[(1+\eta_y^2)\eta_{xx} - 2\eta_{xy}\eta_x\eta_y + (1+\eta_x^2)\eta_{yy}\Big].$$

The additional term $\mathscr{F}$ in (3) represents the nonlinear bending force exerted by the ice sheet onto the fluid surface, as derived by Plotnikov and Toland [34]. It is also a conservative term and thus can be cast into a Hamiltonian formulation as shown below. Two simpler expressions of this bending force have been commonly used in the literature; a linear one based on Euler–Bernoulli theory [3, 24, 33, 37],

$$\mathscr{F} = \eta_{xxxx} + 2\eta_{xxyy} + \eta_{yyyy}\,,$$

and a nonlinear one based on Kirchhoff–Love theory [4, 17, 29, 32]. The system is completed with the boundary condition at the bottom,

$$\Phi_z = 0\,, \qquad \text{at } z = -h\,. \tag{4}$$

In the infinite-depth limit ($h \to \infty$), Eq. (4) is replaced by

$$|\nabla\Phi| \to 0\,, \qquad \text{as } z \to -\infty\,.$$

If $\mathscr{D} = 0$, these are the classical governing equations for the gravity water wave problem [27].

Hereinafter, subscripts are also used as shorthand notation for partial or variational derivatives (e.g. $\Phi_t = \partial_t\Phi$). The vertical bars denote either a vector norm (when applied to a vector) or a complex modulus (when applied to a complex scalar function). The constant $\mathscr{D}$ is the coefficient of flexural rigidity for the ice sheet, ρ the density of the fluid and g the acceleration due to gravity. The dynamic condition (3) stems from the Bernoulli equation [34]. The inertia of the thin elastic plate is neglected, so the plate acceleration term is not considered here [37]. We also assume that the elastic plate is not pre-stressed and neglect plate stretching.

The dispersion relation for the linearized problem with solutions of the form $e^{\mathrm{i}(\mathbf{k}\cdot\mathbf{x}-\omega t)}$ is

$$c^2 = \left(\frac{g}{k} + \frac{\mathscr{D}k^3}{\rho}\right)\tanh(hk)\,, \tag{5}$$

where $k = |\mathbf{k}|$ and $c = \omega/k$ is the phase speed. It can be shown that the phase speed $c(\mathbf{k})$ has a minimum $c_{\min}$ at $\mathbf{k} = \mathbf{k}_{\min}$ for any parameter values [32, 37]. At this minimum, the phase velocity and group velocity are equal. Another critical speed in finite depth is the long-wave limit $c_0 = \sqrt{gh}$ as $k \to 0$.

The total energy

$$H = \frac{1}{2}\iint_{-\infty}^{\infty}\int_{-h}^{\eta} |\nabla\Phi|^2 dz dy dx + \frac{1}{2}\iint_{-\infty}^{\infty}\left[g\eta^2 + \frac{4\mathscr{D}}{\rho}\mathscr{H}^2\sqrt{\mathscr{A}}\right] dy dx\,, \tag{6}$$

together with the impulse (or momentum) vector

$$I = \iint_{-\infty}^{\infty} \int_{-h}^{\eta} \nabla_{\mathbf{x}} \Phi \, dz dy dx ,$$

where $\nabla_{\mathbf{x}} = (\partial_x, \partial_y)^\top$, and the volume (or mass)

$$V = \iint_{-\infty}^{\infty} \eta \, dy dx , \tag{7}$$

are invariants of motion for (1)–(4). The first integral in (6) represents kinetic energy, while the second integral represents potential energy due to gravity and elasticity.

2.2 Hamiltonian Formulation

Following Zakharov [45] and Craig and Sulem [14], we can reduce the dimensionality of the Laplace problem (1)–(4) by introducing $\xi(x, y, t) = \Phi(x, y, \eta(x, y, t), t)$, the trace of the velocity potential on $z = \eta(x, y, t)$, together with the DNO

$$G(\eta)\xi = (-\nabla_{\mathbf{x}}\eta, 1)^\top \cdot \nabla\Phi\big|_{z=\eta} , \tag{8}$$

which is the singular integral operator that takes Dirichlet data ξ on $z = \eta(x, y, t)$, solves the Laplace equation (1) for Φ subject to (4), and returns the corresponding Neumann data (i.e. the normal fluid velocity there).

In terms of these boundary variables, Eqs. (1)–(4) can be rewritten as

$$\eta_t = G(\eta)\xi , \tag{9}$$

$$\xi_t = -\frac{1}{2(1 + |\nabla_{\mathbf{x}}\eta|^2)} \Big[|\nabla_{\mathbf{x}}\xi|^2 - (G(\eta)\xi)^2 - 2(G(\eta)\xi)\nabla_{\mathbf{x}}\xi \cdot \nabla_{\mathbf{x}}\eta + |\nabla_{\mathbf{x}}\xi|^2 |\nabla_{\mathbf{x}}\eta|^2 - (\nabla_{\mathbf{x}}\xi \cdot \nabla_{\mathbf{x}}\eta)^2 \Big] - g\eta - \frac{\mathscr{D}}{\rho}\mathscr{F} , \tag{10}$$

which are Hamiltonian equations for the canonically conjugate variables η and ξ, extending Zakharov's formulation of the water wave problem to flexural-gravity waves [21–23]. Equations (9) and (10) have the canonical form

$$\begin{pmatrix} \eta_t \\ \xi_t \end{pmatrix} = J \begin{pmatrix} H_\eta \\ H_\xi \end{pmatrix} = \begin{pmatrix} 0 & 1 \\ -1 & 0 \end{pmatrix} \begin{pmatrix} H_\eta \\ H_\xi \end{pmatrix} , \tag{11}$$

whose Hamiltonian

$$H = \frac{1}{2} \iint_{-\infty}^{\infty} \Big[\xi G(\eta)\xi + g\eta^2 + \frac{4\mathscr{D}}{\rho} \mathscr{H}^2 \sqrt{\mathscr{A}} \Big] \, dy dx , \tag{12}$$

corresponds to the total energy (6).

The present Hamiltonian formulation involving the DNO can be extended to account for internal waves propagating e.g. along an interface between two fluid regions [5, 9, 11, 12] and for variable topography at the bottom of the fluid domain [6, 7, 20]. However, these effects will not be considered here.

2.3 *Dirichlet–Neumann Operator*

In light of its analyticity properties [13], the DNO can be expressed as a convergent Taylor series expansion in η,

$$G(\eta) = \sum_{j=0}^{\infty} G_j(\eta)\,, \tag{13}$$

where each term G_j can be determined recursively [14, 31, 44]. More specifically, for $j = 2r > 0$,

$$\begin{aligned} G_{2r}(\eta) &= \frac{1}{(2r)!} G_0(|D_{\mathbf{x}}|^2)^{r-1} D_{\mathbf{x}} \cdot \eta^{2r} D_{\mathbf{x}} \\ &- \sum_{s=0}^{r-1} \frac{1}{(2(r-s))!} (|D_{\mathbf{x}}|^2)^{r-s} \eta^{2(r-s)} G_{2s}(\eta) \\ &- \sum_{s=0}^{r-1} \frac{1}{(2(r-s)-1)!} G_0(|D_{\mathbf{x}}|^2)^{r-s-1} \eta^{2(r-s)-1} G_{2s+1}(\eta), \end{aligned} \tag{14}$$

and, for $j = 2r - 1 > 0$,

$$\begin{aligned} G_{2r-1}(\eta) &= \frac{1}{(2r-1)!} (|D_{\mathbf{x}}|^2)^{r-1} D_{\mathbf{x}} \cdot \eta^{2r-1} D_{\mathbf{x}} \\ &- \sum_{s=0}^{r-1} \frac{1}{(2(r-s)-1)!} G_0(|D_{\mathbf{x}}|^2)^{r-s-1} \eta^{2(r-s)-1} G_{2s}(\eta) \\ &- \sum_{s=0}^{r-2} \frac{1}{(2(r-s-1))!} (|D_{\mathbf{x}}|^2)^{r-s-1} \eta^{2(r-s-1)} G_{2s+1}(\eta), \end{aligned} \tag{15}$$

where $D_{\mathbf{x}} = -\mathrm{i}\nabla_{\mathbf{x}}$ and $G_0 = |D_{\mathbf{x}}| \tanh(h|D_{\mathbf{x}}|)$ are Fourier multiplier operators ($D_{\mathbf{x}}$ is defined in such a way that its Fourier symbol is $\mathbf{k}$ and thus $|D_{\mathbf{x}}|$ corresponds to $|\mathbf{k}| = k$). In the infinite-depth limit ($h \to \infty$), G_0 reduces to $|D_{\mathbf{x}}|$ [21]. Similar expansions of the DNO can be derived in the presence of an interface between two fluid layers [5, 12, 19] and for variable bottom topography [6, 20].

3 Modulational Regime

We now present the derivation of weakly nonlinear models for small-amplitude waves in the modulational regime. For this purpose, we apply a Hamiltonian perturbation approach [5, 6, 8, 10], which is especially suitable for the present Hamiltonian formulation of the flexural-gravity wave problem. An advantage of this approach is that it naturally associates a Hamiltonian to the equations of motion at each order of approximation, although we restrict ourselves to approximations up to the cubic order of nonlinearity in the present paper. Long-wave models can be treated in a similar way [23] but they will not be examined here. Changing variables through canonical transformations and expanding the Hamiltonian are the main ingredients of this approach. We distinguish two cases: finite and infinite depth.

3.1 Finite Depth

3.1.1 Canonical Transformations

The first step is a normal decomposition of the first-harmonic waves, and here we extend the approach of Craig et al. [8, 10] by accounting for higher harmonics according to Stokes' expansion, as assumed in the multiple-scale method [3, 13, 15, 16, 30, 38, 40]. This translates into

$$\eta = \frac{1}{\sqrt{2}} a^{-1}(D_{\mathbf{x}})(v + \overline{v}) + \tilde{\eta} + \eta_2 + \dots, \qquad \tilde{\eta} = \mathbb{P}_0 \eta, \qquad \eta_2 = \mathbb{P}_2 \eta,$$

$$\xi = -\frac{\mathrm{i}}{\sqrt{2}} a(D_{\mathbf{x}})(v - \overline{v}) + \tilde{\xi} + \xi_2 + \dots, \qquad \tilde{\xi} = \mathbb{P}_0 \xi, \qquad \xi_2 = \mathbb{P}_2 \xi, \quad (16)$$

where

$$a(D_{\mathbf{x}}) = \sqrt[4]{\frac{g + \mathscr{D}|D_{\mathbf{x}}|^4/\rho}{G_0(D_{\mathbf{x}})}},$$

$(\tilde{\eta}, \tilde{\xi})$ are the zeroth harmonics representing the induced mean flow, and (η_2, ξ_2) the second harmonics. The overbar represents complex conjugation and $\mathbb{P}_0$, $\mathbb{P}_2$ are the projections that associate to (η, ξ) their zeroth- and second-harmonic components respectively. Higher harmonics can be taken into account but it is sufficient to consider only up to the second ones for the purposes of deriving the cubic NLS and DS equations in the present paper. As will be made clearer later, we use the terminology "first harmonics" to refer to the solution's components with wavenumbers centered around the fundamental (or carrier), "second harmonics" to those with wavenumbers centered around twice the fundamental, and so on.

The new variables $(v, \overline{v}, \eta_2, \xi_2, \tilde{\eta}, \tilde{\xi})^\top$ are expressed in terms of $(\eta, \xi)^\top$ as

$$\begin{pmatrix} v \\ \overline{v} \\ \eta_2 \\ \xi_2 \\ \tilde{\eta} \\ \tilde{\xi} \end{pmatrix} = A_1 \begin{pmatrix} \eta \\ \xi \end{pmatrix} = \frac{1}{\sqrt{2}} \begin{pmatrix} a(D_{\mathbf{x}})(\mathbb{I} - \mathbb{P}_0 - \mathbb{P}_2) & \mathrm{i}a^{-1}(D_{\mathbf{x}})(\mathbb{I} - \mathbb{P}_0 - \mathbb{P}_2) \\ a(D_{\mathbf{x}})(\mathbb{I} - \mathbb{P}_0 - \mathbb{P}_2) & -\mathrm{i}a^{-1}(D_{\mathbf{x}})(\mathbb{I} - \mathbb{P}_0 - \mathbb{P}_2) \\ \sqrt{2}\mathbb{P}_2 & 0 \\ 0 & \sqrt{2}\mathbb{P}_2 \\ \sqrt{2}\mathbb{P}_0 & 0 \\ 0 & \sqrt{2}\mathbb{P}_0 \end{pmatrix} \begin{pmatrix} \eta \\ \xi \end{pmatrix},$$

where $\mathbb{I}$ denotes the identity operator. Accordingly, the symplectic structure of the system is changed to $J_1 = A_1 J A_1^\top$ [5, 8] and the equations of motion become

$$\begin{pmatrix} v_t \\ \overline{v}_t \\ \eta_{2t} \\ \xi_{2t} \\ \tilde{\eta}_t \\ \tilde{\xi}_t \end{pmatrix} = J_1 \begin{pmatrix} H_v \\ H_{\overline{v}} \\ H_{\eta_2} \\ H_{\xi_2} \\ H_{\tilde{\eta}} \\ H_{\tilde{\xi}} \end{pmatrix} = \begin{pmatrix} 0 & -\mathrm{i}(\mathbb{I} - \mathbb{P}_0 - \mathbb{P}_2) & 0 & 0 & 0 & 0 \\ \mathrm{i}(\mathbb{I} - \mathbb{P}_0 - \mathbb{P}_2) & 0 & 0 & 0 & 0 & 0 \\ 0 & 0 & 0 & \mathbb{P}_2 & 0 & 0 \\ 0 & 0 & -\mathbb{P}_2 & 0 & 0 & 0 \\ 0 & 0 & 0 & 0 & 0 & \mathbb{P}_0 \\ 0 & 0 & 0 & 0 & -\mathbb{P}_0 & 0 \end{pmatrix} \begin{pmatrix} H_v \\ H_{\overline{v}} \\ H_{\eta_2} \\ H_{\xi_2} \\ H_{\tilde{\eta}} \\ H_{\tilde{\xi}} \end{pmatrix},$$

given the fact that $\mathbb{P}_0^2 = \mathbb{P}_0$ and similarly for $\mathbb{P}_2$. By also decomposing the second harmonics into normal modes,

$$\eta_2 = \frac{1}{\sqrt{2}} a^{-1}(D_{\mathbf{x}})(v_2 + \overline{v}_2), \qquad \xi_2 = -\frac{\mathrm{i}}{\sqrt{2}} a(D_{\mathbf{x}})(v_2 - \overline{v}_2),$$

we obtain

$$\begin{pmatrix} v_t \\ \overline{v}_t \\ v_{2t} \\ \overline{v}_{2t} \\ \tilde{\eta}_t \\ \tilde{\xi}_t \end{pmatrix} = J_2 \begin{pmatrix} H_v \\ H_{\overline{v}} \\ H_{v_2} \\ H_{\overline{v}_2} \\ H_{\tilde{\eta}} \\ H_{\tilde{\xi}} \end{pmatrix},$$

where

$$J_2 = \begin{pmatrix} 0 & -\mathrm{i}(\mathbb{I} - \mathbb{P}_0 - \mathbb{P}_2) & 0 & 0 & 0 & 0 \\ \mathrm{i}(\mathbb{I} - \mathbb{P}_0 - \mathbb{P}_2) & 0 & 0 & 0 & 0 & 0 \\ 0 & 0 & \frac{\mathrm{i}}{2}(a\mathbb{P}_2 a^{-1} - a^{-1}\mathbb{P}_2 a) & -\frac{\mathrm{i}}{2}(a\mathbb{P}_2 a^{-1} + a^{-1}\mathbb{P}_2 a) & 0 & 0 \\ 0 & 0 & \frac{\mathrm{i}}{2}(a\mathbb{P}_2 a^{-1} + a^{-1}\mathbb{P}_2 a) & -\frac{\mathrm{i}}{2}(a\mathbb{P}_2 a^{-1} - a^{-1}\mathbb{P}_2 a) & 0 & 0 \\ 0 & 0 & 0 & 0 & 0 & \mathbb{P}_0 \\ 0 & 0 & 0 & 0 & -\mathbb{P}_0 & 0 \end{pmatrix}.$$

If higher harmonics were to be considered, this would increase the size of the system of equations. These higher harmonics can also be expressed in terms of $(\eta, \xi)^\top$ by using the associated projections as in (16).

The next step introduces the modulational Ansatz

$$v = \varepsilon\, u(\mathbf{X}, t)\mathrm{e}^{\mathrm{i}\mathbf{k}_0\cdot\mathbf{x}}\,, \qquad \overline{v} = \varepsilon\, \overline{u}(\mathbf{X}, t)\mathrm{e}^{-\mathrm{i}\mathbf{k}_0\cdot\mathbf{x}}\,, \tag{17}$$

$$v_2 = \varepsilon^2 u_2(\mathbf{X}, t)\mathrm{e}^{2\mathrm{i}\mathbf{k}_0\cdot\mathbf{x}}\,, \qquad \overline{v}_2 = \varepsilon^2 \overline{u}_2(\mathbf{X}, t)\mathrm{e}^{-2\mathrm{i}\mathbf{k}_0\cdot\mathbf{x}}\,, \tag{18}$$

in the spirit of Stokes' expansion, together with

$$\tilde{\eta} = \varepsilon^\alpha \eta_0(\mathbf{X}, t)\,, \qquad \tilde{\xi} = \varepsilon^\beta \xi_0(\mathbf{X}, t)\,, \tag{19}$$

where the exponents $\beta > 1$ and $\alpha = \beta + 1$ are dependent on whether the depth is finite or infinite [8]. This implies that we look for solutions in the form of quasi-monochromatic waves with nonzero carrier wavenumber $\mathbf{k}_0 = (k_x, k_y)^\top$ and with slowly varying amplitude depending on $\mathbf{X} = \varepsilon \mathbf{x}$. Wave steepness is measured by the small parameter $\varepsilon \sim |\mathbf{k}_0| a_0 \ll 1$ where a_0 is a characteristic wave amplitude. In [8, 10, 21, 23], the second harmonics were assumed to be of higher order than $O(\varepsilon^2)$ and thus did not contribute to the level of approximation considered. Such a regime may be interpreted as that for weakly nonlinear waves which are very close to being monochromatic (or equivalently for a very narrow-banded wave spectrum centered around $\mathbf{k}_0$). In the present case, these second harmonics give contributions, albeit small.

The system is now determined by the slowly varying amplitudes

$$\begin{pmatrix} u \\ \overline{u} \\ u_2 \\ \overline{u}_2 \\ \eta_0 \\ \xi_0 \end{pmatrix} = A_3 \begin{pmatrix} v \\ \overline{v} \\ v_2 \\ \overline{v}_2 \\ \tilde{\eta} \\ \tilde{\xi} \end{pmatrix} = \begin{pmatrix} \varepsilon^{-1}\mathrm{e}^{-\mathrm{i}\mathbf{k}_0\cdot\mathbf{x}} & 0 & 0 & 0 & 0 & 0 \\ 0 & \varepsilon^{-1}\mathrm{e}^{\mathrm{i}\mathbf{k}_0\cdot\mathbf{x}} & 0 & 0 & 0 & 0 \\ 0 & 0 & \varepsilon^{-2}\mathrm{e}^{-2\mathrm{i}\mathbf{k}_0\cdot\mathbf{x}} & 0 & 0 & 0 \\ 0 & 0 & 0 & \varepsilon^{-2}\mathrm{e}^{2\mathrm{i}\mathbf{k}_0\cdot\mathbf{x}} & 0 & 0 \\ 0 & 0 & 0 & 0 & \varepsilon^{-\alpha} & 0 \\ 0 & 0 & 0 & 0 & 0 & \varepsilon^{-\beta} \end{pmatrix} \begin{pmatrix} v \\ \overline{v} \\ v_2 \\ \overline{v}_2 \\ \tilde{\eta} \\ \tilde{\xi} \end{pmatrix},$$

whose evolution equations read

$$\begin{pmatrix} u_t \\ \overline{u}_t \\ u_{2t} \\ \overline{u}_{2t} \\ \eta_{0t} \\ \xi_{0t} \end{pmatrix} = J_3 \begin{pmatrix} H_u \\ H_{\overline{u}} \\ H_{u_2} \\ H_{\overline{u}_2} \\ H_{\eta_0} \\ H_{\xi_0} \end{pmatrix}, \tag{20}$$

where

$$J_3 = \varepsilon^2 A_3 J_2 A_3^\top = \varepsilon^2 \begin{pmatrix} 0 & \mathscr{J}_{12} & 0 & 0 & 0 & 0 \\ \mathscr{J}_{21} & 0 & 0 & 0 & 0 & 0 \\ 0 & 0 & 0 & \mathscr{J}_{34} & 0 & 0 \\ 0 & 0 & \mathscr{J}_{43} & 0 & 0 & 0 \\ 0 & 0 & 0 & 0 & 0 & \varepsilon^{-\alpha-\beta}\mathbb{P}_0 \\ 0 & 0 & 0 & 0 & -\varepsilon^{-\alpha-\beta}\mathbb{P}_0 & 0 \end{pmatrix},$$

and

$$\mathscr{J}_{12} = -\mathrm{i}\varepsilon^{-2}\mathrm{e}^{-\mathrm{i}\mathbf{k}_0\cdot\mathbf{x}}(\mathbb{I} - \mathbb{P}_0 - \mathbb{P}_2)(\mathrm{e}^{\mathrm{i}\mathbf{k}_0\cdot\mathbf{x}}.)\,, \qquad \mathscr{J}_{21} = \overline{\mathscr{J}_{12}}\,,$$

$$\mathscr{J}_{34} = -\mathrm{i}\varepsilon^{-4}\mathrm{e}^{-2\mathrm{i}\mathbf{k}_0\cdot\mathbf{x}}\,\mathbb{P}_2(\mathrm{e}^{2\mathrm{i}\mathbf{k}_0\cdot\mathbf{x}}.)\,, \qquad \mathscr{J}_{43} = \overline{\mathscr{J}_{34}}\,.$$

Note that the additional factor ε^2 in J_3 is due to the spatial rescaling $\mathbf{x} \to \mathbf{X}$ [5, 8]. This new symplectic structure reduces to

$$J_3 = \begin{pmatrix} 0 & -\mathrm{i} & 0 & 0 & 0 & 0 \\ \mathrm{i} & 0 & 0 & 0 & 0 & 0 \\ 0 & 0 & 0 & -\mathrm{i}\varepsilon^{-2} & 0 & 0 \\ 0 & 0 & \mathrm{i}\varepsilon^{-2} & 0 & 0 & 0 \\ 0 & 0 & 0 & 0 & 0 & \varepsilon^{2-\alpha-\beta} \\ 0 & 0 & 0 & 0 & -\varepsilon^{2-\alpha-\beta} & 0 \end{pmatrix},$$

when applied to a homogenized Hamiltonian in terms of functions of $\mathbf{X}$ alone, as described next.

3.1.2 Expansion of the Hamiltonian

The modulational Ansatz (17)–(19) also introduces the small parameter ε in the expression of the Hamiltonian (12) which can then be expanded in powers of ε, by using the Taylor series expansion (13) of the DNO. The mean-flow exponents are set to $\alpha = 2$ and $\beta = 1$, as determined in [8] for finite depth. Up to order $O(\varepsilon^2)$, we find

$$\begin{aligned} H = \iint_{-\infty}^{\infty} \Big[& \frac{1}{2}\overline{u}\left(\omega(\mathbf{k}_0) + \varepsilon\nabla_{\mathbf{k}}\omega(\mathbf{k}_0)\cdot D_{\mathbf{X}} + \frac{\varepsilon^2}{2}\partial^2_{k_j k_\ell}\omega(\mathbf{k}_0) D^2_{X_j X_\ell}\right) u + \text{c.c.} \\ & + \varepsilon^2\omega(2\mathbf{k}_0)|u_2|^2 + \varepsilon^2\alpha_3(\mathbf{k}_0)|u|^4 + \varepsilon^2\alpha_2(\mathbf{k}_0)\left(u^2\overline{u}_2 + \overline{u}^2 u_2\right) \\ & + \varepsilon^2\left(\mathrm{i}\mathbf{k}_0\cdot D_{\mathbf{X}}\xi_0 + \alpha_1(\mathbf{k}_0)\eta_0\right)|u|^2 + \frac{\varepsilon^2}{2}\left(h\xi_0|D_{\mathbf{X}}|^2\xi_0 + g\eta_0^2\right)\Big]\, dYdX + O(\varepsilon^3), \quad (21) \end{aligned}$$

where c.c. stands for the complex conjugate of all the preceding terms on the right-hand side of the equation, and

$$\alpha_1(\mathbf{k}_0) = \frac{1}{2}a^2(\mathbf{k}_0)\Big(|\mathbf{k}_0|^2 - G_0^2(\mathbf{k}_0)\Big),$$

$$\alpha_2(\mathbf{k}_0) = \frac{1}{2\sqrt{2}}a(2\mathbf{k}_0)\Big(2|\mathbf{k}_0|^2 - G_0(\mathbf{k}_0)G_0(2\mathbf{k}_0)\Big) + \frac{1}{4\sqrt{2}}a^{-1}(2\mathbf{k}_0)a^2(\mathbf{k}_0)\Big(|\mathbf{k}_0|^2 + G_0^2(\mathbf{k}_0)\Big),$$

$$\alpha_3(\mathbf{k}_0) = \frac{1}{4}G_0(\mathbf{k}_0)\Big(G_0(\mathbf{k}_0)G_0(2\mathbf{k}_0) - |\mathbf{k}_0|^2\Big) - \frac{5\mathscr{D}}{8\rho}\left(k_x^6 + k_y^6 + 3k_x^4k_y^2 + 3k_x^2k_y^4\right)a^{-4}(\mathbf{k}_0)\,.$$

The coefficient

$$\omega(\mathbf{k}) = \sqrt{G_0(k)(g + \mathscr{D}k^4/\rho)}\,,$$

denotes the linear dispersion relation in terms of the angular frequency and the indices $j, \ell = \{1, 2\}$ refer to the two horizontal directions. The scale separation lemma of Craig et al. [6] is applied to homogenize the fast oscillations in $\mathbf{x}$, so that four-wave resonant terms are retained and non-resonant terms are eliminated. Note the zeroth- and second-harmonic contributions to this order of approximation in (21).

The Hamiltonian (21) can be further simplified by subtracting a multiple of the wave action

$$M = \iint_{-\infty}^{\infty} |u|^2\, dYdX\,,$$

together with a (scalar) multiple of the impulse

$$I = \iint_{-\infty}^{\infty} \eta \nabla_{\mathbf{x}} \xi\, dydx\,,$$
$$= \iint_{-\infty}^{\infty} \Big[\mathbf{k}_0|u|^2 + \frac{\varepsilon}{2}\Big(\overline{u}D_{\mathbf{X}}u + u\overline{D_{\mathbf{X}}u}\Big) + 2\varepsilon^2\mathbf{k}_0|u_2|^2 + i\varepsilon^2\eta_0 D_{\mathbf{X}}\xi_0\Big]\, dYdX\,,$$

so that it takes the "renormalized" form

$$\hat{H} = H - \nabla_{\mathbf{k}}\omega(\mathbf{k}_0)\cdot I - \Big(\omega(\mathbf{k}_0) - \mathbf{k}_0\cdot\nabla_{\mathbf{k}}\omega(\mathbf{k}_0)\Big)M\,,$$
$$= \varepsilon^2 \iint_{-\infty}^{\infty} \Big[\frac{1}{2}\partial^2_{k_jk_\ell}\omega(\mathbf{k}_0)\overline{u}D^2_{X_jX_\ell}u + \Big(\omega(2\mathbf{k}_0) - 2\mathbf{k}_0\cdot\nabla_{\mathbf{k}}\omega(\mathbf{k}_0)\Big)|u_2|^2$$

$$+ \alpha_3(\mathbf{k}_0)|u|^4 + \alpha_2(\mathbf{k}_0)\Big(u^2\overline{u}_2 + \overline{u}^2 u_2\Big) + \Big(\mathrm{i}\mathbf{k}_0 \cdot D_{\mathbf{X}}\xi_0 + \alpha_1(\mathbf{k}_0)\eta_0\Big)|u|^2$$
$$+ \frac{1}{2}h\xi_0|D_{\mathbf{X}}|^2\xi_0 + \frac{1}{2}g\eta_0^2 - \mathrm{i}\nabla_{\mathbf{k}}\omega(\mathbf{k}_0)\cdot\eta_0 D_{\mathbf{X}}\xi_0\Bigg] dYdX + O(\varepsilon^3). \tag{22}$$

The quantities I and M are conserved by the system, at least at the level of approximation considered. Therefore, they Poisson commute with H and do not modify its symplectic structure [5, 8]. The subtraction of a multiple of M from H reflects the fact that our approximation of the problem is phase invariant, while the subtraction of $\nabla_{\mathbf{k}}\omega(\mathbf{k}_0)\cdot I$ is equivalent to changing the coordinate system into a reference frame moving with the group velocity $\nabla_{\mathbf{k}}\omega(\mathbf{k}_0)$.

3.1.3 DS System

By using (22), the equations of motion (20) reduce to

$$\mathrm{i}u_\tau = -\frac{1}{2}\partial^2_{k_jk_\ell}\omega(\mathbf{k}_0)\partial^2_{X_jX_\ell}u + 2\alpha_3(\mathbf{k}_0)|u|^2u + 2\alpha_2(\mathbf{k}_0)\overline{u}\,u_2$$
$$+\Big(\mathrm{i}\mathbf{k}_0\cdot D_{\mathbf{X}}\xi_0 + \alpha_1(\mathbf{k}_0)\eta_0\Big)u\,, \tag{23}$$

$$\varepsilon\eta_{0_\tau} = h|D_{\mathbf{X}}|^2\xi_0 - \mathbf{k}_0\cdot\nabla_{\mathbf{X}}|u|^2 + \nabla_{\mathbf{k}}\omega(\mathbf{k}_0)\cdot\nabla_{\mathbf{X}}\eta_0\,, \tag{24}$$

$$\varepsilon\xi_{0_\tau} = -\Big(g\eta_0 + \alpha_1(\mathbf{k}_0)|u|^2 - \nabla_{\mathbf{k}}\omega(\mathbf{k}_0)\cdot\nabla_{\mathbf{X}}\xi_0\Big)\,, \tag{25}$$

$$\mathrm{i}\varepsilon^2 u_{2_\tau} = \Big(\omega(2\mathbf{k}_0) - 2\mathbf{k}_0\cdot\nabla_{\mathbf{k}}\omega(\mathbf{k}_0)\Big)u_2 + \alpha_2(\mathbf{k}_0)u^2\,, \tag{26}$$

where $\tau = \varepsilon^2 t$. To lowest order in ε, the right-hand sides of (24)–(26) equal zero, hence

$$h|D_{\mathbf{X}}|^2\xi_0 - \mathbf{k}_0\cdot\nabla_{\mathbf{X}}|u|^2 + \nabla_{\mathbf{k}}\omega(\mathbf{k}_0)\cdot\nabla_{\mathbf{X}}\eta_0 = 0\,, \tag{27}$$

and

$$\eta_0 = -\frac{1}{g}\alpha_1(\mathbf{k}_0)|u|^2 + \frac{1}{g}\nabla_{\mathbf{k}}\omega(\mathbf{k}_0)\cdot\nabla_{\mathbf{X}}\xi_0\,, \tag{28}$$

$$u_2 = \frac{\alpha_2(\mathbf{k}_0)}{2\mathbf{k}_0\cdot\nabla_{\mathbf{k}}\omega(\mathbf{k}_0) - \omega(2\mathbf{k}_0)}u^2\,. \tag{29}$$

Then substituting (28)–(29) into (23) and (27) leads to the DS system

$$\mathrm{i}u_\tau = -\frac{1}{2}\partial^2_{k_jk_\ell}\omega(\mathbf{k}_0)\partial^2_{X_jX_\ell}u + \alpha_4(\mathbf{k}_0)|u|^2u + \alpha_5(\mathbf{k}_0)\cdot u\nabla_{\mathbf{X}}\xi_0\,,$$
$$0 = \mathscr{L}\xi_0 - \alpha_5(\mathbf{k}_0)\cdot\nabla_{\mathbf{X}}|u|^2\,, \tag{30}$$

where

$$\alpha_4(\mathbf{k}_0) = \frac{2\alpha_2^2(\mathbf{k}_0)}{2\mathbf{k}_0 \cdot \nabla_{\mathbf{k}}\omega(\mathbf{k}_0) - \omega(2\mathbf{k}_0)} + 2\alpha_3(\mathbf{k}_0) - \frac{1}{g}\alpha_1^2(\mathbf{k}_0),$$

$$\alpha_5(\mathbf{k}_0) = \mathbf{k}_0 + \frac{1}{g}\alpha_1(\mathbf{k}_0)\nabla_{\mathbf{k}}\omega(\mathbf{k}_0),$$

$$\mathscr{L} = -h|\nabla_{\mathbf{X}}|^2 + \frac{1}{g}\Big(\partial_{k_j}\omega(\mathbf{k}_0)\Big)\Big(\partial_{k_\ell}\omega(\mathbf{k}_0)\Big)\partial^2_{X_jX_\ell}.$$

As mentioned earlier, setting $\mathscr{D} = 0$ in (30) and in the subsequent envelope equations reduces them to models for surface gravity water waves.

3.1.4 NLS Equation

In the two-dimensional case, the DS system (30) simplifies to

$$iu_\tau + \frac{1}{2}\partial_k^2\omega(k_0)\partial_X^2 u - \alpha_4(k_0)|u|^2u - \alpha_5(k_0)u\,\partial_X\xi_0 = 0,$$

$$\left[-h + \frac{1}{g}\Big(\partial_k\omega(k_0)\Big)^2\right]\partial_X^2\xi_0 - \alpha_5(k_0)\partial_X|u|^2 = 0.$$

Integrating the second equation above with respect to X by assuming vanishing conditions at infinity (as is the case for solitary waves) gives

$$\partial_X\xi_0 = \frac{\alpha_5(k_0)}{\frac{1}{g}(\partial_k\omega(k_0))^2 - h}|u|^2, \tag{31}$$

and then substituting this into the first equation yields the NLS equation

$$iu_\tau + \lambda\partial_X^2 u + \mu|u|^2u = 0, \tag{32}$$

where

$$\lambda = \frac{1}{2}\partial_k^2\omega(k_0),$$

$$\mu = -\alpha_4(k_0) - \frac{\alpha_5^2(k_0)}{\frac{1}{g}(\partial_k\omega(k_0))^2 - h}.$$

The corresponding Hamiltonian (with respect to τ) reads

$$H = \int_{-\infty}^{\infty}\left(\lambda|\partial_X u|^2 - \frac{1}{2}\mu|u|^4\right)dX, \tag{33}$$

so that $u_\tau = -\mathrm{i}H_{\overline{u}}$ and is obtained by inserting (28), (29), (31) in (22). For convenience, the hat notation is dropped from (33).

Similarly to the classical water wave problem in finite depth [1, 15], the coefficient μ of the nonlinear term may have two singularities at $(\partial_k\omega(k_0))^2 = gh$ and $\omega(2k_0)/(2k_0) = \partial_k\omega(k_0)$ corresponding to resonances between the zeroth and first harmonics and between the first and second harmonics, respectively. The former singularity occurs if the group velocity of the first harmonics equals the phase velocity of the zeroth harmonics (i.e. the long-wave limit c_0), while the latter singularity occurs if the same group velocity equals the phase velocity of the second harmonics. The presence of the first-harmonic group velocity $\partial_k\omega(k_0)$ in these singularities is related to the fact that the reference frame is moving with this velocity, as mentioned above. In the present problem, a natural choice for k_0 is $k_{\min}$. For a given value of h, the corresponding $k_{\min}$ is found numerically where the dispersion relation (5) achieves its minimum $c_{\min}$. Figure 2 reveals that both $c(2k_{\min}) = \omega(2k_{\min})/(2k_{\min})$ and $c_0 = \sqrt{gh}$ tend to $c_{\min} = \partial_k\omega(k_{\min})$ as $h \to 0$. Therefore, the modulational regime becomes inadequate and the long-wave regime should be preferred in the shallow-water limit, as could be expected [23].

According to (16), the ice-sheet deflection can be expressed in terms of u as

$$\eta(X,\tau) = \frac{\varepsilon}{\sqrt{2}}\left[a^{-1}(k_0+\varepsilon D_X)u(X,\tau)e^{\mathrm{i}k_0X/\varepsilon} + \text{c.c.}\right]$$
$$+\frac{\varepsilon^2}{\sqrt{2}}\left[a^{-1}(2k_0+\varepsilon D_X)u_2(X,\tau)e^{2\mathrm{i}k_0X/\varepsilon} + \text{c.c.}\right] + \varepsilon^2\eta_0(X,\tau), \quad (34)$$

where u_2 is given by (29), and

$$\eta_0 = \frac{\alpha_1(k_0)h + k_0\partial_k\omega(k_0)}{(\partial_k\omega(k_0))^2 - gh}|u|^2,$$

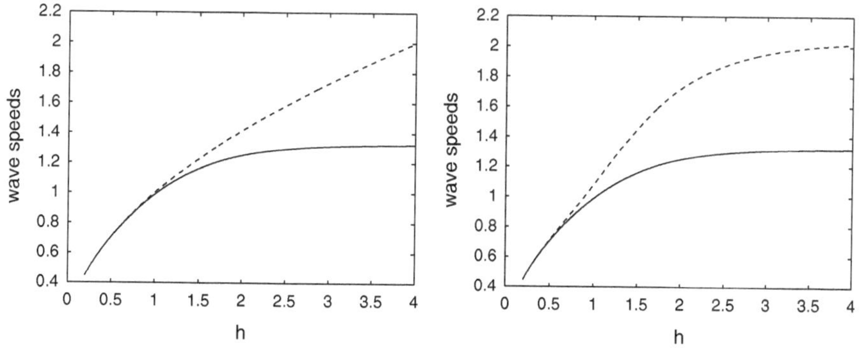

Fig. 2 *Left panel*: $c_{\min}$ (*solid line*) and c_0 (*dashed line*) versus h. *Right panel*: $c_{\min}$ (*solid line*) and $c(2k_{\min})$ (*dashed line*) versus h

by combining (28) with (31). The zeroth and second harmonics add corrections to the coefficients in the envelope equation for the first harmonics but also to the formula recovering the ice-sheet deflection. An expression similar to (34) holds for η in the three-dimensional case, with η_0 determined by (28) and the solution of the DS system. Equation (34) can be evaluated numerically by a pseudo-spectral method, which is a natural choice for computing such Fourier multipliers as a^{-1} [10, 14, 20, 21, 31, 44].

3.1.5 Soliton Solutions

In view of presenting numerical results, we non-dimensionalize the equations by using the characteristic length and velocity scales

$$\mathscr{L} = \left(\frac{\mathscr{D}}{\rho g}\right)^{1/4}, \qquad \mathscr{V} = \left(\frac{\mathscr{D} g^3}{\rho}\right)^{1/8},$$

respectively, so that $g = 1$ and $\mathscr{D}/\rho = 1$ as a consequence [4, 29, 39].

The NLS equation (32) is of focusing type and thus admits stable soliton solutions traveling at the group velocity $\partial_k \omega(k_0)$ if $\lambda\mu > 0$ [18, 38]. The graphs of λ and μ for $k_0 = k_{\min}$ are shown in Fig. 3. We see that λ is increasing and always positive, while μ is decreasing and changes sign at the critical depth $h_c \simeq 36.75$. Accordingly, the NLS equation (32) is of focusing type if $h < h_c$ and defocusing if $h > h_c$. Because $\mu = 0$ at $h = h_c$, this implies that higher-order terms must be included in the equation, but we will not consider this situation here. In the linear Euler–Bernoulli case (by setting $\mathscr{D} = 0$ in α_3), we find $h_c \simeq 5.54$, which is close to the value $h_c \simeq 5.91$ reported by Milewski and Wang [30]. For a Kirchhoff–Love model of the ice sheet, Părău and Dias [32] found $h_c \simeq 7.63$, which is smaller than the present value for the Cosserat model. The fact that $\mu \to \infty$ as $h \to 0$ in Fig. 3 is consistent with the two resonances in the shallow-water limit as discussed in Sect. 3.1.4.

Since we are interested in solitary waves, the key parameters to be examined are the wave speed $c < c_{\min}$ and the water depth $h < h_c$. Figures 4 and 5 present a comparison of solitary wave profiles for various values of (c, h), which are obtained from direct numerical simulations of (1)–(4) and from the exact NLS soliton solution

$$u(X, \tau) = \sqrt{2}\, u_0 \operatorname{sech}\left(u_0 \sqrt{\frac{\mu}{\lambda}}\, X\right) e^{\mathrm{i}\mu u_0^2 \tau}, \tag{35}$$

which corresponds to solitary waves whose crests are stationary relative to their envelopes [2]. In the latter case, the ice-sheet deflection (34) is evaluated as

$$\eta(X,\tau) = \frac{\varepsilon}{\sqrt{2}} \left[\mathrm{FT}^{-1}\left\{a^{-1}(k_0 + \varepsilon K)\, \mathrm{FT}(u)\right\} e^{\mathrm{i}k_0 X/\varepsilon} + \text{c.c.}\right]$$
$$+ \frac{\varepsilon^2}{\sqrt{2}} \left[\mathrm{FT}^{-1}\left\{a^{-1}(2k_0 + \varepsilon K)\, \mathrm{FT}(u_2)\right\} e^{2\mathrm{i}k_0 X/\varepsilon} + \text{c.c.}\right] + \varepsilon^2 \eta_0(X,\tau),$$

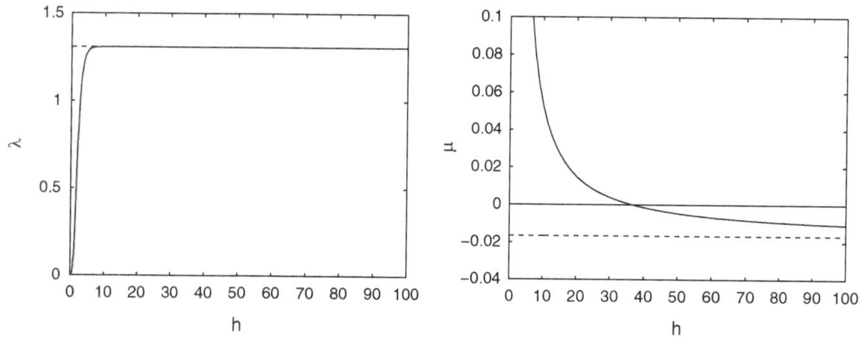

Fig. 3 NLS coefficients λ (*left panel, solid line*) and μ (*right panel, solid line*) versus h. As a reference, the corresponding values in the infinite-depth limit (see Sect. 3.2) are represented by a *dashed line*

where FT denotes the fast Fourier transform [10]. A number of 4096 grid points are typically used in our computations. For convenience, we set $\varepsilon = 1$ and only vary u_0 in (35) to match the fully nonlinear profile as closely as possible (which is equivalent to absorbing ε into u_0). The direct numerical simulations are based on a boundary-integral method with finite-difference approximations and the reader is referred to [21–23] for further details.

Overall there is a good agreement, especially regarding the relative amplitude of the central trough and the wavelength. The agreement is satisfactory even for moderately large wave amplitudes (compared to h, see Fig. 5), which is remarkable given the weakly nonlinear nature of the cubic NLS equation. The NLS prediction is able to capture well the main features, whether the solution is a localized or broader solitary wavepacket. This confirms in particular that the inclusion of the mean-flow component η_0 in (34) is crucial at reproducing well the vertical asymmetry of the solution. The second-harmonic corrections, however, are negligible according to the comparison of the two columns in Figs. 4 and 5. The left column of these figures shows results without second-harmonic contributions as in [23]. Only little improvement due to these second harmonics is noticeable in Fig. 5 for $h = 3.095$. Consistent with statements in Sect. 3.1.4, the agreement between numerical and NLS predictions slowly deteriorates as h decreases. We pay attention to the case $h = 3.095$ because it corresponds to Takizawa's experiments on Lake Saroma (Japan) [39], where the ice thickness was 0.17 m and the water depth was 6.8 m. Waves were generated by moving a load (ski-doo snowmobile) at various speeds on top of the ice sheet. Wavelengths of order $O(10)$ m were observed. For $h = 3.095$, our results resemble some of his observations. We find similar wave profiles for larger depths.

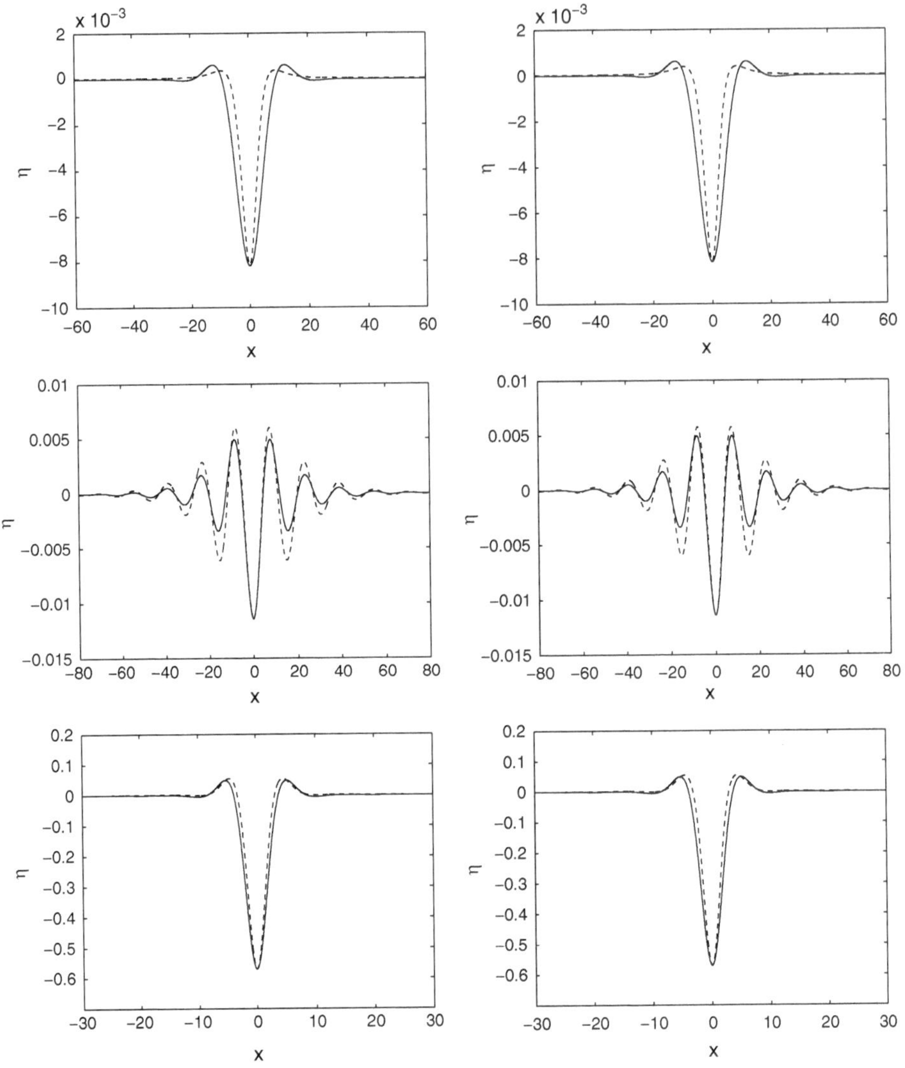

Fig. 4 Comparison of solitary wave profiles obtained from direct numerical simulations (*solid line*) and the NLS soliton (35) (*dashed line*) for $(c, h) = (0.7, 0.5)$, $(0.985, 1)$ and $(0.9, 1.5)$ (from *top* to *bottom*). The *left* and *right columns* show the solutions without and with second-harmonic contributions, respectively

3.2 *Infinite Depth*

In this regime, the mean-flow exponents $\alpha = 3$ and $\beta = 2$ are larger than those in finite depth (see [8] for an explanation). As a consequence, the mean-flow terms do not contribute to the order of approximation being considered, and the renormalized

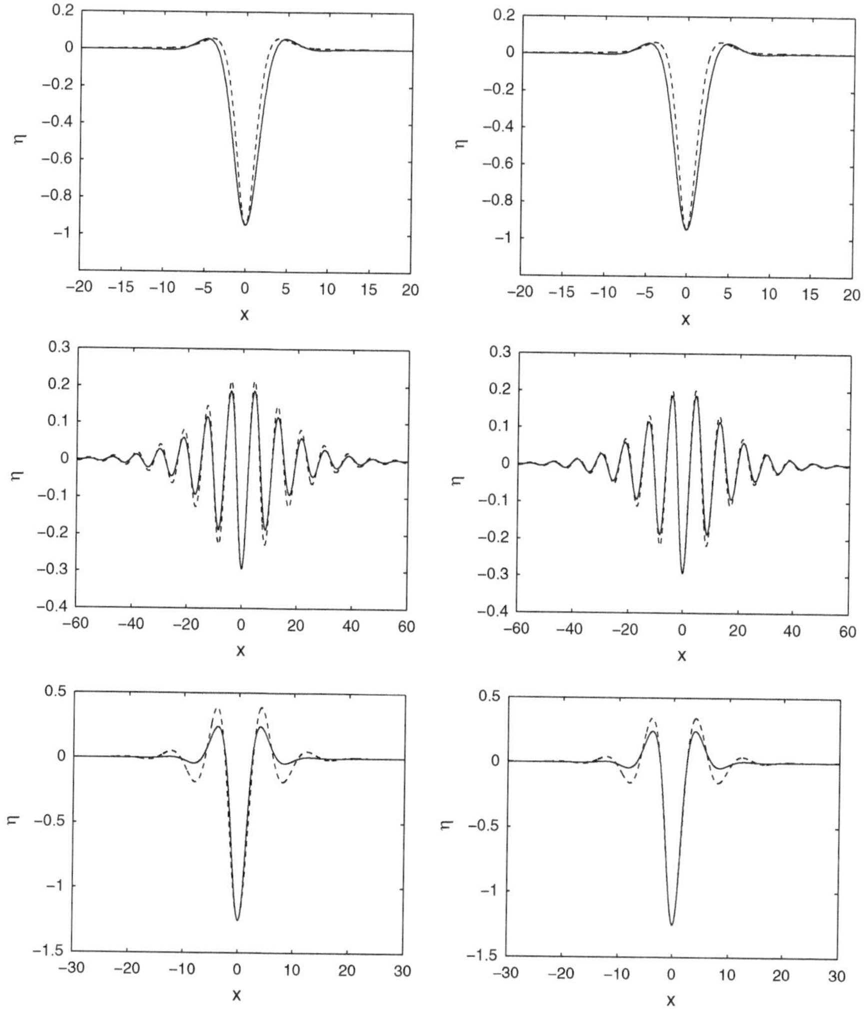

Fig. 5 Comparison of solitary wave profiles obtained from direct numerical simulations (*solid line*) and the NLS soliton (35) (*dashed line*) for $(c, h) = (0.658, 1.5)$, $(1.3, 3.095)$ and $(1.056, 3.095)$ (from *top* to *bottom*). The *left* and *right columns* show the solutions without and with second-harmonic contributions, respectively

Hamiltonian takes the form

$$\hat{H} = \varepsilon^2 \iint_{-\infty}^{\infty} \left[\frac{1}{2} \partial^2_{k_j k_\ell} \omega(\mathbf{k}_0) \overline{u} D^2_{X_j X_\ell} u + \Big(\omega(2\mathbf{k}_0) - 2\mathbf{k}_0 \cdot \nabla_{\mathbf{k}} \omega(\mathbf{k}_0) \Big) |u_2|^2 \right.$$
$$\left. + \left\{ \frac{1}{4} |\mathbf{k}_0|^3 - \frac{5\mathscr{D}}{8\rho} \Big(k_x^6 + k_y^6 + 3k_x^4 k_y^2 + 3k_x^2 k_y^4 \Big) a^{-4}(\mathbf{k}_0) \right\} |u|^4 \right.$$

$$+ \frac{1}{2\sqrt{2}}|\mathbf{k}_0|^2 a^{-1}(2\mathbf{k}_0)a^2(\mathbf{k}_0)\Big(u^2\overline{u}_2 + \overline{u}^2 u_2\Big)\Big] \, dYdX + O(\varepsilon^3) . \tag{36}$$

The equations of motion (20) then become

$$\begin{aligned} \mathrm{i}u_\tau &= -\frac{1}{2}\partial^2_{k_j k_\ell}\omega(\mathbf{k}_0)\partial^2_{X_j X_\ell}u + \frac{1}{\sqrt{2}}|\mathbf{k}_0|^2 a^{-1}(2\mathbf{k}_0)a^2(\mathbf{k}_0)\overline{u}\,u_2 \\ &\quad + \left\{\frac{1}{2}|\mathbf{k}_0|^3 - \frac{5\mathscr{D}}{4\rho}\Big(k_x^6 + k_y^6 + 3k_x^4 k_y^2 + 3k_x^2 k_y^4\Big)a^{-4}(\mathbf{k}_0)\right\}|u|^2 u , \end{aligned} \tag{37}$$

$$\begin{aligned} \mathrm{i}\varepsilon^2 u_{2\tau} &= \Big(\omega(2\mathbf{k}_0) - 2\mathbf{k}_0\cdot\nabla_{\mathbf{k}}\omega(\mathbf{k}_0)\Big)u_2 + \frac{1}{2\sqrt{2}}|\mathbf{k}_0|^2 a^{-1}(2\mathbf{k}_0)a^2(\mathbf{k}_0)u^2, \\ \varepsilon^3 \eta_{0\tau} &= 0, \\ \varepsilon^3 \xi_{0\tau} &= 0, \end{aligned} \tag{38}$$

which confirms that the mean-flow contributions are negligible, while

$$u_2 = \frac{|\mathbf{k}_0|^2 a^{-1}(2\mathbf{k}_0)a^2(\mathbf{k}_0)}{2\sqrt{2}\,(2\mathbf{k}_0\cdot\nabla_{\mathbf{k}}\omega(\mathbf{k}_0) - \omega(2\mathbf{k}_0))}u^2 , \tag{39}$$

to lowest order by virtue of (38). Substituting (39) into (36) and (37) leads to the NLS equation

$$\begin{aligned} \mathrm{i}u_\tau = &-\frac{1}{2}\partial^2_{k_j k_\ell}\omega(\mathbf{k}_0)\partial^2_{X_j X_\ell}u + \left\{\frac{|\mathbf{k}_0|^4 a^{-2}(2\mathbf{k}_0)a^4(\mathbf{k}_0)}{4\,(2\mathbf{k}_0\cdot\nabla_{\mathbf{k}}\omega(\mathbf{k}_0) - \omega(2\mathbf{k}_0))}\right. \\ &\left. + \frac{1}{2}|\mathbf{k}_0|^3 - \frac{5\mathscr{D}}{4\rho}\Big(k_x^6 + k_y^6 + 3k_x^4 k_y^2 + 3k_x^2 k_y^4\Big)a^{-4}(\mathbf{k}_0)\right\}|u|^2 u , \end{aligned} \tag{40}$$

whose Hamiltonian (with respect to τ) is

$$\begin{aligned} H = \iint_{-\infty}^{\infty}\Bigg[&\frac{1}{2}\partial^2_{k_j k_\ell}\omega(\mathbf{k}_0)(\overline{\partial_{X_j}u})(\partial_{X_\ell}u) + \left\{\frac{|\mathbf{k}_0|^4 a^{-2}(2\mathbf{k}_0)a^4(\mathbf{k}_0)}{8\,(2\mathbf{k}_0\cdot\nabla_{\mathbf{k}}\omega(\mathbf{k}_0) - \omega(2\mathbf{k}_0))}\right. \\ &\left. + \frac{1}{4}|\mathbf{k}_0|^3 - \frac{5\mathscr{D}}{8\rho}\Big(k_x^6 + k_y^6 + 3k_x^4 k_y^2 + 3k_x^2 k_y^4\Big)a^{-4}(\mathbf{k}_0)\right\}|u|^4\Bigg] \, dYdX . \end{aligned}$$

In the two-dimensional case, we again obtain an NLS equation of the form (32), with Hamiltonian (33), where

$$\lambda = \frac{1}{2}\partial^2_k\omega(k_0) = \frac{5\mathscr{D}k_0^3}{\rho\sqrt{gk_0 + \mathscr{D}k_0^5/\rho}} - \frac{(g + 5\mathscr{D}k_0^4/\rho)^2}{8(gk_0 + \mathscr{D}k_0^5/\rho)^{3/2}} ,$$

and

$$\mu = \frac{5\mathscr{D}k_0^7}{4\rho(g+\mathscr{D}k_0^4/\rho)} - \frac{k_0^3}{2} - \frac{k_0^3(g+\mathscr{D}k_0^4/\rho)}{4\left(2k_0\partial_k\omega(k_0)-\omega(2k_0)\right)}\sqrt{\frac{2k_0}{g+16\mathscr{D}k_0^4/\rho}}.$$

Without loss of generality, the carrier wavenumber $k_0 = k_x$ is assumed to be positive. Again the coefficient μ may have a singularity at $\omega(2k_0)/(2k_0) = \partial_k\omega(k_0)$ due to the first-second harmonic resonance, but the zeroth-first harmonic resonance is absent here since the mean flow does not come into play. The ice-sheet deflection can be recovered from u by using (34) as well, but without need for the higher-order $O(\varepsilon^3)$ contribution from η_0.

In infinite depth [21, 37], the phase speed is minimum at

$$k_{\min} = \left(\frac{g\rho}{3\mathscr{D}}\right)^{1/4}.$$

After applying the same non-dimensionalization as in Sect. 3.1.5, we find that $\lambda\mu < 0$ if $k_0 = k_{\min}$ since

$$\lambda = \frac{3^{7/8}}{2} \simeq 1.307 > 0,$$

and

$$\mu = -\frac{3^{1/4}(41\sqrt{38}-228)}{912(\sqrt{38}-4)} \simeq -0.016 < 0\,.$$

Incidentally, the above-mentioned denominator in μ

$$\frac{\omega(2k_{\min})}{2k_{\min}} - \partial_k\omega(k_{\min}) = \frac{3^{5/8}(\sqrt{38}-4)}{6} \simeq 0.717\,,$$

does not vanish, as also indicated in Fig. 2 for large h. Therefore, the NLS equation is of defocusing type and no soliton solutions exist in this limit. This result is consistent with the asymptotic behavior of λ and μ for finite depth as $h \to \infty$ (see Sect. 3.1.5). Previous studies using different methods of derivation and different models for the ice sheet (e.g. Kirchhoff–Love theory) also obtained a defocusing NLS equation in this regime [21, 29].

3.3 *Exact Linear Dispersion*

3.3.1 Finite Depth

As suggested in [10, 27, 41], the linear dispersive properties of envelope equations can be improved by retaining the exact linear dispersion relation rather than Taylor expanding it. For finite depth, a counterpart to (22) is

$$
\begin{aligned}
\hat{H} &= H - \omega(\mathbf{k}_0)M\,, \\
&= \varepsilon^2 \iint_{-\infty}^{\infty} \Big[\frac{1}{\varepsilon^2}\overline{u}\Big(\omega(\mathbf{k}_0 + \varepsilon D_{\mathbf{X}}) - \omega(\mathbf{k}_0)\Big)u + \omega(2\mathbf{k}_0)|u_2|^2 + \alpha_3(\mathbf{k}_0)|u|^4 \\
&\quad + \alpha_2(\mathbf{k}_0)\Big(u^2\overline{u}_2 + \overline{u}^2 u_2\Big) + \Big(\mathrm{i}\mathbf{k}_0 \cdot D_{\mathbf{X}}\xi_0 + \alpha_1(\mathbf{k}_0)\eta_0\Big)|u|^2 \\
&\quad + \frac{1}{2\varepsilon^2}\xi_0 G_0(\varepsilon D_{\mathbf{X}})\xi_0 + \frac{1}{2}g\eta_0^2\Big]\, dY dX + O(\varepsilon^3)\,,
\end{aligned}
\tag{41}
$$

and the corresponding evolution equations are

$$
\begin{aligned}
\mathrm{i}u_\tau &= \frac{1}{\varepsilon^2}\Big(\omega(\mathbf{k}_0 + \varepsilon D_{\mathbf{X}}) - \omega(\mathbf{k}_0)\Big)u + 2\alpha_3(\mathbf{k}_0)|u|^2 u + 2\alpha_2(\mathbf{k}_0)\overline{u}\,u_2 \\
&\quad + \Big(\mathrm{i}\mathbf{k}_0 \cdot D_{\mathbf{X}}\xi_0 + \alpha_1(\mathbf{k}_0)\eta_0\Big)u\,, \\
\varepsilon\eta_{0\tau} &= \frac{1}{\varepsilon^2}G_0(\varepsilon D_{\mathbf{X}})\xi_0 - \mathbf{k}_0 \cdot \nabla_{\mathbf{X}}|u|^2\,, \\
\varepsilon\xi_{0\tau} &= -\Big(g\eta_0 + \alpha_1(\mathbf{k}_0)|u|^2\Big)\,, \\
\mathrm{i}\varepsilon^2 u_{2\tau} &= \omega(2\mathbf{k}_0)u_2 + \alpha_2(\mathbf{k}_0)u^2\,.
\end{aligned}
$$

By following the same procedure as in Sect. 3.1.3, we find the modified DS system

$$
\begin{aligned}
\mathrm{i}u_\tau &= \frac{1}{\varepsilon^2}\Big(\omega(\mathbf{k}_0 + \varepsilon D_{\mathbf{X}}) - \omega(\mathbf{k}_0)\Big)u + \left(2\alpha_3(\mathbf{k}_0) - \frac{2\alpha_2^2(\mathbf{k}_0)}{\omega(2\mathbf{k}_0)} - \frac{1}{g}\alpha_1^2(\mathbf{k}_0)\right)|u|^2 u \\
&\quad + u\,\mathbf{k}_0 \cdot \nabla_{\mathbf{X}}\xi_0\,, \\
0 &= \frac{1}{\varepsilon^2}G_0(\varepsilon D_{\mathbf{X}})\xi_0 - \mathbf{k}_0 \cdot \nabla_{\mathbf{X}}|u|^2\,,
\end{aligned}
$$

where

$$
\eta_0 = -\frac{1}{g}\alpha_1(\mathbf{k}_0)|u|^2\,, \qquad u_2 = -\frac{\alpha_2(\mathbf{k}_0)}{\omega(2\mathbf{k}_0)}u^2\,,
$$

to lowest order. In the two-dimensional case, the second equation in this DS system can be solved for $\partial_X\xi_0$, hence

$$
\partial_X\xi_0 = \varepsilon^2 k_0 G_0^{-1}(\varepsilon D_X)\partial_X^2|u|^2\,.
$$

Substituting this into the first equation yields the modified NLS equation

$$iu_\tau = \frac{1}{\varepsilon^2}\Big(\omega(k_0+\varepsilon D_X)-\omega(k_0)\Big)u + \left(2\alpha_3(k_0) - \frac{2\alpha_2^2(k_0)}{\omega(2k_0)} - \frac{1}{g}\alpha_1^2(k_0)\right)|u|^2u$$
$$+\varepsilon^2 k_0^2\, u\, G_0^{-1}(\varepsilon D_X)\partial_X^2|u|^2\,,$$

with Hamiltonian

$$H = \int_{-\infty}^{\infty}\left[\frac{1}{\varepsilon^2}\overline{u}\Big(\omega(k_0+\varepsilon D_X)-\omega(k_0)\Big)u + \frac{1}{2}\left(2\alpha_3(k_0) - \frac{2\alpha_2^2(k_0)}{\omega(2k_0)} - \frac{1}{g}\alpha_1^2(k_0)\right)|u|^4\right.$$
$$\left.+\frac{1}{2}\varepsilon^2 k_0^2\,|u|^2 G_0^{-1}(\varepsilon D_X)\partial_X^2|u|^2\right]dX\,,$$

as derived from (41). Here again, these modified DS and NLS equations can be solved numerically by a pseudo-spectral method which is suitable for handling the Fourier multipliers ω and G_0. Note that the operator $G_0^{-1}(\varepsilon D_X)\partial_X^2$ is well-defined, and in particular it is not singular at $k = 0$ as can be shown by a Taylor series expansion in ε.

3.3.2 Infinite Depth

For infinite depth, the renormalized Hamiltonian is given by

$$\hat{H} = \varepsilon^2\iint_{-\infty}^{\infty}\left[\frac{1}{\varepsilon^2}\overline{u}\Big(\omega(\mathbf{k}_0+\varepsilon D_\mathbf{X})-\omega(\mathbf{k}_0)\Big)u + \omega(2\mathbf{k}_0)|u_2|^2\right.$$
$$+\left\{\frac{1}{4}|\mathbf{k}_0|^3 - \frac{5\mathscr{D}}{8\rho}\Big(k_x^6+k_y^6+3k_x^4k_y^2+3k_x^2k_y^4\Big)a^{-4}(\mathbf{k}_0)\right\}|u|^4$$
$$\left.+\frac{1}{2\sqrt{2}}|\mathbf{k}_0|^2a^{-1}(2\mathbf{k}_0)a^2(\mathbf{k}_0)\Big(u^2\overline{u}_2+\overline{u}^2u_2\Big)\right]dYdX + O(\varepsilon^3)\,,$$

whose dynamics obeys

$$iu_\tau = \frac{1}{\varepsilon^2}\Big(\omega(\mathbf{k}_0+\varepsilon D_\mathbf{X})-\omega(\mathbf{k}_0)\Big)u + \frac{1}{\sqrt{2}}|\mathbf{k}_0|^2a^{-1}(2\mathbf{k}_0)a^2(\mathbf{k}_0)\overline{u}\,u_2$$
$$+\left\{\frac{1}{2}|\mathbf{k}_0|^3 - \frac{5\mathscr{D}}{4\rho}\Big(k_x^6+k_y^6+3k_x^4k_y^2+3k_x^2k_y^4\Big)a^{-4}(\mathbf{k}_0)\right\}|u|^2u\,, \quad (42)$$

$$i\varepsilon^2u_{2\tau} = \omega(2\mathbf{k}_0)u_2 + \frac{1}{2\sqrt{2}}|\mathbf{k}_0|^2a^{-1}(2\mathbf{k}_0)a^2(\mathbf{k}_0)u^2\,. \quad (43)$$

Combining (42) and (43) as in Sect. 3.2 leads to the modified NLS equation

$$
\begin{aligned}
iu_\tau = & \frac{1}{\varepsilon^2}\Big(\omega(\mathbf{k}_0 + \varepsilon D_{\mathbf{X}}) - \omega(\mathbf{k}_0)\Big)u - \left\{ \frac{|\mathbf{k}_0|^4 a^{-2}(2\mathbf{k}_0) a^4(\mathbf{k}_0)}{4\omega(2\mathbf{k}_0)} \right. \\
& \left. - \frac{1}{2}|\mathbf{k}_0|^3 + \frac{5\mathscr{D}}{4\rho}\Big(k_x^6 + k_y^6 + 3k_x^4k_y^2 + 3k_x^2k_y^4\Big)a^{-4}(\mathbf{k}_0) \right\} |u|^2 u \,,
\end{aligned}
$$

such that

$$
u_2 = -\frac{|\mathbf{k}_0|^2 a^{-1}(2\mathbf{k}_0) a^2(\mathbf{k}_0)}{2\sqrt{2}\,\omega(2\mathbf{k}_0)} u^2 \,.
$$

The corresponding Hamiltonian reads

$$
\begin{aligned}
H = & \iint_{-\infty}^{\infty} \left[\frac{1}{\varepsilon^2}\overline{u}\Big(\omega(\mathbf{k}_0 + \varepsilon D_{\mathbf{X}}) - \omega(\mathbf{k}_0)\Big)u - \left\{ \frac{|\mathbf{k}_0|^4 a^{-2}(2\mathbf{k}_0) a^4(\mathbf{k}_0)}{8\omega(2\mathbf{k}_0)} \right.\right. \\
& \left.\left. - \frac{1}{4}|\mathbf{k}_0|^3 + \frac{5\mathscr{D}}{8\rho}\Big(k_x^6 + k_y^6 + 3k_x^4k_y^2 + 3k_x^2k_y^4\Big)a^{-4}(\mathbf{k}_0) \right\} |u|^4 \right] dYdX \,.
\end{aligned}
$$

Their expressions in the two-dimensional case follow directly, namely

$$
\begin{aligned}
iu_\tau = & \frac{1}{\varepsilon^2}\Big(\omega(k_0 + \varepsilon D_X) - \omega(k_0)\Big)u - \left(\frac{k_0^4\, a^{-2}(2k_0) a^4(k_0)}{4\omega(2k_0)} \right. \\
& \left. - \frac{1}{2}k_0^3 + \frac{5\mathscr{D}}{4\rho}k_0^6\, a^{-4}(k_0) \right) |u|^2 u \,,
\end{aligned}
$$

and

$$
\begin{aligned}
H = & \int_{-\infty}^{\infty} \left[\frac{1}{\varepsilon^2}\overline{u}\Big(\omega(k_0 + \varepsilon D_X) - \omega(k_0)\Big)u - \left(\frac{k_0^4\, a^{-2}(2k_0) a^4(k_0)}{8\omega(2k_0)} \right.\right. \\
& \left.\left. - \frac{1}{4}k_0^3 + \frac{5\mathscr{D}}{8\rho}k_0^6\, a^{-4}(k_0) \right) |u|^4 \right] dX \,.
\end{aligned}
$$

4 Conclusions

A Hamiltonian formulation for three-dimensional nonlinear flexural-gravity waves propagating at the surface of an ideal fluid covered by ice is presented. The ice sheet is modeled as a thin elastic plate, based on the special Cosserat theory for hyperelastic shells as proposed by Plotnikov and Toland [34]. Weakly nonlinear models for small-amplitude waves on finite and infinite depth are derived in the modulational regime, by applying the Hamiltonian perturbation approach of

Craig et al. [8, 10]. A new contribution of the present paper is the inclusion of second-harmonic effects, leading to corrections in the cubic coefficient of the envelope equations and in the expression of the ice-sheet deflection. However, comparison with two-dimensional direct numerical simulations reveals no much improvement from these higher-order corrections in the parameter regime considered.

In the future, it would be of interest to further analyze the resulting NLS and DS equations, in particular those incorporating exact linear dispersion, and to compute localized traveling solutions numerically. We also plan to investigate the long-wave regime for this three-dimensional hydroelastic problem within the Hamiltonian framework.

Acknowledgements This work was partially supported by the Simons Foundation through grant No. 246170. The author thanks Walter Craig, Emilian Părău and Catherine Sulem for fruitful discussions. Part of this work was carried out while the author was visiting the Isaac Newton Institute for Mathematical Sciences during the Theory of Water Waves program in the summer 2014.

References

1. Ablowitz, M.J., Segur, H.: Evolution of packets of water waves. J. Fluid Mech. **92**, 691–715 (1979)
2. Akylas, T.R.: Envelope solitons with stationary crests. Phys. Fluids **5**, 789–791 (1993)
3. Alam, M.R.: Dromions of flexural-gravity waves. J. Fluid Mech. **719**, 1–13 (2013)
4. Bonnefoy, F., Meylan, M.H., Ferrant, P.: Nonlinear higher-order spectral solution for a two-dimensional moving load on ice. J. Fluid Mech. **621**, 215–242 (2009)
5. Craig, W., Guyenne, P., Kalisch, H.: Hamiltonian long wave expansions for free surfaces and interfaces. Commun. Pure Appl. Math. **58**, 1587–1641 (2005)
6. Craig, W., Guyenne, P., Nicholls, D.P., Sulem, C.: Hamiltonian long-wave expansions for water waves over a rough bottom. Proc. R. Soc. A **461**, 839–873 (2005)
7. Craig, W., Guyenne, P., Sulem, C.: Water waves over a random bottom. J. Fluid Mech. **640**, 79–107 (2009)
8. Craig, W., Guyenne, P., Sulem, C.: A Hamiltonian approach to nonlinear modulation of surface water waves. Wave Motion **47**, 552–563 (2010)
9. Craig, W., Guyenne, P., Sulem, C.: Coupling between internal and surface waves. Nat. Hazards **57**, 617–642 (2011)
10. Craig, W., Guyenne, P., Sulem, C.: Hamiltonian higher-order nonlinear Schrödinger equations for broader-banded waves on deep water. Eur. J. Mech. B. Fluids **32**, 22–31 (2012)
11. Craig, W., Guyenne, P., Sulem, C.: The surface signature of internal waves. J. Fluid Mech. **710**, 277–303 (2012)
12. Craig, W., Guyenne, P., Sulem, C.: Internal waves coupled to surface gravity waves in three dimensions. Commun. Math. Sci. **13**, 893–910 (2015)
13. Craig, W., Schanz, U., Sulem, C.: The modulational regime of three-dimensional water waves and the Davey–Stewartson system. Ann. Inst. H. Poincaré (C) Nonlinear Anal. **14**, 615–667 (1997)
14. Craig, W., Sulem, C.: Numerical simulation of gravity waves. J. Comput. Phys. **108**, 73–83 (1993)

15. Craig, W., Sulem, C., Sulem, P.-L.: Nonlinear modulation of gravity waves: a rigorous approach. Nonlinearity **5**, 497–522 (1992)
16. Düll, W.-P., Schneider, G., Wayne, C.E.: Justification of the Nonlinear Schrödinger equation for the evolution of gravity driven 2D surface water waves in a canal of finite depth. (2013, submitted)
17. Forbes, L.K.: Surface waves of large amplitude beneath an elastic sheet. Part 1. High-order series solution. J. Fluid Mech. **169**, 409–428 (1986)
18. Grillakis, M., Shatah, J., Strauss, W.: Stability theory of solitary waves in the presence of symmetry. I. J. Funct. Anal. **74**, 160–197 (1987)
19. Guyenne, P., Lannes, D., Saut, J.-C.: Well-posedness of the Cauchy problem for models of large amplitude internal waves. Nonlinearity **23**, 237–275 (2010)
20. Guyenne, P., Nicholls, D.P.: A high-order spectral method for nonlinear water waves over moving bottom topography. SIAM J. Sci. Comput. **30**, 81–101 (2007)
21. Guyenne, P., Părău, E.I.: Computations of fully nonlinear hydroelastic solitary waves on deep water. J. Fluid Mech. **713**, 307–329 (2012)
22. Guyenne, P., Părău, E.I.: Forced and unforced flexural-gravity solitary waves. In: Proceedings of IUTAM Nonlinear Interfacial Wave Phenomena from the Micro- to the Macro-Scale, vol. 11, pp. 44–57. Limassol (2014)
23. Guyenne, P., Părău, E.I.: Finite-depth effects on solitary waves in a floating ice sheet. J. Fluids Struct. **49**, 242–262 (2014)
24. Hărăguş-Courcelle, M., Ilichev, A.: Three-dimensional solitary waves in the presence of additional surface effects. Eur. J. Mech. B. Fluids **17**, 739–768 (1998)
25. Hegarty, G.M., Squire, V.A.: A boundary-integral method for the interaction of large-amplitude ocean waves with a compliant floating raft such as a sea-ice floe. J. Eng. Math. **62**, 355–372 (2008)
26. Korobkin, A., Părău, E.I., Vanden-Broeck, J.-M.: The mathematical challenges and modelling of hydroelasticity. Philos. Trans. R. Soc. Lond. A **369**, 2803–2812 (2011)
27. Lannes, D.: The Water Waves Problem: Mathematical Analysis and Asymptotics. American Mathematical Society, Providence (2013)
28. Marko, J.R.: Observations and analyses of an intense waves-in-ice event in the Sea of Okhotsk. J. Geophys. Res. **108**, 3296 (2003)
29. Milewski, P.A., Vanden-Broeck, J.-M., Wang, Z.: Hydroelastic solitary waves in deep water. J. Fluid Mech. **679**, 628–640 (2011)
30. Milewski, P.A., Wang, Z.: Three dimensional flexural-gravity waves. Stud. Appl. Math. **131**, 135–148 (2013)
31. Nicholls, D.P.: Traveling water waves: Spectral continuation methods with parallel implementation. J. Comput. Phys. **143**, 224–240 (1998)
32. Părău, E., Dias, F.: Nonlinear effects in the response of a floating ice plate to a moving load. J. Fluid Mech. **460**, 281–305 (2002)
33. Părău, E.I., Vanden-Broeck, J.-M.: Three-dimensional waves beneath an ice sheet due to a steadily moving pressure. Phil. Trans. Roy. Soc. Lond. A **369**, 2973–2988 (2011)
34. Plotnikov, P.I., Toland, J.F.: Modelling nonlinear hydroelastic waves. Philos. Trans. R. Soc. Lond. A **369**, 2942–2956 (2011)
35. Schneider, G., Wayne, C.E.: Justification of the NLS approximation for a quasilinear water wave model. J. Differ. Equ. **251**, 238–269 (2011)
36. Squire, V.A.: Past, present and impendent hydroelastic challenges in the polar and subpolar seas. Philos. Trans. R. Soc. A **369**, 2813–2831 (2011)
37. Squire, V.A., Hosking, R.J., Kerr, A.D., Langhorne, P.J.: Moving Loads on Ice Plates. Kluwer, Dordrecht (1996)
38. Sulem, C., Sulem, P.-L.: The Nonlinear Schrödinger Equation: Self-focusing and Wave Collapse. Springer, New York (1999)
39. Takizawa, T.: Deflection of a floating sea ice sheet induced by a moving load. Cold Reg. Sci. Technol. **11**, 171–180 (1985)

40. Totz, N., Wu, S.: A rigorous justification of the modulation approximation to the 2D full water wave problem. Commun. Math. Phys. **310**, 817–883 (2012)
41. Trulsen, K., Kliakhandler, I., Dysthe, K.B., Velarde, M.G.: On weakly nonlinear modulation of waves on deep water. Phys. Fluids **12**, 2432–2437 (2000)
42. Vanden-Broeck, J.-M., Părău, E.I.: Two-dimensional generalised solitary waves and periodic waves under an ice sheet. Philos. Trans. R. Soc. Lond. A **369**, 2957–2972 (2011)
43. Xia, X., Shen, H.T.: Nonlinear interaction of ice cover with shallow water waves in channels. J. Fluid Mech. **467**, 259–268 (2002)
44. Xu, L., Guyenne, P.: Numerical simulation of three-dimensional nonlinear water waves. J. Comput. Phys. **228**, 8446–8466 (2009)
45. Zakharov, V.E.: Stability of periodic waves of finite amplitude on the surface of a deep fluid. J. Appl. Mech. Tech. Phys. **9**, 190–194 (1968)

Dissipation of Narrow-Banded Surface Water Waves

Diane Henderson, Girish Kumar Rajan, and Harvey Segur

We dedicate this paper to our friend and colleague, Walter Craig.

Abstract Our overall objective is to find mathematical models that describe accurately how waves in nature propagate and evolve. One process that affects evolution is dissipation (Segur et al., J Fluid Mech 539:229–271, 2005), so in this paper we explore several models in the literature that incorporate various dissipative physical mechanisms. In particular, we seek theoretical models that (1) agree with measured dissipation rates in laboratory and field experiments, and (2) have the mathematical properties required to be of use in weakly nonlinear models of the evolution of waves with narrow-banded spectra, as they propagate over long distances on deep water.

1 Introduction

The equations for water waves were first proposed by Stokes [21] as an energy-conserving system. More recently it was discovered [10, 11, 17, 20, 25] that dissipation is important in the long-time evolution of narrow-banded wavetrains. So, our objective is to find physically realistic models of dissipation of gravity-driven surface water waves. In this paper we report on comparisons of measured vs. predicted dissipation rates of such waves, under a variety of conditions. The laboratory generated waves have frequencies in the range of 1–4 Hz, and the waves propagate on water of either finite or infinite depth, when the air/water interface is either cleaned or else contaminated with olive oil, or with plastic wrap spread

D. Henderson (✉) • G.K. Rajan
Department of Mathematics, Penn State University, 218 McAllister Building, University Park, PA 16802, USA
e-mail: dmh@math.psu.edu; girish@psu.edu

H. Segur
Department of Applied Mathematics, University of Colorado, Boulder, CO 80309-0526, USA
e-mail: segur@colorado.edu

P. Guyenne et al. (eds.), *Hamiltonian Partial Differential Equations and Applications*, Fields Institute Communications 75,
DOI 10.1007/978-1-4939-2950-4_6

across the interface, or from ambient conditions in the laboratory. We also consider previously published observations of dissipation rates for ocean swell, with much lower frequencies. These measured dissipation rates are compared to dissipation rates predicted by several theories in the published literature. The various theories assume: (1) a clean interface in either a fluid/vacuum system or a fluid/fluid system; (2) two limits of a surfactant-covered interface in a fluid/vacuum system—either infinite elasticity or else maximum dissipation rate due to a resonance; or (3) a thin layer of a viscoelastic fluid at the interface of a fluid/vacuum system.

Interest in dissipation of the energy of ocean swell goes back at least to Pliny the Elder in the first century AD (as referenced by Miles [18], who reviewed the origins of observations and modeling). In addition to understanding mechanisms for energy dissipation during wave propagation, we also are concerned with the effects of dissipation on the nonlinear evolution of both laboratory waves and narrow-banded ocean waves, such as ocean swell. In particular, narrow-banded spectra of deep-water, surface gravity waves are known to undergo a modulational instability, or Benjamin-Feir [1] instability, if modeled using inviscid, nonlinear Schrödinger (NLS)—type equations, first developed for water waves and other applications in [2, 19, 26, 27] and extended to higher orders by Dysthe [8] and others (e.g. [9, 23].) However, the inclusion of non-zero linear dissipation stabilizes this instability [20]. Predictions of the growth of perturbation amplitudes from dissipative NLS models that use *measured* dissipation rates agree quantitatively with measurements from laboratory experiments on waves with both one-dimensional [17, 20] and two-dimensional [11] surface patterns, and have been shown to be relevant for understanding the propagation of ocean swell [10]. Measured values of damping rates are sometimes significantly larger than values typically predicted by published theories. For example, Dias et al. [6] derived a dissipative NLS model following Lamb's approach in §349 of [15] that assumed a shear-free boundary condition at the vacuum/water interface. The resulting prediction for dissipation rate agrees fairly well with laboratory experiments for which care is taken to "clean" the air/water interface, but is *four orders of magnitude smaller* than a typical value measured for ocean swell [10]. On the other hand, Lamb's "inextensible film model" in §351 of [15] makes no assumption on the shear stress at the interface, but assumes that the tangential velocities are zero there. This model results in a prediction of dissipation rates that agrees with measurements of dissipation of ocean swell [10], but has the drawback that its boundary condition cannot be used for nonlinear models.

In the remainder of the paper, we present in Sect. 2 several models from the published literature that predict dissipation rates. These models assume either a clean interface between a fluid and a vacuum or between two fluids; or they assume some type of contamination at the interface. The models are listed in Table 1 with references, the equation number in this paper for the formula for dissipation rate, distinguishing characteristics of the model, and a list of the rheological parameters required for its application. The model name given in the table is for reference

Table 1 List of dissipation models considered herein

Reference	Model name equation number	Distinguishing feature(s)	Material parameters required
Van Dorn [24]	Sidewall & bottom(1)	Damping from sidewall and bottom boundary layers Does not account for dissipation at the interface	Fluid viscosity
Lamb [15] §349	Clean-surface (2)	Shear-free vacuum/fluid interface	Fluid viscosity
Dore [7]	Two-fluid (3)	Shear-free fluid/fluid interface	Viscosities of 2 fluids
Lamb [15] §351	Inextensible-surface (4)	Zero-tangential velocities at vacuum/fluid interface	Fluid viscosity
Miles [18]	Surfactant-surface (5)	Surfactant film at the vacuum/fluid interface	1 Elasticity coefficient 2 Surface viscosities 1 Surface tension 1 Fluid viscosity Surfactant solubility
Huhnerfuss et al. [13]	Surfactant-max (9)	Surfactant film at the vacuum/fluid interface Fluid wave - Marangoni wave resonance	Fluid viscosity
Jenkins & Jacobs [14]	Thin-layer(10)	Thin-fluid/fluid interface Vacuum/thin-fluid interface	Elasticity coefficients of 2 interfaces Viscosities of 2 fluids Surface tensions at 2 interfaces Shear viscosities at 2 interfaces 1 Thin-fluid thickness

in this paper. In some cases, the reference listed is not the original reference, but instead one that provides a literature review of the theoretical development as well as the formulae used in this paper. The theoretical models under consideration in this paper share a common assumption, that dissipation is a linear effect. We consider no models that include nonlinear dissipation. In Sect. 3 we describe laboratory experiments using waves with frequencies from 1–4 Hz, in which we vary the condition of the air/water interface. The measurements of resulting dissipation rates are reported in Sect. 4 with comparisons from the various models. The experiments include (1) clean-surface experiments, in which we clean contamination off of the air/water interface and measure dissipation rates as a function of frequency; (2) exposed-surface experiments, in which we clean the air/water interface and then leave it exposed to the ambient atmosphere, measuring dissipation rates as a function of surface age for two frequencies; (3) oil-contaminated experiments, in which we clean the air/water interface and then add known amounts of olive oil, measuring dissipation rates as a function of concentration of oil for two frequencies; and (4) cling-wrap experiments, in which we clean the air/water interface and then coat it with sheets of thin plastic wrap, measuring dissipation rates as a function of frequency. Our conclusions are listed in Sect. 5. In brief, we find that the clean-interface models are adequate for laboratory waves with frequencies less than about 3 Hz if the air/water interface is cleaned; the model that assumes a thin layer of viscoelastic fluid at the vacuum/fluid interface is required to predict the larger dissipation rates observed when the air/water interface is exposed to the ambient atmosphere for several days, but is inadequate to predict the largest dissipation rates, measured with waves on an oil-contaminated interface; and the surfactant-model limit that assumes a resonance between elastic waves and the underlying fluid best predicts the dissipation rates of waves when the air/water interface is covered with plastic wrap. Dissipation rates for ocean swell are most accurately predicted by the surfactant model in the limit of infinite elasticity.

2 Models of Dissipation Rates

Here we present formulae for the predicted spatial rate, δ, of amplitude dissipation from the models listed in Table 1, which assume various interfacial boundary conditions. So, assume the amplitude, a, decays like $a = a_0 e^{-\delta x}$. For comparisons of predictions with measured results from the laboratory wavetank, we include the contributions from the sidewall and bottom boundary layers.

2.1 *Sidewall and Bottom Boundary-Layer Model*

Dissipation due to boundary layers along the sidewalls and bottom of the fluid domain do not affect ocean waves on deep water. However, these boundary layers are an important source of dissipation for waves in wavetanks, so we include the contribution, worked out by van Dorn [24], here. He showed that the dissipation rate due to boundary layers at the sidewalls and bottom of the tank for waves with wavenumber, k, and frequency, ω, at a vacuum/fluid interface in a tank of width W and fluid depth h is

$$\delta_{sb} = \left(\frac{\nu}{2\omega}\right)^{1/2}\left(\frac{2|k|}{W}\right)\left(\frac{|k|W + \sinh(2|k|h)}{2|k|h + \sinh(2|k|h)}\right), \tag{1}$$

where ν is the kinematic viscosity of the fluid. The wavenumber, $k \in \mathbb{R}$, can be positive or negative, depending on the direction of propagation, while $\omega > 0$.

In the remainder of this section, we present models for dissipation that are due to conditions at the air/water interface. For laboratory experiments, the total dissipation rate is the sum of δ_{sb} and the rate due to the movable surface.

2.2 *Clean-Surface Model*

The clean-surface model was first worked out by Lamb [15], §349, and predicts the damping rate of waves at an interface between a fluid with weak viscosity and a vacuum. The name "clean surface" comes from the dynamic boundary conditions at the interface, which are that both the normal and tangential components of stress are zero there. The resulting spatial dissipation rate is

$$\delta_c = 4\frac{\nu|k|^3}{\omega(k)}. \tag{2}$$

This model is commonly used in practice. For example, Dias et al. [6] used it to derive a dissipative nonlinear Schrödinger equation from the Navier-Stokes equations. Lo and Mei [16] stated that it is the model for waves "in the open ocean or a very wide tank." However, Henderson and Segur [10] (HS) used previously published observations of ocean data to show that predictions using (2) do not agree with observed dissipation rates of ocean swell, as noted in Sect. 1. For laboratory data, for which the frequencies are typically 1 Hz and higher, this model does reasonably well if the air/water interface is cleaned (as described, for example in Sect. 3). However, as shown in Sect. 4, when the interface ages, or if it has a contaminating film, then this model significantly underpredicts observed dissipation rates.

2.3 *Two-Fluid Model*

The clean-surface model in Sect. 2.2 ignores the dynamics of the air above the water. Dore [7] took into account energy loss to the air in the two-fluid model, which predicts the damping rate of waves at an interface between two fluids with weak viscosities. The dynamic boundary conditions at the air/water interface are that both the normal and tangential components of stress are continuous there. Dore found that for waves with lengths appropriate for ocean swell, the effects of air are significant, increasing the dissipation rate over that predicted by the clean-surface model by two orders of magnitude. Dore's prediction for dissipation rate is

$$\delta_{2fluid} = \left[\sqrt{2} \frac{\rho_a}{\rho_w} \left(\nu_a k^2 \right)^{1/2} \left(g|k| \right)^{1/4} + 2\nu_w k^2 \right] \left(2\sqrt{\frac{|k|}{g}} \right), \tag{3}$$

where $\rho_{a/w}$ is the density of the air/water, and $\nu_{a/w}$ is the kinematic viscosity of the air/water. Although the prediction of dissipation rate from the two-fluid model is a substantial improvement over predictions from the clean-surface model for waves on the scale of ocean swell, HS [10] found that it nevertheless also underpredicts observed dissipation rates of ocean swell.

2.4 Inextensible-Surface Model

Recognizing that in nature the interface between air and water rarely behaves as if "clean" (as required in Sects. 2.2 and 2.3) without special care to approximate that condition, Lamb [15], §351, also derived a model that assumed a "rigid-lid" surface, which has also been called a "fully contaminated" or "inextensible" surface. In this model, the vacuum/water interface can oscillate vertically as a rigid lid; it cannot stretch horizontally. So the dynamic boundary conditions at the inextensible surface are that the normal stress vanishes, but the shear stress is unconstrained. Instead, there is a no-slip condition that causes the tangential velocity to vanish. HS [10] generalized Lamb's result to include finite depth; their prediction for dissipation rate is

$$\delta_{in} = \sqrt{\frac{\nu}{2\omega}} \left[\frac{2k^2 \cosh^2(|k|h)}{2|k|h + \sinh(2|k|h)} \right]. \tag{4}$$

HS [10] showed that this model agrees well with observed dissipation rates of ocean swell, but pointed out that it cannot be used in a model that allows for nonlinearity. The condition of zero tangential velocity at the interface is inconsistent with wave motion with finite amplitude.

2.5 Surfactant-Surface Model

The inextensible-surface model of Sect. 2.4 describes a fully-contaminated surface—one for which further contamination does not change the dissipation rate. The surfactant-surface model allows for partial contamination. It describes a system in which a viscoelastic, massless film with its own constitutive law is at the vacuum/fluid interface. Then the stress balance at the interface has to account for the rheological properties of the film. Miles [18] reviewed the development of the corresponding prediction for the resulting dissipation rate. In its generality, the prediction takes into account the elasticity of the surfactant surface, the shear and dilational viscosities of the surfactant surface and the solubility of the surfactant. The elasticity is due to gradients in surface tension that arise when the film is stretched and compressed due to the presence of waves, so that the coefficient of elasticity is proportional to the gradient in surface tension with respect to the rest-concentration of the surfactant. The prediction for dissipation rate that takes into account a surfactant film is

$$\delta_s = \frac{|k|}{g}\sqrt{\frac{\nu\omega^3}{2}}\,\frac{\xi(\xi+\eta)+\zeta(\zeta+2)}{(\xi-1)^2+(1+\eta)^2+\zeta(\zeta+2)}, \tag{5}$$

where

$$\xi = \sqrt{2}\Big(\frac{k^2\chi}{\rho\sqrt{\nu\omega^3}}\Big), \tag{6}$$

$$\zeta = \sqrt{2}\Big(\frac{k^2(\nu_1+\nu_2)}{\rho\sqrt{\nu\omega}}\Big), \tag{7}$$

$$\eta = \frac{(2D/\omega)^{1/2}}{(d\Gamma/d\gamma)_0}. \tag{8}$$

Here, ξ is a dimensionless elasticity parameter, ζ is a dimensionless viscous parameter, and η is a dimensionless solubility parameter. Then ν is the kinematic viscosity of the fluid; ρ is the fluid density; $\chi = -\Gamma_0(dT/d\Gamma)_0$ is a measure of the elasticity, which arises from gradients in (dynamic) surface tension, T, due to changes in surfactant concentration, Γ, on the vacuum/water interface; the subscript, 0, indicates the interface at rest; ν_1 and ν_2 are the dilational and shear viscosities of the surfactant film; D, the bulk diffusion coefficient, is a measure of solubility of the surfactant, and γ is a measure of the bulk concentration. This model assumes that $D \ll \nu$, so that bulk diffusion is confined to a thin boundary layer; and $\eta\zeta \ll 1$, which is valid for both capillary and gravity waves.

The prediction, (5), for dissipation rate works fairly well in a controlled laboratory setting when one is able to clean the air/water interface and apply known quantities of a monomolecular film. For example, see Davies and Vose [5] or [12].

There are two limits that are of particular interest. One limit corresponds to the inextensible model discussed in Sect. 2.4. It is obtained by considering an infinite elasticity for which $\xi \to \infty$. The other limit corresponds to a resonance between the water wave and elastic waves in the surfactant film. This limit gives the maximum decay rate possible for a water/vacuum system with a surfactant-film surface and is discussed in Sect. 2.6.

2.6 Surfactant-Maximum Model

Miles [18] showed that the maximum dissipation rate possible from the surfactant-surface model described in Sect. 2.5 is twice that obtained from the inextensible surface model described in Sect. 2.4. This result was also found by others. In the notation of Huhnerfuss et al. [13], the prediction for maximum dissipation rate of gravity waves due to a surfactant film at the vacuum/water interface is

$$\delta_{sm} = \sqrt{2}\left[\frac{\nu^2|k|^7}{g}\right]^{1/4}. \tag{9}$$

This model underpredicts measured rates in laboratory experiments using aged interfaces and oil-covered interfaces, but predicts decay rates in reasonable agreement with experiments that used a cling-wrap covered interface as discussed in Sect. 4.

2.7 Thin-Layer Model

Jenkins and Jacobs [14] generalized the surfactant-surface model discussed in Sect. 2.5 to consider a two-fluid system that consists of two interfaces: an interface between a weakly viscous fluid and a thin fluid layer (with mass and dynamics); and an interface between the thin fluid layer and a vacuum above it. Each interface has its own elasticity and shear viscosity. The dynamic boundary conditions at the two interfaces are continuity of normal and tangential stresses. The prediction of dissipation rate due to this model is

$$\delta = \sqrt{g|k|}\,\mathrm{Re}\left(\frac{\delta_1+\delta_2}{\delta_3}\right)/C_g(k), \tag{10}$$

where the (nondimensional) parameters are

$$\delta_1 = 2\epsilon^2(k) + \mathrm{i}\,d(k)\Big[\frac{\gamma(k)(1-\rho_f)-\gamma_w(k)}{\sqrt{\Gamma(k)}}\Big] + \Big[\frac{\nu_T(k)\rho_f d(k)\Gamma^{1/4}(k)}{2\epsilon(k)\sqrt{\mathrm{i}}}\Big], \tag{11}$$

$$\delta_2 = \frac{1}{2}\nu_T(k) + \frac{1}{2\epsilon(k)}\sqrt{\mathrm{i}}\,\rho_f^2 d^2(k)\Gamma^{3/4}(k)(R^2(k)-1), \tag{12}$$

$$\delta_3 = 1 + \frac{\sqrt{\mathrm{i}}\,\nu_T(k)}{\epsilon(k)\Gamma^{1/4}(k)} + \frac{\rho_f d(k)\Gamma^{1/4}(k)}{\sqrt{\mathrm{i}}\,\epsilon(k)} \tag{13}$$

$$\epsilon(k) = \sqrt{\frac{|k|^{3/2}\hat{\nu}_w}{g^{1/2}}}, \tag{14}$$

$$\gamma_f(k) = \hat{\gamma}_f \frac{k^2}{g\hat{\rho}_w}, \tag{15}$$

$$\gamma_w(k) = \hat{\gamma}_w \frac{k^2}{g\hat{\rho}_w}, \tag{16}$$

$$n(k) = -\mathrm{i}\sqrt{1+\gamma(k)} - 2\epsilon^2(k), \tag{17}$$

$$\nu_{sf}(k) = \hat{\nu}_{sf} \frac{|k|^{5/2}}{g^{1/2}\hat{\rho}_w}, \tag{18}$$

$$\nu_{sw}(k) = \hat{\nu}_{sw} \frac{|k|^{5/2}}{g^{1/2}\hat{\rho}_w}, \tag{19}$$

$$\nu_T(k) = \left(\frac{\chi_w(k) + \chi_f(k)}{n(k)}\right) + \nu_{sf}(k) + \nu_{sw}(k) + (4\rho_f \nu_f(k) d(k)) \tag{20}$$

$$+\left[\frac{\nu_{Ef}(k)\nu_{Ew}(k)d(k)}{\rho_f \nu_f(k)}\right], \tag{21}$$

$$\gamma(k) = \gamma_w(k) + \gamma_f(k), \tag{22}$$

$$\Gamma(k) = \sqrt{1+\gamma(k)}, \tag{23}$$

$$R(k) = \frac{\rho_f + \gamma_f(k)}{\rho_f \Gamma(k)}, \tag{24}$$

$$\chi_f(k) = \hat{\chi}_f \frac{k^2}{g\hat{\rho}_w}, \tag{25}$$

$$\chi_w(k) = \hat{\chi}_w \frac{k^2}{g\hat{\rho}_w}, \tag{26}$$

$$\nu_{Ef}(k) = \frac{\chi_f(k)}{n(k)} + \nu_{sf}, \tag{27}$$

$$\nu_{Ew}(k) = \frac{\chi_w(k)}{n(k)} + \nu_{sw}, \tag{28}$$

$$\sqrt{\mathrm{i}} = +\frac{\sqrt{2}}{2}(1+\mathrm{i}). \tag{29}$$

Here Re indicates the real part; $d(k) = \hat{d}k$ is a nondimensional thickness of the thin layer with dimensional thickness $\hat{d}$; $\hat{\nu}_w$ and $\hat{\nu}_f$ are the dimensional kinematic viscosities of the bottom fluid and thin layer, respectively; $\hat{\rho}_w$ and $\hat{\rho}_f$ are the dimensional densities of the bottom fluid and thin layer; and $\rho_f = \hat{\rho}_f/\hat{\rho}_w$ and $\rho_w = \hat{\rho}_w/\hat{\rho}_w = 1$. The interface between the bottom fluid and thin layer has a dimensional (dynamic) surface tension $\hat{\gamma}_w$, a dimensional elasticity $\hat{\chi}_w$, and a dimensional shear viscosity $\hat{\nu}_{sw}$. The interface between the thin layer and the vacuum above has a dimensional surface tension $\hat{\gamma}_f$ a dimensional elasticity $\hat{\chi}_f$, and a dimensional shear viscosity $\hat{\nu}_{sf}$. To make (10) a spatial decay rate, we have divided by the group velocity, $C_g(k) = d\omega(k)/dk$. For computations used in sect. 4, we used the inviscid dispersion relation that includes surface tension, so that $C_g(k) = (g + 3k^2\hat{\gamma}_w/\hat{\rho}_w)/(2\sqrt{gk + k^3\hat{\gamma}_w/\hat{\rho}_w})$.

This model is difficult to apply unless one is able to measure all of the rheological properties of the fluids and surfaces. Nevertheless, one can ask—does this model provide the capability to predict the large dissipation rates that are observed in the experiments discussed in Sect. 4?

3 Experimental Facility

Experiments were conducted in a wave channel that is 50 ft long, $W = 10$ in wide, and 1 ft deep. The tank was cleaned with alcohol and filled with tap water to a depth of $h > 20$ cm. Typically, the air/water interface was then cleaned by blowing wind on the water surface at one end of the tank. The wind set up a surface current that pushed a thin, top layer of the water down the tank. This top layer was vacuumed with a wet-vac at the other end of the tank (50 ft away from the source) to a depth of $h = 20$ cm, which was measured with a Lory Type C point gage.

Waves were generated with a plunger-type paddle with a triangular cross-section attached to a ball-screw, which is attached to a motor. The motor was programmed using LabView software to oscillate at a particular frequency. Time series of surface displacement were measured using fifteen, capacitance-type surface wave gages separated by 43 cm in the direction of wave propagation. This coverage corresponds to about 600 cm, which was chosen to avoid reflections from the end wall. Each gage consisted of a coated-wire probe connected to an oscillator. The difference frequency between this oscillator and a fixed oscillator was read by a field programmable gate array (FPGA), NI PCI-7833R. Thus, no D/A conversion, filtering, or A/D conversion was required. The surface capacitance gage was held in a rack on wheels that are attached to a programmable belt. We calibrated the capacitance gage by traversing the rack at a known speed over a precisely machined, trapezoidally shaped speed bump that is 1 cm high. The amplitudes of waves generated on a clean surface as measured at the first gage site and the corresponding wave slopes are listed in Table 2.

Table 2 Frequencies, amplitudes and corresponding slopes of waves measured at the first gage site from the experiments using a clean surface

f_0 (Hz)	a_0 (cm)	a_0k
4.00	0.050	0.032
3.50	0.077	0.038
3.33	0.12	0.053
3.00	0.082	0.030
2.50	0.12	0.030
2.00	0.14	0.022
1.50	0.11	0.010

We used a Fourier Transform of the time-series to obtain its L_2 norm. We fit a line to the logarithm of these values as a function of distance from the wavemaker. The dissipation rate then corresponds to half the line's slope.

In Sect. 4 we present measurements of dissipation rates and comparisons with predictions discussed in Sect. 2 for four types of experiments: (1) clean-surface experiments; (2) aged-surface experiments; (3) oil-surface experiments and (4) experiments in which a plastic wrap (Glad Cling Wrap) was laid on the air/water interface.

For the clean-surface experiments discussed in Sect. 4.1, the dissipation rates were measured within two hours of cleaning during which time the frequency of the waves was varied.

For the aged-surface experiments discussed in Sect. 4.2, the dissipation rates were measured for 2 and 3.33 Hz waves during a time period of about a week. The water surface was exposed to the laboratory environment—nothing was done to treat the air/water interface. At about 100 h, we cleaned the air/water interface to see if the measured dissipation rates would return to the values measured for the clean interface. Then the interface was allowed to age again without further cleaning the water or tank.

For the oil-surface experiments discussed in Sect. 4.3, the dissipation rates were measured for a clean air/water interface and then for interfaces that had olive oil added. We used a micropipette to add to the air/water interface drops of oil with known (equal) volumes at fifteen evenly spaced locations that spanned the long dimension of the tank along the centerline of the tank. We measured dissipation rates, and then added another set of fifteen drops. We did this five times. The addition of olive oil decreases the surface tension at the interface—a gradient in surface tension causes a flow along the interface, so theoretically, the oil should have spread to a uniform thickness. However, after the first about three sets of drops were added, we could see nonuniformities in the oil-film thickness.

For the cling wrap-surface experiments discussed in Sect. 4.4, the air/water interface was cleaned and then covered with pieces of cling wrap that each spanned the width of the tank. So, there were small gaps between the pieces of cling wrap, and each sheet had some wrinkles. The sheets were not taut. The wave gages had to fit between sheets, so we could not spread one long sheet of cling wrap that spanned the tank width, down the length of the air/water interface.

4 Results

Here we present measurements of dissipation rates of surface waves with frequencies 1–4 Hz and compare them with the models discussed in Sect. 2. In the experiments, we varied the properties of the air/water interface to observe the corresponding effect on dissipation rates. The first set of experiments used a "clean" air/water interface, where we defined "clean" in Sect. 3. The second set used a clean interface that was allowed to age with no additional treatment. The third set used a clean interface to which measured amounts of olive oil were added. The fourth set used a clean interface that had sheets of cling wrap laid on top. In all cases, the predictions include two parts: a part that incorporates sidewall and bottom boundary layers, (1), and a part that takes into account one of the models of surface dissipation (2)–(10). We also present observations and comparisons with models of dissipation rates from the ocean.

In experiments with an aged surface (Sect. 4.2), oil-contaminated surface (Sect. 4.3), and cling-wrap surface, (Sect. 4.4), the measured dissipation rates were larger than those predicted by the models that assume either a clean surface (Sects. 2.2 and 2.3) or a surfactant-covered surface (Sects. 2.4 and 2.6). For these comparisons, we also used the model that assumes a thin layer of fluid at the surface (Sect. 2.7). The values used for the rheological properties required by that model are listed in Table 3.

4.1 Clean Surface

Figure 1 shows measured and predicted dissipation rates when the surface was cleaned as described in Sect. 3. (The data are also discussed in [10].) The models that are most likely to apply to this situation are the clean-surface model discussed in Sect. 2.2 and the (clean-surface) two-fluid model discussed in Sect. 2.3. The clean-surface model (2) and the two-fluid model (3) predict almost the same rates for these frequencies, so one assumes that for these frequencies, dissipation due to air dynamics is not significant. These models' assumption of a clean surface underpredicts somewhat the measured dissipation rates, with the discrepancy increasing with increasing frequency. Presumably, there is some contamination on the surface, regardless of our effort at cleaning, but not enough to cause the extreme condition of no tangential flow required by the inextensible surface model (4), which over-predicts the dissipation rates.

Table 3 List of parameters used in (10) to obtain the dotted curves in Figs. 2b and 4b

Parameter	Value for surface age experiments	Value for surface oil experiments
$\hat{d}$ (cm)	0.012	0–0.0005 (black, dashed-dotted curve in Fig. 4b)
		0–0.013 (gray, dashed-dotted curve in Fig. 4b)
$\hat{\nu}_w$ (m^2/s)	1×10^{-6}	1×10^{-6}
$\hat{\nu}_f$ (m^2/s)	1×10^{-4}	8.7×10^{-6}
$\hat{\rho}_w$ (kg/m^3)	998	998
$\hat{\rho}_f$ (kg/m^3)	900	915
$\hat{\gamma}_w$ (mN/m)	25–73	30
$\hat{\gamma}_f$ (mN/m)	30	32
$\hat{\chi}_w$ (mN/m)	30	30
$\hat{\chi}_f$ (mN/m)	30	30
$\hat{\upsilon}_{sw}$ (kg/s)	0	0
$\hat{\upsilon}_{sf}$ (kg/s)	0	0

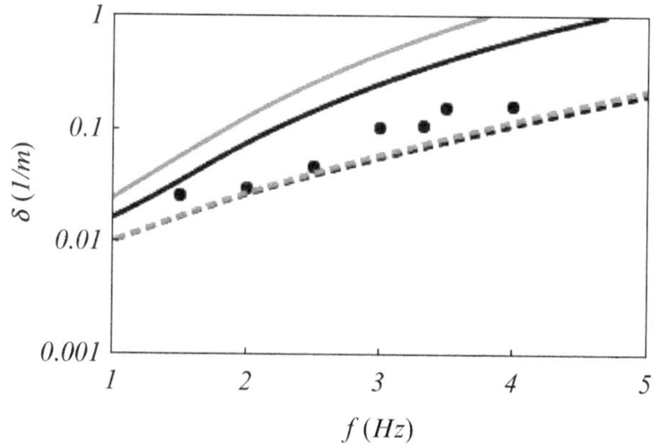

Fig. 1 Dissipation rates as a function of frequency for the clean-surface experiments. The *dots* are measurements. The *black dashed curve* is from the clean-surface model, (2), plus (1). The *gray dashed curve* (almost identical to the *black dashed curve*) is from the two-fluid model, (3), plus (1). The *black solid curve* is from the inextensible surface model, (4), plus (1). The *gray solid curve* is from the surfactant maximum model, (9), plus (1)

4.2 Exposed Surface

Figure 2 shows measured and predicted dissipation rates for 2 and 3.33 Hz waves when the surface was cleaned as described in Sect. 3 and then exposed to the laboratory atmosphere for several days. The models that are most likely to apply to this situation are the clean-surface model discussed in Sect. 2.2 and the (clean-surface) two-fluid model discussed in Sect. 2.3 for small surface ages; the surfactant-surface model discussed in Sect. 2.5 and the surfactant-maximum model discussed in Sect. 2.6 for larger surface ages; and the inextensible-surface model discussed in Sect. 2.4 for large surface ages when the contamination becomes saturated in some sense.

Figure 2a shows that the dissipation rate for the 2 Hz waves was fairly insensitive to surface contamination. However, Fig. 2b shows that after about 1 day of exposure, the dissipation rate for the 3.33 Hz waves increased by an order of magnitude to a value much larger than predictions from all of these models.

Does the thin-layer model discussed in Sect. 2.7 apply to this experimental configuration? One imagines that for small surface ages, the surface is contaminated with dust and particulate matter that rises from the bulk of the fluid. However, after a surface age of about 2 days, there appears to be a continuous film at the surface, which could be biological in nature and could behave as a thin layer of fluid. After several days, the film is quite visible. If one pushes the surface horizontally with a vertical plate in the water, one sees short-wavelength (a few millimeters) waves in the film. If the plate is removed, the film bounces back. If one continues to push the plate, the film buckles and may tear. Thus, we were interested to determine if we could find rheological properties that when used in (10) would result in a prediction from the thin-layer model that is about as large as the largest dissipation rate measured.

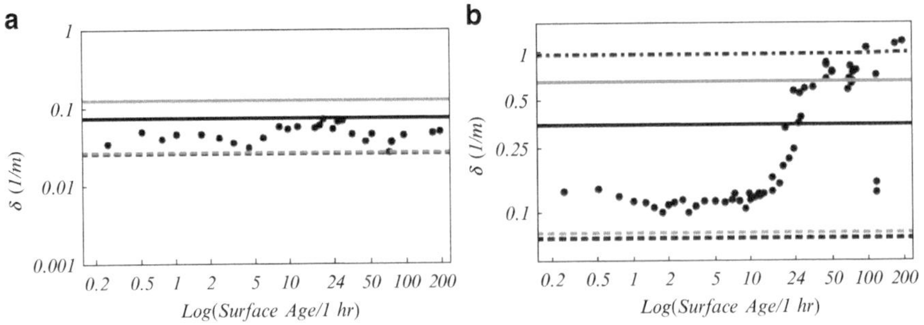

Fig. 2 Dissipation rates of a (**a**) 2Hz wave and (**b**) 3.33 Hz wave as a function of the time a cleaned surface was exposed to ambient contamination. The *dots* are measurements. The *black dashed curve* is from the clean-surface model, (2), plus (1). The *gray dashed curve* (almost identical to the *black dashed curve*) is from the two-fluid model, (3), plus (1). The *black solid curve* is from the inextensible-surface model, (4), plus (1). The *gray solid curve* is from the surfactant-maximum model, (9), plus (1). In (**b**) the *dashed-dotted curve* is from the thin-layer model, (10), plus (1) that assumes an interfacial fluid thickness of 0.12 mm. See Table 3 for other values used. At about 100 h, the surface was cleaned

We do not know the rheological properties of the contamination at the air/water interface when it is exposed to the atmosphere. We did measure the surface tension of the air/water interface from water that was in a beaker in the same room as the wavetank—these measurements are shown in Fig. 3. Each time we measured dissipation rates, we also measured the surface tension from the beaker system using a DuNuoy-type tensiometer. One cannot conclude that the surface tension from the beaker system was the same as that from the wavetank system. For example, the film observed on the air/water interface in the tank did not grow on the air/water interface in the beaker. The beaker surface was initially cleaned by vacuuming off the surface with a pipette attached to a pump, and the measured value of surface tension agrees with the theoretical value for a clean surface at the measured temperature. The value decreased with age to about 58 mN/m at 114.75 h. When we cleaned the surface again at 115.25 h, the measured value of surface tension again agreed with the theoretical value for a clean surface. Then after waiting another 116.42 h, the surface tension dropped to about 68 mN/m, not as low as the previous value of 58 mN/m. Apparently, the film that grows with surface age is not reproducible, so that after about 10 h, the measured value of surface tension shows scatter.

It is not clear how to translate the surface tension measurements into coefficients of elasticity to apply the thin-layer model; however, Jenkins and Jacobs [14] give a range of surface elasticities of 10–50 mN/m for surfactant materials on water. To compute a dissipation rate using (10), we looked at this range of elasticities: the maximum dissipation rate we were able to compute did not depend significantly on the choice of $\hat{\chi}_f$. It did depend significantly on the choice of $\hat{\chi}_w$. The value of $\hat{\chi}_w$ that gives the largest dissipation rate is typically not at the limits of this elasticity range, but at the mid-values. The dissipation rate is fairly insensitive to the choice of surface tension at both interfaces, and following Jenkins and Jacobs,

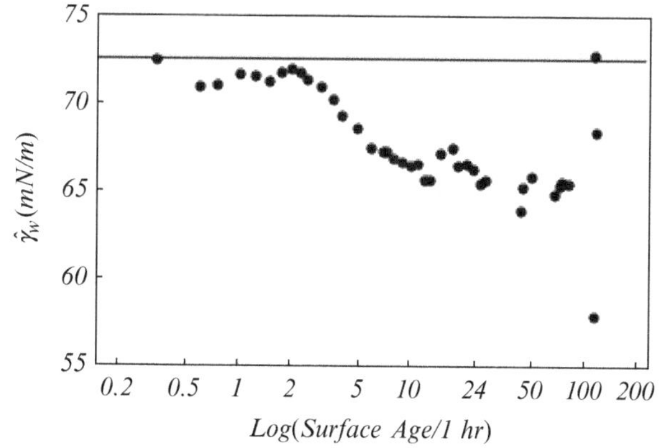

Fig. 3 Measured values of surface tension (*dots*) as a function of surface age at the surface of air and the water in a beaker. The *horizontal line* is the value corresponding to a clean surface at 21.3° C. At about 100 h, the surface was cleaned

we neglected the interfacial viscosities, $\hat{\upsilon}_{sf}$ and $\hat{\upsilon}_{sw}$. However, varying the thickness of the layer makes a significant difference. For the values of viscosities, elasticities, surface tensions, and densities listed in Table 3, we found a layer thickness (also listed in Table 3), at which a maximum dissipation rate resulted. Figure 2b shows that this maximum is quite close to the largest dissipation rate measured when the surface age was large.

To obtain the dashed-dotted curve in Fig. 2b, which shows predictions from the thin-layer model (10), we varied the value of surface tension at the fluid/water interface, $\hat{\gamma}_w$, used in (10). The variable $\hat{\gamma}_w$ allows for the changing surface tension with surface age that occurs in the experiments. However, the prediction of dissipation rate is insensitive to the variable surface tension value, as shown by the almost constant dashed-dot curve.

The discontinuous jump in the measured dissipation rate in Fig. 2b at about 100 h corresponds to the surface's being cleaned at that time. This return to the value measured for the initially clean surface provides further evidence that the enhanced dissipation is due to surface contamination, as opposed to bulk fluid properties or sidewall boundary layers.

4.3 Oil-Contaminated Surface

Figure 4 shows measured and predicted dissipation rates for 2 and 3.33 Hz waves when the surface was cleaned as described in Sect. 3 and then oil was added as described in Sect. 3. The models that are most likely to apply to this situation are the clean-surface model discussed in Sect. 2.2 and the (clean-surface) two-fluid model discussed in Sect. 2.3 for the clean-surface before oil was added; the surfactant-

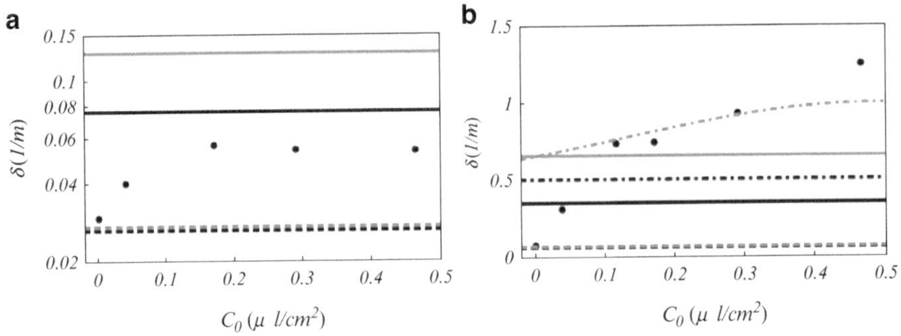

Fig. 4 Dissipation rates of a (**a**) 2-Hz wave and (**b**) 3.33 Hz wave as a function of the amount of oil added per unit area of cleaned surface. The *dots* are measurements. The *black dashed curve* is from the clean-surface model, (2), plus (1). The *gray dashed curve* (almost identical to the *black dashed curve*) is from the two-fluid model, (3), plus (1). The *black solid curve* is from the inextensible-surface model, (4), plus (1). The *gray solid curve* is from the surfactant-maximum model, (9), plus (1). In (**b**) the *dashed-dotted black curve* is from the thin-layer model, (10), plus (1) that assumes a thin layer with varying thickness that corresponded to that of the experiment; the *dashed-dotted gray curve* is from the thin-layer model, (10), plus (1) that assumes a thin layer with varying thickness that was chosen to increase the predicted dissipation rate. See Table 3 for the values used

model discussed in Sect. 2.5 and the surfactant-maximum model discussed in Sect. 2.6 for the surfaces with non-saturated levels of oil; the inextensible-model discussed in Sect. 2.4 for the oil-saturated surfaces, which occur when the oil that is added no longer spreads; and the thin-fluid model discussed in Sect. 2.7 for the addition of any amount of oil.

Figure 4a shows that the dissipation rate for the 2 Hz waves increased by about a factor of 2 due to the addition of the oil. Figure 4b shows that the dissipation rate for the 3.33 Hz waves increased by an order of magnitude to a value much larger than predictions from all of the models except perhaps the one that takes into account a thin layer of fluid at the interface between the water and a vacuum (Sect. 2.7); although, even this model could not account for the largest dissipation rate measured.

Figure 4b shows two curves from the thin-layer model. The black, dot-dashed curve was computed using different film thicknesses, $\hat{d}$, equal to the ratio of the volume of the fifteen drops and the surface area in the tank. (See Table 3.) However, as discussed in Sect. 3, after the first set of fifteen drops were applied to the surface, the additional sets of drops did not spread uniformly in thickness. Thus, the gray, dot-dashed curve was computed using different layer thicknesses, $\hat{d}$, from zero to just above the value, $\hat{d} = 0.012$ cm, that gives the maximum dissipation rate that we found for this model. The thin-layer model assumes an interfacial layer of fluid with a uniform thickness, which we do not think was the case for the experiments; so comparing its predictions with our data is pushing the assumptions of the model. We used a value of viscosity for olive oil at room temperature consistent with

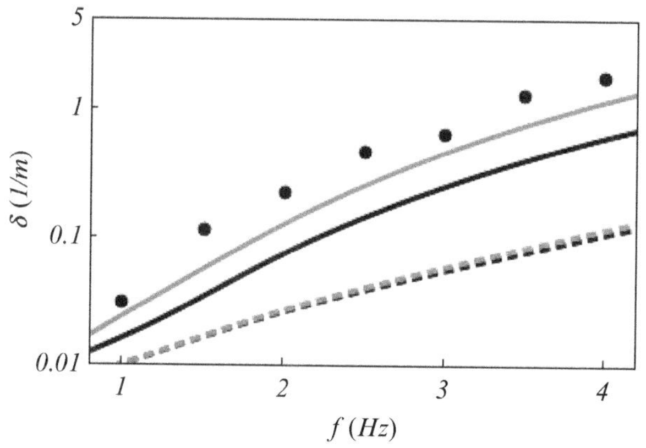

Fig. 5 Dissipation rates as a function of frequency for the cling wrap surface experiments. The *dots* are measurements. The *black dashed curve* is from the clean-surface model, (2), plus (1). The *gray dashed curve* (almost identical to the *black dashed curve*) is from the two-fluid model, (3), plus (1). The *black solid curve* is from the inextensible surface model, (4), plus (1). The *gray solid curve* is from the surfactant maximum model, (9), plus (1)

measurements reported in [3]. We used a value of density for olive oil at room temperature consistent with values reported in

http://www.engineeringtoolbox.com/liquids-densities-d_743.html. We used a value of surface tension between the olive oil and air at room temperature consistent with values reported in

http://physics.about.com/od/physicsexperiments/a/surfacetension_5.htm.

4.4 Cling Wrap Covered Surface

Figure 5 shows measured and predicted dissipation rates for waves with seven frequencies spread evenly between 1 and 4 Hz when the surface was cleaned as described in Sect. 3 and then covered with sheets of cling wrap. The models that are most likely to apply to this situation are the inextensible-surface model discussed in Sect. 2.4 and the surfactant-maximum model discussed in Sect. 2.6. Both of these models underpredict the measured dissipation rate, but predictions from the surfactant-maximum model are fairly close. This model corresponds to a resonance between elastic waves of a surface film and the underlying water wave. Since the cling wrap had tiny wrinkles, it was allowed to stretch and compress, so that this resonance could potentially occur. The thin-layer model (Sect. 2.7) would not seem to apply to this experimental configuration. Nevertheless, we did try to find rheological parameters, along with measurements of the cling wraps' density and thickness, in (10) to see if that model could predict measured dissipation rates. We could not find a set of parameters that gave even qualitative agreement.

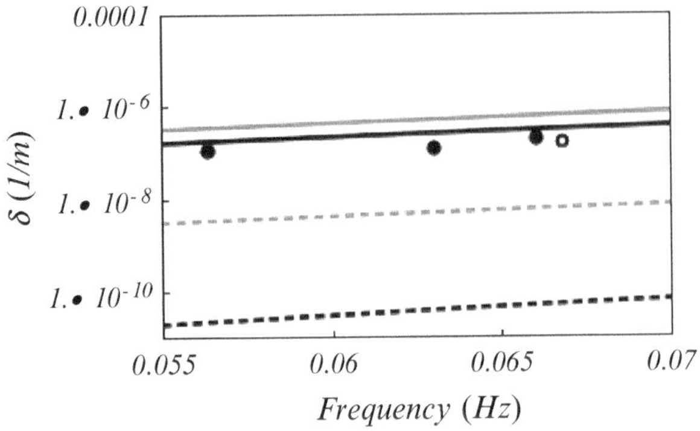

Fig. 6 Dissipation rates as a function of frequency for oceanographic observations. The *solid circles* are based on measurements from [22]. The *hollow circle* is a measurement from [4]. The *black dashed curve* is from the clean-surface model, (2). The *gray dashed curve* is from the two-fluid model, (3). The *black solid curve* is from the inextensible surface model, (4). The *gray solid curve* is from the surfactant maximum model, (9)

4.5 Oceanographic Observations

Figure 6 shows measured and predicted dissipation rates for ocean swell as reported in [10] from data measured by Snodgrass et al [22], and Collard et al [4]. For the surfactant maximum model, we used the deep-water limit of (Sect. 2.4), which is

$$\delta_{in\infty} = k^2 \sqrt{\frac{\nu}{2\omega(k)}}. \tag{30}$$

This inextensible-surface model (Sect. 2.4) does the best job of predicting the measurements. So, it can be useful for taking into account dissipation in linear models of ocean swell. It does not generalize to nonlinear models, but it is a limit of both the surfactant-surface model (Sect. 2.5) and the thin-layer model (Sect. 2.7), which can be generalized to nonlinear models. Of note is that, as pointed out in [10], the observed dissipation rates of ocean swell are orders of magnitude larger than predicted by the clean-surface, vacuum/water model, so that model (Sect. 2.2), simply does not apply to ocean swell. The two-fluid model, (Sect. 2.3), shows that on the scale of ocean swell, the dynamics of the air are important, but even this model underpredicts measurements by a couple of orders of magnitude. So, surface dynamics are important in modeling dissipation of ocean swell.

5 Summary and Conclusions

In this paper we present several models for the dissipation rates of surface waves and compare their predictions with laboratory measurements and oceanographic observations. We make the following summary and conclusions.

- The clean-surface model is adequate for laboratory waves with "small enough" frequencies (less than about 3 Hz) if the surface is adequately cleaned (for example, as described in Sect. 3).
- The clean-surface two-fluid model predicts dissipation rates a couple of orders of magnitude larger than those of the (fluid/vacuum) clean surface model for waves on the scale of ocean swell. Thus, it is likely that air plays an important role in understanding the dynamics of ocean swell. Nevertheless, this model underpredicts observed dissipation rates of ocean swell. It is not clear what the "correct" boundary conditions are at the ocean-atmosphere surface, or what the "best" conditions are that could be used in practice to predict both dissipative behaviour and nonlinear evolution.
- We did not apply the surfactant model (of an infinitely thin, massless film), but instead looked at predictions from two of its limits.
  - The inextensible surface-model corresponds to the limit of infinite elasticity of the surfactant film. Its predictions agreed well with observations of dissipation rates of ocean swell. However, this model cannot be generalized to nonlinear models of ocean swell.
  - The surfactant-maximum model corresponds to a resonance between elastic waves in the surfactant film and the water waves. It can occur for surfactant concentrations less than saturation. It did the best job of predicting dissipation rates that occurred when the surface was covered with cling wrap that was allowed to wrinkle—thereby compressing and expanding.
- The thin-layer model is the most general of all of the models considered here. All of the other models discussed here except for Dore's [7] model in Sect. 2.3, which includes the effects of air, are limits that can be obtained from the thin-layer model. This model was able to predict the large damping rate observed in the 3.33 Hz waves propagating on a air/water interface that had been exposed to the ambient atmosphere for several days. Under this laboratory condition, a film grows on the air/water interface that may then be modeled as a thin viscoelastic fluid. This model was inadequate in predicting the largest dissipation rates observed with waves propagating on a air/water interface that was contaminated with variable amounts of olive oil. One possible explanation for the discrepancies is that the olive oil did not spread uniformly. Other possibilities are also being investigated. We point out, though, that the thin-layer model has numerous free parameters, so it is difficult to use in practice.

The goal of this work is to find a model for dissipation that predicts all of the observed dissipation rates from the laboratory and the ocean, and could be used in nonlinear models of wave evolution. We find that different models are useful for different situations, as discussed above, and that none of the models is adequate when we add over-saturated amounts of olive oil to the surface.

Acknowledgements This work was supported in part by the National Science Foundation, DMS-1107379 and DMS-1107354.

References

1. Benjamin, B., Feir, F.: The disintegration of wavetrains in deep water. Part 1. J. Fluid Mech. **27**, 417–430 (1967)
2. Benney, D., Newell, A.: The propagation of nonlinear wave envelopes. J. Math. Phys. Stud. Appl. Math. **46**, 133–139 (1967)
3. Bonnet, J.-P., Devesvre, L., Artaud, J., Moulin, P.: Dynamic viscosity of olive oil as a function of composition and temperature: a first approach. Eur. J. Lipid Sci. Technol. **113**, 1019–1025 (2011)
4. Collard, F., Ardhuin, F., Chapron, B: Monitoring and analysis of ocean swell fields from space: new methods for routine observations. J. Geophys. Res. **114**, C07023 (2009)
5. Davies, J.T., Vose, R.M.: On the damping of capillary waves by surface films. Proc. Roy. Soc. Lond. **A286**, 218 (1965)
6. Dias, F., Dyachenko, A., Zakharov, V.: Theory of weakly damped free-surface flows: a new formulation based on potential flow solutions. Phys. Lett. **A372**, 1297–1302 (2008)
7. Dore, B.: Some effects of the air-water interface on gravity waves. Geophys. Astophys. Fluid Dyn. **10**, 213–230 (1978)
8. Dysthe, K.: Note on a modification to the nonlinear Schrödinger equation for application to deep water waves. Proc. Roy. Soc. Lond. **A369**, 105–114 (1979)
9. Gramstad, O., Trulsen, K.: Fourth-order coupled nonlinear Schrödinger equations for gravity waves on deep water. Phys. Fluids. **23**, 062102–062111 (2011)
10. Henderson, D., Segur, H.: The role of dissipation in the evolution of ocean swell. J. Geo. Res: Oceans **118**, 5074–5091 (2013)
11. Henderson, D., Segur, H., Carter, J.: Experimental evidence of stable wave patterns on deep water. J. Fluid Mech. **658**, 247–278 (2010)
12. Henderson, D.M.: Effects of surfactants on Faraday-wave dynamics. J. Fluid Mech. **365**, 89–107 (1998)
13. Huhnerfuss, H., Lange, P.A., Walters, W.: Relaxation effects in monolayers and their contribution to water wave damping. II. The Marangoni phenomenon and gravity wave attenuation. J. Coll. Int. Sci. **108**, 442–450 (1985)
14. Jenkins, A.D., Jacobs, S.J.: Wave damping by a thin layer of viscous fluid. Phys. Fluids **9**, 1256–1264 (1997)
15. Lamb, H.: Hydrodynamics, 6th edn. Dover, New York (1932)
16. Lo, E., Mei, C.C.: A numerical study of water-wave modulations based on a higher-order nonlinear Schrödinger equation. J. Fluid Mech. **150**, 395–415 (1985)
17. Ma, Y., Dong, G., Perlin, M., Ma, X., Wang, G.: Experimental investigation on the evolution of the modulation instability with dissipation. J. Fluid Mech. **711**, 101–121 (2012)
18. Miles, J.: Surface-wave damping in closed basins. Proc. Roy. Soc. Lond. **A297**, 459–475 (1967)
19. Ostrovsky, L.: Propagation of wave packets and space-time self-focussing in a nonlinear medium. Sov. J. Exp. Theor. Phys. **24**, 797–800 (1967)
20. Segur, H., Henderson, D., Carter, J., Hammack, J., Li, C.-M., Pheiff, D., Socha, K.: Stabilizing the Benjamin-Feir instability. J. Fluid Mech. **539**, 229–271 (2005)
21. Stokes, G.G.: On the theory of oscillatory waves. Trans. Camb. Phil. Soc **8**, 441 (1847)

22. Snodgrass, F.E., Groves, G.W., Hasselmann, K., Miller, G.R., Munk, W.H., Powers, W.H.: Propagation of ocean swell across the Pacific. Phil. Trans. Roy. Soc. Lond. **A259**, 431–497 (1966)
23. Trulsen, K., Dysthe, K.: A modified nonlinear Schrödinger equation for broader bandwidth gravity waves on deep water. Wave Motion **24**, 281–289 (1996)
24. van Dorn, W.G.: Boundary dissipation of oscillatory waves. J. Fluid Mech. **24**, 769–779 (1966)
25. Wu, G., Liu, Y., Yue, D.: A note on stabilizing the Benjamin-Feir instability. J. Fluid Mech. **556**, 45–54 (2006)
26. Zakharov, V.: The wave stability in nonlinear media. Sov. J. Exp. Theor. Phys. **24**, 455–459 (1967)
27. Zakharov, V.: Stability of periodic waves of finite amplitude on the surface of a deep fluid. J. Appl. Mech. Tech. Phys. **2**, 190–194 (1968)

The Kelvin-Helmholtz Instabilities in Two-Fluids Shallow Water Models

David Lannes and Mei Ming

To Walter Craig, for his 60th birthday, with friendship and admiration.

Abstract The goal of this paper is to describe the formation of Kelvin-Helmholtz instabilities at the interface of two fluids of different densities and the ability of various shallow water models to reproduce correctly the formation of these instabilities.

Working first in the so called rigid lid case, we derive by a simple linear analysis an explicit condition for the stability of the low frequency modes of the interface perturbation, an expression for the critical wave number above which Kelvin-Helmholtz instabilities appear, and a condition for the stability of all modes when surface tension is present. Similar conditions are derived for several shallow water asymptotic models and compared with the values obtained for the full Euler equations. Noting the inability of these models to reproduce correctly the scenario of formation of Kelvin-Helmholtz instabilities, we derive new models that provide a perfect matching. A comparisons with experimental data is also provided.

Moreover, we briefly discuss the more complex case where the rigid lid is replaced by a free surface. In this configuration, it appears that some frequency modes are stable when the velocity jump at the interface is large enough; we explain why such stable modes do not appear in the rigid lid case.

D. Lannes (✉)
IMB, Université de Bordeaux et CNRS UMR 5251, 351 Cours de la Libération, 33405 Talence Cedex, France
e-mail: David.Lannes@math.u-bordeaux1.fr

M. Ming
School of Mathematics and Computational Science, Sun Yat-Sen University, 135 Xingangxi Street, Guangzhou 510275, China
e-mail: mingm@mail.sysu.edu.cn

P. Guyenne et al. (eds.), *Hamiltonian Partial Differential Equations and Applications*, Fields Institute Communications 75,
DOI 10.1007/978-1-4939-2950-4_7

1 Introduction

Describing the motion of the *surface* of a non viscous fluid of constant density is known as the water waves problems and is now quite well-understood. The equations are known to be well-posed for smooth data under the Rayleigh-Taylor criterion stating that the vertical derivative of the pressure must be negative at the surface

$$\text{(Rayleigh-Taylor criterion)} \qquad -\partial_z P_{|\text{surf}} > 0,$$

which is equivalent to saying that the downward acceleration of the fluid must not exceed gravity; away from singularities, this condition is known to be always satisfied [31, 42]. The solutions of the water waves equations are also known to be well-approximated by simpler asymptotic models in several asymptotic regimes. In particular, in the so called shallow water regime which is relevant for most applications to coastal oceanography or tsunami propagation for instance, many asymptotic models have been derived and justified, one of the pioneer works being Walter Craig's justification of the KdV approximation [17]. The so-called Nonlinear Shallow Water (or Saint-Venant) equations and the Green-Naghdi (or fully nonlinear Boussinesq, or Serre) equations are two examples of widely used models in the shallow water regime (in the sense that the depth is much smaller than the typical horizontal length); we refer to [33] for more details on these aspects.

The related problem consisting in describing the *interface* between two fluids of different density happens to be much more complicated due to the presence of a new kind of instabilities that do not exist in the water waves problem, the so-called Kelvin-Helmholtz instabilities, created by the discontinuity of the tangential velocity at the interface. If interfacial waves can be observed in spite of these instabilities, this is because some extra mechanism is involved which prevents the growth of Kelvin-Helmholtz instabilities. It has for instance been shown in [32] that a small amount of surface tension may be enough to control these instabilities, leading to the following generalization of the Rayleigh-Taylor criterion for the stability of two-fluids interfaces,

$$-[\![\partial_z P^{\pm}_{|\text{interf}}]\!] > \frac{1}{4}\frac{(\rho^+\rho^-)^2}{\sigma(\rho^+ + \rho^-)^2}\mathfrak{c}(\zeta)\big|[\![\mathbf{U}^{\pm}_{|\text{interf}}]\!]\big|^4_\infty, \tag{1}$$

where $\rho^{\pm}$ denote the densities of both fluids, σ the surface tension coefficient, $\mathfrak{c}(\zeta)$ a constant of little importance, and $[\![\mathbf{U}^{\pm}_{|\text{interf}}]\!]$ the velocity jump at the interface. This criterion is the result of two mechanisms: gravity stabilizes the low frequency components of the interfacial waves, while surface tension stabilizes the high frequency modes. The stability of the low frequency modes is not always granted: the *relative gravity* $g' = g\frac{\rho^+ - \rho^-}{\rho^+ + \rho^-}$ must be large enough. From the analysis of [32] this can be expressed as follows,

$$g' > \frac{\rho^+\rho^-}{(\rho^+ + \rho^-)^2} C(\zeta) |[\![\mathbf{U}^\pm_{|\text{interf}}]\!]|^2_\infty, \tag{2}$$

for some constant $C(\zeta)$ depending on ζ (and homogeneous to the inverse of a length).

If this condition is satisfied, then there is a critical wave number k_{KH} such that Kelvin-Helmholtz instabilities appear above this wave number in the absence of surface tension. If k_{KH} is large enough, then we may expect a regularization by small scale processes: capillarity (as in [32]), viscosity, etc. A precise description of k_{KH} is therefore important.

A precise evaluation of k_{KH} in the general nonlinear case seems out of reach because the computations based on the linearization around any solution is extremely technical. It is however possible to get some interesting insight on the formation of Kelvin-Helmholtz instabilities by considering the linearization around a constant shear, and this is the approach we shall use in this paper. For instance, we shall show that a necessary and sufficient condition for the stability of the low frequency modes for the linearized system around a shear flow with constant speeds c^+ and c^- is

$$|[\![c^\pm]\!]|^2 < \Omega_{KH} := \frac{g' H_0}{\underline{\rho}^+ \underline{\rho}^-}, \qquad \text{with } H_0 = \underline{\rho}^+ H^- + \underline{\rho}^- H^+ \tag{3}$$

(with $\underline{\rho}^\pm = \frac{\rho^\pm}{\rho^+ + \rho^-}$ denoting the relative densities of the fluids), a condition which is both consistent with (2), and more explicit. Moreover, the technical simplifications brought by linearizing around a constant state allow us to give a formula for k_{KH} when the above condition is satisfied. We are also able to derive a condition for the stability of *all modes*, which is therefore a linear but more precise version of (1),

$$[\![c^\pm]\!]^2 \leq \Omega^{cr} \quad \text{with} \quad \Omega^{cr} \sim \frac{2}{\sqrt{\mathrm{Bo}}} \Omega_{KH}, \tag{4}$$

and where Bo is the Bond number

$$\mathrm{Bo} = \frac{(\rho^+ + \rho^-) g' H_0^2}{\sigma}. \tag{5}$$

Since the Bond number is very large in most cases, this condition is much more restrictive than (3).

As for the one fluid (water waves) case, simpler asymptotic models are often used to describe the propagation of interfacial waves. We will stick here to the shallow water regime (in which the depth of both fluids are small compared to the typical horizontal length); we refer to [8] for a systematic derivation of asymptotic models in this context (see also [25] for a spectacular application of such models for the explanation of the dead water phenomenon).

Since shallow water models somehow only keep the low frequency part of the wave, they behave differently with respect to Kelvin-Helmholtz instabilities and may be well posed without stabilizing phenomenons such as surface tension. Their well-posedness is however subject to additional conditions on the data, and it turns out that these conditions are actually what remains of the Kelvin-Helmholtz instabilities for these models. The singularity corresponding to the violation of these extra conditions must therefore be interpreted as the prediction of a Kelvin-Helmholtz instability. One of the questions we raise here is: how well do the various shallow water models approximate the scenario of formation of Kelvin-Helmholtz instabilities of the full Euler equations? This is to our knowledge the first time this issue is addressed.

Our strategy to answer this question is to consider the same problem as for the full Euler equations and that we described above. Namely, we consider the stability of the linearization of the various shallow water models around a constant shear. We exhibit a condition for the low frequency stability that we compare to (3), and a critical wave number $k_{KH}^{\rm app}$ for the apparition of the Kelvin-Helmholtz instabilities, and that we compare to k_{KH}. We also compute a condition for the stability of *all* modes of the form $[\![c^{\pm}]\!]^2 \leq \Omega^{cr,app}$ that we compare to (4).

The asymptotic models we consider here are the so-called Shallow Water/Shallow Water (SW/SW) model which is the model obtained at first order in the shallow water limit. We show that this model *underestimates* the Kelvin-Helmholtz instabilities. We then consider the Green-Naghdi/Green-Naghdi model which is obtained in the same regime as the SW/SW model, but which is precise up to second order. This model *overestimates* the Kelvin-Helmholtz instabilities. Moreover, we derive two new families of regularized GN/GN model (one of which being a generalization of the models derived in [13]), and we show that they are able to reproduce *exactly* the singularity formation scenario of the full Euler equations. These results are summed up in the following table, and used to interpret some experimental measurements (see Sect. 4.3).

All the results described above are obtained in the framework of the *rigid lid* approximation for which the fluid above is bounded from above by a rigid bottom. We also consider here the *free surface* case where the upper boundary is now, like the interface, a free surface. This framework is of course relevant for many applications and has been considered for instance in [1, 3, 11, 19–21, 24, 26, 40]. We follow the same approach as above: we exhibit a condition for the stability of low frequency modes, and a stronger condition for the stability of all modes. The problem is however much more technical than for the rigid lid case, and we resort to numerical computations. As noticed in [1, 11, 35, 40], the free surface case is also marked by a peculiar phenomenon: low frequency modes are stable for small shears as in the rigid lid case, but also for large enough shears (Table 1).

As in the rigid lid case, we compare the formation of Kelvin-Helmholtz instabilities for the full Euler equations and for a Shallow Water/Shallow Water model with free surface. Except for a special configuration (the depth of both fluids is the same) where we can carry the computations out, this comparison is numerical. The investigation of the next order models (GN/GN) is not done in this

Table 1 Stability criteria for Euler system and different models

	LF stability	Critical wave number	Stability of all modes
Euler	$[\![c^\pm]\!] < \Omega_{KH}$	$0 < k_{KH} < \infty$	$\Omega^{cr} \sim \frac{2}{\sqrt{\mathrm{Bo}}}\Omega_{KH} < \Omega_{KH}$
SW/SW	idem	$k_{SW} = \infty$	$\Omega^{cr}_{SW} = \Omega_{KH} > \Omega^{cr}$
GN/GN	idem	$k_{GN,\sigma} < k_{KH}$	$\Omega^{cr}_{GN_\sigma} < \Omega^{cr}$
GN/GN$_{\mathrm{reg}}$	idem	$k_{GN_r} = k_{KH}$ possible	$\Omega^{cr}_{GN_r} = \Omega^{cr}$ possible

paper because it is highly technical, and because it can be done following the same approach as in the rigid lid case. Note that GN/GN type models have been studied in the free surface case in [1], and that it is possible to generalize this study by deriving regularized models with the techniques used here in the rigid lid case in order to obtain a better description of the formation of Kelvin-Helmholtz instabilities.

The last point we address in this paper is a study of the behavior of the additional stability area observed for large shears in the free surface case. More precisely, we want to understand how this stability area behaves in the so called rigid lid limit. We infer from this study that the rigid lid approximation can only be true for very small density contrast between both fluids, and that the large shear stability area disappears in the rigid lid case because it corresponds only to infinite shears.

1.1 Organization of the Paper

We first recall in Sect. 2 the equations of motions, both in the rigid lid and in the free surface cases. We also derive the reduced formulations that are used throughout the paper for the computations. The Kelvin-Helmholtz instabilities are then studied in Sect. 3 for the full equations in the rigid lid case. In Sect. 4, we study several shallow water models, focusing our attention on their ability to describe the Kelvin-Helmholtz instabilities. An experimental application is also given. We then turn to study, in Sect. 5, the formation of Kelvin-Helmholtz instabilities for the full Euler equations in the free surface case. The shallow water behavior of these instabilities and the behavior of the large shear stability area is then investigated in Sect. 6.

1.2 Notations

- If A^+ and A^- are two quantities (real numbers, functions, etc.), the notations $[\![A^\pm]\!]$ and $\langle A^\pm \rangle$ stand for

$$[\![A^\pm]\!] = A^+ - A^- \quad \text{and} \quad \langle A^\pm \rangle = \frac{A^+ + A^-}{2}.$$

- We use relative densities

$$\underline{\rho}^+ = \frac{\rho^+}{\rho^+ + \rho^-}, \qquad \underline{\rho}^- = \frac{\rho^-}{\rho^+ + \rho^-}$$

as the ratios of densities ρ^+, ρ^- for the lower and upper layers respectively.
- Parameters δ and γ stand for the depth ratio $\delta = H^-/H^+$ and the density ratio $\gamma = \underline{\rho}^-/\underline{\rho}^+$ respectively.
- The notations $\mathrm{th}^\pm$, $\mathrm{cth}^\pm$, $\mathrm{sh}^\pm$ and $\mathrm{ch}^\pm$ stand for the hyperbolic Fourier multipliers $\tanh(H^\pm|D|)$, $\coth(H^\pm|D|)$, $\sinh(H^\pm|D|)$ and $\cosh(H^\pm|D|)$, with $iD = \nabla$ and $H^\pm$ the average depths for the upper and lower fluids. When no misunderstanding is possible, we also take $\mathrm{th}^\pm$, $\mathrm{cth}^\pm$, $\mathrm{sh}^\pm$ and $\mathrm{ch}^\pm$ as their Fourier modes $\tanh(H^\pm|\mathbf{k}|)$, $\coth(H^\pm|\mathbf{k}|)$, $\sinh(H^\pm|\mathbf{k}|)$ and $\cosh(H^\pm|\mathbf{k}|)$.

2 The Equations of Motion

In this section, we introduce briefly the equations of motion for both the rigid lid case and the free surface case.

2.1 *The Rigid Lid Case*

We consider first the case of a rigid lid; in such a configuration (see Fig. 1), the inferior fluid domain Ω_t^+ is delimited from below by a flat bottom Γ^+ at height $z = -H^+$ and from above by the interface $\Gamma_t = \{(X, z) \in \mathbb{R}^d \times \mathbb{R}, z = \zeta(t, X)\}$, and the superior fluid is bounded from below by Γ_t and from above by a rigid lid Γ^- at height $z = H^-$.

2.1.1 Basic Equations

We denote by $\mathbf{U}^\pm$ the velocity field in $\Omega_t^\pm$; the horizontal component of $\mathbf{U}^\pm$ is $V^\pm$ and its vertical one $w^\pm$. The pressure is denoted by $P^\pm$. The equations of motion are then the following,

- Equations in the fluid layers. In both fluid layers, the velocity field $\mathbf{U}^\pm$ and the pressure $P^\pm$ satisfy the equations

$$\operatorname{div} \mathbf{U}^\pm(t,\cdot) = 0, \qquad \operatorname{curl} \mathbf{U}^\pm(t,\cdot) = 0, \qquad \text{in } \Omega_t^\pm \quad (t \geq 0), \tag{6}$$

which express the incompressibility and irrotationality assumptions, and

$$\rho^\pm(\partial_t \mathbf{U}^\pm + (\mathbf{U}^\pm \cdot \nabla_{X,z})\mathbf{U}^\pm) = -\nabla_{X,z} P^\pm - g\mathbf{e}_z \qquad \text{in } \Omega_t^\pm \quad (t \geq 0), \tag{7}$$

which expresses the conservation of momentum (Euler equation).

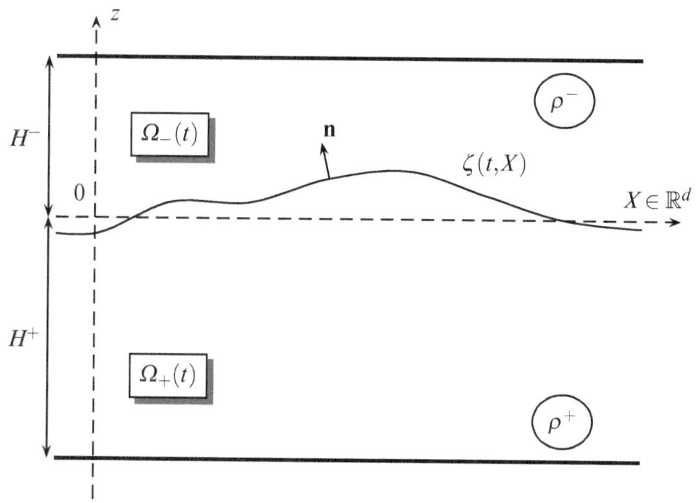

Fig. 1 The rigid lid configuration

- Boundary conditions at the rigid bottom and lid. Impermeability of these two boundaries is classically rendered by

$$w^+(t,\cdot)_{|\Gamma^+} = 0, \text{ and } w^-(t,\cdot)_{|\Gamma^-} = 0 \qquad (t \geq 0). \tag{8}$$

- Boundary conditions at the moving interface. The fact that the interface is a bounding surface (the fluid particles do not cross it) yields the equations

$$\partial_t \zeta - \sqrt{1 + |\nabla\zeta|^2}\mathbf{U}_n^{\pm} = 0, \qquad (t \geq 0), \tag{9}$$

where $\mathbf{U}_n^{\pm} := \mathbf{U}^{\pm}{}_{|\Gamma_t} \cdot \mathbf{n}$, $\mathbf{n}$ being the upward unit normal vector to the interface Γ_t. A direct consequence of (9) is that *there is no jump of the normal component of the velocity at the interface*. Finally, the continuity of the stress tensor at the interface gives

$$[\![P^{\pm}(t,\cdot)_{|\Gamma_t}]\!] = \sigma\kappa(\zeta), \qquad (t \geq 0), \tag{10}$$

where $\sigma \geq 0$ is the surface tension coefficient and

$$\kappa(\zeta) = -\nabla \cdot \Big(\frac{\nabla\zeta}{\sqrt{1 + |\nabla\zeta|^2}}\Big)$$

is the mean curvature of the interface.

2.1.2 Reduction to the Interface

As stressed above, the normal component of the velocity field $\mathbf{U}^\pm \cdot \mathbf{n}$ is continuous at the interface. This is not the case for the tangential component of the velocity. Since the vertical component of this vector can be deduced from its horizontal components, we will focus our attention on these horizontal components and more specifically on the quantity $U_\parallel^\pm$ defined as

$$\mathbf{U}^\pm_{|z=\zeta} \times \mathbf{N} = \begin{pmatrix} -(U_\parallel^\pm)^\perp \\ -(U_\parallel^\pm)^\perp \cdot \nabla\zeta \end{pmatrix}, \qquad \text{with} \quad \mathbf{N} = \sqrt{1+|\nabla\zeta|^2}\mathbf{n}.$$

Taking the vector product of the trace of (7) at the interface with $\mathbf{N}$, and considering only the horizontal components of the resulting equation, one readily gets

$$\partial_t U_\parallel^\pm + g\nabla\zeta + \frac{1}{2}\nabla|U_\parallel^\pm|^2 - \frac{1}{2}\nabla\big((1+|\nabla\zeta|^2)|\underline{w}^\pm|^2\big) = -\frac{1}{\rho^\pm}\nabla\underline{P}^\pm,$$

where we denoted

$$\underline{P}^\pm := P^\pm_{|z=\zeta}, \qquad \underline{\mathbf{U}}^\pm = (\underline{V}^\pm, \underline{w}^\pm) := \mathbf{U}^\pm_{|z=\zeta} = (V^\pm_{|z=\zeta}, w^\pm_{|z=\zeta}).$$

The interfacial waves equations (6)–(10) can therefore be recast as two sets of two equations on the fixed domain $\mathbb{R}_t^+ \times \mathbb{R}_X^d$, i.e. on the interface

$$\begin{cases} \partial_t\zeta - \underline{\mathbf{U}}^\pm \cdot \mathbf{N} = 0, \\ \partial_t U_\parallel^\pm + g\nabla\zeta + \frac{1}{2}\nabla|U_\parallel^\pm|^2 - \frac{1}{2}\nabla\big((1+|\nabla\zeta|^2)|\underline{w}^\pm|^2\big) = -\frac{1}{\rho^\pm}\nabla\underline{P}^\pm, \end{cases} \tag{11}$$

with the jump condition

$$[\![P^\pm(t,\cdot)_{|\Gamma_t}]\!] = -\sigma\nabla\cdot\Big(\frac{\nabla\zeta}{\sqrt{1+|\nabla\zeta|^2}}\Big). \tag{12}$$

Remark 1. i- In (11), the incompressibility and irrotationality conditions (6) imply[1] that one can express $\underline{\mathbf{U}}^\pm \cdot \mathbf{N}$ in terms of ζ and the tangential component $U_\parallel^\pm$.

ii- It is possible to reduce further (11) to *one* set of two equations. Let us define

$$U_\parallel = \underline{\rho}^+ U_\parallel^+ - \underline{\rho}^- U_\parallel^-;$$

---

[1]It has been proved recently [12] in the particular case of one single fluid ($\rho^- = 0$) that the irrotationality condition is not needed and that (11), together with the standard evolution equation for the vorticity, form a closed, well-posed, set of equations.

multiplying the second equation of $(11)_\pm$ by $\underline{\rho}^+$ and subtracting the resulting + and − equations, we obtain

$$\begin{cases} \partial_t \zeta - \underline{\mathbf{U}}^\pm \cdot \mathbf{N} = 0, \\ \partial_t U_\parallel + g' \nabla \zeta + \frac{1}{2} \nabla [\![\underline{\rho}^\pm |U_\parallel^\pm|^2]\!] - \frac{1}{2} \nabla \big((1 + |\nabla \zeta|^2) [\![\underline{\rho}^\pm |\underline{w}^\pm|^2]\!] \big) \\ \qquad = -\frac{\sigma}{\rho^+ + \rho^-} \nabla \kappa(\zeta), \end{cases} \tag{13}$$

where $g' = g(\underline{\rho}^+ - \underline{\rho}^-)$ is the reduced gravity. It is possible to show that $\underline{\mathbf{U}}^+ \cdot \mathbf{N} = \underline{\mathbf{U}}^- \cdot \mathbf{N}$, $U_\parallel^\pm$ and $\underline{w}^\pm$ can be expressed in terms of ζ and $U_\parallel$, so that (13) is a closed set of equations on ζ and $U_\parallel$.

iii- From the incompressibility and irrotationality assumptions (6), we infer that the velocity fields $\mathbf{U}^\pm$ derive from a scalar harmonic potential $\Phi^\pm$ (i.e. $\mathbf{U}^\pm = \nabla_{X,z}\Phi^\pm$, with $\Delta_{X,z}\Phi^\pm = 0$ in $\Omega_t^\pm$). Denoting by $\psi^\pm$ the traces of $\Phi^\pm$ at the interface, $\psi^\pm = \Phi^\pm_{|_{z=\zeta}}$, one can check that $U_\parallel^\pm = \nabla \psi^\pm$. Introducing

$$\psi = \underline{\rho}^+ \psi^+ - \underline{\rho}^- \psi^-,$$

one has $U_\parallel = \nabla \psi$ and the second equation in (13) can be replaced by a scalar equation on ψ, leading to

$$\begin{cases} \partial_t \zeta - \sqrt{1 + |\nabla \zeta|^2} \mathbf{U}_n^\pm = 0, \\ \partial_t \psi + g' \zeta + \frac{1}{2} [\![\underline{\rho}^\pm |U_\parallel^\pm|^2]\!] - \frac{1}{2} (1 + |\nabla \zeta|^2) [\![\underline{\rho}^\pm |\underline{w}^\pm|^2]\!] = -\frac{\sigma}{\rho^+ + \rho^-} \kappa(\zeta). \end{cases} \tag{14}$$

This is the formulation used in [32] to establish the nonlinear well posedness of the two-fluids interfaces equations. In the case of one fluid (water waves), these equations coincide with the Zakharov-Craig-Sulem formulation of the water waves equations [22, 43]. One of the main features of this formulation is its Hamiltonian structure; it is remarkable that the two-fluid generalization of the Zakharov-Craig-Sulem formulation is also Hamiltonian, as remarked in [4] (see in particular Sect. 5 of that reference for considerations on Kelvin-Helmholtz instabilities). Using asymptotic expansions of the Hamiltonian, Craig, Guyenne and Kalisch derived in [19] a family of Hamiltonian approximations to the two-fluids equations (14).

2.2 *The Free Surface Case*

We consider here the case of a free surface (see Fig. 2). The configuration is the same as in Sect. 2.1 except that the superior fluid is now bounded from above by a free surface $\Gamma_t^s = \{(X,z) \in \mathbb{R}^d \times \mathbb{R}, z = \zeta^s(t,X)\}$.

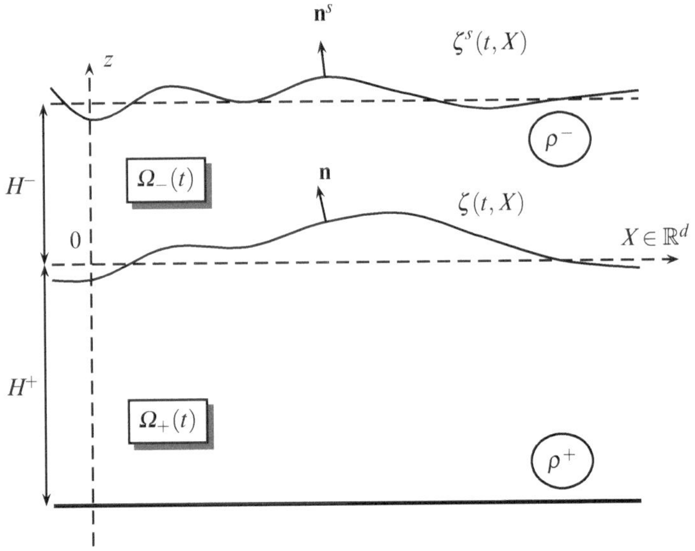

Fig. 2 The free surface configuration

2.2.1 Basic Equations

The only difference between the free surface and rigid lid cases is that the boundary condition (8) must be replaced by the kinematic condition

$$\partial_t \zeta^s - \sqrt{1 + |\nabla \zeta^s|} \mathbf{U}^-_{n^s} = 0, \tag{15}$$

where $\mathbf{U}^-_{n^s} := \mathbf{U}^-_{|_{\Gamma_t^s}} \cdot \mathbf{n}^s$, $\mathbf{n}^s$ being the upward unit normal vector to the free surface Γ_t^s, and that the value of the pressure at the free surface must be prescribed; neglecting the effects of surface tension at the surface, the pressure is constant, and since it is defined up to a constant, we take it equal to zero,

$$P_{|_{z=\zeta^s}} = 0. \tag{16}$$

2.2.2 Reduction to the Interface

Consistently with the notations previously used, we define the quantity $U_\parallel^s$ as

$$\mathbf{U}^-_{|_{z=\zeta^s}} \times \mathbf{N}^s = \begin{pmatrix} -(U_\parallel^s)^\perp \\ -(U_\parallel^s)^\perp \cdot \nabla \zeta^s \end{pmatrix}, \quad \text{with} \quad \mathbf{N}^s = \sqrt{1 + |\nabla \zeta^s|^2} \mathbf{n}^s.$$

Taking the vector product of the trace of (7) at the free surface with $\mathbf{N}^s$, and considering only the horizontal components of the resulting equation, one readily gets

$$\partial_t U_\parallel^s + g\nabla\zeta^s + \frac{1}{2}\nabla|U_\parallel^s|^2 - \frac{1}{2}\nabla\big((1+|\nabla\zeta^s|^2)|\underline{w}^s|^2\big) = 0,$$

where we denoted

$$\underline{\mathbf{U}}^s = (\underline{V}^s, \underline{w}^s) := \mathbf{U}^-_{|_{z=\zeta^s}} = (V^-_{|_{z=\zeta^s}}, w^-_{|_{z=\zeta^s}})$$

and the surface tension at the top free surface is neglected. The interfacial waves equations (6)–(10) can therefore be recast as three sets of two equations on the fixed domain $\mathbb{R}_t^+ \times \mathbb{R}_X^d$, namely,

$$\begin{cases} \partial_t\zeta^s - \underline{\mathbf{U}}^s\cdot\mathbf{N}^s = 0, \\ \partial_t U_\parallel^s + g\nabla\zeta^s + \frac{1}{2}\nabla|U_\parallel^s|^2 - \frac{1}{2}\nabla\big((1+|\nabla\zeta^s|^2)|\underline{w}^s|^2\big) = 0, \end{cases} \tag{17}$$

and

$$\begin{cases} \partial_t\zeta - \underline{\mathbf{U}}^\pm\cdot\mathbf{N} = 0, \\ \partial_t U_\parallel^\pm + g\nabla\zeta + \frac{1}{2}\nabla|U_\parallel^\pm|^2 - \frac{1}{2}\nabla\big((1+|\nabla\zeta|^2)|\underline{w}^\pm|^2\big) = -\frac{1}{\rho^\pm}\nabla\underline{P}^\pm, \end{cases} \tag{18}$$

with the jump condition

$$[\![P^\pm(t,\cdot)_{|_{\Gamma_t}}]\!] = \sigma\kappa(\zeta). \tag{19}$$

Remark 2. Proceeding as in Remark 1, it is possible to reduce this set of seven equations into a set of four equations on $(\zeta^s, U_\parallel^s, \zeta, U_\parallel)$, with $U_\parallel = \underline{\rho}^+U_\parallel^+ - \underline{\rho}^-U_\parallel^-$ These equations are

$$\begin{cases} \partial_t\zeta^s - \underline{\mathbf{U}}^s\cdot\mathbf{N}^s = 0, \\ \partial_t U_\parallel^s + g\nabla\zeta^s + \frac{1}{2}\nabla|U_\parallel^s|^2 - \frac{1}{2}\nabla\big((1+|\nabla\zeta^s|^2)|\underline{w}^s|^2\big) = 0, \\ \partial_t\zeta - \underline{\mathbf{U}}^\pm\cdot\mathbf{N} = 0, \\ \partial_t U_\parallel + g'\nabla\zeta + \frac{1}{2}\nabla[\![\underline{\rho}^\pm|U_\parallel^\pm|^2]\!] - \frac{1}{2}\nabla\big((1+|\nabla\zeta|^2)[\![\underline{\rho}^\pm|\underline{w}^\pm|^2]\!]\big) \\ \qquad\qquad = -\frac{\sigma}{\rho^+ + \rho^-}\nabla\kappa(\zeta), \end{cases} \tag{20}$$

where $g' = g(\underline{\rho}^+ - \underline{\rho}^-)$ is the reduced gravity. Giving a version of these equations in terms of velocity potential similar to (14) in the rigid lid case is of course possible.

3 The Kelvin-Helmholtz Instability in the Rigid Lid Case

The formation of Kelvin-Helmholtz instabilities for two-fluids interfaces and in the case of a rigid lid is investigated here. In this setting, it is known [32] that the full nonlinear equations are well posed under the generalized Rayleigh-Taylor criterion (1). This criterion is a sufficient condition ensuring the stability of all frequency modes. By sticking to the much simpler linear theory, we want to get additional information that will also be used to compare with the formation of Kelvin-Helmholtz instabilities in shallow water asymptotic models and in the free surface case, where the nonlinear analysis is still an open problem. More precisely, we consider the linear stability of perturbations to a constant shear flow. We insist on three important aspects: (1) the existence of a critical value of the shear flow below which *low frequency modes* are stable (2) the existence for such flows of a critical wavenumber delimiting this stability region, and (3) the possible existence of a second critical shear flow below which *all modes* are stable. These stability criteria are studied in detail, and explicit formulas or accurate approximations are provided, which allows one to comment on the influence of the density ratio $\gamma = \rho^-/\rho^+$ and of the depth ratio $\delta = H^-/H^+$ for instance.

3.1 Linearization Around a Constant Shear $(\mathbf{c}^+, \mathbf{c}^-)$

We consider here the linear equations governing small perturbations of the constant horizontal shear $\zeta = 0$, $V^\pm = \mathbf{c}^\pm$, $w^\pm = 0$ (and therefore $U_\parallel^\pm = \mathbf{c}^\pm$).

Linearizing (10)–(11) around $\zeta = 0$, $V^\pm = \mathbf{c}^\pm$ and $w^\pm = 0$ we find the following linear equations for the perturbation $(\dot\zeta, \dot V^\pm, \dot w^\pm)$

$$\begin{cases} \partial_t \dot\zeta + \mathbf{c}^\pm \cdot \nabla \dot\zeta - \dot w^\pm = 0, \\ \partial_t \dot V^\pm + g\nabla\dot\zeta + \mathbf{c}^\pm \cdot \nabla \dot V^\pm = -\frac{1}{\rho^\pm}\nabla \underline{P}^\pm, \end{cases}$$

with

$$[\![\underline{P}^\pm]\!] = -\sigma \Delta \dot\zeta.$$

In the same spirit as in Remark 1, we can derive a reduced formulation for these equations; denoting $\dot V = \underline{\rho}^+ \dot V^+ - \underline{\rho}^- \dot V^-$ and remarking that

$$[\![\underline{\rho}^\pm \mathbf{c}^\pm \cdot \dot V^\pm]\!] = \langle \mathbf{c}^\pm \rangle \cdot \dot V + [\![\mathbf{c}^\pm]\!] \cdot \langle \underline{\rho}^\pm \dot V^\pm \rangle,$$

we first get

$$\begin{cases} \partial_t \dot{\zeta} + \mathbf{c}^{\pm} \cdot \nabla \dot{\zeta} - \dot{w}^{\pm} = 0, \\ \partial_t \dot{V} + \langle \mathbf{c}^{\pm} \rangle \cdot \nabla \dot{V} + g' \nabla \dot{\zeta} + [\![\mathbf{c}^{\pm}]\!] \cdot \nabla \langle \underline{\rho}^{\pm} \dot{V}^{\pm} \rangle = \dfrac{\sigma}{\rho^{+} + \rho^{-}} \nabla \Delta \dot{\zeta}. \end{cases} \tag{21}$$

In order to prove that these equations are in closed form, we still need to express $\dot{V}^{\pm}$ and $\dot{w}^{\pm}$ in terms of $\dot{\zeta}$ and $\dot{V}$. This is done in the following lemma.

Lemma 1. *One has*

$$\dot{V}^{+} = \frac{\underline{\rho}^{-}}{\underline{\rho}^{-}\mathrm{th}^{+} + \underline{\rho}^{+}\mathrm{th}^{-}} \frac{[\![\mathbf{c}^{\pm}]\!] \cdot \nabla}{|D|} \nabla \dot{\zeta} + \frac{\mathrm{th}^{-}}{\underline{\rho}^{-}\mathrm{th}^{+} + \underline{\rho}^{+}\mathrm{th}^{-}} \dot{V}$$

$$\dot{V}^{-} = \frac{\underline{\rho}^{+}}{\underline{\rho}^{-}\mathrm{th}^{+} + \underline{\rho}^{+}\mathrm{th}^{-}} \frac{[\![\mathbf{c}^{\pm}]\!] \cdot \nabla}{|D|} \nabla \dot{\zeta} - \frac{\mathrm{th}^{+}}{\underline{\rho}^{-}\mathrm{th}^{+} + \underline{\rho}^{+}\mathrm{th}^{-}} \dot{V},$$

and $\dot{w}^{\pm} = \mp \mathrm{th}^{\pm} \frac{\nabla}{|D|} \cdot \dot{V}^{\pm}$.

Proof. From the first equation of (21), we get

$$\dot{w}^{+} - \dot{w}^{-} = [\![\mathbf{c}^{\pm}]\!] \cdot \nabla \dot{\zeta}$$

(this is the linear version of the continuity of the normal component of the velocity). For the same reason as in the third point of Remark 1, we can also write $\dot{V}^{\pm} = \nabla \dot{\psi}^{\pm}$ and $\dot{V} = \nabla \dot{\psi}$, with $\dot{\psi} = \underline{\rho}^{+} \dot{\psi}^{+} - \underline{\rho}^{-} \dot{\psi}^{-}$, and where

$$\dot{\psi}^{\pm} = \Phi^{\pm}_{|z=0} \quad \text{with} \quad \Delta_{X,z} \Phi^{\pm} = 0 \quad \text{in } -H^{\pm} < \pm z < 0 \quad \text{and } \partial_z \Phi^{\pm}_{|z=\mp H^{\pm}} = 0.$$

From a simple linear analysis, we get therefore

$$\dot{w}^{\pm} = \pm |D| \mathrm{th}^{\pm} \dot{\psi}^{\pm}.$$

Since moreover, one has by definition $\dot{\psi} = \underline{\rho}^{+} \dot{\psi}^{+} - \underline{\rho}^{-} \dot{\psi}^{-}$, one can express $\dot{\psi}^{\pm}$ (and therefore $\dot{V}^{\pm} = \nabla \dot{\psi}^{\pm}$ and $\dot{w}^{\pm}$ through the above relation) in terms of $\dot{V} = \nabla \dot{\psi}$ and $\dot{\zeta}$ by solving the system

$$\begin{cases} |D| \mathrm{th}^{+} \dot{\psi}^{+} + |D| \mathrm{th}^{-} \dot{\psi}^{-} = [\![\mathbf{c}^{\pm}]\!] \cdot \nabla \dot{\zeta}, \\ \underline{\rho}^{+} \dot{\psi}^{+} - \underline{\rho}^{-} \dot{\psi}^{-} = \dot{\psi}, \end{cases}$$

which leads to the formula of the lemma. □

Thanks to the lemma, we can rewrite (21) under the form

$$\begin{cases} \partial_t \dot{\zeta} + \mathbf{c}(D)\cdot\nabla\dot{\zeta} + \frac{\mathrm{th}^+\mathrm{th}^-}{\underline{\rho}^-\mathrm{th}^+ + \underline{\rho}^+\mathrm{th}^-}\frac{\nabla}{|D|}\cdot\dot{V} = 0, \\ \partial_t \dot{V} + \mathbf{c}(D)\cdot\nabla\dot{V} + (g' - e(D) - \frac{\sigma}{\rho^+ + \rho^-}\Delta)\nabla\dot{\zeta} = 0, \end{cases} \tag{22}$$

with

$$\mathbf{c}(D) = \frac{\mathbf{c}^+\underline{\rho}^+\mathrm{th}^- + \mathbf{c}^-\underline{\rho}^-\mathrm{th}^+}{\underline{\rho}^-\mathrm{th}^+ + \underline{\rho}^+\mathrm{th}^-}, \qquad e(D) = \frac{\underline{\rho}^+\underline{\rho}^-}{\underline{\rho}^-\mathrm{th}^+ + \underline{\rho}^+\mathrm{th}^-}\frac{([\![\mathbf{c}^\pm]\!]\cdot D)^2}{|D|}. \tag{23}$$

3.2 Kelvin-Helmholtz Instabilities for the Linearized Two-Fluids Equations

A brief look at (22) shows that the Fourier modes $(\mathscr{F}\dot{\zeta}(\mathbf{k}), \mathscr{F}\dot{\psi}(\mathbf{k}))$ are stable (i.e. they are not exponentially[2] amplified) if and only if

$$g' - e(\mathbf{k}) + \frac{\sigma}{\rho^+ + \rho^-}|\mathbf{k}|^2 > 0.$$

From the explicit expression of $e(\mathbf{k})$ stemming from (23), the most unstable direction corresponds to $\mathbf{k}$ parallel to $[\![c^\pm]\!]$, and we therefore restrict our stability analysis to wave numbers $\mathbf{k}$ oriented along this most unstable direction. Without loss of generality, we can assume that

$$[\![\mathbf{c}^\pm]\!] = [\![c^\pm]\!]\mathbf{e}_x, \qquad ([\![c^\pm]\!] = |[\![\mathbf{c}^\pm]\!]|),$$

and we consider therefore the stability of the Fourier modes $(\mathscr{F}\dot{\zeta}(\mathbf{k}), \mathscr{F}\dot{\psi}(\mathbf{k}))$ corresponding to $\mathbf{k} = k\mathbf{e}_x$, with $k = |\mathbf{k}|$.

From (23), we deduce easily that for all $\mathbf{k} = k\mathbf{e}_x$,

$$g' - e(\mathbf{k}) + \frac{\sigma}{\rho^+ + \rho^-}|\mathbf{k}|^2 > 0 \iff [\![c^\pm]\!]^2 < \Omega_{KH}\alpha_\sigma(k), \tag{24}$$

where we used the notations

$$\Omega_{KH} = \frac{g'H_0}{\underline{\rho}^+\underline{\rho}^-} \qquad (\text{and } H_0 = \underline{\rho}^-H^+ + \underline{\rho}^+H^-), \tag{25}$$

$$\alpha_\sigma(k) = \Big(1 + \frac{\sigma}{g'(\rho^+ + \rho^-)}k^2\Big)\alpha(k), \tag{26}$$

[2] The endpoint case $g' - e(\mathbf{k}) = 0$ actually corresponds to a *linear* amplification.

$$\alpha(k) = \underline{\rho}^{-}\frac{\mathrm{th}^{+}}{H_0 k} + \underline{\rho}^{+}\frac{\mathrm{th}^{-}}{H_0 k}. \tag{27}$$

Since $\alpha(\cdot)$ is a decreasing function on $\mathbb{R}^+$ with maximal value 1 at the origin, *the modes with low wavenumber are stable* if and only if

$$\text{(Stab 1)} \qquad [\![c^{\pm}]\!]^2 < \Omega_{KH};$$

if it is satisfied, there is a unique $k_{KH} \in \mathbb{R}_*^+ \cup \{+\infty\}$ for each $[\![c^{\pm}]\!]$, such that all the modes corresponding to $0 \leq k < k_{KH}$ are stable,

$$\text{(Critical wavenumber)} \qquad k_{KH} = \min \alpha_\sigma^{-1}\Big(\frac{[\![c^{\pm}]\!]^2}{\Omega_{KH}}\Big);$$

the quantity k_{KH} is therefore the critical wavenumber above which Kelvin-Helmholtz instabilities appear for the perturbation of the constant shear flow. Since one obviously has

$$\alpha(k) \leq \alpha_\infty(k) := \frac{1}{kH_0},$$

one obtains the following upper bound for k_{KH} (see Fig. 3a),

$$k_{KH} \leq k_{KH}^{\infty} := \min \alpha_{\sigma,\infty}^{-1}\Big(\frac{[\![c^{\pm}]\!]^2}{\Omega_{KH}}\Big),$$

where $\alpha_{\sigma,\infty}(k) = (1 + \frac{\sigma}{g'(\rho^+ + \rho^-)}k^2)\alpha_\infty(k)$. Direct computations yield

$$k_{KH}^{\infty} = \frac{[\![c^{\pm}]\!]^2}{\Omega_{KH}}\frac{\mathrm{Bo}}{2H_0}\Big(1 - \sqrt{1 - 4\frac{\Omega_{KH}^2}{\mathrm{Bo}[\![c^{\pm}]\!]^4}}\Big) \qquad \text{(with surface tension)},$$

$$k_{KH}^{\infty} = \frac{\Omega_{KH}}{[\![c^{\pm}]\!]^2}\frac{1}{H_0} \qquad \text{(without surface tension)}.$$

Note that this upper bound coincides with the critical wavenumber k_{KH} when both fluid layers are of infinite depth ($H^+ = H^- = \infty$).

It is straightforward to analyze the influence on the stable area (i.e. the set all the couples $([\![c^{\pm}]\!], k)$ such that (24) is satisfied) of the depth H^+, the depth ratio $\delta = H^-/H^+$ and the density ratio $\gamma = \underline{\rho}^-/\underline{\rho}^+$. One can see from (24) that, when the surface tension is neglected ($\sigma = 0$), the interface problem is more stable when H^+ increases, δ increases, or γ decreases (see Fig. 3b–d in the case without surface tension).

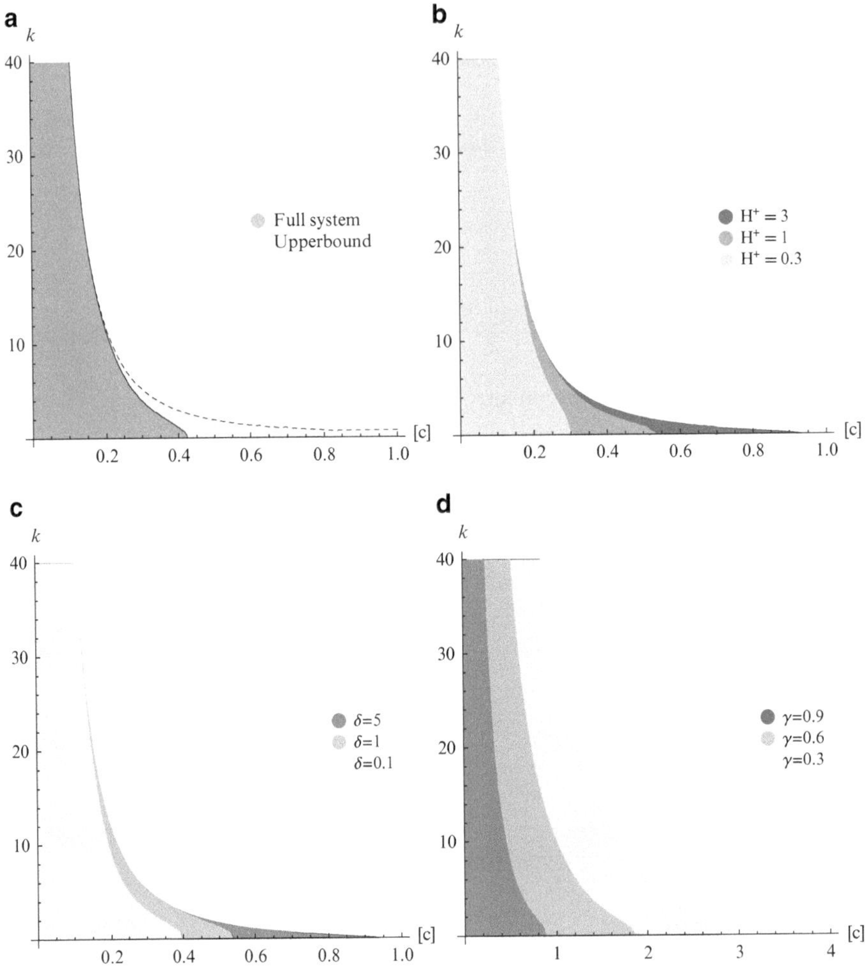

Fig. 3 Surface tension is neglected ($\sigma = 0$). Unless otherwise specified, the parameters are $\delta = 0.15/0.62$, $H^+ = 0.62$, $\gamma = 0.494/0.506$ and $g = 9.81$. We write $[\![c^\pm]\!]$ simply as $[c]$ in all the figures through this paper. (**a**) An upper bound for the stable area. (**b**) Influence of H^+ on the stable area. (**c**) Influence of the depth ratio δ. (**d**) Influence of the density ratio γ

Remark 3. We chose to present the linear stability criteria derived in this paper in terms of $[\![c^\pm]\!]$ to make the comparison with the nonlinear criterion (1) more transparent. Another possibility, often used in the literature, would have been to use instead the (dimensionless) Froude number

$$\mathrm{Fr} = \frac{[\![c^\pm]\!]}{\sqrt{g' H_0}}.$$

Using this notation, the criterion (Stab 1) derived above can be restated as

$$\mathrm{Fr} < (\underline{\rho}^{+}\underline{\rho}^{-})^{-1/2};$$

the right-hand-side depends only on the density ratio of the two layers, and is always greater than 2. This indicates that this stability criterion is quite reasonable, even when the density ratio goes to 1, even though g' goes to zero. This approach is consequently used in Sect. 6.2 to study the rigid limit in the case of two layers with a free surface.

In absence of surface tension ($\sigma = 0$), $\alpha_\sigma(\cdot) = \alpha(\cdot)$ is a decreasing function and goes to zero at infinity. One then easily deduces that in presence of a nonzero shear $[\![\mathbf{c}^{\pm}]\!]$ there are always unstable modes. In presence of surface tension, the situation is different. If (Stab 1) is satisfied, then there the modes with low wave numbers are stable; since $\alpha_\sigma(\cdot)$ grow to infinity at infinity, one directly infers from (24) that modes with high wavenumbers are stable. It is then possible to have stability of all Fourier modes if the shear does not exceed a critical value $[\![c^{\pm}]\!] = \sqrt{\Omega^{cr}}$; more precisely, *all modes are stable* if and only if

$$\text{(Stab 2)} \qquad [\![c^{\pm}]\!]^2 \leq \Omega^{cr} := \Omega_{KH} \min_{\mathbb{R}^+} \alpha_\sigma$$

(note that this criterion can be understood as the linear approximation of (1)). When $\sigma = 0$, one gets $\Omega^{cr} = 0 < \Omega_{KH}$; in presence of surface tension, $\Omega^{cr} \leq \Omega_{KH}$ and equality holds if the capillary effects are strong enough (see Fig. 4), in which case the existence of *some* stable modes implies that *all* modes are stable.

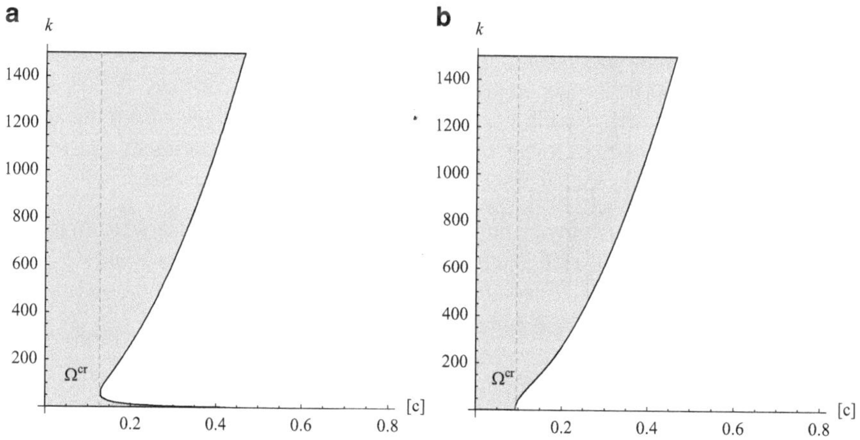

Fig. 4 Stable area with surface tension. Parameters are $\delta = 0.15/0.62$, $\gamma = 0.494/0.506$, $\rho^+ + \rho^- = 2021$, $\sigma = 0.073$ and $g = 9.81$. (**a**) $\Omega^{cr} < \Omega_{KH}$ (here, $H^+ = 0.62$). (**b**) $\Omega^{cr} = \Omega_{KH}$ (here, $H^+ = 0.03$)

There does not seem to be any simple explicit formula for the critical value Ω^{cr}. However, it is possible to give upper and lower bounds for this quantity. Note that these bounds are very precise in most situations (for the experiment studied in Sect. 4.3 below for instance, there is an error of only 3 % between Ω_-^{cr} and Ω_+^{cr}). We also point out that the upper bound given in the proposition coincides with Kelvin's criterion derived in the setting $H^+ = H^- = \infty$ [23].

Proposition 1. *One has*

$$\Omega_-^{cr} < \Omega^{cr} < \Omega_+^{cr},$$

with

$$\Omega_-^{cr} = \Omega_{KH}\,\alpha\Big(\frac{1}{2H_0}\sqrt{\mathrm{Bo}}\Big), \qquad \Omega_+^{cr} = \Omega_{KH}\frac{2}{\sqrt{\mathrm{Bo}}},$$

where $\alpha(\cdot)$ *is as in* (27) *and the Bond number* Bo *as defined in* (5).

Remark 4. Note that the approximation $\Omega^{cr} \sim \Omega_+^{cr}$ is in perfect agreement with the nonlinear criterion (1) derived in [32], namely,

$$-[\![\partial_z P^{\pm}_{|\mathrm{interf}}]\!] > \frac{1}{4}\frac{(\rho^+\rho^-)^2}{\sigma(\rho^+ + \rho^-)^2}\mathfrak{c}(\zeta)\big|[\![\mathbf{U}^{\pm}_{|\mathrm{interf}}]\!]\big|_\infty^4.$$

Indeed, replacing $\mathbf{U}^{\pm}_{|\mathrm{interf}}$ by $c^{\pm}$ in the above formula, and replacing the pressure by its hydrostatic approximation $\partial_z P^{\pm} = -\rho^{\pm} g$, this nonlinear criterion yields

$$[\![c^{\pm}]\!]^2 < \mathfrak{c}(\zeta)^{-1/2}\Omega_+^{cr}.$$

Sticking to the linear theory as we do in this article allows us to carry the computations further and to get an explicit expression for the constant $\mathfrak{c}(\zeta)$ (or more precisely of its equivalent for the linear theory); we get in particular from the above proposition some upper and lower bounds for $\mathfrak{c}(\zeta)$ (for the experiment considered in Sect. 4.3, one deduces that $\mathfrak{c}(\zeta) \sim 1$).

Proof. Let us first prove a lower bound α_σ^- to α_σ. Let $k_0 > 0$. One has obviously $\alpha_\sigma(k) \geq \alpha(k)$ for all $k \geq 0$ and, since $\alpha(\cdot)$ is a decreasing function, we get

$$\forall 0 \leq k \leq k_0, \quad \alpha(k) \geq \alpha(k_0).$$

For $k \geq k_0$, let us remark first that

$$g' + \frac{\sigma}{\rho^+ + \rho^-}k^2 \geq \beta k,$$

if $\beta > 0$ is small enough. The largest possible value for β is $\beta = 2\frac{\sqrt{g'\sigma}}{\sqrt{\rho^+ + \rho^-}}$, so that

$$\alpha_\sigma(k) = (1 + \frac{\sigma}{g'(\rho^+ + \rho^-)}k^2)\alpha(k) \geq 2\frac{\sqrt{\sigma}}{\sqrt{g'(\rho^+ + \rho^-)}}k\alpha(k).$$

Since the function $k \mapsto k\alpha(k)$ is increasing on $\mathbb{R}^+$, we get that

$$\forall k \geq k_0, \quad \alpha_\sigma(k) \geq 2\frac{\sqrt{\sigma}}{\sqrt{g'(\rho^+ + \rho^-)}}k_0\alpha(k_0).$$

It follows from the above that

$$\forall k \geq 0, \quad \alpha_\sigma(k) \geq \min\{\alpha(k_0), 2\frac{\sqrt{\sigma}}{\sqrt{g'(\rho^+ + \rho^-)}}k_0\alpha(k_0)\}.$$

We now choose k_0 such that the two quantities in the right-hand-side are equal, namely,

$$k_0 = \frac{1}{2}\frac{\sqrt{g'(\rho^+ + \rho^-)}}{\sqrt{\sigma}},$$

to get

$$\forall k \geq 0, \quad \alpha_\sigma(k) \geq \alpha_\sigma^-, \quad \text{with} \quad \alpha_\sigma^- = \alpha\Big(\frac{1}{2}\frac{\sqrt{g'(\rho^+ + \rho^-)}}{\sqrt{\sigma}}\Big).$$

We now turn to derive an upper bound α_σ^+ for the minimum of α_σ. It is obtained simply by observing that $\alpha(k) \leq (kH_0)^{-1}$. This yields

$$\alpha_\sigma(k) \leq \frac{1}{kH_0}\Big(1 + \frac{\sigma}{g'(\rho^+ + \rho^-)}k^2\Big)$$

and an upper bound for the minimum of α_σ is therefore given by the minimum of the right-hand-side

$$\forall k \geq 0, \quad \alpha_\sigma(k) \leq \alpha_\sigma^+, \quad \text{with} \quad \alpha_\sigma^+ = \frac{2}{H_0}\frac{\sqrt{\sigma}}{\sqrt{g'(\rho^+ + \rho^-)}}.$$

The expressions for $\Omega_\pm^{cr}$ follow easily from $\alpha_\sigma^\pm$. □

4 Shallow Water Approximations in the Rigid lid Case

The goal of this section is to investigate whether shallow water models are able to give a good account of the formation of Kelvin-Helmholtz instabilities for two-fluids interfaces, in the rigid lid case. We investigate several shallow water models (some of them new) and study the same stability issues as in Sect. 3 for the full Euler equations. This allows us to give some insight on the ability of these asymptotic models to describe the Kelvin-Helmholtz instabilities correctly.

Notation: When the fluid layers are both of finite depth, the first equation in (11) can equivalently be written under the form (conservation of mass)

$$\partial_t \zeta \pm \nabla \cdot (H^{\pm}(\zeta)\overline{V}^{\pm}) = 0, \qquad \text{with} \quad H^{\pm}(\zeta) = H^{\pm} \pm \zeta, \tag{28}$$

and where $\overline{V}^{\pm}$ is the vertically averaged horizontal velocity,

$$\overline{V}^{\pm}(t,X) = \pm \frac{1}{H^{\pm}(\zeta)} \int_{\mp H^{\pm}}^{\zeta} V^{\pm}(t,X,z)dz.$$

Shallow water approximations consist in writing the second equation of (11) as an approximate evolution equation on $\overline{V}^{\pm}$. In order to do so, $U_{\parallel}^{\pm}$ is related to $\overline{V}^{\pm}$ through an asymptotic expansion in terms of the shallowness parameter $\mu^{\pm} = (H^{\pm})^2/L^2$, where L is the typical horizontal length of the interfacial waves (see for instance [8] for a systematic approach). With a first order expansion, one obtains the Shallow water/Shallow water system (SW/SW) described in Sect. 4.1, while a second order expansion gives the Green-Naghdi/Green-Naghdi system (GN/GN) in Sect. 4.2. We also consider in Sects. 4.2.4 and 4.2.5 two classes of "regularized" models. We show in particular that these news models allow one to reproduce the same values for the critical shear as for the full Euler equations (which is not the case with the standard GN/GN model, even in presence of surface tension). Finally, we apply in Sect. 4.3 some of our results to experiments reported in [29].

4.1 *First Order Approximation*

We consider here the first order approximation under the Shallow Water/Shallow Water regime. We derive the so called Shallow Water/Shallow Water system first, and then the system is linearized near a constant shear flow. The stability criterion is obtained as in the full system case.

4.1.1 The Shallow Water/Shallow Water equations

Up to terms that are $O(\mu)$ times smaller, one can write

$$U_{\parallel}^{\pm} \sim \overline{V}^{\pm} \quad \text{and} \quad \underline{w}^{\pm} \sim 0$$

(see for instance [8, 14, 15, 34, 41]), and the second equation of (11) can be written at leading order under the form

$$\partial_t \overline{V}^{\pm} + g\nabla\zeta + \frac{1}{2}\nabla|\overline{V}^{\pm}|^2 = -\frac{1}{\rho^{\pm}}\nabla\underline{P}^{\pm}.$$

The first order shallow water approximation of (11) is then given by

$$\begin{cases} \partial_t\zeta \pm \nabla\cdot(H^{\pm}(\zeta)\overline{V}^{\pm}) = 0, \\ \partial_t\overline{V}^{\pm} + g\nabla\zeta + \frac{1}{2}\nabla|\overline{V}^{\pm}|^2 = -\frac{1}{\rho^{\pm}}\nabla\underline{P}^{\pm}, \end{cases} \tag{29}$$

with

$$H^{\pm}(\zeta) = H^{\pm} \pm \zeta.$$

Remark 5. As in Remark 1 for the full equations, one can derive a reduced formulation of (11) in terms of ζ and $\overline{V} = \underline{\rho}^{+}\overline{V}^{+} - \underline{\rho}^{-}\overline{V}^{-}$. In dimension $d = 1$ (writing $\overline{v} = \overline{V}$) this system is

$$\begin{cases} \partial_t\zeta + \partial_x\big[H(\zeta)\overline{v}\big] = 0 \\ \partial_t\overline{v} + g'\partial_x\zeta + \frac{1}{2}\partial_x\big[H'(\zeta)\overline{v}^2\big] = -\frac{\sigma}{\rho^{+}+\rho^{-}}\kappa(\zeta), \end{cases} \tag{30}$$

with

$$H(\zeta) = \frac{H^{-}(\zeta)H^{+}(\zeta)}{\underline{\rho}^{+}H^{-}(\zeta) + \underline{\rho}^{-}H^{+}(\zeta)};$$

in dimension $d = 2$, one can also obtain a reduced formulation, but it is more complicated and involves non local operators [8, 30]. Note also that this model can be simplified under the so called Boussinesq approximation when the two densities are very close [9].

4.1.2 Linearization Around a Constant Shear $(\mathbf{c}^+, \mathbf{c}^-)$

We consider here the linear equations governing small perturbations of the constant horizontal shear $\zeta = 0$, $V^\pm = \mathbf{c}^\pm$ (and therefore $\overline{V}^\pm = \mathbf{c}^\pm$).

Linearizing (29) around $\zeta = 0$, $V^\pm = \mathbf{c}^\pm$, we find the following equations for the perturbations $(\dot\zeta, \dot V^\pm)$,

$$\begin{cases} \partial_t \dot\zeta + \mathbf{c}^\pm \cdot \nabla\dot\zeta \pm H^\pm \nabla \cdot \dot V^\pm = 0, \\ \partial_t \dot V^\pm + g\nabla\dot\zeta + \mathbf{c}^\pm \cdot \nabla \dot V^\pm = -\frac{1}{\rho^\pm}\nabla \underline{P}^\pm; \end{cases}$$

note that since the flow is assumed to be potential at leading order in order to derive the nonlinear shallow water equations (30), we restrict to perturbations $\dot V^\pm$ such that $\nabla^\perp \cdot \dot V^\pm = 0$.

In the same spirit as in Remark 1, we can derive a reduced formulation for these equations; denoting $\dot V = \underline{\rho}^+ \dot V^+ - \underline{\rho}^- \dot V^-$, we first get

$$\begin{cases} \partial_t \dot\zeta + \mathbf{c}^\pm \cdot \nabla\dot\zeta \pm H^\pm \nabla \cdot \dot V^\pm = 0, \\ \partial_t \dot V + \langle \mathbf{c}^\pm \rangle \cdot \nabla \dot V + g'\nabla\dot\zeta + [\![\mathbf{c}^\pm]\!] \cdot \nabla \langle \underline{\rho}^\pm \dot V^\pm \rangle = \frac{\sigma}{\rho^+ + \rho^-}\Delta\nabla\dot\zeta; \end{cases} \tag{31}$$

in order to put this system in a closed system of equations on $(\dot\zeta, \dot V)$, we need the following lemma.

Lemma 2. *One has, with* $H_0 = \underline{\rho}^+ H^- + \underline{\rho}^- H^+$,

$$\dot V^+ = \frac{H^-}{H_0}\dot V - \frac{\underline{\rho}^-}{H_0}\Pi\big([\![\mathbf{c}^\pm]\!]\dot\zeta\big), \qquad \dot V^- = -\frac{H^+}{H_0}\dot V - \frac{\underline{\rho}^+}{H_0}\Pi\big([\![\mathbf{c}^\pm]\!]\dot\zeta\big),$$

where $\Pi = \frac{\nabla\nabla^T}{\Delta}$ *is the projector onto gradient vector fields.*

Proof. From the first equation of (31), we get

$$H^+\nabla \cdot \dot V^+ + H^-\nabla \cdot \dot V^- = -[\![\mathbf{c}]\!] \cdot \nabla\dot\zeta;$$

recalling that the perturbations $\dot V^\pm$ are such that $\nabla^\perp \cdot \dot V^\pm = 0$, we get after integrating in space that

$$H^+\dot V^+ + H^-\dot V^- = -\Pi\big([\![\mathbf{c}]\!]\dot\zeta\big).$$

Together with the definition of $\dot V$, namely,

$$\underline{\rho}^+ \dot V^+ - \underline{\rho}^- \dot V^- = \dot V,$$

this yields a system on $\dot V^\pm$ whose solution is provided by the formulas stated in the lemma. □

Using the lemma, we can rewrite (31) under the form as a system of $(\dot{\zeta}, \dot{V})$

$$\begin{cases} \partial_t \dot{\zeta} + \Big(\mathbf{c}^+ - \underline{\rho}^- \frac{H^+}{H_0} [\![\mathbf{c}^\pm]\!]\Big) \cdot \nabla \dot{\zeta} + \frac{H^+ H^-}{H_0} \nabla \cdot \dot{v} = 0, \\ \partial_t \dot{V} + \Big(\mathbf{c}^+ - \underline{\rho}^- \frac{H^+}{H_0} [\![\mathbf{c}^\pm]\!]\Big) \cdot \nabla \dot{V} \\ \qquad\qquad + \Big(g' - \frac{\underline{\rho}^+ \underline{\rho}^-}{H_0} \frac{([\![\mathbf{c}^\pm]\!] \cdot D)^2}{D^2} - \frac{\sigma}{\rho^+ + \rho^-} \Delta\Big) \nabla \dot{\zeta} = 0. \end{cases} \tag{32}$$

4.1.3 The SW/SW Model and the Kelvin-Helmholtz Instabilities

In Sect. 3.2 we derived a stability condition (Stab 1) ensuring the existence of stable modes with low wave number, derived the expression of the critical wave number at the boundary of the stability area, and derived a stronger stability condition (Stab 2) ensuring the stability of *all* modes. We derive here similar conditions for the SW/SW model, which allows us to conclude that this model underestimates the Kelvin-Helmholtz instabilities.

A brief look at (32) shows that the Fourier modes $(\mathscr{F}\dot{\zeta}(\mathbf{k}), \mathscr{F}\dot{\psi}(\mathbf{k}))$ are stable if and only if

$$g' - \frac{\underline{\rho}^+ \underline{\rho}^-}{H_0} \frac{([\![\mathbf{c}^\pm]\!] \cdot \mathbf{k})^2}{|\mathbf{k}|^2} + \frac{\sigma}{\rho^+ + \rho^-} |\mathbf{k}|^2 > 0.$$

As for the full equations in Sect. 3.2, the most unstable modes are those for which $\mathbf{k}$ is parallel to the shear, and we again focus on these most unstable modes. As in Sect. 3.2, we write $[\![\mathbf{c}^\pm]\!] = [\![c^\pm]\!] \mathbf{e}_x$ and $\mathbf{k} = k\mathbf{e}_x$, with $[\![c^\pm]\!] = |[\![\mathbf{c}^\pm]\!]|$ and $k = |\mathbf{k}|$. We readily get that a Fourier mode corresponding to the wavenumber $\mathbf{k}$ is stable for (32) if and only if

$$g'\Big(1 - \frac{[\![c^\pm]\!]^2}{\Omega_{KH}}\Big) + \frac{\sigma}{\rho^+ + \rho^-} k^2 > 0 \iff [\![c^\pm]\!]^2 < \Omega_{KH} \left(1 + \frac{\sigma}{g'(\rho^+ + \rho^-)} k^2\right). \tag{33}$$

It follows easily that *the modes with low wavenumber are stable* if and only if

$$(\text{Stab } 1)_{SW} \qquad [\![c^\pm]\!]^2 < \Omega_{KH},$$

which is exactly the same as the condition (Stab 1) derived for the full equations in Sect. 3.2. If it is satisfied, then all the modes are stable, so that in the SW/SW model, we have for each $[\![c^\pm]\!]$

$$(\text{Critical wavenumber})_{SW} \qquad k_{SW} = +\infty,$$

and the critical shear $\sqrt{\Omega_{SW}^{cr}}$ below which all modes are stable coincides with $\sqrt{\Omega_{KH}}$: *all modes are stable* if and only if

$$(\text{Stab } 2)_{SW} \qquad [\![c^{\pm}]\!]^2 \leq \Omega_{SW}^{cr} \quad \text{with} \quad \Omega_{SW}^{cr} = \Omega_{KH}.$$

This condition differs in general from (Stab 2) derived in Sect. 3.2 for the full equation since we always have $\Omega^{cr} \leq \Omega_{KH}$ and, in general, $\Omega^{cr} < \Omega_{KH}$ (this is the case for small capillary effects, see Fig. 3). Therefore, *the SW/SW model underestimates Kelvin-Helmholtz instabilities.*

Remark 6. The full, nonlinear, analysis can be carried out for the SW/SW equations. This has been done in [30] and improved in [10]; in particular, the SW/SW equations (29) (or equivalently, (30) when $d = 1$) are locally well-posed provided that the initial condition $(\zeta^0, \overline{V}^{\pm,0})$ satisfies

$$[\![\overline{V}^{\pm,0}]\!]^2 < \Omega_{KH}(\zeta^0) \quad \text{with} \quad \Omega_{KH}(\zeta^0) = \frac{g'(\underline{\rho}^- H^-(\zeta^0) + \underline{\rho}^+ H^+(\zeta^0))}{\underline{\rho}^+ \underline{\rho}^-},$$

and $H^{\pm}(\zeta^0) = H^{\pm} \pm \zeta^0$. The stability condition $(\text{Stab } 1)_{SW}$ is therefore the condition one naturally expects from the condition above when linearizing around the constant shear $(\zeta = 0, \overline{V}^{\pm} = \mathbf{c}^{\pm})$.

4.2 *Second Order Approximation*

This section is devoted to the second order approximation under the SW/SW regime, which is the so called Green-Naghdi/Green-Naghdi (or GN/GN) system. We linearize the system again near a constant shear flow. A different stability criterion is then obtained based on the linearized system. Besides, we also derive and discuss two "regularized" GN/GN models in order to get a better approximation on the critical shear and therefore a better description of the formation of Kelvin-Helmholtz instabilities.

4.2.1 The Green-Naghdi/Green-Naghdi Equations

We proceed as in Sect. 4.1, but go one step further in the asymptotic expression of $\dot{V}^{\pm}$ in terms of $\overline{V}^{\pm}$, namely

$$U_{\parallel}^{\pm} \sim \big(1 + \mathcal{T}^{\pm}\big)\overline{V}^{\pm} \quad \text{and} \quad \underline{w}^{\pm} \sim \mp H^{\pm}(\zeta)\partial_x \overline{V}^{\pm}$$

where

$$\mathcal{T}^{\pm}\bullet = -\frac{1}{3H^{\pm}(\zeta)}\nabla\big(H^{\pm}(\zeta)^3\nabla\cdot\bullet\big) \quad \text{with} \quad H^{\pm}(\zeta) = H^{\pm} \pm \zeta \tag{34}$$

(we refer to [15, 34] for the derivation of this asymptotic expansion).

Putting the first equation of (11) under the equivalent form (28), and plugging the above approximations for $U_{\parallel}^{\pm}$ and $\underline{w}^{\pm}$ into the second equation of (11), we obtain the second order shallow water approximation [15, 18, 36] from which many other asymptotic models can be derived [27],

$$\begin{cases} \partial_t\zeta \pm \nabla\cdot(H^{\pm}(\zeta)\overline{V}^{\pm}) = 0, \\ (1+\mathcal{T}^{\pm})\partial_t\overline{V}^{\pm} + g\nabla\zeta + \frac{1}{2}\nabla|\overline{V}^{\pm}|^2 + Q^{\pm} = -\frac{1}{\rho^{\pm}}\nabla\underline{P}^{\pm}, \end{cases} \tag{35}$$

with $[\![\underline{P}^{\pm}]\!] = \sigma\kappa(\zeta)$ and

$$Q^{\pm}(\overline{V}^{\pm}) = -\frac{1}{3H^{\pm}(\zeta)}\nabla\Big(H^{\pm}(\zeta)^3\big(\overline{V}^{\pm}\cdot\nabla(\nabla\cdot\overline{V}^{\pm}) - |\nabla\cdot\overline{V}^{\pm}|^2\big)\Big).$$

In the one fluid case $\rho^{-} = 0$, these equations coincide with the Green-Naghdi (or Serre) equations; adopting the terminology of [8], we therefore refer to these equations as the Green-Naghdi/Green-Naghdi (or GN/GN) equations. As shown in [15] these equations have a (non-canonical) Hamiltonian structure.

4.2.2 Linearization Around a Constant Shear (c^{+}, c^{-})

Though it is possible to work with the two dimensional case $d = 2$ as in Sect. 4.1, we focus our interest here on the one dimensional case $d = 1$ for the sake of clarity. We therefore denote by v the horizontal velocity instead of V (and similarly, we write $\overline{v}$, etc). Linearizing (35) around $\zeta = 0$, $\overline{v}^{\pm} = c^{\pm}$, we find the following equations,

$$\begin{cases} \partial_t\dot{\zeta} + c^{\pm}\partial_x\dot{\zeta} \pm H^{\pm}\partial_x\dot{v}^{\pm} = 0, \\ \Big(1 - \frac{1}{3}(H^{\pm})^2\partial_x^2\Big)(\partial_t\dot{v}^{\pm} + c^{\pm}\partial_x\dot{v}^{\pm}) + g\partial_x\dot{\zeta} = -\frac{1}{\rho^{\pm}}\partial_x\underline{P}^{\pm}. \end{cases}$$

Using the formulas of Lemma 2, we can rewrite the second equation under the form

$$\begin{cases} \partial_t\dot{\zeta} + \Big(c^{+} - \underline{\rho}^{-}\frac{H^{+}}{H_0}[\![c^{\pm}]\!]\Big)\partial_x\dot{\zeta} + \frac{H^{+}H^{-}}{H_0}\partial_x\dot{v} = 0, \\ \Big(1 - \frac{1}{3}(H^{\pm})^2\partial_x^2\Big)\Big((\partial_t + c^{\pm}\partial_x)\big(\pm\frac{H^{\mp}}{H_0}\dot{v} - \frac{\underline{\rho}^{\mp}}{H_0}[\![c^{\pm}]\!]\dot{\zeta}\big)\Big) + g\partial_x\dot{\zeta} = -\frac{1}{\rho^{\pm}}\partial_x\underline{P}^{\pm}. \end{cases}$$

Eliminating the pressure from these equations, we get after some computations,

$$\begin{cases} \partial_t \dot{\zeta} + c_0 \partial_x \dot{\zeta} + \frac{H^+ H^-}{H_0} \partial_x \dot{v} = 0, \\ (1 - \alpha H_0^2 \partial_x^2)(\partial_t \dot{v} + c_0 \partial_x \dot{v}) + \gamma [\![c^\pm]\!] H_0^2 \partial_x^3 \dot{v} \\ \qquad + \Big(g' - \frac{\underline{\rho}^+ \underline{\rho}^-}{H_0} [\![c^\pm]\!]^2 (1 - \beta H_0^2 \partial_x^2) - \frac{\sigma}{\rho^+ + \rho^-} \partial_x^2 \Big) \partial_x \dot{\zeta} = 0, \end{cases} \tag{36}$$

with

$$c_0 = \Big(c^+ - \underline{\rho}^- \frac{H^+}{H_0} [\![c^\pm]\!] \Big), \qquad \alpha = \frac{1}{3} (\underline{\rho}^+ H^+ + \underline{\rho}^- H^-) \frac{H^+ H^-}{H_0^3},$$

$$\beta = \frac{\underline{\rho}^- (H^+)^3 + \underline{\rho}^+ (H^-)^3}{3 H_0^3}, \qquad \gamma = -\frac{2}{3} \frac{\underline{\rho}^+ \underline{\rho}^-}{H_0} ((H^+)^2 - (H^-)^2) \frac{H^+ H^-}{H_0^3}.$$

4.2.3 The GN/GN Model and the Kelvin-Helmholtz Instabilities

As in Sect. 4.1.3 with the SW/SW model, we want to study the linear stability of (36) and compare it with the behavior of the full equations investigated in Sect. 3.2.

With a simple analysis, one gets that the Fourier mode $(\mathscr{F}\zeta(k), \mathscr{F}\dot{v}(k))$ is stable (in the sense of Sect. 3.2) for the equations (36) if and only if

$$g' \Big(1 + \frac{\sigma}{g'(\rho^+ + \rho^-)} k^2 \Big)(1 + \alpha H_0^2 k^2) \\ > \underline{\rho}^+ \underline{\rho}^- \frac{1}{H_0} [\![c^\pm]\!]^2 \Big(1 + \frac{1}{3} (H^+)^2 k^2 \Big) \Big(1 + \frac{1}{3} (H^-)^2 k^2 \Big),$$

or equivalently

$$[\![c^\pm]\!]^2 < \Omega_{KH} \, \alpha_{GN_\sigma}(k),$$

with

$$\alpha_{GN_\sigma}(k) = (1 + \frac{1}{\mathrm{Bo}} H_0^2 k^2) \alpha_{GN}(k)$$

and

$$\alpha_{GN}(k) = \frac{1}{H_0} \Big(\frac{\underline{\rho}^+ H^-}{(1 + \frac{1}{3}(H^-)^2 k^2)} + \frac{\underline{\rho}^- H^+}{(1 + \frac{1}{3}(H^+)^2 k^2)} \Big).$$

It follows immediately that *the modes with low wavenumber are stable* if and only if

$$(\text{Stab } 1)_{GN_\sigma} \qquad [\![c^\pm]\!]^2 < \Omega_{KH},$$

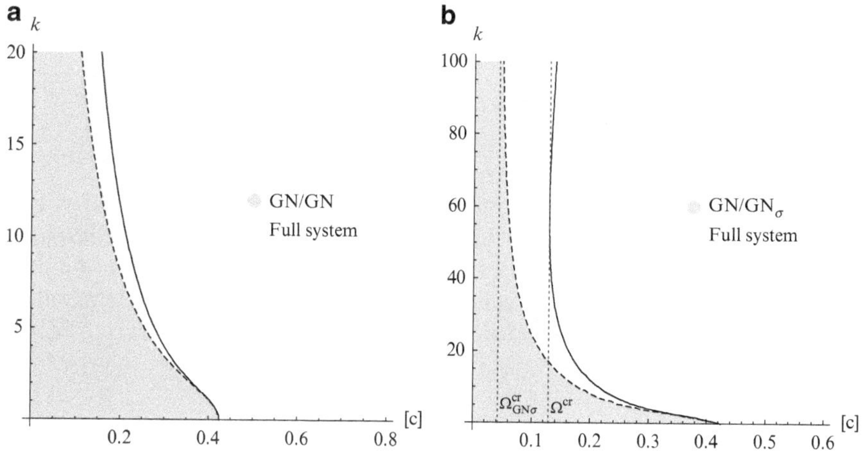

Fig. 5 Stables areas for the full system (Euler) and the GN/GN model without or with surface tension ($\sigma = 0.073$). Parameters $\rho^+ + \rho^- = 2021$, $\gamma = 0.494/0.506$, $H^+ = 0.62$, $\delta = 0.15/0.62$ and $g = 9.81$. (**a**) Without surface tension. (**b**) With surface tension

which is exactly the same as the condition (Stab 1) and (Stab 1)$_{SW}$ derived respectively for the full equations and the SW/SW model. If it is satisfied, there is a unique $k_{GN_\sigma} \in \mathbb{R}^+_* \cup \{+\infty\}$ for each $[\![c^\pm]\!]$, such that all the modes corresponding to $\xi \in \mathbb{R}^d$ with $|\xi| < k_{GN_\sigma}$ are stable,

$$\text{(Critical wavenumber)}_{GN_\sigma} \qquad k_{GN_\sigma} = \min \alpha^{-1}_{GN_\sigma}\Big(\frac{|[\![c^\pm]\!]|^2}{\Omega_{KH}}\Big).$$

In fact, one always has (at least for reasonable values of σ as in Footnote 3) that $\alpha_{GN_\sigma}(k) < \alpha_\sigma(r)$ for $\sigma \geq 0$, with α_σ as in (26). Therefore we find $k_{GN_\sigma} < k_{KH}$ (see Fig. 5b), and *the GN/GN model* (35) *therefore overestimates Kelvin-Helmholtz instabilities* (see Fig. 5).

In absence of surface tension ($\sigma = 0$), $\alpha_{GN_\sigma}(\cdot) = \alpha_{GN}(\cdot)$ is a decreasing function and goes to zero at infinity. As for the full equations, there are therefore always unstable modes in presence of a nonzero shear $[\![c^\pm]\!]$, as noted in [13]. In presence of surface tension, the function $\alpha_{GN_\sigma}(\cdot)$ is still decaying[3] but tends to a non zero limit

[3]If σ is very large, $\alpha_{GN_\sigma}(\cdot)$ is no longer decreasing over $\mathbb{R}^+$. However, if σ satisfies

$$\sigma \leq \frac{1}{3} g'(\rho^+ + \rho^-)\min\{(H^+)^2, (H^-)^2\},$$

which is always satisfied in realistic physical configurations, then $\alpha_{GN_\sigma}(\cdot)$ is indeed a decreasing function. We always assume that we are in such a regime.

at infinity. There exists therefore a critical value $\Omega^{cr}_{GN_\sigma} := \Omega_{KH} \min_{\mathbb{R}^+} \alpha_{GN_\sigma}$ such that *all modes are stable* if and only if

$$(\text{Stab }2)_{GN_\sigma} \qquad [\![c^\pm]\!]^2 \leq \Omega^{cr}_{GN_\sigma} = \Omega_{KH}\frac{3}{\text{Bo}}\Big(1+\underline{\rho}^+\underline{\rho}^-\frac{(H^+-H^-)^2}{H^+H^-}\Big);$$

the same analysis as for the discussion on the critical wave number shows that $\Omega^{cr}_{GN_\sigma} < \Omega^{cr}$; the GN/GN model being unable to reproduce correctly the threshold for Kelvin-Helmholtz instabilities, we are led to derive other models in the next sections. Note that contrary to the standard GN/GN model which has a Hamiltonian structure [15], these regularized models are in general not Hamiltonian (see Sect. 4 of [6] for a discussion on this aspect in the weakly nonlinear case).

4.2.4 A First Class of Regularized Models

As shown in Sect. 4.2.3, the second order shallow water model (or Green-Naghdi/Green-Naghdi) is always unstable for large wave numbers in absence of surface tension (even if the shear $[\![c^\pm]\!]$ is very small). This led Choi, Barros and Jo [13] to derive an asymptotically equivalent model with better high frequency behavior in the sense that for a small enough shear $[\![c^\pm]\!]$, *all* Fourier modes are stable.

Their idea was to use a technique commonly used in the one fluid case (water waves) to improve the dispersion relation of asymptotic models; it consists in working with a velocity variable that differs from the averaged velocity $\overline{v}$; this can typically be the velocity at some fixed depth or a given level line, of the fluid domain (see [5, 7, 39] and more generally Sect. 5.2 of [33]). This approach has also been used in the context of interfacial waves [8, 38]; the idea in [13] (see also [2], and [16] for a related approach) was to use it to remove the Kelvin-Helmholtz instabilities from the standard GN/GN model. The authors rewrote the GN/GN equations (35) in the variables $(\zeta, \hat{v}_r^\pm)$ instead of $(\zeta, \overline{v}^\pm)$, where $\hat{v}_r^\pm$ is the horizontal velocity at the fixed height $\hat{z}_r^\pm$, with

$$\hat{z}_r^\pm = \mp H^\pm\big(1-\sqrt{-2r+1/3}\big), \qquad -1/3 \leq r \leq 1/6$$

(so that $\hat{z}_r^\pm = 0$ if $r = -1/3$ and $\hat{z}_r^\pm = \mp H^\pm$ if $r = 1/6$). Using a standard shallow water expansion of the velocity field in the fluid domain (see for instance [41], Sect. 13.11 or [33], Sect. 5.6.2), $\hat{v}_r^\pm$ can be related to $\overline{v}^\pm$ through the relation

$$\overline{v}^\pm \sim \hat{v}_r^\pm - \frac{1}{6}H^\pm(\zeta)^2\partial_x^2\hat{v}_r^\pm + \Big(-r+\frac{1}{6}\Big)(H^\pm)^2\partial_x^2\hat{v}_r^\pm \qquad (H^\pm(\zeta) = H^\pm \pm \zeta).$$

The regularized equations of [13] are obtained by replacing $\overline{v}^\pm$ by this approximation in (35) and dropping smallest order terms. For $r = 0$, the linearization of these equations around the rest state coincides with the linearization of the standard GN/GN model, but the nonlinear terms differ. Inspired by Sect. 5.2 of [33] we therefore propose a slight modification (and generalization to the two dimensional case $d = 2$) to the approach of [13] which consists in working with the velocity $V_r^{\pm}$ defined as

$$V_r^{\pm} = (1 + 3r^{\pm}\mathscr{T}^{\pm})^{-1}\overline{V}^{\pm} \qquad (r \geq 0) \qquad (\text{with } r^{\pm} = \frac{H_0^2}{(H^{\pm})^2}r \quad \text{and } r \geq 0),$$

so that, with $\mathscr{T}^{\pm}$ as in (34),

$$\overline{V}^{\pm} = V_r^{\pm} + 3r^{\pm}\mathscr{T}^{\pm}V_r^{\pm}.$$

Replacing $\overline{V}^{\pm}$ by this approximation in (35) and dropping the $O((\mu^{\pm})^2)$ terms (with $\mu^{\pm} = (H^{\pm})^2/L^2$, L being the typical horizontal scale), one obtains the following family of regularized GN/GN models indexed by the parameter $r \geq 0$,

$$\begin{cases} \partial_t \zeta \pm \nabla \cdot \big[H^{\pm}(\zeta)(1 + 3r^{\pm}\mathscr{T}^{\pm})V_r^{\pm}\big] = 0, \\ \big[1 + (1 + 3r^{\pm})\mathscr{T}^{\pm}\big]\partial_t V_r^{\pm} + g\nabla\zeta + \frac{1}{2}\nabla|V_r^{\pm}|^2 + Q_r^{\pm} = -\frac{1}{\rho^{\pm}}\nabla \underline{P}^{\pm} \end{cases} \tag{37}$$

where

$$Q_r^{\pm} := Q^{\pm}(V_r^{\pm}) + 3r^{\pm}\nabla(V_r^{\pm} \cdot \mathscr{T}^{\pm}V_r^{\pm}) + 3r^{\pm}(\partial_t \mathscr{T}^{\pm})V_r^{\pm},$$

where $Q^{\pm}(\cdot)$ as defined before in original GN/GN model and

$$(\partial_t \mathscr{T}^{\pm})V_r^{\pm} = -\frac{1}{3H^{\pm}(\zeta)^2}\nabla \cdot \big(H^{\pm}(\zeta)V_r^{\pm}\big)\nabla\big(H^{\pm}(\zeta)^3\nabla \cdot V_r^{\pm}\big)$$
$$+\frac{1}{H^{\pm}(\zeta)}\nabla\Big(H^{\pm}(\zeta)^2\nabla \cdot \big(H^{\pm}(\zeta)V_r^{\pm}\big)(\nabla \cdot V_r^{\pm})\Big).$$

In its one dimensional form $d = 1$, this model is a small variant of the system (3.7) and (3.8) of [13]; the first interest of working with (37) is that the standard GN/GN model (35) is a particular case (corresponding to $r = 0$) of (37), while it does not belong to any of the regularized systems derived in [13]. The second and main advantage is that the definition of $v_r^{\pm}$ makes sense[4] for all $r \geq 0$ while the definition of $\hat{v}_r$ in [13] requires that $r \leq 1/6$.

[4]When $r \geq 1/6$, there is no obvious physical meaning for $v_r^{\pm}$. For $0 \leq r \leq 1/6$, and for small amplitude waves $v_r^{\pm}$ is the horizontal velocity evaluated on the level line $\{z = \hat{z}_r^{\pm}(1 \pm \frac{\zeta}{H^{\pm}}) + \zeta\}$ (see [33], Sect. 5.6.2).

As for the previous examples, we now restrict to the case $d = 1$ to investigate how this regularized model handles the Kelvin-Helmholtz instabilities. Linearizing around the constant shear $(\zeta, v_r^\pm) = (0, c^\pm)$, we get

$$\begin{cases} \partial_t\dot{\zeta} + c^\pm\partial_x\dot{\zeta} \pm H^\pm(1 - rH_0^2\partial_x^2)\partial_x\dot{v}_r^\pm = 0, \\ \Big(1 - \frac{1}{3}(3rH_0^2 + (H^\pm)^2)\partial_x^2\Big)(\partial_t\dot{v}_r^\pm + c^\pm\partial_x\dot{v}_r^\pm) + g\partial_x\dot{\zeta} = -\frac{1}{\rho^\pm}\partial_x\underline{P}^\pm. \end{cases}$$

The formulas of Lemma 2 must now be replaced by

$$\dot{v}_r^+ = \frac{H^-}{H_0}\dot{v}_r - \frac{\underline{\rho}^-}{H_0}[\![c^\pm]\!](1 - rH_0^2\partial_x^2)^{-1}\dot{\zeta}, \qquad \dot{v}_r^- = -\frac{H^+}{H_0}\dot{v}_r - \frac{\underline{\rho}^+}{H_0}[\![c^\pm]\!](1 - rH_0^2\partial_x^2)^{-1}\dot{\zeta}$$

so that, proceeding as for (36), we are led (in absence of surface tension) to the system

$$\begin{cases} \partial_t\dot{\zeta} + c_0\partial_x\dot{\zeta} + \frac{H^+H^-}{H_0}(1 - rH_0^2\partial_x^2)\partial_x\dot{v}_r = 0, \\ \Big(1 - (\alpha + r)H_0^2\partial_x^2\Big)(\partial_t\dot{v}_r + c_0\partial_x\dot{v}_r) + \gamma[\![c^\pm]\!]H_0^2\partial_x^3\dot{v}_r \\ \qquad + \Big(g' - \frac{\underline{\rho}^+\underline{\rho}^-}{H_0}[\![c^\pm]\!]^2(1 - rH_0^2\partial_x^2)^{-1}\big(1 - (\beta + r)H_0^2\partial_x^2\big)\Big)\partial_x\dot{\zeta} = 0, \end{cases} \tag{38}$$

where $\dot{v} = \underline{\rho}^+\dot{v}^+ - \underline{\rho}^-\dot{v}^-$. The Fourier mode $(\mathscr{F}\dot{\zeta}(k), \mathscr{F}\dot{v}(k))$ is therefore stable for the system (38) if and only if

$$|[\![c^\pm]\!]|^2 \leq \Omega_{KH}\alpha_{GN_r}(\xi)$$

where, for all $r \geq 0$,

$$\alpha_{GN_r}(k) = \underline{\rho}^+\frac{H^-}{H_0}\frac{1 + rH_0^2k^2}{1 + \big(\frac{1}{3}(H^-)^2 + rH_0^2\big)k^2} + \underline{\rho}^-\frac{H^+}{H_0}\frac{1 + rH_0^2k^2}{1 + \big(\frac{1}{3}(H^+)^2 + rH_0^2\big)k^2}.$$

Comparing α_{GN_r} with α_{GN} from the standard GN/GN model, one of course finds that $\alpha_{GN_r}(k) = \alpha_{GN}(k)$ when $r = 0$, since the standard GN/GN model then exactly coincides with (37). Both α_{GN_r} and α_{GN} are decreasing functions but when $r > 0$, $\alpha_{GN_r}(k)$ tends to a nonzero value as $k \to \infty$ while $\alpha_{GN}(k)$ tends to 0 as $k \to \infty$. It follows similarly as before that *there exists stable modes for low wavenumber if and only if*

$$(\text{Stab } 1)_{GN_r} \qquad |[\![c^\pm]\!]|^2 < \Omega_{KH}.$$

With this condition, the critical wave number corresponding to $[\![c^\pm]\!]$ is

$$\text{(Critical wavenumber)}_{GN_r} \qquad k_{GN_r} = \min \alpha_{GN_r}^{-1}\Big(\frac{|[\![c^\pm]\!]|^2}{\Omega_{KH}}\Big),$$

and there is a critical value $\Omega_{GN_r}^{cr}$ such that *all modes are stable* if and only if

$$\text{(Stab 2)}_{GN_r} \qquad [\![c^\pm]\!]^2 \leq \Omega_{GN_r}^{cr};$$

one has $\Omega_{GN_r}^{cr} = \Omega_{KH} \min \alpha_{GN,r}$, that is

$$\Omega_{GN_r}^{cr} = \Omega_{KH}\Big[\underline{\rho}^+\frac{H^-}{H_0}\frac{rH_0^2}{\frac{1}{3}(H^-)^2 + rH_0^2} + \underline{\rho}^-\frac{H^+}{H_0}\frac{rH_0^2}{\frac{1}{3}(H^+)^2 + rH_0^2}\Big].$$

In [13], the authors choose $r = 1/6$, which gives the largest possible value for $\Omega_{GN_r}^{cr}$ (namely, $\Omega_{GN_r}^{cr} = 1/3\Omega_{KH}$) for a system written in $(\zeta, \hat{v}_r^\pm)$ and therefore subject to the constraint $r \geq 1/6$. Writing the regularized GN/GN model in terms of $(\zeta, V_r^\pm)$ as suggested here, it is possible to take any $r \geq 0$, and therefore $\Omega_{GN_r}^{cr}$ can take any value in $[0, \Omega_{KH})$. The lower bound of this interval is achieved when $r = 0$; *the standard GN/GN is therefore the most unstable of the whole family of regularized models* (37). The fact that the whole range $[0, \Omega_{KH})$ can be covered with our approach implies in particular that it is possible to find some $r_0 \geq 0$ such that the regularized model (37) has exactly the critical shear as the full equations (11)–(12) with surface tension, i.e. such that $\Omega_{GN_{r_0}}^{cr} = \Omega^{cr}$. One readily finds that

$$r_0 = \frac{-b + \sqrt{b^2 - 4ac}}{2a}, \tag{39}$$

with

$$a = \Big(1 - \frac{\Omega_{cr}}{\Omega_{KH}}\Big)H_0^4,$$

$$b = \frac{H_0^2}{3}\Big(H^+H^-\frac{\underline{\rho}^+H^+ + \underline{\rho}^-H^-}{H_0} - \frac{\Omega_{cr}}{\Omega_{KH}}(((H^-)^2 + (H^+)^2)\Big),$$

$$c = -\frac{1}{9}\frac{\Omega_{cr}}{\Omega_{KH}}(H^+H^-)^2.$$

Remark 7. The lines above show that it is possible to reproduce the same behavior with respect to the apparition of Kelvin-Helmholtz instabilities as for the full Euler equations with surface tension, even though surface tension is not taken into account in (37). This is because the regularization used here can behave as a control mechanism for high frequencies that can be tuned to reproduce the surface tension effects in the full Euler equations. Taking surface tension into account in (37) would lead to a stability condition of the form

$$|[\![c^{\pm}]\!]|^2 \leq \Omega_{KH}\alpha_{GN_{\sigma,r}}(k) \quad \text{with} \quad \alpha_{GN_{\sigma,r}}(k) = \Big(1 + \frac{1}{\mathrm{Bo}}H_0^2k^2\Big)\alpha_{GN_r}(k).$$

4.2.5 A Second Class of Regularized Models

In the previous section, we extended the approach of [13] and obtained a very flexible approach to describe the formation of Kelvin-Helmholtz instabilities. There is however a drawback when working with the models of [13] and more generally with the regularized models (37). Indeed, the first equation in (35) is *exact* (conservation of mass), while the first equation in (37) is only valid up to order $O((\mu^{\pm})^2)$ when $r \neq 0$ (we recall that $\mu^{\pm} = (H^{\pm})^2/L^2$ where L is the typical horizontal scale).

We propose here a second family of regularized models, that keeps the same advantages as (37) to handle Kelvin-Helmholtz instabilities, but for which the conservation of mass is also exact.

We recall that at first order in the shallowness parameter $\mu^{\pm}$, the full equations (11) can be approximated by the SW/SW equations (29) and that the GN/GN equations (35) are a second order approximation. In particular, the second equation of (35) can be written under the form

$$\partial_t\overline{V}^{\pm} = -g\nabla\zeta - \frac{1}{2}\nabla|\overline{V}^{\pm}|^2 - \frac{1}{\rho^{\pm}}\nabla\underline{P}^{\pm} + O(\mu^{\pm}).$$

We now use a generalization of the so called BBM trick, namely, we write

$$\begin{aligned}\mathscr{T}^{\pm}\partial_t\overline{V}^{\pm} &= (1+3s^{\pm})\mathscr{T}^{\pm}\partial_t\overline{V}^{\pm} + 3s^{\pm}g\mathscr{T}^{\pm}\nabla\zeta + 3s^{\pm}\frac{1}{2}\mathscr{T}^{\pm}\nabla|\overline{V}^{\pm}|^2\\ &\quad +3s^{\pm}\frac{1}{\rho^{\pm}}\mathscr{T}^{\pm}\nabla\underline{P}^{\pm} + O(\mu^{\pm})\end{aligned}$$

(with $s^{\pm} = \frac{H_0^2}{(H^{\pm})^2}s$ where $s \in \mathbb{R}$), and plug this into the second equation of (35) to obtain the following family of regularized GN/GN systems indexed by the parameters $s^{\pm}$ as

$$\begin{cases}\partial_t\zeta \pm \nabla\cdot(H^{\pm}(\zeta)\overline{V}^{\pm}) = 0,\\ \big(1+(1+3s^{\pm})\mathscr{T}^{\pm}\big)\partial_t\overline{V}^{\pm} + g\nabla\zeta + \frac{1}{2}\nabla|\overline{V}^{\pm}|^2 + \mathcal{Q}_s^{\pm} = -\frac{1}{\rho^{\pm}}(1+3s^{\pm}\mathscr{T}^{\pm})\nabla\underline{P}^{\pm}\end{cases} \tag{40}$$

with

$$Q_s^{\pm} = Q^{\pm} + 3s^{\pm}\Big(g\mathcal{T}^{\pm}\nabla\zeta + \frac{1}{2}\mathcal{T}^{\pm}\nabla|\overline{V}^{\pm}|^2\Big).$$

Focusing on the 1-D case and linearizing this system around the constant shear (c^+, c^-), one finds the following linear system

$$\begin{cases} \partial_t\dot{\zeta} + c^{\pm}\partial_x\dot{\zeta} \pm H^{\pm}\partial_x\dot{v}^{\pm} = 0, \\ \Big(1 - (\frac{1}{3} + s^{\pm})(H^{\pm})^2\partial_x^2\Big)(\partial_t\dot{v}^{\pm} + c^{\pm}\partial_x\dot{v}^{\pm}) + g\big(1 - s^{\pm}(H^{\pm})^2\partial_x^2\big)\partial_x\dot{\zeta} \\ \quad = -\frac{1}{\rho^{\pm}}\big(1 - s^{\pm}(H^{\pm})^2\partial_x^2\big)\nabla \underline{P}^{\pm}. \end{cases}$$

Using Lemma 2 again to express $\dot{v}^{\pm}$ in terms of $\dot{v} = \underline{\rho}^+\dot{v}^+ - \underline{\rho}^-\dot{v}^-$ and $\dot{\zeta}$, and proceeding as for (36), this system can be rewritten as

$$\begin{cases} \partial_t\dot{\zeta} + c_0\partial_x\dot{\zeta} + \frac{H^+H^-}{H_0}\partial_x\dot{v} = 0, \\ \Big(1 - (\alpha + s)H_0^2\partial_x^2\Big)\big(\partial_t\dot{v} + c_0\partial_x\dot{v}\big) + \gamma[\![c^{\pm}]\!]H_0^2\partial_x^3\dot{v} \\ \quad + \Big(g'(1 - sH_0^2\partial_x^2) - \frac{\underline{\rho}^+\underline{\rho}^-}{H_0}[\![c^{\pm}]\!]^2\big(1 - (\beta + s)H_0^2\partial_x^2\big)\Big)\partial_x\dot{\zeta} = 0, \end{cases} \tag{41}$$

where α, β, γ and c_0 are the same as before.

With a simple analysis, one gets that the Fourier mode $(\mathcal{F}\dot{\zeta}(k), \mathcal{F}\dot{v}(k))$ is stable (in the sense of Sect. 3.2) for the equations (41) if and only if

$$[\![c^{\pm}]\!]^2 < \Omega_{KH}\,\alpha_{GN_s}(k),$$

with

$$\alpha_{GN_s}(k) = \underline{\rho}^+\frac{H^-}{H_0}\frac{1 + sH_0^2k^2}{1 + \big(\frac{1}{3}(H^-)^2 + sH_0^2\big)k^2} + \underline{\rho}^-\frac{H^+}{H_0}\frac{1 + sH_0^2k^2}{1 + \big(\frac{1}{3}(H^+)^2 + sH_0^2\big)k^2};$$

this is exactly the same as the function α_{GN_r} derived for the first regularization (37). Consequently, keeping an *exact* equation for the conservation of mass, *the second regularization* (40), *provides the same description of the formation of Kelvin-Helmholtz instabilities as* (37).

4.3 Application

We apply here some of the results derived above to analyze some experimental data that can be found in [29], where the propagation of interfacial solitary waves is studied; it is in particular shown that large amplitude solitary waves are destroyed by Kelvin-Helmholtz instabilities.

First of all, in the case of solitary waves, one can express the shear speed $[\![c^{\pm}]\!]$ with the wave speed c and the amplitude of solitary wave a. In fact, plugging the transverse solution $\zeta(x-ct)$ into the first equation of (35) and integrating with respect to x in 1-D case results in

$$-c\zeta \pm H^{\pm}(\zeta)\overline{v}^{\pm} = 0$$

where we assume ζ, $\overline{v} \to 0$ when x goes to ∞. This implies that at the bottom point of the solitary wave, i.e. $\zeta = a$ (with $a < 0$ for a depression solitary wave), one has

$$\overline{v}^{\pm} = \pm \frac{c\,a}{H^{\pm} \mp a}.$$

If we define the jump of speed $[\![c^{\pm}]\!]$ as $[\![c^{\pm}]\!] := \overline{v}^{+} - \overline{v}^{-}$, we can have

$$[\![c^{\pm}]\!] = \frac{ca(H^{+} + H^{-})}{(H^{+} - a)(H^{-} + a)}.$$

One can also find this formula in [13].

Now we want to fix the value of surface tension coefficient σ for the linearized full system (22) according to this expression, the critical value Ω^{cr} for (22) and the experimental data in [29].

One has from [29] that $\rho^{+} = 1022\,\mathrm{g/m^3}$, $\rho^{-} = 999\,\mathrm{kg/m^3}$ (hence $\underline{\rho}^{+} = 0.506$, $\underline{\rho}^{-} = 0.494$), $H^{+} = 0.62\,\mathrm{m}$ and $H^{-} = 0.15\,\mathrm{m}$. Taking the experimental values $c/c_0 - 1 = 0.24$ and $a/H^{-} = 1.23$ (corresponding to the last experiment before the apparition of Kelvin-Helmholtz instabilities in Fig. 7 in [29]), where $c_0^2 = \frac{g' H^{+} H^{-}}{\underline{\rho}^{+} H^{-} + \underline{\rho}^{+} H^{+}}$ is the linear long wave speed from KdV equation, one can compute that $[\![c^{\pm}]\!] = 0.20\,\mathrm{m/s}$ at the onset of Kelvin-Helmholtz instabilities. This means that, with the notations of Sect. 3.2, one has $\Omega^{cr} = [\![c^{\pm}]\!]^2 = 0.04\,(\mathrm{m/s})^2$. Using Proposition 1, we use the approximation $\Omega_{+}^{cr} \sim \Omega^{cr}$ and the definition of Ω_{+}^{cr} to get

$$\frac{1}{\mathrm{Bo}} = \frac{1}{4}\Big(\frac{\Omega_{+}^{cr}}{\Omega_{KH}}\Big)^2 \sim \frac{1}{4}\Big(\frac{0.04}{\Omega_{KH}}\Big)^2;$$

from the definition (5) of the Bond number, we deduce that

$$\sigma \sim \frac{1}{4g'}(0.04)^2 \frac{(\rho^{+}\rho^{-})^2}{(\rho^{+} + \rho^{-})^3} = 0.45$$

(using this value of σ, on can compute that the lower bound estimate for Ω_{cr} provided by Proposition 1 is $\Omega_{cr}^{-} = 0.038$ so that the error made by approximating Ω_{cr} by Ω_{cr}^{+} to compute σ is only 3 %). This value of σ is significantly larger than the air-water surface tension ($\sigma \sim 0.07$), which was to be expected because we treat here the surface tension as a model to describe the various different stabilizing effects (such as the mixing layer) between two miscible fluids. Note however that even though σ is larger than usual, it does not significantly affect the behavior of the wave while Kelvin-Helmholtz instabilities do not destroy it. Indeed, surface tension in (36) plays a role through a term of the form $-\sigma/(\rho^+ + \rho^-)\partial_x^3\zeta$ while the dispersive term coming from non-hydrostatic terms is $-\alpha H_0^2\partial_x^2\partial_t v \sim \alpha g' H_0^2\partial_x^3\zeta$. Using the above data then shows that the dispersive effects coming from surface tension are only about 5.9 % of non hydrostatic ones. This confirms the general picture exhibited in [32], namely, that surface tension is necessary for the wave to exist, but that it does not affect significantly its propagation.

As shown in Sect. 4.2.4, it is also possible to find some r_0 such that the regularized model (37) reproduces the same critical value for the apparition of Kelvin-Helmholtz instabilities (i.e. let $\Omega^{cr}_{GN_r} = \Omega^{cr}$); according to (39), one gets $r_0 = 0.12$. Note in particular that $0 < r_0 < 1/6$ and is therefore in the range allowed by the regularized models of [13], but close to its upper range. Slightly more stable configurations would therefore fall out of the range of validity of [13], but would stay within the range allowed for (37); Since the stability criterion for the second family of regularized systems (40) is the same as for (37), these comments hold for these models also.

5 The Kelvin-Helmholtz Instability in the Free Surface Case

We investigate here the formation of (linear) Kelvin-Helmholtz instabilities for two-fluids interfaces in the case of a free top surface. In fact, we consider again the linear stability of perturbations near a constant shear flow as we did in the rigid lid case, and we follow the formulation above.

With a free top surface, we get a more complex system of $(\zeta_s, U^s_\parallel, \zeta, U^\pm_\parallel)$ (see Sect. 2.2), and the linear stability analysis becomes more difficult. We perform some numerical computations, and are able to find the expressions of the foot points for the stable area in some special cases. A comparison with the rigid lid case is also done in this section, where we discuss in particular the presence of a new range of stable frequencies.

5.1 Linearization Around a Constant Shear $(\mathbf{c}^+, \mathbf{c}^-)$

We consider here the linear equations governing small perturbations of the constant horizontal shear $\zeta^s = \zeta = 0$, $V^\pm = \mathbf{c}^\pm$, $w^\pm = 0$ (and therefore $U^\pm_\parallel = \mathbf{c}^\pm$, $U^s_\parallel = \mathbf{c}^-$).

Linearizing (17)–(19) around $\zeta^s = \zeta = 0$, $\underline{V}^s = \mathbf{c}^-$, $\underline{V}^\pm = \mathbf{c}^\pm$ we find the following linear equations for the perturbation $(\dot{\zeta}^s, \dot{V}^s, \dot{\zeta}, \dot{V}^\pm)$

$$\begin{cases} \partial_t \dot{\zeta}^s + \mathbf{c}^- \cdot \nabla \dot{\zeta}^s - \dot{w}^s = 0, \\ \partial_t \dot{V}^s + g\nabla \dot{\zeta}^s + \mathbf{c}^- \cdot \nabla \dot{V}^s = 0, \\ \partial_t \dot{\zeta} + \mathbf{c}^\pm \cdot \nabla \dot{\zeta} - \dot{w}^\pm = 0, \\ \partial_t \dot{V}^\pm + g\nabla \dot{\zeta} + \mathbf{c}^\pm \cdot \nabla \dot{V}^\pm = -\frac{1}{\rho^\pm} \nabla \underline{P}^\pm \end{cases}$$

(we refer to Lemma 3 for an expression of $\dot{w}^\pm$), with

$$[\![\underline{P}^\pm]\!] = -\sigma \Delta \dot{\zeta}.$$

As for (21), this can be further reduced by introducing $\dot{V} = \underline{\rho}^+ \dot{V}^+ - \underline{\rho}^- \dot{V}^-$; one gets the following free surface version of (21)

$$\begin{cases} \partial_t \dot{\zeta}^s + \mathbf{c}^- \cdot \nabla \dot{\zeta}^s - \dot{w}^s = 0, \\ \partial_t \dot{V}^s + g\nabla \dot{\zeta}^s + \mathbf{c}^- \cdot \nabla \dot{V}^s = 0, \\ \partial_t \dot{\zeta} + \mathbf{c}^\pm \cdot \nabla \dot{\zeta} - \dot{w}^\pm = 0, \\ \partial_t \dot{V} + \langle \mathbf{c}^\pm \rangle \cdot \nabla \dot{V} + g' \nabla \dot{\zeta} + [\![\mathbf{c}^\pm]\!] \cdot \nabla \langle \underline{\rho}^\pm \dot{V}^\pm \rangle = \frac{\sigma}{\rho^+ + \rho^-} \nabla \Delta \dot{\zeta}. \end{cases} \tag{42}$$

We then need the following lemma (similar to Lemma 1) to handle the free surface case.

Lemma 3. *One has*

$$\dot{V}^+ = \frac{\underline{\rho}^-}{b(D)} ([\![\mathbf{c}^\pm]\!] \cdot \nabla) \nabla \dot{\zeta} + \frac{\mathrm{cth}^- |D|}{b(D)} \dot{V} + \frac{\underline{\rho}^- |D|}{\mathrm{sh}^- b(D)} \dot{V}^s$$

$$\dot{V}^- = \frac{\underline{\rho}^+}{b(D)} ([\![\mathbf{c}^\pm]\!] \cdot \nabla) \nabla \dot{\zeta} - \frac{\mathrm{th}^+ |D|}{b(D)} \dot{V} + \frac{\underline{\rho}^+ |D|}{\mathrm{sh}^- b(D)} \dot{V}^s,$$

and

$$\dot{w}^+ = -\mathrm{th}^+ \frac{\nabla}{|D|} \cdot \dot{V}^+, \qquad \dot{w}^- = \mathrm{cth}^- \frac{\nabla}{|D|} \cdot \dot{V}^- - \frac{1}{\mathrm{sh}^-} \frac{\nabla}{|D|} \cdot \dot{V}^s,$$

and

$$\dot{w}^s = \frac{1}{\mathrm{sh}^-} \frac{\nabla}{|D|} \cdot \dot{V}^- - \mathrm{cth}^- \frac{\nabla}{|D|} \cdot \dot{V}^s$$

with the notation $b(D) = (\underline{\rho}^+ \mathrm{cth}^- + \underline{\rho}^- \mathrm{th}^+)|D|$.

Proof. As in the proof of Lemma 1, we write $\dot{V}^{\pm} = \nabla\dot{\psi}^{\pm}$, $\dot{V} = \nabla\dot{\psi}$ with $\dot{\psi} = \underline{\rho}^{+}\dot{\psi}^{+} - \underline{\rho}^{-}\dot{\psi}^{-}$, we also write $\dot{V}^{s} = \nabla\dot{\psi}^{s}$.

Let $\dot{\phi}^{-}$ be the harmonic potential in the upper strip $0 < z < H^{-}$ and such that $\dot{\phi}^{-}_{|z=0} = \dot{\psi}^{-}$ and $\dot{\phi}^{-}_{|z=H^{-}} = \dot{\psi}^{s}$. Explicit computations show that

$$\dot{\phi}^{-} = \frac{\sinh(z|D|)}{\sinh(H^{-}|D|)}\dot{\psi}^{s} - \frac{\sinh((z-H^{-})|D|)}{\sinh(H^{-}|D|)}\dot{\psi}^{-},$$

from which we deduce that

$$\underline{w}^{-} = \partial_z\dot{\phi}^{-}_{|z=0} = \frac{1}{\mathrm{sh}^{-}}|D|\dot{\psi}^{s} - \mathrm{cth}^{-}|D|\dot{\psi}^{-}.$$

Proceeding as in the proof of Lemma 1, we are therefore led to solve

$$\begin{cases} |D|\mathrm{th}^{+}\dot{\psi}^{+} + |D|\mathrm{cth}^{-}\dot{\psi}^{-} = [\![\mathbf{c}^{\pm}]\!]\cdot\nabla\dot{\zeta} + \frac{1}{\mathrm{sh}^{-}}|D|\dot{\psi}^{s}, \\ \underline{\rho}^{+}\dot{\psi}^{+} - \underline{\rho}^{-}\dot{\psi}^{-} = \dot{\psi}; \end{cases}$$

this leads to

$$\begin{aligned} \dot{\psi}^{+} &= \frac{\underline{\rho}^{-}}{b(D)}[\![\mathbf{c}^{\pm}]\!]\cdot\nabla\dot{\zeta} + \frac{\mathrm{cth}^{-}|D|}{b(D)}\dot{\psi} + \frac{\underline{\rho}^{-}|D|}{\mathrm{sh}^{-}b(D)}\dot{\psi}^{s} \\ \dot{\psi}^{-} &= \frac{\underline{\rho}^{+}}{b(D)}[\![\mathbf{c}^{\pm}]\!]\cdot\nabla\dot{\zeta} - \frac{\mathrm{th}^{+}|D|}{b(D)}\dot{\psi} + \frac{\underline{\rho}^{+}|D|}{\mathrm{sh}^{-}b(D)}\dot{\psi}^{s}, \end{aligned}$$

and the formulas of the lemma follow easily after remarking that

$$\dot{w}^{s} = \mathrm{cth}^{-}|D|\psi^{s} - \frac{|D|}{\mathrm{sh}^{-}}\dot{\psi}^{-}.$$

This ends the proof. □

Thanks to the lemma, we can rewrite (42) under the form

$$\partial_t W + A(D)W = 0 \tag{43}$$

where $W = (\dot{\zeta}^{s}, \dot{V}^{s}, \dot{\zeta}, \dot{V})$ and

$$A(D) = \begin{pmatrix} \mathbf{c}^{-}\cdot\nabla & \frac{b_0(D)\nabla}{b(D)\mathrm{th}^{-}}\cdot & \frac{\underline{\rho}^{+}|D|[\![\mathbf{c}^{\pm}]\!]\cdot\nabla}{b(D)\mathrm{sh}^{-}} & \frac{\mathrm{th}^{+}\nabla}{b(D)\mathrm{sh}^{-}}\cdot \\ g\nabla & \mathbf{c}^{-}\cdot\nabla & 0 & 0 \\ 0 & \frac{\underline{\rho}^{-}\mathrm{th}^{+}\nabla}{b(D)\mathrm{sh}^{-}}\cdot & \bar{\mathbf{c}}(D)\cdot\nabla & \frac{\mathrm{th}^{+}\mathrm{cth}^{-}\nabla}{b(D)}\cdot \\ 0 & \frac{\underline{\rho}^{+}\underline{\rho}^{-}|D|}{b(D)\mathrm{sh}^{-}}([\![\mathbf{c}^{\pm}]\!]\cdot\nabla) & a(D)\nabla & \bar{\mathbf{c}}(D)\cdot\nabla \end{pmatrix}$$

where $b(D) = (\underline{\rho}^+\mathrm{cth}^- + \underline{\rho}^-\mathrm{th}^+)|D|$ is already defined in Lemma 3, $b_0(D) = \underline{\rho}^+ \mathrm{th}^- + \underline{\rho}^-\mathrm{th}^+$, and moreover

$$\bar{\mathbf{c}}(D) = \frac{(\mathbf{c}^+\underline{\rho}^+\mathrm{cth}^- + \mathbf{c}^-\underline{\rho}^-\mathrm{th}^+)|D|}{b(D)}, \qquad a(D) = g' + \frac{\underline{\rho}^+\underline{\rho}^-}{b(D)}(\llbracket c^\pm \rrbracket \cdot \nabla)^2 - \frac{\sigma}{\rho^+ + \rho^-}\Delta.$$

5.2 *Kelvin-Helmholtz Instabilities for the Linearized Two-Fluids Equations with Free Surface*

In this section we only focus on the $d = 1$ case for the sake of clarity. As in Sect. 3.2, we write $\llbracket \mathbf{c}^\pm \rrbracket = \llbracket c^\pm \rrbracket \mathbf{e}_x$ and $\mathbf{k} = k\mathbf{e}_x$, with $\llbracket c^\pm \rrbracket = |\llbracket \mathbf{c}^\pm \rrbracket|$ and $k = |\mathbf{k}|$.

In order to find out the stability criterion for the linearized system (43), we need to explore the eigenvalue ω for the Fourier mode $A(k)$ of operator $A(D)$. Indeed, ω satisfies the equation

$$a_0 x^4 + a_1 x^3 + a_2 x^2 + a_3 x + a_4 = 0$$

where $x = c^- - \frac{\omega}{ik}$ and

$$a_0 = 1, \qquad a_1 = \frac{2\underline{\rho}^+\mathrm{cth}^- k\llbracket c^\pm \rrbracket}{b(k)}$$

$$a_2 = \frac{\mathrm{cth}^-}{b(k)}\Big(\underline{\rho}^+ k\llbracket c^\pm \rrbracket^2 - g b_0(k) - (g' + \frac{\sigma}{\rho^+ + \rho^-}k^2)\mathrm{th}^+\Big),$$

$$a_3 = -\frac{2g\underline{\rho}^+\llbracket c^\pm \rrbracket}{b(k)}, \qquad a_4 = -\frac{g\underline{\rho}^+\llbracket c^\pm \rrbracket^2}{b(k)} + \frac{g\,\mathrm{th}^+}{b(k)k}\Big(g' + \frac{\sigma}{\rho^+ + \rho^-}k^2\Big),$$

and we write $b = b(k)$ and $b_0 = b_0(k)$ for short in the following text. The linear system (43) is stable if and only if all the eigenvalues ω are pure imaginary, that is, all the roots x for the equation above are real. As in [1], we use [28] to express the stability criterion under the form

$$\Delta_3 > 0,\ \Delta_5 > 0,\ \Delta_7 > 0 \quad \text{or} \quad \Delta_3 \geq 0,\ \Delta_5 = 0,\ \Delta_7 = 0 \tag{44}$$

where

$$\Delta_7 = \begin{vmatrix} a_0 & a_1 & a_2 & a_3 & a_4 & 0 & 0 \\ 0 & a_0 & a_1 & a_2 & a_3 & a_4 & 0 \\ 0 & 0 & a_0 & a_1 & a_2 & a_3 & a_4 \\ 0 & 0 & 0 & 4a_0 & 3a_1 & 2a_2 & a_3 \\ 0 & 0 & 4a_0 & 3a_1 & 2a_2 & a_3 & 0 \\ 0 & 4a_0 & 3a_1 & 2a_2 & a_3 & 0 & 0 \\ 4a_0 & 3a_1 & 2a_2 & a_3 & 0 & 0 & 0 \end{vmatrix}$$

and Δ_3, Δ_5 are the 3×3, 5×5 inner center determinants.

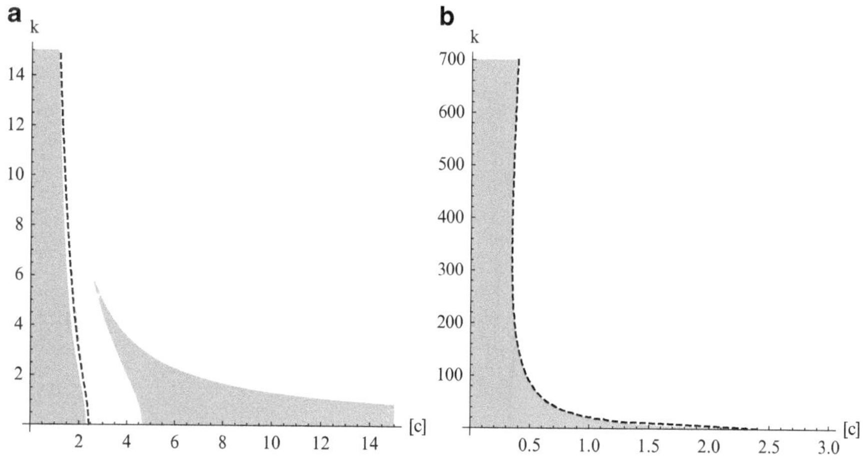

Fig. 6 The stable area for the linear system (43) in free surface case with surface tension on the interface. The *dashed line* represents the bound for the stable area in the corresponding rigid lid case. Parameters $H^+ = 0.62$, $\delta = 0.15/0.62$, $\rho^+ + \rho^- = 2021$, $\gamma = 0.4$, $\sigma = 0.073$ and $g = 9.81$. (**a**) the two parts of the stable area. (**b**) a closer look at the left part

Since these conditions are far more complicated than the stability criterion for the rigid lid case, we cannot write down the exact stability expressions and we prefer to use numerical computations.

Compared to the stable area for the rigid lid case in Sect. 3.2, Fig. 6 tells us that there are two parts for the stable area of system (43). The left part is similar and close to the stable area for the rigid lid case, while there is a new stable part in the free surface case on the right-hand side. *This means that the linear system can also be stable for high shear and low frequency, which is completely different from the rigid lid case.*

When surface tension is neglected, i.e. $\sigma = 0$, there are also two parts of stable area for the system (43). Similarly as the surface tension case above, the left part is also close to the stable area of the corresponding rigid lid case. In fact, the stable area is pretty much the same as the surface tension case in the low frequency part.

Summing up the numerical analysis, the stability condition for system (43) can be described roughly as

$$|[\![c^\pm]\!]|^2 \leq \Omega_{KH,1}(k) \quad \text{or} \quad \Omega_{KH,2}(k) \leq |[\![c^\pm]\!]|^2 \leq \Omega_{KH,3}(k)$$

where $\Omega_{KH,i}(k)$ $(i = 1, 2, 3)$ can be expressed in terms of k, and $\Omega_{KH,1}(k)$ is close to $\Omega_{KH}\alpha_\sigma(k)$ from the rigid lid case. So *the modes with low wave number are stable* if and only if

$$\text{(Stab 1)} \qquad |[\![c^\pm]\!]|^2 \leq \Omega_{KH,1}(0) \quad \text{or} \quad |[\![c^\pm]\!]|^2 \geq \Omega_{KH,2}(0).$$

We shall provide in Sect. 6.1.3 exact values of $\Omega_{KH,1}(0)$ and $\Omega_{KH,2}(0)$ that will allow us to show another striking difference with the rigid lid case: *one has* $\Omega_{KH,1}(0) \neq \Omega_{KH}$ (in fact $\Omega_{KH,1}(0) < \Omega_{KH}$ in our numerical computations), *that is, the threshold for the stability of small wavenumbers is not the same as in the rigid lid model* (even if we restrict it to its left part corresponding to small shears. One can have the condition (Stab 2) for the stability of *all* wave numbers:

$$\text{(Stab 2)} \qquad |[\![c^{\pm}]\!]|^2 \leq \Omega_{FS}^{cr} := \min_{\mathbb{R}+} \Omega_{KH,1}(k) > 0,$$

where we know from the numeral computations that Ω_{FS}^{cr} is close to the one Ω^{cr} in the corresponding rigid lid case and therefore can be estimated approximately by Proposition 1.

Remark 8. In absence of surface tension, i.e. $\sigma = 0$, we obtain that the stability condition for *all* wave numbers holds only for a zero shear, and there is therefore no difference with the rigid lid case as far as the stability condition (Stab 2) is concerned.

6 Shallow Water Approximations in the Free Surface Case

This section investigates the (linear) Kelvin-Helmholtz instabilities of the shallow water approximation when the upper bound is a free surface. We consider here only a first-order approximation leading to a free surface version of the SW/SW system, where we are able to derive the exact expressions for stability criterion in the case when the two layers have the same average depths ($H^+ = H^-$). Based on this observation, we can understand more about the differences between the rigid lid case and the free surface case.

Notation: We need to adapt the definition of $\overline{V}^{\pm}$ in Sect. 4 to the free surface case as follows

$$\overline{V}^{+} = \frac{1}{H^{+}(\zeta)} \int_{-H^{+}}^{\zeta} V^{+}(t,X,z)dz,$$

$$\overline{V}^{-} = \frac{1}{H^{-}(\zeta^{s},\zeta)} \int_{\zeta}^{H^{-}+\zeta^{s}} V^{-}(t,X,z)dz,$$

where $H^{+}(\zeta) = H^{+} + \zeta$ and $H^{-}(\zeta^{s},\zeta) = H^{-} + \zeta^{s} - \zeta$.

6.1 First Order Approximation

A first order approximation is discussed here in this section, similarly as in the SW/SW part in the rigid lid case. We linearize the system around a constant shear as before, and study numerically its stability; we also derive an explicit criterion in the particular case where both fluid layers have the same depth at rest.

6.1.1 The Shallow Water/Shallow Water Equations with Free Surface

At first order, one writes

$$U^s_{\parallel} \sim \overline{V}^-, \quad U^{\pm}_{\parallel} \sim \overline{V}^{\pm}, \quad \text{and} \quad \underline{w}^s \sim 0, \quad \underline{w}^{\pm} \sim 0.$$

Plugging this into system (17) and (18) one has the nonlinear SW/SW system as

$$\begin{cases} \partial_t \zeta^s + \nabla \cdot (H^-(\zeta^s, \zeta)\overline{V}^-) + \nabla \cdot (H^+(\zeta)\overline{V}^+) = 0, \\ \partial_t \overline{V}^- + g\nabla\zeta^s + \frac{1}{2}\nabla|\overline{V}^-|^2 = 0, \\ \partial_t \zeta + \nabla \cdot (H^+(\zeta)\overline{V}^+) = 0, \\ \partial_t \overline{V}^{\pm} + g\nabla\zeta + \frac{1}{2}\nabla|\overline{V}^{\pm}|^2 = -\frac{1}{\rho^{\pm}}\nabla \underline{P}^{\pm}|_{z=\zeta}; \end{cases} \tag{45}$$

this system has been derived and studied in [3, 14, 19], and justified in [24].

6.1.2 Linearization Around a Constant Shear $(\mathbf{c}^+, \mathbf{c}^-)$

For the sake of clarity, we only consider the 1-D case here, so we write $\mathbf{c}^{\pm}$ to be $c^{\pm}$. After linearizing around the constant shear as before one obtains

$$\begin{cases} \partial_t \zeta^s + c^- \partial_x \zeta^s + H^- \partial_x \dot{v}^- + [\![c^{\pm}]\!]\partial_x \zeta + H^+ \partial_x \dot{v}^+ = 0, \\ \partial_t \dot{v}^- + g\partial_x \zeta^s + c^- \partial_x \dot{v}^- = 0, \\ \partial_t \zeta + c^+ \partial_x \zeta + H^+ \partial_x \dot{v}^+ = 0, \\ \partial_t \dot{v} + g'\nabla\zeta + [\![\underline{\rho}^{\pm} c^{\pm} \partial_x \dot{v}^{\pm}]\!] = 0, \end{cases}$$

where as usual $\dot{v} = \underline{\rho}^+ \dot{v}^+ - \underline{\rho}^- \dot{v}^-$. Proceeding as for Lemma 3 to derive a free surface version of Lemma 2, one readily obtains that the free surface version of the linearized SW/SW system (32) takes the following form,

$$\partial_t W + A_{SW}(D)W = 0 \tag{46}$$

where $W = (\zeta^s, \dot{v}^-, \zeta, \dot{v})$ with the notation

$$A_{SW}(D) = \begin{pmatrix} c^- & \frac{H_0}{\underline{\rho}^+} & [\![c^\pm]\!] & \frac{H^+}{\underline{\rho}^+} \\ g & c^- & 0 & 0 \\ 0 & \frac{\underline{\rho}^- H^+}{\underline{\rho}^+} & c^+ & \frac{H^+}{\underline{\rho}^+} \\ 0 & \underline{\rho}^- [\![c^\pm]\!] & g' - \frac{\sigma}{\rho^+ + \rho^-}\partial_x^2 & c^+ \end{pmatrix} \partial_x.$$

6.1.3 The Free Surface SW/SW Model and the Kelvin-Helmholtz Instabilities

Similar as for the full system case, the stability criterion for (46) is equivalent to the fact that the Fourier mode $A_{SW}(k)$ has purely imaginary eigenvalues for all k, which leads to the following condition

$$x^4 + 2[\![c^\pm]\!]x^3 + \Big([\![c^\pm]\!]^2 - g(H^+ + H^-) - \frac{\sigma H^+}{\underline{\rho}^+(\rho^+ + \rho^-)}k^2\Big)x^2 - 2gH^-[\![c^\pm]\!]x$$

$$-gH^-[\![c^\pm]\!]^2 + g\frac{H^+H^-}{\underline{\rho}^+}\big(g' + \frac{\sigma}{\rho^+ + \rho^-}k^2\big) = 0 \quad \text{has four real roots.}$$

With the same notations as in Sect. 5.2, this equation has four real roots if and only if

$$\Delta_3^{SW} > 0,\ \Delta_5^{SW} > 0,\ \Delta_7^{SW} > 0 \quad \text{or} \quad \Delta_3^{SW} \geq 0,\ \Delta_5^{SW} = 0,\ \Delta_7^{SW} = 0.$$

Since the surface tension at the interface affects only the high frequency part, we omit the surface tension i.e. set $\sigma = 0$ in the following computations. Direct computations show that

$$\Delta_3^{SW} = 4[\![c^\pm]\!]^2 + 8g(H^+ + H^-) > 0,$$

and

$$\Delta_5^{SW} = 8g\Big[(H^+ + H^-)\big([\![c^\pm]\!]^2 - g(H^+ - H^-)\big)^2 + 4[\![c^\pm]\!]^2 g(3H^+ - H^-)H^-\Big]$$
$$+16g^2\underline{\rho}^- H^+H^-\big([\![c^\pm]\!]^2 + 2g(H^+ + H^-)\big)/\underline{\rho}^+.$$

We can tell from the expression above that $\Delta_5^{SW} > 0$ at least when $3H^+ > H^-$ (our data in figures fit the case), therefore the stability is governed by the condition $\Delta_7^{SW} > 0$; but the sign of Δ_7^{SW} in general can only be assessed numerically. Figure 7 shows how the free surface SW/SW model behaves compared to the full system.

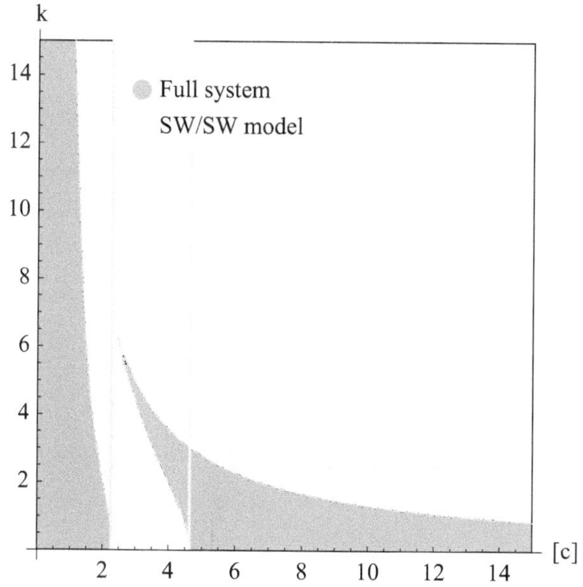

Fig. 7 Stable areas of the free surface SW/SW model compared to the full system ($\sigma = 0$). Parameters $H^+ = 0.62$, $\delta = 0.15/0.62$, $\gamma = 0.4$ and $g = 9.81$

In the particular case where $H^+ = H^- = H$, the expression for the inner determinants Δ_5^{SW} and Δ_7^{SW} can be simplified considerably and one finds

$$\Delta_5^{SW} = 16[\![c^\pm]\!]^4 gH + 16[\![c^\pm]\!]^2(4 + \underline{\rho}^-/\underline{\rho}^+)g^2H^2 + 64(\underline{\rho}^-/\underline{\rho}^+)g^3H^3 > 0,$$

and

$$\Delta_7^{SW} = 16g^2H^2\Big([\![c^\pm]\!]^2 + (\underline{\rho}^-/\underline{\rho}^+)gH\Big)^2\Big([\![c^\pm]\!]^4 - 8[\![c^\pm]\!]^2 gH + 16(1 - \underline{\rho}^-/\underline{\rho}^+)g^2H^2\Big).$$

Therefore the stability criterion for the linear SW/SW system (46) generated from $\Delta_7^{SW} > 0$ reads

$$[\![c^\pm]\!]^4 - 8[\![c^\pm]\!]^2 gH + 16(1 - \underline{\rho}^-/\underline{\rho}^+)g^2H^2 > 0.$$

So the stability is achieved if and only if

$$[\![c^\pm]\!]^2 < 4gH(1 - \sqrt{\gamma}) \quad \text{or} \quad [\![c^\pm]\!]^2 > 4gH(1 + \sqrt{\gamma}),$$

where we recall that γ is the density ratio: $\gamma = \rho^-/\rho^+ \in [0, 1)$. Remarking that the threshold Ω_{KH} obtained in the rigid lid case for the stability of low wavenumbers can be written $\Omega_{KH} = gH\frac{1-\gamma^2}{\gamma}$, we obtain the following condition for the stability of the free surface SW/SW model,

$$(\text{Stab } 1)_{SW} \qquad [\![c^\pm]\!]^2 < \Omega_{KH} \times 4\gamma\frac{1-\sqrt{\gamma}}{1-\gamma^2} \quad \text{or} \quad [\![c^\pm]\!]^2 > \Omega_{KH} \times 4\gamma\frac{1+\sqrt{\gamma}}{1-\gamma^2};$$

this is consistent with [37] where it is proved the free surface SW/SW model is hyperbolic[5] if and only if $[\![c^\pm]\!] < \Omega^-_{KH}$ or $[\![c^\pm]\!] > \Omega^+_{KH}$ where $\Omega^\pm_{KH}$ are two non explicit constants). This is of course equivalent to the condition for the stability of low wavenumbers derived in Sect. 5.2 for the full system. As in the rigid lid case, stability of low wavenumbers for the SW/SW model is equivalent to the stability of *all* wavenumbers, i.e.

$$(\text{Stab } 2)_{SW} \quad \Longleftrightarrow \quad (\text{Stab } 1)_{SW}.$$

6.2 Behavior of the Kelvin-Helmholtz Instabilities in the Rigid Lid Limit

We have seen in the previous section that there are at least two major differences for the behavior of Kelvin-Helmholtz instabilities in the rigid lid and in the free surface cases. Namely,

(1) In both cases, the linearized equations are stable for small shears, but the (Stab 1) condition for the stability of low frequencies differ. These conditions are given by

$$[\![c^\pm]\!] < \Omega_{KH} \quad (\text{ rigid lid }) \quad \text{and} \quad [\![c^\pm]\!] < \Omega_{KH,1}(0) \neq \Omega_{KH} \quad (\text{ free surface })$$

(the fact that $\Omega_{KH,1}(0) \neq \Omega_{KH}$ is either checked numerically or, in the particular case where $H^+ = H^-$, using the explicit expression computed in Sect. 6.1.3).

(2) There exists in the free surface case a zone of stability for low frequency modes when the shear is large. This stability area represented in Fig. 6 and also noticed in [1, 11, 35, 40] does not exist in the rigid lid case.

Since the rigid lid approximation is much simpler and often used in applications, we discuss here whether the two differences described above are compatible with the rigid lid approximation.

From (1), one can infer that the rigid lid approximation can only be correct if $\Omega_{KH,1}(0) \to \Omega_{KH}$ in the rigid lid limit (i.e. when the amplitude of the surface perturbations is much smaller than the interface displacement). From the explicit expression of $\Omega_{KH,1}(0)$ computed in Sect. 6.1.3, this implies that one necessarily has $\gamma \to 1$, i.e. that the densities $\rho^\pm$ of both fluids become very similar (which

[5]Hyperbolicity does not imply well posedness of the nonlinear system (45); a stronger *sufficient* well-posedness condition is derived in [24, 37]; contrary to $(\text{Stab } 1)_{SW}$, this condition is satisfied only for small shears.

is the case for oceanographic applications). This is in this context that the rigid lid approximation for the SW/SW model (45) has been justified in [26]; note also that it is possible to derive a simpler model in this framework by making the so called Boussinesq approximation [9].

The question raised by (2) is: what happens to the second stability area in the rigid lid limit? This question is best answered by working in dimensionless variables

$$\tilde{X} = \frac{X}{L}, \quad \tilde{\zeta}^s = \alpha\frac{\zeta^s}{H_0}, \quad \tilde{\zeta} = \frac{\zeta}{H_0}, \quad \tilde{v} = \frac{v}{v_0}, \quad \tilde{t} = t\frac{v_0}{L},$$

where v stand here for all the velocities involved in the equations ($v = v^{\pm}, v^s, c^{\pm}, \ldots$), L is the typical horizontal length, $H_0 = \underline{\rho}^+ H^- + \underline{\rho}^- H^+$, and v_0 is the typical speed of the linearized equations in shallow water and in the rigid lid case, $v_0^2 = g' H_0$ (and the rigid lid limit corresponds to $\alpha \to 0$). In these dimensionless variables, the (Stab 1)$_{SW}$ condition for the free surface SW/SW model derived in Sect. 6.1.3 is written

$$\text{(Stab 1)}_{SW} \qquad [\![\tilde{c}^{\pm}]\!]^2 < 4(1+\gamma)^2\frac{1-\sqrt{\gamma}}{1-\gamma^2} \quad \text{or} \quad [\![\tilde{c}^{\pm}]\!]^2 > 4(1+\gamma)^2\frac{1+\sqrt{\gamma}}{1-\gamma^2}$$

with $[\![\tilde{c}^{\pm}]\!] = [\![c^{\pm}]\!]/\sqrt{g' H_0}$.

As seen above, the rigid lid limit can only be relevant if $\epsilon := 1 - \gamma$ goes to zero. The above stability criterion is therefore asymptotically equivalent to

$$\text{(Stab 1)}_{SW} \qquad [\![\tilde{c}^{\pm}]\!]^2 < 4 \quad \text{or} \quad [\![\tilde{c}^{\pm}]\!]^2 > \frac{16}{\epsilon}.$$

In the rigid lid limit, the behavior of the two stability areas observed in the free surface case is the following: the left part (small shear) converges to the stability area of the rid lid model, while the right part (large shear) is shifted more to the right (see Fig. 8). In the rigid lid limit, the large shear stability condition is therefore trivial in the sense that it requires an infinite shear to be satisfied.

6.3 *Remarks on Other Interesting Limits*

6.3.1 The One-Fluid Limit $\gamma \to 0$

It is proved in [32] that, in the rigid-lid case, the full (nonlinear) two-fluids equations converge to the one-fluid (water waves) equation as $\gamma \to 0$ (it is in particular shown that it is possible to consider simultaneously $\gamma \to 0$ and $\text{Bo} \to \infty$ and to get convergence to the water waves equations without surface tension).

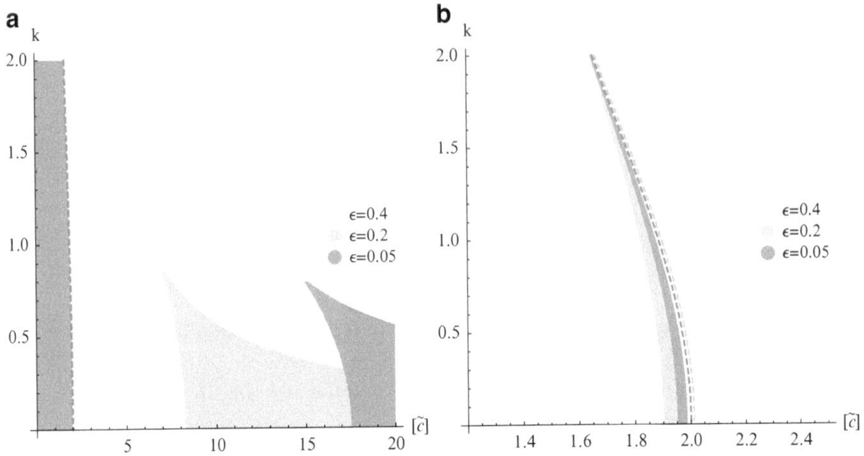

Fig. 8 The stable areas for the linear system (43) with different parameter $\gamma = 1 - \epsilon$. The horizontal axis is rescaled to be $[\![\tilde{c}^{\pm}]\!]$ (written as $[\tilde{c}]$ in the figures) in order to compare with the rigid lid case. The *dashed lines* represent the bounds for the stable area in the corresponding rigid lid cases. Parameters $\rho^{+} + \rho^{-} = 2021$, $H^{+} = 0.62 = H^{-}$, $\sigma = 0.073$ and $g = 9.81$. (**a**) the stable areas with different ϵ. (**b**) a closer look at the left part

It is natural to wonder what happens in the free surface case for which it is expected that the stability region grows larger as $\gamma \to 0$ since for the linearized water waves equations, all wave numbers are stable (there is no Kelvin-Helmholtz instability). As shown in Fig. 9, we numerically get that when $\gamma \to 0$, the left and the right parts of the stable area in the free surface case become closer and closer and get united as one piece in the limit. For the small numbers, this phenomenon becomes transparent by looking into the condition $(\text{Stab } 1)_{SW}$ and setting $\gamma \to 0$.

6.3.2 The Small Aspect Ratio Limit $\delta \to 0$

To different configurations lead to a small aspect ratio $\delta = H^{-}/H^{+}$:

- When H^{+} is fixed and H^{-} gets smaller and smaller. In this case, the one fluid case is expected to be recovered. Consequently, the stable area in the corresponding two-fluids case with free surface should grow to cover the whole range of wave-numbers. This scenario is numerically confirmed, as shown in Fig. 10. The right part of the stable area gets larger and larger. Meanwhile, the left part becomes closer and closer to the stable area in the rigid lid case.
- When H^{-} is kept fixed and H^{+} grows larger and larger. In this case, the rigid lid limit is expected to be valid, and it is interesting to look at the behavior of the high frequency stable area specific to the free surface case. This behavior is represented in Fig. 11.

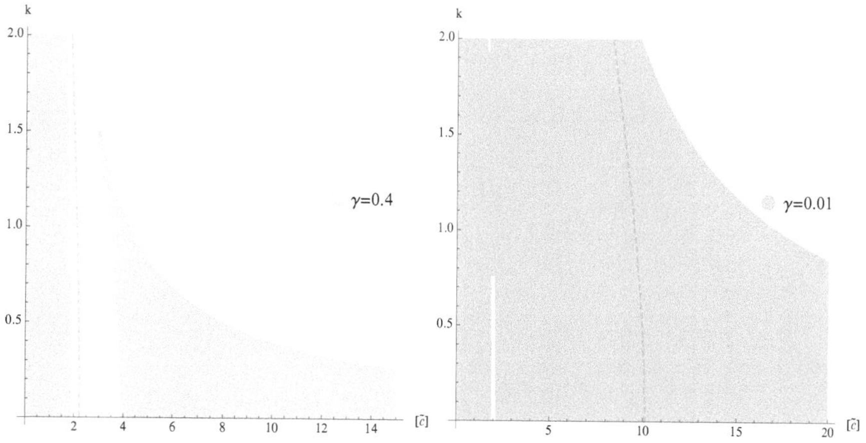

Fig. 9 The stable areas for the linear system (43) with parameter $\gamma \rightarrow 0$. The *dashed lines* represent the bounds for the stable area in the corresponding rigid lid cases. Parameters $\rho^{+}+\rho^{-} = 2021$, $H^{+} = 0.62 = H^{-}$, $\sigma = 0.073$ and $g = 9.81$. (**a**) Stable area when $\gamma = 0.4$. (**b**) Stable area when $\gamma = 0.01$

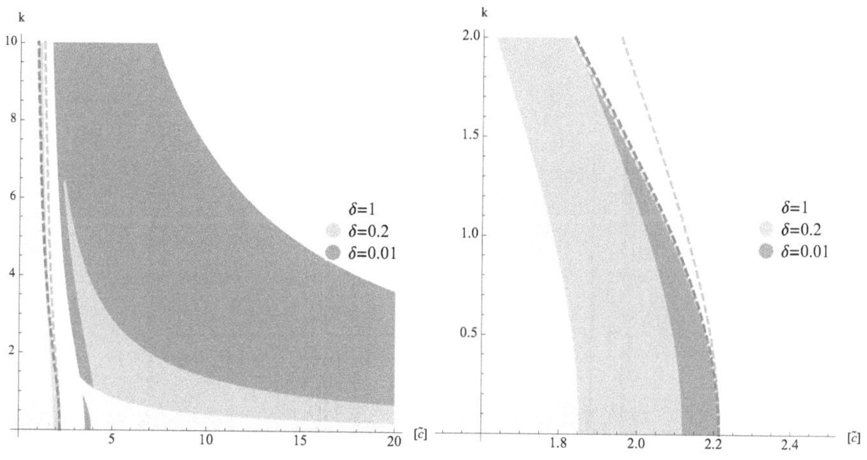

Fig. 10 The stable areas for the linear system (43) with different parameter δ where $H_{+} = 0.62$ kept fixed. The *dashed lines* represent the bounds for the stable area in the corresponding rigid lid cases. Parameters $\rho^{+} + \rho^{-} = 2021$, $\gamma = 0.4$, $\sigma = 0.073$ and $g = 9.81$. (**a**) the stable areas with different δ. (**b**) a closer look at the left part

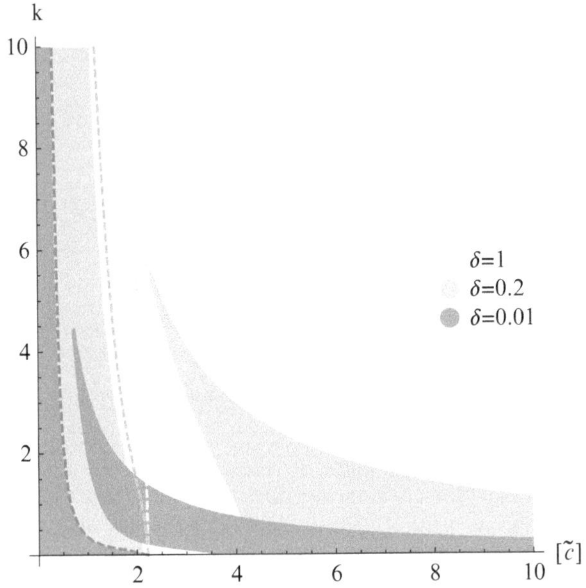

Fig. 11 The stable areas for the linear system (43) with different parameter δ where $H_- = 0.15$ kept fixed. The *dashed lines* represent the bounds for the stable area in the corresponding rigid lid cases again. Parameters $\rho^+ + \rho^- = 2021$, $\gamma = 0.4$, $\sigma = 0.073$ and $g = 9.81$

Acknowledgements The authors want to address their warmest thanks to the referee for his/her careful reading and valuable suggestions. D. L. acknowledges support from the ANR-13-BS01-0003-01 DYFICOLTI and the ANR BOND. M.. This work was supported by Fondation Sciences Mathématiques de Paris (FSMP) when M. Ming was a postdoc with D. L. at DMA, l'École Normale Supérieure in 2012.

References

1. Barros, R., Choi, W.: Inhibiting shear instability induced by large amplitude internal solitary waves in two-layer flows with a free surface. Stud. Appl. Math. **122**, 325–346 (2009)
2. Barros, R., Choi, W.: On regularizing the strongly nonlinear model for two-dimensional internal waves. Phys. D **264**, 27–34 (2013)
3. Barros, R., Gavrilyuk, S., Teshukov, V.M.: Dispersive nonlinear waves in two-layer flows with free surface. I. Model derivation and general properties. Stud. Appl. Math. **119**, 191–211 (2007)
4. Benjamin, T.B., Bridges, T.J.: Reappraisal of the Kelvin-Helmholtz problem. II. Interaction of the Kelvin-Helmholtz, superharmonic and Benjamin-Feir instabilities. J. Fluid Mech. **333**, 327–373 (1997)
5. Bona, J.L., Chen, M., Saut, J.-C.: Boussinesq equations and other systems for small-amplitude long waves in nonlinear dispersive media. I. Derivation and linear theory. J. Nonlinear Sci. **12**, 283–318 (2002)
6. Bona, J.L., Chen, M., Saut, J.-C.: Boussinesq equations and other systems for small amplitude long waves in nonlinear dispersive media. II. The nonlinear theory. Nonlinearity **17**, 925–952 (2004)

7. Bona, J.L., Colin, T., Lannes, D.: Long wave approximations for water waves. Arch. Ration. Mech. Anal. **178**, 373–410 (2005)
8. Bona, J.L., Lannes, D., Saut, J.-C.: Asymptotic models for internal waves. J. Math. Pures Appl. **89**, 538–566 (2008)
9. Boonkasame, A., Milewski, P.A.: The stability of large-amplitude shallow interfacial non-Boussinesq flows. Studies in Applied Math **128**(1), 40–58 (2012)
10. Bresch, D., Renardy, M.: Well-posedness of two-layer shallow-water flow between two horizontal rigid plates. Nonlinearity **24**, 1081–1088 (2011)
11. Bresch, D., Renardy, M.: Kelvin-Helmholtz instability with a free surface, Z. Angew. Math. Phys. **64**, 905–915 (2013)
12. Castro, A., Lannes, D.: Well-posedness and shallow-water stability for a new Hamiltonian formulation of the water waves equations with vorticity. to appear in Indiana University Math. Journal http://arxiv.org/abs/1402.0464
13. Choi, W., Barros, R., Jo, T.-C.: A regularized model for strongly nonlinear internal solitary waves. J. Fluid Mech. **629**, 73–85 (2009)
14. Choi, W., Camassa, R.: Weakly nonlinear internal waves in a two-fluid system. J. Fluid Mech. **313**, 83–103 (1996)
15. Choi, W., Camassa, R.: Fully nonlinear internal waves in a two-fluid system. J. Fluid Mech. **396**, 1–36 (1999)
16. Cotter, C.J., Holm, D.D., Percival, J.R.: The square root depth wave equations. Proc. R. Soc. Lond. Ser. A Math. Phys. Eng. Sci. **466**, 3621–3633 (2010)
17. Craig, W.: An existence theory for water waves and the Boussinesq and Korteweg-de Vries scaling limits. Commun. Partial Differ. Equ. **10**, 787–1003 (1985)
18. Craig, W., Guyenne, P., Kalisch, H.: A new model for large amplitude long internal waves. C. R. Mecanique **332**, 525–530 (2004)
19. Craig, W., Guyenne, P., Kalisch, H.: Hamiltonian long-wave expansions for free surfaces and interfaces. Commun. Pure Appl. Math. **58**, 1587–1641 (2005)
20. Craig, W., Guyenne, P., Sulem, C.: Coupling between internal and surface waves. Nat. Hazards **57**, 617–642 (2010)
21. Craig, W., Guyenne, P., Sulem, C.: The surface signature of internal waves. J. Fluid Mech. **710**, 277–303 (2012)
22. Craig, W., Sulem, C.: Numerical simulation of gravity waves. J. Comput. Phys. **108**, 73–83 (1993)
23. Drazin, P.G., Reid, W.H.: Hydrodynamic stability. Cambridge Monographs on Mechanics and Applied Mathematics. Cambridge University Press, Cambridge (1982)
24. Duchêne, V.: Asymptotic shallow water models for internal waves in a two-fluid system with a free surface. SIAM J. Math. Anal. **42**, 2229–2260 (2010)
25. Duchêne, V.: Asymptotic models for the generation of internal waves by a moving ship, and the dead-water phenomenon. Nonlinearity **24**, 2281–2323 (2011)
26. Duchêne, V.: On the rigid-lid approximation for two shallow layers of immiscible fluids with small density contrast. J. Nonlinear Sci. **24**, 579–632 (2014)
27. Duchêne, V., Israwi, S., Talhouk, R.: Shallow water asymptotic models for the propagation of internal waves. Discrete Contin. Dyn. Syst. Ser. S **7**, 239–269 (2014)
28. Fuller, A.T.: Root location criteria for quartic equations. IEEE Trans. Automat. Contr. **AC-26**, 777–782 (1981)
29. Grue, J., Jensen, A., Rusås, P.-O., Sveen, J.K.: Properties of large-amplitude internal waves. J. Fluid Mech. **380**, 257–278 (1999)
30. Guyenne, P., Lannes, D., Saut, J.-C.: Well-posedness of the Cauchy problem for models of large amplitude internal waves. Nonlinearity **23**, 237–275 (2010)
31. Lannes, D.: Well-posedness of the water-waves equations. J. Am. Math. Soc. **18**, 605–654 (2005)
32. Lannes, D.: A stability criterion for two-fluid interfaces and applications. Arch. Ration. Mech. Anal. **208**, 481–567 (2013)

33. Lannes, D.: The Water Waves Problems. Mathematical Analysis and Asymptotics. Mathematical Surveys and Monographs vol. 188. AMS, Providence (2013)
34. Lannes, D., Bonneton, P.: Derivation of asymptotic two-dimensional time-dependent equations for surface water wave propagation. Phys. Fluids **21**, 016601 (2009)
35. Liska, R., Margolin, L., Wendroff, B.: Nonhydrostatic two-layer models of incompressible flow. Comput. Math. Appl. **29**, 25–37 (1995)
36. Miyata, M.: Long internal waves of large amplitude. In: Horikawa, H., Maruo, H. (ed.) Proceedings of the IUTAM Symposium on Nonlinear Water Waves, pp. 399–406. Springer, Berlin (1988)
37. Monjarret, R.: Local well-posedness of the two-layer shallow water model with free surface, submitted.
38. Nguyen, H.Y., Dias, F.: A boussinesq system for two-way propagation of interfacial waves. Phys. D **237**, 2365–2389 (2008)
39. Nwogu, O.: Alternative form of boussinesq equations for nearshore wave propagation. J. Waterw. Port, Coast. Ocean Eng. **119**, 616–638 (1993)
40. Stewart, A.L., Dellar, P.J.: Multilayer shallow water equations with complete coriolis force. part 3. hyperbolicity and stability under shear. J. Fluid Mech. **723**, 289–317 (2013)
41. Whitham, G.B.: Linear and nonlinear waves. Pure and Applied Mathematics. Wiley, New York (1999) [Reprint of the 1974 original, A Wiley-Interscience Publication]
42. Wu, S.: Well-posedness in sobolev spaces of the full water wave problem in 3-d. J. Am. Math. Soc. **12**, 445–495 (1999)
43. Zakharov, V.E.: Stability of periodic waves of finite amplitude on the surface of a deep fluid. J. Appl. Mech. Tech. Phys. **9**, 190–194 (1968)

Some Analytic Results on the FPU Paradox

Dario Bambusi, Andrea Carati, Alberto Maiocchi, and Alberto Maspero

Dedicato a Walter Craig il cui entusiasmo è sempre contagioso.

Abstract We present some analytic results aiming at explaining the lack of thermalization observed by Fermi Pasta and Ulam in their celebrated numerical experiment. In particular we focus on results which persist as the number N of particles tends to infinity. After recalling the FPU experiment and some classical heuristic ideas that have been used for its explanation, we concentrate on more recent rigorous results which are based on the use of (i) canonical perturbation theory and KdV equation, (ii) Toda lattice, (iii) a new approach based on the construction of functions which are adiabatic invariants with large probability in the Gibbs measure.

1 Introduction

In their celebrated paper of the year 1954 Fermi Pasta and Ulam (see [20]) studied the dynamics of a chain of nonlinear oscillators by numerical integration of the equations of motion, with the aim of testing the dynamical foundation of equilibrium statistical mechanics. They looked at the evolution of the normal mode energies and of their time averages. FPU considered initial data with all the energy in the first Fourier mode and observed that, for the initial data and the ranges of time considered, (1) the harmonic energies seem to have a recurrent behaviour, and (2) the time averages of the harmonic energies quickly relax to a distribution which is exponentially decreasing with the wave number (the so called FPU packet of modes). This was quite surprising since the solution was expected to explore a

D. Bambusi (✉) • A. Carati • A. Maspero
Dipartimento di Matematica, Università degli Studi di Milano,
Via Saldini 50, 20133 Milano, Italy
e-mail: dario.bambusi@unimi.it; andrea.carati@unimi.it; alberto.maspero@unimi.it

A. Maiocchi
Laboratoire de Mathématiques, Université de Cergy-Pontoise,
2 avenue Adolphe Chauvin, Cergy-Pontoise, France

P. Guyenne et al. (eds.), *Hamiltonian Partial Differential Equations and Applications*, Fields Institute Communications 75,
DOI 10.1007/978-1-4939-2950-4_8

whole energy surface in phase space in such a way that the normal mode energies would relax to equipartition, according to the prescription of equilibrium statistical mechanics. For this reason the FPU result is sometimes referred to as the FPU paradox.

A qualitatively new result was then obtained by Izrailev and Chirikov in the year 1966 [29] (confirmed by Bocchieri et al. [16]), who discovered that the paradox disappears if initial data with sufficiently high energy are considered. In the same period Zabuski and Kruskal [39] used the KdV equation in order to try to describe analytically the recurrent behaviour observed by FPU.

Thus two kind of analytic problems naturally arise. The first one is to describe the FPU recurrent behaviour, maybe along the lines of Zabuski and Kruskal; the second one is to establish whether the FPU paradox persists in the thermodynamic limit, i.e. the limit in which the number N of particles in the chain tends to infinity while keeping the specific energy E/N fixed, which is the relevant limit for the foundations of statistical mechanics.

The aim of the present paper is to present a short review of the status of the research, focusing only on analytic results and in particular on a couple of results recently obtained by the authors [5, 31].

The paper is organized as follows: in Sect. 2 we present some numerical computations which essentially coincide with those by FPU. We also add a further numerical computation showing the existence of an energy threshold above which the paradox disappears. In Sect. 3 we will discuss some theoretical heuristic ideas which have been used in order to try to explain and to understand the FPU paradox. In particular, in Sect. 3.1 we will discuss the relation between FPU lattice and KdV equation, while in Sect. 3.2 we discuss the use of KAM theory and canonical perturbation theory (and Nekhoroshev's theorem) in the context of FPU dynamics. In Sect. 4 we present some rigorous results that have been obtained in the last ten years on the problem and which give some explanation of the existence of the so called FPU packet of modes. The limitation of all these results is that they apply to a regime in which the specific energy goes to zero as $N \to \infty$. The section is split into three subsection, the first one deals with a result based on the KdV approximation, the second one deals with a result based on multifrequency expansion and the third one deals with a result based on the approximation by Toda lattice. The subsection on Toda lattice contains the best results now available on the FPU packet of modes. In Sect. 5 we will present an averaging theorem for the FPU chain valid in the thermodynamic limit. This last result in particular deals with a slightly different problem, namely the exchange of energy among the different degrees of freedom when one starts with an initial datum belonging to a set of large Gibbs measure. We conclude the paper with a short discussion in Sect. 6.

2 The FPU Paradox

The Hamiltonian of the FPU–chain can be written, in suitably rescaled variables, as

$$H_{FPU} = H_0 + H_1 + H_2 \tag{1}$$

where

$$H_0 \stackrel{\text{def}}{=} \sum_j \left(\frac{p_j^2}{2} + \frac{(q_{j+1} - q_j)^2}{2} \right) ,$$

$$H_1 \stackrel{\text{def}}{=} \frac{1}{3!} \sum_j (q_{j+1} - q_j)^3$$

$$H_2 \stackrel{\text{def}}{=} \frac{A}{4!} \sum_j (q_{j+1} - q_j)^4 ,$$

where (p, q) are canonically conjugated variables. We will consider either periodic boundary conditions or Dirichlet boundary conditions: the index j runs from 0 to N in the case of Dirichlet boundary conditions, namely $q_0 = q_{N+1} = 0 = p_0 = p_{N+1}$, while it runs from $-(N + 1)$ to N in the case of periodic boundary conditions, i.e. $q_{-N-1} = q_{N+1}$ and $p_{-N-1} = p_{N+1}$.

In order to introduce the Fourier basis consider the vectors

$$\hat{e}_k(j) = \begin{cases} \frac{\delta_{PD}}{\sqrt{N+1}} \sin\left(\frac{jk\pi}{N+1}\right), & k = 1, \dots, N, \\ \frac{1}{\sqrt{N+1}} \cos\left(\frac{jk\pi}{N+1}\right), & k = -1, \dots, -N, \\ \frac{1}{\sqrt{2N+2}}, & k = 0, \\ \frac{(-1)^j}{\sqrt{2N+2}}, & k = -N - 1. \end{cases} \tag{2}$$

Then the Fourier basis is formed by $\hat{e}_k$, $k = 1, \dots, N$ and $\delta_{PD} = \sqrt{2}$ in the case of Dirichlet boundary conditions, and by $\hat{e}_k$, $k = -N - 1, \dots, N$ and $\delta_{PD} = 1$ in the case of periodic boundary conditions.

Introducing the Fourier variables $(\hat{p}_k, \hat{q}_k)$ by

$$p_j = \sum_k \hat{p}_k \hat{e}_k(j) , \qquad q_j = \sum_k \hat{q}_k \hat{e}_k(j) \tag{3}$$

with

$$\omega_k = 2 \sin\left(\frac{|k|\pi}{2(N + 1)} \right) , \tag{4}$$

the system takes the form

$$H = H_0 + H_1 + H_2 \tag{5}$$

where

$$H_0(\hat{p},\hat{q}) = \sum_k \frac{\hat{p}_k^2 + \omega_k^2 \hat{q}_k^2}{2} , \quad H_1 = H_1(\hat{q}) , \quad H_2 = H_2(\hat{q}) . \tag{6}$$

We also introduce the harmonic energies

$$E_k = \frac{\hat{p}_k^2 + \omega_k^2 \hat{q}_k^2}{2} ,$$

and their time averages

$$\overline{E_k}(T) := \frac{1}{T} \int_0^T E_k(t)dt . \tag{7}$$

We will often use also the *specific harmonic energies* defined by

$$\mathrm{e}_k := \frac{E_k}{N} . \tag{8}$$

We recall that according to the principles of classical statistical mechanics, at equilibrium at temperature T, each of the harmonic oscillators should have an energy equal to β^{-1}, where $\beta = (k_B T)^{-1}$ is the standard parameter entering in the Gibbs measure (and k_B is the Boltzmann constant). Furthermore, if the system has good statistical properties, the time averages of the different quantities should quickly relax to their equilibrium value.

Fermi Pasta and Ulam studied the time evolution of E_k and of the corresponding time averages $\overline{E_k}$ under Dirichlet boundary conditions. Figure 1 shows the results of the numerical computations by FPU[1]; the initial data are chosen with $E_1(0) \neq 0$ and $E_k(0) = 0$ for any $|k| > 1$.

From Fig. 1 one sees that the energy flows quickly to some modes of low frequency, but after a short period it returns almost completely to the first mode; in the right part of the figure the final values of $\overline{E_k}(t)$ are plotted in a linear scale. The final distribution turns out to be exponentially decreasing with k.

If one continues the integration one sees that the return phenomenon repeats itself almost identically for a very long time (see Fig. 2). The distribution of the $\overline{E_k}(t)$, too, is almost unchanged: one usually says that a packet of modes has formed.

In Fig. 3 the time averages $\overline{E_k}(t)$ are plotted versus time in a semi-log scale. One sees that the quantities $\overline{E_k}(t)$ quickly relax to well defined values which, as shown in Fig. 2 (right figure), decay exponentially with the wave number. To describe the situation with the words by Fermi Pasta and Ulam "The result shows very little, if

[1]Actually these figures where obtained more recently by Antonio Giorgilli by repeating the computations of FPU.

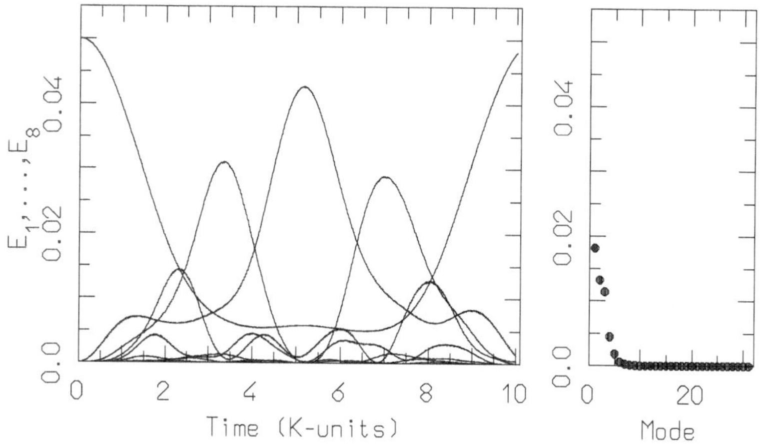

Fig. 1 Mode energies vs time (*left*) and final values of their time averages vs mode number k (*right*)

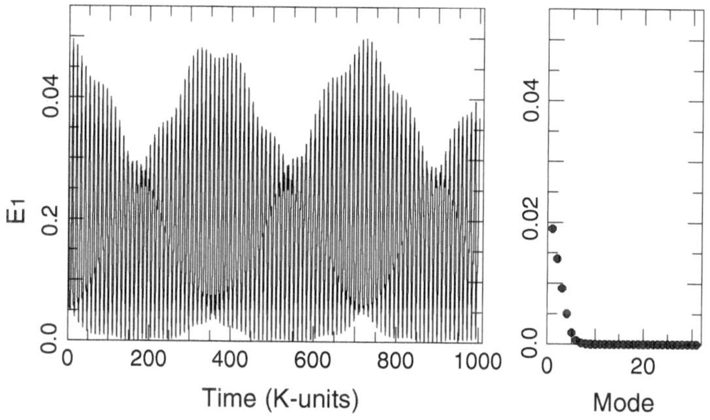

Fig. 2 Energy of the first mode and final values of $\overline{E_k}(t)$ at longer time scales

any, tendency towards equipartition of energy among the degrees of freedom." This is what is usually known as the Fermi Pasta Ulam paradox.

All the above results correspond to initial data with small energy. It was however discovered by Izrailev and Chirikov [29] that the results qualitatively change when the energy per particle is increased. This is illustrated in Fig. 4 from which one sees that the FPU paradox disappears in this regime, because equipartition is quickly attained.

The FPU numerical experiment originated a huge amount of scientific research and in particular subsequent numerical computations have established the shape of the packet of modes to which energy flows (see e.g. [13]) into FPU regime and have put into evidence that the FPU packet is only metastable [21], namely that after a quite long time, whose precise length is not yet precisely established, the system relaxes to equipartition (see e.g. [12, 14]).

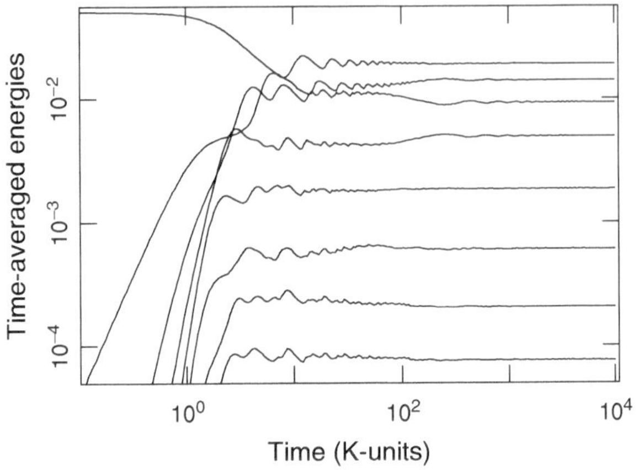

Fig. 3 $\overline{E_k}(t)$ versus time

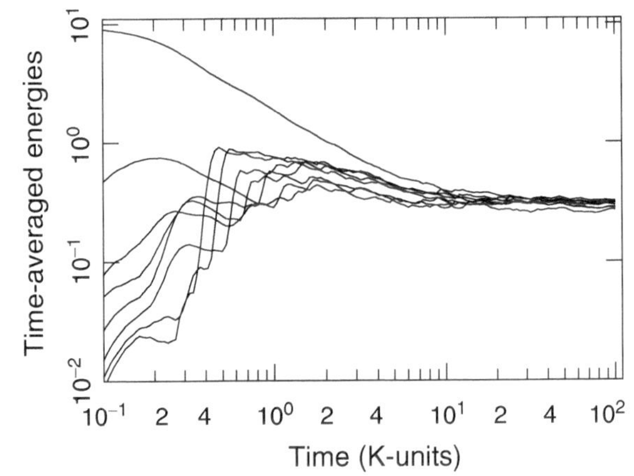

Fig. 4 $\overline{E_k}(t)$ versus time at large energy

3 Heuristic Theoretical Analysis

We remark that the theoretical understanding of the FPU paradox would be absolutely fundamental: indeed it is clear that the phenomenon would have a strong relevance for the foundations of statistical mechanics if it were proven to persist in the thermodynamic limit, i.e., in the limit in which the number of particles $N \to \infty$ while the energy per mode, namely $\sum_k E_k/N$, is kept fixed. Of course numerics can just give some indications, while a definitive result can only come from a theoretical result, which is the only one able to attain the limit $N = \infty$.

3.1 KdV

One of the first attempts to explain the FPU paradox was based on the use of the Korteweg de Vries equation (KdV). The point is that on the one hand KdV is known to approximate the FPU and on the other one KdV is also known to be integrable, so that it displays a recurrent behaviour.

We now recall briefly the way KdV is introduced as a modulation equation for the FPU. We consider the case of periodic boundary conditions and confine the study to the subspace

$$\sum_j q_j = 0 = \sum_j p_j \tag{9}$$

which is invariant under the dynamics. The idea is to consider initial data with large wavelength and small amplitude, namely to interpolate the difference $q_j - q_{j+1}$ through a smooth small function slowly changing in space (and time). This is obtained through an Ansatz of the form

$$q_j - q_{j+1} = \epsilon u(\mu j, t), \quad \mu := \frac{1}{N}, \quad \epsilon \ll 1 \tag{10}$$

with u periodic of period 2. It turns out that in order to fulfill the FPU equations, the function u should have the form

$$u(x,t) = f(x-t, \mu^3 t) + g(x+t, \mu^3 t)$$

with $f(y, \tau)$ and $g(y, \tau)$ fulfilling the equations

$$f_\tau + \frac{\mu^2}{\epsilon} f_{yyy} + f f_y = O(\mu^2), \quad g_\tau - \frac{\mu^2}{\epsilon} g_{yyy} - g g_y = O(\mu^2), \tag{11}$$

namely, up to higher order corrections, the system is described by a couple of KdV equations with dispersion of order μ^2/ϵ. The origin of this group of ideas is the celebrated paper by Zabuski and Kruskal [39] on the dynamics of the KdV equation, which was the starting point of soliton theory and led in particular to the understanding that KdV is integrable. Thus, the enthusiasm for the discovery of such a beautiful and important phenomenology, led to the idea that also the FPU paradox might be an integrable phenomenon, or more precisely could be the shadow of the fact that KdV is an integrable system nearby FPU.

In order to transform such a heuristic idea into a theorem one should fill two gaps. The first one consists in showing that in the KdV equation a phenomenon of the kind of the formation and persistence of the packet of modes occurs; the second one consisting in showing that the solutions of the KdV equation actually describe well the dynamics of the FPU, namely that the higher order corrections neglected in (11), are actually small.

As discussed below, both problems can be solved in the case $\epsilon = \mu^2$, in which the KdV equation turns out to be the standard one.

In particular, in this case one can exploit some analytic properties of action angle variables for KdV (see [30]) in order to show that if one puts all the energy in the first Fourier mode, then the energy remains forever localized in an exponentially localized packet of Fourier modes. However, if one wants to take the limit $N \to \infty$ while keeping ϵ fixed (as needed in order to get a result valid in the thermodynamic limit), one has to study the dispersionless limit of the KdV equation and very little is known on the behaviour of action angle variables in such a limit. Thus we can say that, **in the KdV equation**, the phenomenon of formation and persistence of the packet is not explained in the limit which corresponds to the thermodynamic limit of the FPU lattice.

The second problem (justification of KdV as an approximation of FPU) is far from trivial, since FPU is a singular perturbation of KdV, namely the $O(\mu^2)$ terms in (11) contain higher order derivatives: the proof of theorems connecting the solutions of KdV and the solutions of FPU have only recently been obtained [6, 36], and only in the case $\epsilon = \mu^2$.

3.2 KAM Theory and Canonical Perturbation Theory

Izrailev and Chirikov [29] in 1966 suggested an explanation of the FPU paradox through KAM theory. We recall that KAM theory deals with perturbations of integrable systems and ensures that, provided the perturbation is small enough, most of the invariant tori in which the phase space of the unperturbed system is foliated persist in the complete system. In the case of FPU the simplest integrable system is the linearized chain for which the perturbation is provided by the nonlinearity. So the size of the nonlinearity increases with the energy of the initial datum and KAM theory should apply for energy smaller than some N-dependent threshold ϵ_N. This approach has the advantage of potentially explaining the FPU paradox *and also* to predict that it should disappear for energy larger then some threshold (as actually observed numerically). From the argument of Izrailev and Chirikov (based on Chirikov's criterion of overlapping of resonances) one can extract also an explicit estimate of ϵ_N which should go to zero like $N^{-4} \equiv \mu^4$. Such an estimate is derived by Izrailev and Chirikov by considering initial data on high frequency Fourier modes, while they do not deduce any explicit estimate for the case of initial data on low frequency modes. Their argument was extended to initial data on low frequency Fourier modes by Shepeliansky [37] leading to the claim that also in correspondence to such a kind of initial data the FPU phenomenon should disappear as $N \to \infty$. However a subsequent reanalysis of the problem led Ponno [33] to different conclusions, so, we can at least say that the situation is not yet clear.

We emphasize that the actual application of KAM theory to the FPU lattice is quite delicate since the hypotheses of KAM theory involve a Diophantine type nonresonance condition and also a nondegeneracy condition. The two conditions have been verified only much later by Rink [35] (see also [26, 32]). Then one has to estimate the dependence of the threshold ϵ_N on N and it turns out that a rough estimate gives that ϵ_N goes to zero exponentially with N (essentially due to the Diophantine type nonresonance condition). This is the main reason which led Izrailev and Chirikov to conjecture that the FPU paradox disappears in the thermodynamic limit.

In order to weaken this condition on ϵ_N, Benettin, Galgani, Giorgilli and collaborators [1, 8–11, 22] started to investigate the possibility of using averaging theory and Nekhoroshev's theorem to explain the FPU paradox. This a quite remarkable change of point of view, since at variance with KAM theory averaging theory and Nekhoroshev's theorem give results controlling the dynamics over long, but finite times, so that such a point of view leaves open the possibility that the FPU paradox disappears after a finite but long time, which is what is actually observed in numerical investigations (see also the remarkable theoretical paper [21]). Results along this line have been obtained for chains of rotators ([1, 9]) and FPU chains with alternate masses [1, 22]. An application to the true FPU model is given in the next section.

4 Some Rigorous Results at Vanishing Specific Energy

4.1 KdV and FPU

The unification of the two points of view illustrated above was obtained in the paper [6]. In that paper, first of all canonical perturbation theory is used in order to deduce a couple of KdV equations as resonant normal form for the FPU lattice and, secondly, the KdV equations are used in order to describe the phenomenon of formation and metastability of the FPU packet. We briefly recall the result of [6].

We consider here the case of periodic boundary conditions. Consider a state of the form (10) and write the equation for the evolution of the function u. Then it turns out that such an equation is a Hamiltonian perturbation of the wave equation, so one can use canonical perturbation theory for PDEs in order to simplify the equation. Passing to the variables f, g the normal form turns out to be the Hamiltonian of a couple of non interacting KdV equations. In [6] a rigorous theory estimating the error was developed, and the main results of that paper are contained in Theorem 1 and Corollary 1 below.

Consider the KdV equation

$$f_\tau + f_{yyy} + ff_y = 0 ;$$

it is well known [30] that if the initial datum extends to a function analytic in a complex strip of width σ, then the solution (as a function of the space variable y) is also analytic (in general in a smaller complex strip).

Consider now a couple of solutions f, g of KdV with analytic initial data and let $q_j^{KdV}(t)$ be the unique sequence such that

$$q_j^{KdV}(t) - q_{j+1}^{KdV}(t) = \mu^2 \left[f\left(\mu(j-t), \mu^3 t\right) + g\left(-\mu(j+t), \mu^3 t\right)\right],$$
$$\sum_j q_j^{KdV}(t) \equiv 0 , \tag{12}$$

where, as above, $\mu := N^{-1}$. Then the result of Theorem 1 below is that q_j^{KdV} approximates well the true solution of the FPU lattice.

Let $q_j(t)$ be the solution of the FPU equations with initial data $q_j(0) = q_j^{KdV}(0)$, $\dot{q}_j(0) = \dot{q}_j^{KdV}(0)$; denote by $E_k(t)$ the energy in the kth Fourier mode of the solution of the FPU with such initial data and $\mathrm{e}_k := E_k/N$.

The following theorem holds

Theorem 1 ([6]). *Fix an arbitrary $T_f > 0$. Then there exists μ_* such that, if $\mu < \mu_*$ then for all times t fulfilling*

$$|t| \leq \frac{T_f}{\mu^3} \tag{13}$$

one has

$$\sup_j \left|r_j(t) - r_j^{KdV}(t)\right| \leq C\mu^3 , \tag{14}$$

where $r_j := q_j - q_{j+1}$ and similarly for r_j^{KdV}. Furthermore, there exists $\sigma > 0$ s.t., for the same times, one has

$$\mathrm{e}_k(t) \leq C\mu^4 e^{-\sigma|k|} + C\mu^5 . \tag{15}$$

Exploiting known results on the dynamics of KdV (and of Hill's operators [34]) one gets the following corollary which is directly relevant to the FPU paradox.

Corollary 1. *Fix a positive R and a positive T_f; then there exists a positive constant μ_*, with the following property: assume $\mu < \mu_*$ and consider the FPU system with an initial datum fulfilling*

$$\mathrm{e}_1(0) = \mathrm{e}_{-1}(0) = R^2\mu^4 , \quad \mathrm{e}_k(0) \equiv \mathrm{e}_k(t)\big|_{t=0} = 0 , \quad \forall |k| \neq 1 . \tag{16}$$

Then, along the corresponding solution, Eq. (15) *holds for the times* (13).

Furthermore there exists a sequence of almost periodic functions $\{F_k(t)\}$ such that one has

$$\left|\mathrm{e}_k(t) - \mu^4 F_k(t)\right| \leq C_2 \mu^5 , \quad |t| \leq \frac{T_f}{\mu^3} . \tag{17}$$

Remark 1. One can show that the following limit exists

$$\bar{F}_k := \lim_{T\to\infty} \frac{1}{T} \int_0^T F_k(t) dt . \tag{18}$$

It follows that up to a small error the time average of $\mathrm{e}_k(t)$ relaxes to the limit distribution obtained by rescaling $\bar{F}_k$. Of course $\bar{F}_k$ is exponentially decreasing with k, but one can also show that actually one has $\bar{F}_k \neq 0 \; \forall k \neq 0$

The strong limitation of the above results rests in the fact that they only apply to initial data with specific energy of order μ^4, thus they do not apply to the thermodynamic limit.

4.2 Longer Time Scales at Smaller Energy

We present here a result by Hairer and Lubich [24] which is valid in a regime of specific energy smaller then that considered above, but controls the dynamics for longer time scales. The proof of the result is based on the technique of modulated Fourier expansion developed by the authors and their collaborators. In some sense such a technique can be considered as a variant of classical perturbation theory. The key tool that they use for the proof is an accurate analysis of the small denominators entering in the perturbative construction.

To be precise [24] deals with the case of periodic boundary conditions.

Theorem 2. *There exist positive constants R_*, N_*, T, with the following property: consider the FPU system with an initial datum fulfilling* (16) ***with* $\mathbf{R} < \mathbf{R}_*$**. *Then, along the corresponding solution, one has*

$$\mathrm{e}_k(t) \leq R^2 \mu^4 R^{2(|k|-1)} , \quad \forall\, 1 \leq |k| \leq N , \quad \forall |t| \leq \frac{T}{\mu^2 R^5} , \tag{19}$$

where as above $\mathrm{e}_k := E_k/N$.

It is interesting to compare the time scale covered by this theorem with the time scale of Corollary 1. It is clear that the time scale (19) is longer than (13) as far as

$$R < N^{-1/5} \tag{20}$$

(where we made the choice $T_f := T$), namely in a regime where the specific energy goes to zero faster than in Theorem 1.

One has also to remark that in Theorem 2 one gets an exponential decay of the Fourier modes valid for all k's (the term of order μ^5 present in (15) is here absent).

4.3 Toda Lattice

It is well known that close to the FPU lattice there exists a remarkable integrable system, namely the Toda lattice [25, 38] whose Hamiltonian is given by

$$H_{Toda}(p, q) = \frac{1}{2}\sum_j p_j^2 + \sum_j e^{q_j - q_{j+1}}, \tag{21}$$

so that one has

$$H_{FPU}(p, q) = H_{Toda}(p, q) + (A - 1)H_2(q) + H^{(3)}(q),$$

where

$$H_l(q) := \sum_j \frac{(q_j - q_{j+1})^{l+2}}{(l+2)!}, \quad \forall l \geq 2,$$

$$H^{(3)} := -\sum_{l\geq 3} H_l,$$

which shows the vicinity of H_{FPU} and H_{Toda}.

The idea of exploiting the Toda lattice in order to deduce information on the dynamics of the FPU chain is an old one; however in order to make it effective, one has first to deduce information on the dynamics of the Toda lattice itself, and this is far from trivial. The most obvious way to proceed consists in constructing action angle coordinates for the Toda lattice and using them to study the dynamics of the Toda lattice itself. An important result along this program was obtained by Henrici and Kappeler [26, 27] who constructed action angle coordinates and Birkhoff coordinates (a kind of cartesian action angle coordinates) showing that, for any N, such coordinates are globally analytic (see Theorem 3 below for a precise statement). However the construction by Henrici and Kappeler is not uniform in the number N of particles, thus it is not possible to exploit it directly in order to get results in the limit $N \to \infty$.

Results on the behaviour of the integrable structure of Toda for large N have been recently obtained in a series of papers [2–5, 15]. In particular in [2–4], exploiting ideas from [15], it has been shown that, as $N \to \infty$, the actions and the frequencies of the Toda lattice are well described by the actions and the frequencies of a couple of KdV equations, at least in a regime equal to that of Theorem 1, namely of specific energy of order μ^4.

Further results (exploiting some ideas from [2–4]) directly applicable to the FPU metastability problem have been obtained in [5] and now we are going to present them. In [5] the regularity properties of the Birkhoff map, namely the map introducing Birkhoff coordinates for the FPU lattice, have been studied and lower and upper bounds to the radius of the ball over which such a map is analytic have been given.

To come to a precise statement we start by recalling the result by Henrici and Kappeler.

Consider the Toda lattice in the submanifold (9) and introduce the linear Birkhoff variables

$$X_k = \frac{\hat{p}_k}{\sqrt{\omega_k}} , \quad Y_k = \sqrt{\omega_k}\hat{q}_k , \quad |k| = 1, \ldots, N ; \tag{22}$$

using such coordinates, H_0 takes the form

$$H_0 = \sum_{|k|=1}^{N} \omega_k \frac{X_k^2 + Y_k^2}{2} . \tag{23}$$

With an abuse of notations, we re-denote by H_{Toda} the Hamiltonian (21) written in the coordinates (X, Y).

Theorem 3 ([28]). *For any integer $N \geq 2$ there exists a global real analytic canonical diffeomorphism $\Phi_N : \mathbb{R}^{2N} \times \mathbb{R}^{2N} \to \mathbb{R}^{2N} \times \mathbb{R}^{2N}$, $(X, Y) = \Phi_N(x, y)$ with the following properties:*

(i) The Hamiltonian $H_{Toda} \circ \Phi_N$ is a function of the actions $I_k := \frac{x_k^2+y_k^2}{2}$ only, i.e. (x_k, y_k) are Birkhoff variables for the Toda Lattice.
(ii) The differential at the origin is the identity: $d\Phi_N(0, 0) = \mathbb{1}$.

In order to state the analyticity properties fulfilled by the map Φ_N as $N \to \infty$ we need to introduce suitable norms: for any $\sigma \geq 0$ define

$$\|(X, Y)\|_\sigma^2 := \frac{1}{N} \sum_k e^{2\sigma|k|} \omega_k \frac{|X_k|^2 + |Y_k|^2}{2} . \tag{24}$$

We denote by $B^\sigma(R)$ the ball in $\mathbb{C}^{2N} \times \mathbb{C}^{2N}$ of radius R and center 0 in the topology defined by the norm $\|.\|_\sigma$. We will also denote by $B^\sigma_{\mathbb{R}} := B^\sigma(R) \cap (\mathbb{R}^{2N} \times \mathbb{R}^{2N})$ the *real* ball of radius R.

Remark 2. We are particularly interested in the case $\sigma > 0$ since, in such a case, states with finite norm are exponentially decreasing in Fourier space.

The main result of [5] is the following Theorem.

Theorem 4 ([5]). *Fix* $\sigma \geq 0$ *then there exist* $R, R' > 0$ *s.t.* Φ_N *is analytic on* $B^\sigma\left(\frac{R}{N^\alpha}\right)$ *and fulfills*

$$\Phi_N\left(B^\sigma\left(\frac{R}{N^\alpha}\right)\right) \subset B^\sigma\left(\frac{R'}{N^\alpha}\right), \quad \forall N \geq 2 \tag{25}$$

if and only if $\alpha \geq 2$. *The same is true for the inverse map* Φ_N^{-1}.

Remark 3. A state (X, Y) is in the ball $B^\sigma(R/N^2)$ if and only if there exist interpolating periodic functions (β, α), namely functions s.t.

$$p_j = \beta\left(\frac{j}{N}\right), \quad q_j - q_{j+1} = \alpha\left(\frac{j}{N}\right), \tag{26}$$

which are analytic in a strip of width σ and have an analytic norm of size R/N^2. Thus we are in the same regime to which Theorem 1 apply.

Theorem 4 shows that the Birkhoff coordinates are analytic only in a ball of radius of order N^{-2}, which corresponds to initial data with specific energy of order N^{-4}.

We think this is a strong indication of the fact that standard integrable techniques cannot be used beyond such a regime.

As a corollary of Theorem 4, one immediately gets that in the Toda Lattice the analogous of the FPU metastable packet of modes is actually stable, namely it persists for infinite times. Precisely one has the following result.

Corollary 2. *Consider the Toda lattice* (21). *Fix* $\sigma > 0$, *then there exist constants* R_0, C_1, *such that the following holds true. Consider an initial datum fulfilling* (16) ***with*** $\mathbf{R} < \mathbf{R_0}$. *Then, along the corresponding solution, one has*

$$\mathrm{e}_k(t) \leq R^2(1 + C_1R)\mu^4 e^{-2\sigma|k|}, \quad \forall\, 1 \leq |k| \leq N, \quad \forall t \in \mathbb{R}. \tag{27}$$

We recall that this was observed numerically by Benettin and Ponno [7, 12]. One has to remark that according to the numerical computations of [12], the packet exists and is stable over infinite times also in a regime of finite specific energy (which would correspond to the case $\alpha = 0$ in Theorem 4). The understanding of this behaviour in such a regime is still a completely open problem.

Concerning the FPU chain, Theorem 4 yields the following result.

Theorem 5. *Consider the FPU system. Fix* $\sigma \geq 0$; *then there exist constants* R_0', C_2, T, *such that the following holds true. Consider a real initial datum fulfilling* (16) ***with*** $\mathbf{R} < \mathbf{R_0'}$, *then, along the corresponding solution, one has*

$$\mathrm{e}_k(t) \leq 16R^2\mu^4 e^{-2\sigma|k|}, \quad \forall\, 1 \leq |k| \leq N, \quad |t| \leq \frac{T}{R^2\mu^4} \cdot \frac{1}{|A-1| + C_2R\mu^2}. \tag{28}$$

Furthermore, for $1 \le |k| \le N$, consider the action $I_k := \frac{x_k^2+y_k^2}{2}$ of the Toda lattice and let $I_k(t)$ be its evolution according to the FPU flow. Then one has

$$\frac{1}{N}\sum_{|k|=1}^{N} e^{2\sigma|k|}\omega_k|I_k(t)-I_k(0)| \le C_3R^2\mu^5 \qquad \textit{for } t \textit{ fulfilling } (28). \tag{29}$$

So this theorem gives a result which covers times one order of magnitude longer then those covered by Theorem 1. Furthermore the small parameter controlling the time scale is the distance between the FPU and the Toda

This is particularly relevant in view of the fact that, according to Theorem 1 the time scale of formation of the packet is μ^{-3}, thus the present theorem shows that the packet persists at least over a time scale one order of magnitude longer then the time needed for its formation.

5 An Averaging Theorem in the Thermodynamic Limit

In this section we discuss a different approach to the study of the dynamics of the FPU system, which allows to give some results valid in the thermodynamic limit. Such a method is a development of the one introduced in [17] in order to deal with a chain of rotators (see also [19]), and developed in [18] in order to study a Klein Gordon chain.

We consider here the case of Dirichlet boundary conditions and endow the phase space with the Gibbs measure μ_β at inverse temperature β, namely

$$\mathrm{d}\mu_\beta(p,q) \stackrel{\text{def}}{=} \frac{\mathrm{e}^{-\beta H_{FPU}(p,q)}}{Z(\beta)}\mathrm{d}p\mathrm{d}q\ ; \tag{30}$$

where as usual

$$Z(\beta) := \int \mathrm{e}^{-\beta H_{FPU}(p,q)}\mathrm{d}p\mathrm{d}q$$

is the partition function (the integral is over the whole phase space). In the following we will omit the index β from μ. Given a function F on the phase space, we define

$$\langle F\rangle \stackrel{\text{def}}{=} \int F\mathrm{d}\mu\ , \tag{31}$$

$$\|F\|^2 \stackrel{\text{def}}{=} \langle F^2\rangle \equiv \int |F|^2\mathrm{d}\mu\ , \tag{32}$$

$$\sigma_F^2 \stackrel{\text{def}}{=} \|F-\langle F\rangle\|^2\ , \tag{33}$$

which are called respectively the average, the L^2 norm and the variance of F. The time autocorrelation function C_F of a dynamical variable F is defined by

$$C_F(t) := \langle F(t)F \rangle - \langle F^2 \rangle , \tag{34}$$

where $F(t) := F \circ G^t$ and G^t is the flow of the FPU system.

Remark that the Gibbs measure is asymptotically concentrated on the energy surface of energy N/β. Thus, when studying the system in the above setting one is typically considering data with specific energy equal to β^{-1}.

Let $g \in C^2([0,1], \mathbb{R}^+)$ be a twice differentiable function; we are interested in the time evolution of quantities of the form

$$\Phi_g \stackrel{\text{def}}{=} \sum_{k=1}^{N} E_k \, g\left(\frac{k}{N+1}\right) .$$

We are thinking of a function g with a small support close to a fixed wave vector $\bar{k}/(N+1)$, so that the quantity Φ_g represents the energy of a packet of modes centered at $\bar{k}/(N+1)$.

The following theorem was proved in [31]

Theorem 6. *Let $g \in C^2([0,1]; \mathbb{R}^+)$ be a function fulfilling $g'(0) = 0$. There exist constants $\beta^* > 0$, $N^* > 0$ and $C > 0$ s.t., for any $\beta > \beta^*$ and for any $N > N^*$, any $\delta_1, \delta_2 > 0$ one has*

$$\mu\left(\left|\Phi_g(t) - \Phi_g(0)\right| \geq \delta_1 \sigma_{\Phi_g}\right) \leq \delta_2 , \quad |t| \leq \frac{\delta_1 \sqrt{\delta_2}}{C} \beta \tag{35}$$

where, as above, $\Phi_g(t) = \Phi_g \circ G^t$.

This theorem shows that, with large probability, the energy of the packet of modes with profile defined by the function g remains constant over a time scale of order β^{-1}. We also emphasize that the change in the quantity Φ_g is small compared to its variance, which establishes the order of magnitude of the difference between the biggest and the smallest value of Φ_g on the energy surface.

Theorem 6 is actually a corollary of a result controlling the evolution of the time autocorrelation function of Φ_g. We point out that, in some sense the time autocorrelation function is a more important object, at least if one is interested in the problem of the dynamical foundations of thermodynamics. Indeed, it is known by Kubo linear response theory, that the specific heat of system in contact with a thermostat is the time autocorrelation function of its energy. Of course we are here dealing with an isolated system, so the previous theorem is not directly relevant to the problem of foundations of statistical mechanics.

Remark 4. Of course one can repeat the argument for different choices of the function g. For example one can partition the interval $[0,1]$ of the variable $k/(N+1)$ in K sub-intervals and define K different functions $g^{(1)}, g^{(2)}, \ldots, g^{(K)}$, with disjoint support, each one fulfilling the assumptions of Theorem 6, so that one gets that the quantities $\Phi_{g^{(l)}} \stackrel{\text{def}}{=} \sum_k g^{(l)}\left(\frac{k}{N+1}\right) E_k$ are adiabatic invariants, i.e. the energy essentially does not move from one packet to another one.

The scheme of the proof of Theorem 6 is as follows: first, following standard ideas in perturbation theory (see [23]), one performs a formal construction of an integral of motion as a power series in the phase space variables. As usual, already at the first step one has to solve the so called homological equation in order to find the third order correction of the quadratic integral of motion. The solution of such an equation involves some small denominators which are usually the source of one of the problems arising when one wants to control the behaviour of the system in the thermodynamical limit. We show that, if one takes as the quadratic part of the integral the quantity Φ_g, then every small denominator appears with a numerator which is also small, so that the ratio is bounded. The subsequent step consists in adding rigorous estimates on the variance of the time derivative of the so constructed approximate integral of motion. This allows to conclude the proof.

We emphasize that this procedure completely avoids to impose the time invariance of the domain in which the theory is developed, which is the requirement that usually prevents the applicability of canonical perturbation theory to systems in the thermodynamic limit. Indeed in the probabilistic framework the relevant estimates are global in the phase space.

6 Conclusions

Summarizing the above results, we can say that all analytical results available nowadays can be split into two groups: the first group consisting of those which describe the formation of the packet observed by FPU and give some estimates on its time of persistence. Such results do not survive in the thermodynamic limit; indeed they are all confined to the regime in which the specific energy is order N^{-4}. We find particularly surprising the fact that very different methods lead to the same regime and of course this raises the suspect that there is some reality in this limitation. However one has to say that numerics do not provide any evidence of changes in the dynamics when energy is increased beyond this limit.

A few more comments on this point are the following ones: the limitations appearing in constructing the Birkhoff variables in the Toda lattice (which are the source of the limitations in the applicability of Theorem 5) are related to the fact that one is implicitly looking for an integrable behaviour of the system, namely a behaviour in which the system is essentially decoupled into non interacting oscillators. On the contrary the construction leading to Theorem 1 is based on a resonant perturbative construction in which the small denominators are not present. The main limitation for the applicability of Theorem 1 comes from the need of considering the zero dispersion limit of the KdV equation. So, it is surprising that the regime at which the two results apply is the same.

So the question on whether the phenomenon of formation of a metastable packet persists in the thermodynamic limit or not is still completely open. An even more open question is that of the length of the time interval over which it persists. Up to now the best result we know is that of Theorem 5, but from the numerical experiments one would expect longer time scales (furthermore in the thermodynamic limit). How to reach them is by now not known.

At present the only known result valid in the thermodynamic limit is that of Theorem 6. However we think that this should be considered only as a preliminary one. Indeed it leaves open many important questions. The first one is the optimality of the time scale of validity: the technique used for its proof does not extend to higher order constructions. This is due to the fact that at order four new kinds of small denominators appear and up to now we were unable to control them. Furthermore there is no numerical evidence of the optimality of the time scale controlled by such a theorem.

An even more important question is the relevance of the result for the foundations of statistical mechanics. Indeed, one expects that the existence of many integrals of motion independent of the energy should have some influence on the measurement of thermodynamic quantities, for example the specific heat. In particular, since the time needed to exchange energy among different packets of modes increases as one lower the temperature, one would expect that some new behaviour appears as the temperature is lowered towards the absolute zero. However up to now we were unable to put into evidence some clear effect of the mathematical phenomenon described by Theorem 6. This is one of the main goals of our group for the next future.

Acknowledgements We thank Luigi Galgani for many interesting discussions and for his careful reading of the manuscript. This research was founded by the Prin project 2010–2011 "Teorie geometriche e analitiche dei sistemi Hamiltoniani in dimensioni finite e infinite".

References

1. Bambusi, D., Giorgilli, A.: Exponential stability of states close to resonance in infinite-dimensional Hamiltonian systems. J. Statist. Phys. **71**, 569–606 (1993)
2. Bambusi, D., Kappeler, T., Paul, T.: De Toda à KdV. C. R. Math. Acad. Sci. Paris, **347**, 1025–1030 (2009)
3. Bambusi, D., Kappeler, T., Paul, T.: Dynamics of periodic Toda chains with a large number of particles (2013) (ArXiv e-prints, arXiv:1309.5441 [math.AP])
4. Bambusi, D., Kappeler, T., Paul, T.: From Toda to KdV. ArXiv e-prints, arXiv:1309.5324 [math.AP], Sept. (2013).
5. Bambusi D, Maspero A. Birkhoff coordinates for the Toda lattice in the limit of infinitely many particles with an application to FPU. Preprint (2014)
6. Bambusi, D., Ponno A.: On metastability in FPU. Comm. Math. Phys. **264**, 539–561 (2006)

7. Benettin, G., Christodoulidi, H., Ponno, A.: The Fermi-Pasta-Ulam problem and its underlying integrable dynamics. J. Stat. Phys. **152**, 195–212 (2013)
8. Benettin, G., Galgani, L., Giorgilli, A.: Classical perturbation theory for systems of weakly coupled rotators. Nuovo Cimento B (11) **89**, 89–102 (1985)
9. Benettin, G., Galgani, L., Giorgilli, A.: Numerical investigations on a chain of weakly coupled rotators in the light of classical perturbation theory. Nuovo Cimento B (11), **89**, 103–119 (1985)
10. Benettin, G., Galgani, L., Giorgilli, A.: Realization of holonomic constraints and freezing of high frequency degrees of freedom in the light of classical perturbation theory. I. Comm. Math. Phys. **113**, 87–103 (1987)
11. Benettin, G., Galgani, L., Giorgilli, A.: Realization of holonomic constraints and freezing of high frequency degrees of freedom in the light of classical perturbation theory. II. Comm. Math. Phys. **121**, 557–601 (1989)
12. Benettin, G., Ponno, A.: Time-scales to equipartition in the Fermi-Pasta-Ulam problem: finite-size effects and thermodynamic limit. J. Stat. Phys. **144**, 793–812 (2011)
13. Berchialla, L., Galgani, L., Giorgilli, A.: Localization of energy in FPU chains. Discrete Contin. Dyn. Syst. **11**, 855–866 (2004)
14. Berchialla, L., Giorgilli, A., Paleari, S.: Exponentially long times to equipartition in the thermodynamic limit. Phys. Lett. A **321**, 167–172 (2004)
15. Bloch, A., Golse, F, Paul T., Uribe, A.: Dispersionless toda and toeplitz operators. Duke Math. J. **117**, 157–196 (2003)
16. Bocchieri, P., Scotti, A., Bearzi, B., Loinger, A.: Anharmonic chains with Lenard–Jones interactions. Phys. Rev. A **2**, 2013–2019 (1970)
17. Carati, A.: An averaging theorem for Hamiltonian dynamical systems in the thermodynamic limit. J. Stat. Phys. **128**, 1057–1077 (2007)
18. Carati, A., Maiocchi, A.M.: Exponentially long stability times for a nonlinear lattice in the thermodynamic limit. Comm. Math. Phys. **314**, 129–161 (2012)
19. De Roeck, W., Huveneers, F.: Asymptotic localization of energy in non-disordered oscillator chains (2013) [arXiv:1305.512]
20. Fermi, E., Pasta, J., Ulam, S.: Studies of nonlinear problems. In Collected works of E. Fermi, vol.2. Chicago University Press, Chicago (1965)
21. Fucito, F., Marchesoni, F., Marinari, E., Parisi, G., Peliti, L., Ruffo, S., Vulpiani, A.: Approach to equilibrium in a chain of nonlinear oscillators. J. de Physique **43**, 707–713 (1982)
22. Galgani, L., Giorgilli, A., Martinoli, A., Vanzini, S.: On the problem of energy equipartition for large systems of the Fermi-Pasta-Ulam type: analytical and numerical estimates. Phys. D **59**, 334–348 (1992)
23. Giorgilli, A., Galgani, L.: Formal integrals for an autonomous Hamiltonian system near an equilibrium point. Celestial Mech. **17**, 267–280 (1978)
24. Hairer, E., Lubich, C.: On the energy distribution in Fermi-Pasta-Ulam lattices. Arch. Ration. Mech. Anal. **205**, 993–1029 (2012)
25. Hénon, M.: Integrals of the Toda lattice. Phys. Rev. B (3), **9**, 1921–1923 (1974)
26. Henrici, A., Kappeler, T.: Birkhoff normal form for the periodic Toda lattice. In: Integrable systems and random matrices, Contemporary Math, vol. 458, pp. 11–29. American Mathematical Society, Providence (2008)
27. Henrici, A., Kappeler, T.: Global action-angle variables for the periodic Toda lattice. Int. Math. Res. Not. IMRN **11**, 52 (2008). (Art. ID rnn031)
28. Henrici, A., Kappeler, T.: Global Birkhoff coordinates for the periodic Toda lattice. Nonlinearity **21**, 2731–2758 (2008)
29. Izrailev, F.M., Chirikov, B.V.: Statistical properties of a nonlinear string. Sov. Phys. Dokl. **11**, 30–32 (1966)
30. Kappeler, T., Pöschel, J.: KdV & KAM. Ergebnisse der Mathematik und ihrer Grenzgebiete. 3. Folge, vol. 45 (A Series of Modern Surveys in Mathematics) [Results in Mathematics and Related Areas, 3rd Series. A Series of Modern Surveys in Mathematics]. Springer, Berlin (2003)

31. Maiocchi, A.M., Bambusi, D., Carati, A.: An averaging theorem for fpu in the thermodynamic limit. J. Stat. Phys. **155**, 300–322 (2014)
32. Nishida T. A note on an existence of conditionally periodic oscillation in a one-dimensional anharmonic lattice. Mem. Fac. Eng. Kyoto Univ. **33**, 27–34 (1971)
33. Ponno, A.: The Fermi-Pasta-Ulam problem in the thermodynamic limit. In: Chaotic dynamics and transport in classical and quantum systems, NATO Science Series II Mathematics Physics and Chemistry, vol. 182, pp. 431–440. Kluwer Acadamic, Dordrecht (2005)
34. Pöschel, J.: Hill's potentials in weighted Sobolev spaces and their spectral gaps. Math. Ann. **349**, 433–458 (2011)
35. Rink, B.: Symmetry and resonance in periodic FPU chains. Comm. Math. Phys. **218**, 665–685 (2001)
36. Schneider, G., Wayne, C.E.: Counter-propagating waves on fluid surfaces and the continuum limit of the Fermi-Pasta-Ulam model. In: International Conference on Differential Equations (Berlin, 1999), vols. 1, 2, pp. 390–404. World Science, River Edge (2000)
37. Shepelyansky, D.L.: Low-energy chaos in the Fermi–Pasta–Ulam problem. Nonlinearity **10**, 1331–1338 (1997)
38. Toda, T.: Vibration of a chain with nonlinear interaction. J. Phys. Soc. Jpn. **22**, 431 (1967)
39. Zabusky, N.J., Kruskal, M.D.: Interaction of solitons in a collisionless plasma and the recurrence of initial states. Phys. Rev. Lett. **15**, 240–243 (1965)

A Nash-Moser Approach to KAM Theory

Massimiliano Berti and Philippe Bolle

Dedicated to Walter Craig for his 60th birthday.

Abstract Any finite dimensional embedded invariant torus of a Hamiltonian system, densely filled by quasi-periodic solutions, is isotropic. This property allows us to construct a set of symplectic coordinates in a neighborhood of the torus in which the Hamiltonian is in a generalized KAM normal form with angle-dependent coefficients. Based on this observation we develop an approach to KAM theory via a Nash-Moser implicit function iterative theorem. The key point is to construct an approximate right inverse of the differential operator associated to the linearized Hamiltonian system at each approximate quasi-periodic solution. In the above symplectic coordinates the linearized dynamics on the tangential and normal directions to the approximate torus are approximately decoupled. The construction of an approximate inverse is thus reduced to solving a quasi-periodically forced linear differential equation in the normal variables. Applications of this procedure allow to prove the existence of finite dimensional Diophantine invariant tori of autonomous PDEs.

1 Introduction

In the last years much work has been devoted to the existence theory of quasi-periodic solutions of PDEs. The main strategies developed to overcome the well known "*small divisors*" problem are:

1. KAM (Kolmogorov-Arnold-Moser) theory,
2. Newton-Nash-Moser implicit function theorems.

M. Berti (✉)
SISSA, Via Bonomea 265, Trieste 34136, Italy
e-mail: berti@sissa.it

P. Bolle
Université d'Avignon, Laboratoire de Mathématiques d'Avignon (EA2151),
F-84018 Avignon, France
e-mail: philippe.bolle@univ-avignon.fr

P. Guyenne et al. (eds.), *Hamiltonian Partial Differential Equations and Applications*, Fields Institute Communications 75,
DOI 10.1007/978-1-4939-2950-4_9

The KAM approach consists in generating iteratively a sequence of canonical transformations of the phase space which bring, up to smaller and smaller remainders, the Hamiltonian system into a normal form with an invariant torus "at the origin": in the new coordinates, the invariant torus is the zero section of the trivial linear bundle $\mathbb{T}^\nu \times \mathbb{R}^\nu \times E \stackrel{\pi}{\longrightarrow} \mathbb{T}^\nu$, see Sect. 2. In the usual KAM scheme the normal form has constant coefficients (see (30)). This iterative procedure requires, at each step, to solve the so called linear "homological equations". The normal form depends on a frequency vector $\omega \in \mathbb{R}^\nu$, whose components are the "tangential frequencies" of the invariant torus, and the normal frequencies Ω_i at the invariant torus. Solving the homological equations requires lower bounds for the moduli of $\omega \cdot k + \Omega_i$ (first order Melnikov non-resonance conditions) and of $\omega \cdot k + \Omega_i \pm \Omega_j$ (second order Melnikov conditions), for $k \in \mathbb{Z}^\nu$. The final KAM torus is linearly stable.

This scheme has been effectively implemented by Kuksin [20] and Wayne [28] for 1-d nonlinear wave (NLW) and Schrödinger (NLS) equations with Dirichlet boundary conditions. The required second order Melnikov non resonance conditions are violated in presence of multiple normal frequencies, for example, already for periodic boundary conditions (two eigenvalues of ∂_{xx} are equal).

Thus a more direct bifurcation approach was proposed by Craig and Wayne [13], see also [12], for 1-d NLW and NLS equations with periodic boundary conditions. After a Lyapunov-Schmidt decomposition, the search of the invariant torus is reduced to solve a functional equation by some Newton-Nash-Moser implicit function theorem in Banach scales of analytic functions of time and space. The main advantage of this approach is to require only the "first order Melnikov" non-resonance conditions for inverting the linearized operators obtained at each step of the iteration: such conditions mean that the eigenvalues of these linear (self-adjoint) operators are bounded away from zero. On the other hand, the main difficulty is that the linearized equations are PDEs with non-constant coefficients, represented by differential operators that are small perturbations of a diagonal operator having arbitrarily small eigenvalues. Hence it is hard to estimate their inverses in high norms. Craig-Wayne [13] solved this problem for periodic solutions and Bourgain [9] also for quasi-periodic solutions. This approach is particularly useful for PDEs in higher dimension due to the large (possibly unbounded) multiplicities of the normal frequencies. It has been effectively implemented by Bourgain [10, 11], for analytic NLS and NLW with Fourier multipliers on $\mathbb{T}^d$, $d \geq 2$, and by Berti-Bolle [4, 5] for forced NLS and NLW equations with a multiplicative potential and finite differentiable nonlinearities. In Berti-Corsi-Procesi [8] this scheme has been then generalized into an abstract Nash-Moser implicit function theorem with applications to NLW and NLS on general compact Lie groups and homogeneous spaces.

We remark that in the above papers the transformations used to prove estimates in high norms for the inverse linearized operators at the approximate solutions are *not* maps of the phase space, as in the usual KAM approach, and therefore the dynamical system structure of the transformed system is lost.

The aim of this Note is to present a new strategy to KAM theory for PDEs which contains the advantages of both the previous approaches. It is a Nash-Moser implicit function iterative scheme for the search of a torus embedding with the advantages of the normal form KAM approach, as developed in [15, 23], and for PDEs in [16, 20, 21, 25, 28] (and references therein). Instead of applying directly a sequence of canonical maps of the phase space which, at the limit, conjugate the Hamiltonian to another one which possesses an invariant torus "at the origin" (this is the usual normal form approach), we construct an embedded invariant torus by a Nash-Moser iterative scheme in scales of Banach spaces. Then Theorem 1 proves that, in a neighborhood of an invariant Diophantine torus (a torus supporting quasi-periodic orbits of Diophantine frequency vector), it is always possible to construct a set of analytic symplectic variables in which the Hamiltonian system assumes the KAM normal form (28). Note that the quadratic terms of the normal form (28) are, in general, angle dependent. In these coordinates the linearized equations in the tangential and normal directions are decoupled, see (31). Theorem 1 is relevant for implementing a Nash-Moser iterative scheme which constructs better and better approximate invariant tori of the PDE (Sects. 3–4). Actually the existence of an approximate KAM normal form, near an approximate invariant torus, as in (28), simplifies the analysis of the linearized system at such approximate solution: in a set of symplectic coordinates, adapted to each approximate torus, the linearized equations in the tangential and normal directions are approximately decoupled. This reduces the problem to the study of the quasi-periodically forced linearized equation in the normal directions. An advantage is that, when working with autonomous PDEs, one can essentially decouple the problems coming from the study of the bifurcation equation from those coming from the small divisors. The present strategy applies well to NLW on $\mathbb{T}^d$ with multiplicative potential (see [6]) and to quasi-linear PDEs (see [3]); in both cases the usual standard normal form approach does not seem to apply.

The above symplectic construction preserves the Hamiltonian dynamical structure of the equations. Thus it is a decomposition of tangential/normal dynamics sharper than the usual Lyapunov-Schmidt reduction based on the splitting into bifurcation/range equations. The present KAM approach applies well also to PDEs whose flow is ill-posed.

We now present more in detail the main ideas of this KAM strategy. As already mentioned, the main difficulty for implementing a Newton-Nash-Moser iterative scheme is to solve the (non homogeneous) linearized equations at each approximate quasi-periodic solution. This is a difficult matter for the presence of small divisors and because the tangential and normal components to the torus of the linearized equations are coupled by the nonlinearity.

It was noted by Zehnder [29] that, in order to get a "quasi-Newton-Nash-Moser" scheme with still quadratic speed of convergence, it is sufficient to invert the linearized equation only approximately. In [29] Zehnder introduced the precise notion of approximate right inverse linear operator, see (43). Its main feature is to be an *exact* inverse of the linearized equation at an exact solution.

In this approach we construct an approximate right inverse for the functional equation satisfied by the embedding of an invariant torus of a Hamiltonian system, see (16), or (34)–(35). The first observation is that an embedded invariant torus supporting a non-resonant rotation is isotropic. This is classical for finite dimensional Hamiltonian systems, see [19] or [17]-Lemma 33. Actually this property is also true for *infinite dimensional* Hamiltonian systems (Lemma 1) since it requires only that the Hamiltonian flow is well defined on the invariant torus and preserves the symplectic 2-form on it. Near an isotropic torus it is then possible to introduce the symplectic set of coordinates (23) in which the torus is "at the origin". It follows that the existence of an invariant torus and a nearby normal form like (28) are equivalent statements, see Theorem 1. Clearly, with second order Melnikov non-resonance conditions, it is also possible to obtain a constant coefficients normal form as (30), i.e. to prove the reducibility of the torus.

The introduction of such symplectic coordinates adapted to the torus is the bridge with the usual KAM proof based on normal forms transformations. The point is that the normal form (28) means more than the existence of the invariant torus, since it also provides a control of the linearized equations in the normal bundle of the torus. Actually, in the normal form coordinates, the linearized equations at the torus simplify, see (31). In particular the second component in (31) is decoupled from the others and the system (31) can be solved in a "triangular" way.

Of course, there is little interest in inverting the linearized equation at a torus that is already a solution. The point is that, at an approximate invariant torus, it is still possible to construct an approximate right inverse of the linearized equation, which is enough to get a "quasi-Nash-Moser" scheme à la Zehnder.

With this aim, in Sect. 4 we extend the symplectic construction developed in Sect. 2 for an invariant torus, to an approximate solution. Needless to say, an approximate invariant torus is only approximately isotropic (Lemma 5). Thus the first step is to deform it into a nearby exactly isotropic torus (Lemma 6). This enables to define the set of symplectic coordinates (66) in which the isotropic torus is "at the origin". In these new coordinates also the linearized equations (73) simplify and we may invert them approximately, namely solve only (74) and (75). Such system is obtained by neglecting in (73) the terms which are zero at an exact solution, see Lemmas 8 and 3. The linear system (74) and (75) may be solved in a triangular way, first inverting the action-component equation (76) (see (77) and (78)), which is decoupled from the other equations.

In the case of a Lagrangian (finite dimensional) torus, there is not the last normal component in (75), and one may immediately solve equation (82) for the angle component, see (83) and (84). This completes the construction of an approximate inverse. This is another way to recover the classical results of Zehnder [29] and Salomon-Zehnder [27] for maximal dimensional tori, and it is closely related to the method in [14] by De la Llave, Gonzalez, Jorba, Villanueva.

On the contrary, in the general case of an isotropic torus, the present strategy has reduced the search of an approximate right inverse for the embedded torus equation, to the problem of solving the linear equation (79). This is a quasi-periodically forced linear PDE which is a small perturbation of the original linearized PDE, restricted to the normal directions. The existence of a solution for such an equation is very simple for partially hyperbolic (whiskered) tori, because there is no resonance in the normal directions. In the more difficult case of elliptic tori, where small divisors appear, this equation has the same feature as the quasi-periodically forced linear PDE restricted to the normal directions. It makes possible to exploit KAM results that have already been proved for the corresponding forced PDEs, as, for example, [2, 4, 5, 8].

For finite dimensional systems, this construction is deeply related to the Herman-Fejoz KAM normal form theorem used in [17] to prove the existence of elliptic invariant tori in the planetary solar system. Actually le "Théorème de conjugaison tordue" of Herman (Theorem 38 in [17]) is deduced by a Nash-Moser implicit functions theorem in Fréchet spaces.

This scheme may be effectively implemented for autonomous Hamiltonian PDEs, like, for example,

1. **(NLW)** Nonlinear wave equation

$$y_{tt} - \Delta y + V(x)y = f(x, y)\,, \quad x \in \mathbb{T}^d\,, \quad y \in \mathbb{R}\,, \tag{1}$$

   with a real valued multiplicative potential (we may clearly consider also a convolution potential $V * y$ as in [11]) and a real valued nonlinearity f.
2. **(NLS)** Hamiltonian nonlinear Schrödinger equation

$$\mathrm{i}u_t - \Delta u + V(x)u = f(x, u)\,, \quad x \in \mathbb{T}^d\,,\ u \in \mathbb{C}\,, \tag{2}$$

   where $f(x, u) = \partial_{\bar{u}} F(x, u)$ and the potential $F(x, u) \in \mathbb{R}$, $\forall u \in \mathbb{C}$, is real valued. For $u = a + ib$, $a, b \in \mathbb{R}$, the operator $\partial_{\bar{u}}$ is by definition $\frac{1}{2}(\partial_a + \mathrm{i}\partial_b)$.
3. **(KdV)** Quasi-linear Hamiltonian perturbed KdV equations

$$u_t + u_{xxx} + \partial_x u^2 + \mathcal{N}(x, u, u_x, u_{xx}, u_{xxx}) = 0\,, \quad x \in \mathbb{T}\,, \tag{3}$$

   where $\mathcal{N}(x, u, u_x, u_{xx}, u_{xxx}) := -\partial_x\big[(\partial_u f)(x, u, u_x) - \partial_x((\partial_{u_x} f)(x, u, u_x))\big]$ is the most general Hamiltonian (local) nonlinearity, see (4).

The NLW and NLS equations are studied in [6] and the quasi-linear KdV in [3].

All the above PDEs are infinite dimensional Hamiltonian systems. Also in view of the abstract setting of Sect. 2, we present their Hamiltonian formulation:

1. The NLW equation (1) can be written as the Hamiltonian system

$$\frac{d}{dt}\begin{pmatrix} y \\ p \end{pmatrix} = \begin{pmatrix} p \\ \Delta y - V(x)y + f(x, y) \end{pmatrix} = \begin{pmatrix} 0 & Id \\ -Id & 0 \end{pmatrix} \begin{pmatrix} \delta_y H(y, p) \\ \delta_p H(y, p) \end{pmatrix}$$

where $\delta_y H$, $\delta_p H$ denote the $L^2(\mathbb{T}^d_x)$-gradient of the Hamiltonian

$$H(y,p) := \int_{\mathbb{T}^d} \frac{p^2}{2} + \frac{1}{2}\big(|\nabla y|^2 + V(x)y^2\big) + F(x,y)\, dx$$

and $\partial_y F(x,y) = -f(x,y)$. The variables (y,p) are "Darboux coordinates".

2. The NLS equation (2) can be written as the infinite dimensional complex system

$$u_t = \mathrm{i}\delta_{\bar u} H(u)\,, \quad H(u) := \int_{\mathbb{T}^d} |\nabla u|^2 + V(x)|u|^2 - F(x,u)\, dx\,.$$

Actually (2) is a real Hamiltonian PDE in the variables $(a,b) \in \mathbb{R}^2$, real and imaginary part of u. Denoting the real valued potential $W(a,b) := F(x, a+ib)$, so that

$$\partial_{\bar u} F(x, a+ib) := \frac{1}{2}(\partial_a + \mathrm{i}\partial_b)W(a,b) = f(x, a+ib)\,,$$

the NLS equation (2) reads

$$\frac{d}{dt}\begin{pmatrix} a \\ b \end{pmatrix} = \begin{pmatrix} \Delta b - V(x)b + \frac{1}{2}\partial_b W(a,b) \\ -\Delta a + V(x)a - \frac{1}{2}\partial_a W(a,b) \end{pmatrix} = \frac{1}{2}\begin{pmatrix} 0 & -Id \\ Id & 0 \end{pmatrix}\begin{pmatrix} \delta_a H(a,b) \\ \delta_b H(a,b) \end{pmatrix}$$

with real valued Hamiltonian $H(a,b) := H(a+ib)$, δ_a, δ_b denoting the L^2-real gradients.

3. The KdV equation (3) is the Hamiltonian PDE

$$u_t = \partial_x \delta H(u)\,, \qquad H(u) = \int_{\mathbb{T}} \frac{u_x^2}{2} - \frac{u^3}{3} + f(x, u, u_x)\, dx\,, \tag{4}$$

where δH denotes the $L^2(\mathbb{T}_x)$ gradient. A natural phase space for (4) is

$$H^1_0(\mathbb{T}_x) := \Big\{ u(x) \in H^1(\mathbb{T}, \mathbb{R}) \ : \ \int_{\mathbb{T}} u(x) dx = 0 \Big\}\,.$$

In the present paper we shall focus on the geometric construction of the approximate right inverse for the equation satisfied by an embedded torus of a Hamiltonian system, stressing the algebraic aspects of the proof. In the papers [3, 6] the analytic estimates, and small technical variations, may disturb the geometric clarity of the approach.

2 Normal Form Close to an Invariant Torus

We consider the toroidal phase space

$$\mathcal{P} := \mathbb{T}^\nu \times \mathbb{R}^\nu \times E \qquad \text{where} \qquad \mathbb{T}^\nu := \mathbb{R}^\nu/(2\pi\mathbb{Z})^\nu$$

is the standard flat torus and E is a real Hilbert space with scalar product $\langle\ ,\ \rangle$. We denote by $u := (\theta, I, z)$ the variables of $\mathcal{P}$. We call (θ, I) the "action-angle" variables and z the "normal" variables. We assume that E is endowed with a constant exact symplectic 2-form

$$\Omega_E(z, w) = \langle \bar{J}z, w\rangle\,, \quad \forall z, w \in E\,, \tag{5}$$

where $\bar{J} : E \to E$ is an antisymmetric bounded linear operator with trivial kernel. Thus $\mathcal{P}$ is endowed with the symplectic 2-form

$$\Omega := (dI \wedge d\theta) \oplus \Omega_E \tag{6}$$

which is exact, namely

$$\Omega = d\lambda \tag{7}$$

where d denotes the exterior derivative and λ is the 1-form on $\mathcal{P}$ defined by

$$\lambda_{(\theta,I,z)} : \mathbb{R}^\nu \times \mathbb{R}^\nu \times E \to \mathbb{R}\,,$$

$$\lambda_{(\theta,I,z)}[\hat\theta, \hat I, \hat z] := I \cdot \hat\theta + \frac{1}{2}\langle \bar{J}z, \hat z\rangle\,, \quad \forall(\hat\theta, \hat I, \hat z) \in \mathbb{R}^\nu \times \mathbb{R}^\nu \times E\,. \tag{8}$$

The dot " · " denotes the usual scalar product of $\mathbb{R}^\nu$.

Remark 1. For the NLW equation $E = L^2 \times L^2$ with $L^2 := L^2(\mathbb{T}^d, \mathbb{R})$, the operator defining the symplectic structure is $\bar{J} = \begin{pmatrix} 0 & -Id \\ Id & 0 \end{pmatrix}$ and $\langle\ ,\ \rangle$ is the L^2 real scalar product on E. The transposed operator $\bar{J}^T = -\bar{J}$ (with respect to $\langle\ ,\ \rangle$) and the inverse $\bar{J}^{-1} = \bar{J}^T$. The same setting holds for the NLS equation with real valued Hamiltonian, writing it as a real Hamiltonian system in the real and imaginary part. For the KdV equation $E = L^2_0(\mathbb{T}, \mathbb{R}) := \{z \in L^2(\mathbb{T}, \mathbb{R}) : \int_{\mathbb{T}} z(x)\,dx = 0\}$ the operator $\bar{J} = \partial_x^{-1}$ and $\langle\ ,\ \rangle$ is the L^2 real scalar product. Here the transposed operator $\bar{J}^T = -\bar{J}$ and the inverse $\bar{J}^{-1} = \partial_x$ is unbounded.

Given a Hamiltonian function $H : \mathcal{D} \subset \mathcal{P} \to \mathbb{R}$, we consider the Hamiltonian system

$$u_t = X_H(u)\,, \qquad \text{where} \qquad dH(u)[\cdot] = -\Omega(X_H(u), \cdot) \tag{9}$$

formally defines the Hamiltonian vector field X_H. For infinite dimensional systems (i.e. PDEs) the Hamiltonian H is, in general, well defined and smooth only on a dense subset $\mathcal{D} = \mathbb{T}^\nu \times \mathbb{R}^\nu \times E_1 \subset \mathcal{P}$ where $E_1 \subset E$ is a dense subspace of E. We require that, for all $(\theta, I) \in \mathbb{T}^\nu \times \mathbb{R}^\nu$, $\forall z \in E_1$, the Hamiltonian H admits a gradient $\nabla_z H$, defined by

$$d_z H(\theta, I, z)[h] = \langle \nabla_z H(\theta, I, z), h \rangle\,, \quad \forall h \in E_1\,, \tag{10}$$

and that $\nabla_z H(\theta, I, z) \in E$ is in the space of definition of the (possibly unbounded) operator $J := -\bar{J}^{-1}$. Then by (9), (5), (6), (10) the Hamiltonian vector field X_H : $\mathbb{T}^\nu \times \mathbb{R}^\nu \times E_1 \mapsto \mathbb{R}^\nu \times \mathbb{R}^\nu \times E$ writes

$$X_H = (\partial_I H, -\partial_\theta H, J\nabla_z H)\,, \qquad J := -\bar{J}^{-1}\,. \tag{11}$$

A continuous curve $[t_0, t_1] \ni t \mapsto u(t) \in \mathbb{T}^\nu \times \mathbb{R}^\nu \times E$ is a solution of the Hamiltonian system (9) if it is C^1 as a map from $[t_0, t_1]$ to $\mathbb{T}^\nu \times \mathbb{R}^\nu \times E_1$ and $u_t(t) = X_H(u(t))$, $\forall t \in [t_0, t_1]$. For PDEs, the flow map Φ_H^t may not be well-defined everywhere. The next arguments, however, will not require to solve the initial value problem, but only a functional equation in order to find solutions which are quasi-periodic, see (16).

We refer to [21] for a general functional setting of Hamiltonian PDEs on scales of Hilbert spaces.

Remark 2. In the example of remark 1 for NLW and NLS, we can take $E_1 := H^2 \times H^2$. Then the Hamiltonian vector field $J\nabla_z H : H^2 \times H^2 \to L^2 \times L^2$. For KdV we can take $E_1 = H_0^3(\mathbb{T})$ so that $J\nabla_z H = -\partial_x \nabla_z H : H_0^3(\mathbb{T}) \mapsto L_0^2(\mathbb{T})$.

We suppose that (9) possesses an embedded invariant torus

$$\varphi \mapsto i(\varphi) := (\theta_0(\varphi), I_0(\varphi), z_0(\varphi))\,, \tag{12}$$

$$i \in C^1(\mathbb{T}^\nu, \mathcal{P}) \cap C^0(\mathbb{T}^\nu, \mathcal{P} \cap \mathbb{T}^\nu \times \mathbb{R}^\nu \times E_1)\,, \tag{13}$$

which supports a quasi-periodic solution with non-resonant frequency vector $\omega \in \mathbb{R}^\nu$, more precisely

$$i \circ \Psi_\omega^t = \Phi_H^t \circ i\,, \quad \forall t \in \mathbb{R}\,, \tag{14}$$

where Φ_H^t denotes the flow generated by X_H and

$$\Psi_\omega^t : \mathbb{T}^\nu \to \mathbb{T}^\nu\,, \quad \Psi_\omega^t(\varphi) := \varphi + \omega t\,, \tag{15}$$

is the translation flow of vector ω on $\mathbb{T}^\nu$. Since $\omega \in \mathbb{R}^\nu$ is non-resonant, namely $\omega \cdot k \neq 0$, $\forall k \in \mathbb{Z}^\nu \setminus \{0\}$, each orbit of (Ψ^t_ω) is *dense* in $\mathbb{T}^\nu$. Note that (14) *only* requires that the flow Φ^t_H is well defined and smooth on the compact manifold $\mathcal{T} := i(\mathbb{T}^\nu) \subset \mathcal{P}$ and $(\Phi^t_H)_{|\mathcal{T}} = i \circ \Psi^t_\omega \circ i^{-1}$. This remark is important because, for PDEs, the flow could be ill-posed in a neighborhood of $\mathcal{T}$. From a functional point of view (14) is equivalent to the equation

$$\omega \cdot \partial_\varphi i(\varphi) - X_H(i(\varphi)) = 0 \,. \tag{16}$$

Remark 3. In the sequel we will formally differentiate several times the torus embedding i, so that we assume more regularity than (13). In the framework of a Nash-Moser scheme, the approximate torus embedding solutions i are indeed regularized at each step.

We require that $\theta_0 : \mathbb{T}^\nu \to \mathbb{T}^\nu$ is a diffeomorphism of $\mathbb{T}^\nu$ isotopic to the identity. Then the embedded torus $\mathcal{T} := i(\mathbb{T}^\nu)$ is a smooth graph over $\mathbb{T}^\nu$. Moreover the lift on $\mathbb{R}^\nu$ of θ_0 is a map

$$\theta_0 : \mathbb{R}^\nu \to \mathbb{R}^\nu, \quad \theta_0(\varphi) = \varphi + \Theta_0(\varphi)\,, \tag{17}$$

where $\Theta_0(\varphi)$ is 2π-periodic in each component φ_i, $i = 1, \ldots, \nu$, with invertible Jacobian matrix $D\theta_0(\varphi) = Id + D\Theta_0(\varphi)$, $\forall \varphi \in \mathbb{T}^\nu$. In the usual applications $D\Theta_0$ is small and ω is a Diophantine vector, namely

$$|\omega \cdot k| \geq \frac{\gamma}{|k|^\tau}, \qquad \forall k \in \mathbb{Z}^\nu \setminus \{0\}\,. \tag{18}$$

In such a case we say that the invariant torus embedding $\varphi \mapsto i(\varphi)$ is Diophantine. The torus $\mathcal{T}$ is the graph of the function (see (12) and (17))

$$j := i \circ \theta_0^{-1}, \quad j : \mathbb{T}^\nu \to \mathbb{T}^\nu \times \mathbb{R}^\nu \times E, \quad j(\theta) := (\theta, \tilde{I}_0(\theta), \tilde{z}_0(\theta))\,, \tag{19}$$

namely

$$\mathcal{T} = \big\{(\theta, \tilde{I}_0(\theta), \tilde{z}_0(\theta))\,;\, \theta \in \mathbb{T}^\nu\big\}, \quad \text{where} \quad \tilde{I}_0 := I_0 \circ \theta_0^{-1}\,,\ \tilde{z}_0 := z_0 \circ \theta_0^{-1}\,. \tag{20}$$

We want to introduce a symplectic set of coordinates (ψ, y, w) near the invariant torus $\mathcal{T} := i(\mathbb{T}^\nu)$ such that $\mathcal{T}$ is described by $\{y = 0, w = 0\}$ and the restricted flow is simply $\psi(t) = \varphi + \omega t$. We look for a diffeomorphism of the phase space of the form

$$\begin{pmatrix} \theta \\ I \\ z \end{pmatrix} = G \begin{pmatrix} \psi \\ y \\ w \end{pmatrix} := \begin{pmatrix} \theta_0(\psi) \\ I_0(\psi) + B_1(\psi)y + B_2(\psi)w \\ z_0(\psi) + w \end{pmatrix}$$

where $B_1(\psi) : \mathbb{R}^\nu \to \mathbb{R}^\nu$, $B_2(\psi) : E \to \mathbb{R}^\nu$ are linear operators. Note that in the first component G is just the diffeomorphism of $\mathbb{T}^\nu$ induced by the torus embedding and that G is linear in y, w (actually the third component of G is a translation in w).

Remark 4. The above change of variables G is a particular class of those used by Moser in [23], which also allow to "rotate" linearly the third component as $z_0(\psi) + C_1(\psi)y + C_2(\psi)w$.

In order to find a symplectic set of coordinates as above, namely to find $B_1(\psi)$, $B_2(\psi)$ such that G is symplectic, we exploit the isotropy of the invariant torus $i(\mathbb{T}^\nu)$, i.e. the fact that the 2-form Ω vanishes on the tangent space to $i(\mathbb{T}^\nu) \subset \mathcal{P}$,

$$0 = i^*\Omega = i^*d\lambda = d(i^*\lambda)\,. \tag{21}$$

In other words, the 1-form $i^*\lambda$ on $\mathbb{T}^\nu$ is closed. It is natural to use such property: also in the proof of the classical Arnold-Liouville theorem (see e.g. [24]), the first step for the construction of the symplectic action-angle variables is to show that a maximal torus supporting a non-resonant rotation is Lagrangian. We first prove the isotropy of an invariant torus as in [17, 19] (Lemma 5 will provide a more precise result).

Lemma 1. *The invariant torus $i(\mathbb{T}^\nu)$ is isotropic.*

Proof. By (14) the pullback

$$(i \circ \Psi^t_\omega)^*\Omega = (\Phi^t_H \circ i)^*\Omega = i^*\Omega\,. \tag{22}$$

For smooth Hamiltonian systems in finite dimension (22) is true because the 2-form Ω is invariant under the flow map Φ^t_H (i.e. $(\Phi^t_H)^*\Omega = \Omega$). In our setting, the flow (Φ^t_H) may not be defined everywhere, but Φ^t_H is well defined on $i(\mathbb{T}^\nu)$ by the assumption (14), and still preserves Ω on the manifold $i(\mathbb{T}^\nu)$, see the proof of Lemma 5 for details.

Next, denoting the 2-form $(i^*\Omega)(\varphi) = \sum_{i<j} A_{ij}(\varphi)d\varphi_i \wedge d\varphi_j$, we have

$$(i \circ \Psi^t_\omega)^*\Omega = (\Psi^t_\omega)^* \circ i^*\Omega = \sum\nolimits_{i<j} A_{ij}(\varphi + \omega t)d\varphi_i \wedge d\varphi_j\,,$$

and so (22) implies that $A_{ij}(\varphi + \omega t) = A_{ij}(\varphi)$, $\forall t \in \mathbb{R}$. Since the orbit $\{\varphi + \omega t\}$ is dense on $\mathbb{T}^\nu$ (ω is non-resonant) and each function A_{ij} is continuous, it implies that

$$A_{ij}(\varphi) = c_{ij}\,,\ \forall \varphi \in \mathbb{T}^\nu\,, \quad i.e.\ i^*\Omega = \sum\nolimits_{i<j} c_{ij}d\varphi_i \wedge d\varphi_j$$

is constant. But, by (7), the 2-form $i^*\Omega = i^*d\lambda = d(i^*\lambda)$ is also exact. Thus each $c_{ij} = 0$ namely $i^*\Omega = 0$. □

We now consider the diffeomorphism of the phase space

$$\begin{pmatrix} \theta \\ I \\ z \end{pmatrix} = G \begin{pmatrix} \psi \\ y \\ w \end{pmatrix} := \begin{pmatrix} \theta_0(\psi) \\ I_0(\psi) + [D\theta_0(\psi)]^{-T} y - \big[D\tilde{z}_0(\theta_0(\psi))\big]^T \bar{J} w \\ z_0(\psi) + w \end{pmatrix} \tag{23}$$

where $\tilde{z}_0(\theta) := (z_0 \circ \theta_0^{-1})(\theta)$, see (20). The transposed operator $\big[D\tilde{z}_0(\theta)\big]^T : E \to \mathbb{R}^\nu$ is defined by the duality relation

$$\big[D\tilde{z}_0(\theta)\big]^T w \cdot \hat{\theta} = \langle w, D\tilde{z}_0(\theta)[\hat{\theta}] \rangle, \quad \forall w \in E, \ \hat{\theta} \in \mathbb{R}^\nu.$$

Lemma 2. *Let i be an isotropic torus embedding. Then G is symplectic.*

Proof. We may see G as the composition $G := G_2 \circ G_1$ of the diffeomorphisms

$$\begin{pmatrix} \theta \\ I \\ z \end{pmatrix} = G_1 \begin{pmatrix} \psi \\ y \\ w \end{pmatrix} := \begin{pmatrix} \theta_0(\psi) \\ [D\theta_0(\psi)]^{-T} y \\ w \end{pmatrix}$$

and

$$\begin{pmatrix} \theta \\ I \\ z \end{pmatrix} \mapsto G_2 \begin{pmatrix} \theta \\ I \\ z \end{pmatrix} := \begin{pmatrix} \theta \\ \tilde{I}_0(\theta) + I - \big[D\tilde{z}_0(\theta)\big]^T \bar{J} z \\ \tilde{z}_0(\theta) + z \end{pmatrix} \tag{24}$$

where $\tilde{I}_0 := I_0 \circ \theta_0^{-1}$, $\tilde{z}_0 := z_0 \circ \theta_0^{-1}$, see (20). We claim that both G_1, G_2 are symplectic, whence the lemma follows.

G_1 IS SYMPLECTIC. Since G_1 is the identity in the third component, it is sufficient to check that $(\psi, y) \mapsto (\theta_0(\psi), [D\theta_0(\psi)]^{-T} y)$ is a symplectic diffeomorphism on $\mathbb{T}^\nu \times \mathbb{R}^\nu$, which is a direct calculus.

G_2 IS SYMPLECTIC. We prove that $G_2^*\lambda - \lambda$ is closed and so (see (7)) $G_2^*\Omega = G_2^* d\lambda = dG_2^*\lambda = d\lambda = \Omega$. By (24) and the definition of pullback we have

$$\begin{aligned} (G_2^*\lambda)_{(\theta,I,z)}[\hat{\theta}, \hat{I}, \hat{z}] &= \big(\tilde{I}_0(\theta) + I - \big[D\tilde{z}_0(\theta)\big]^T \bar{J} z\big) \cdot \hat{\theta} \\ &\quad + \frac{1}{2} \langle \bar{J}(\tilde{z}_0(\theta) + z), \hat{z} + D\tilde{z}_0(\theta)[\hat{\theta}] \rangle . \end{aligned}$$

Therefore (recall (8))

$$\begin{aligned} \big((G_2^*\lambda)_{(\theta,I,z)} - \lambda_{(\theta,I,z)}\big)[\hat{\theta}, \hat{I}, \hat{z}] &= \big(\tilde{I}_0(\theta) - \big[D\tilde{z}_0(\theta)\big]^T \bar{J} z\big) \cdot \hat{\theta} + \frac{1}{2} \langle \bar{J}\tilde{z}_0(\theta), \hat{z} \rangle \\ &\quad + \frac{1}{2} \langle \bar{J}\tilde{z}_0(\theta), D\tilde{z}_0(\theta)[\hat{\theta}] \rangle + \frac{1}{2} \langle \bar{J} z, D\tilde{z}_0(\theta)[\hat{\theta}] \rangle \\ &= \tilde{I}_0(\theta) \cdot \hat{\theta} + \frac{1}{2} \langle \bar{J}\tilde{z}_0(\theta), D\tilde{z}_0(\theta)[\hat{\theta}] \rangle \\ &\quad + \frac{1}{2} \langle \bar{J}\tilde{z}_0(\theta), \hat{z} \rangle + \frac{1}{2} \langle \bar{J} D\tilde{z}_0(\theta)[\hat{\theta}], z \rangle , \end{aligned} \tag{25}$$

having used that $\bar{J}^T = -\bar{J}$. We first note that the 1-form

$$(\hat{\theta}, \hat{I}, \hat{z}) \mapsto \langle \bar{J}\tilde{z}_0(\theta), \hat{z}\rangle + \langle \bar{J}D\tilde{z}_0(\theta)[\hat{\theta}], z\rangle = d(\langle \bar{J}\tilde{z}_0(\theta), z\rangle)[\hat{\theta}, \hat{I}, \hat{z}] \tag{26}$$

is exact. Moreover

$$\tilde{I}_0(\theta) \cdot \hat{\theta} + \frac{1}{2}\langle \bar{J}\tilde{z}_0(\theta), D\tilde{z}_0(\theta)[\hat{\theta}]\rangle = (j^*\lambda)_\theta[\hat{\theta}] \tag{27}$$

(recall (8)) where $j := i \circ \theta_0^{-1}$ see (19). Hence (25), (26), (27) imply

$$(G_2^*\lambda)_{(\theta,I,z)} - \lambda_{(\theta,I,z)} = \pi^*(j^*\lambda)_{(\theta,I,z)} + d\Big(\frac{1}{2}\langle \bar{J}\tilde{z}_0(\theta), z\rangle\Big),$$

where $\pi : \mathbb{T}^\nu \times \mathbb{R}^\nu \times E \to \mathbb{T}^\nu$ is the canonical projection.

Since the torus $j(\mathbb{T}^\nu) = i(\mathbb{T}^\nu)$ is isotropic (Lemma 1) the 1-form $j^*\lambda$ on $\mathbb{T}^\nu$ is closed (as $i^*\lambda$, see (21)). This concludes the proof. □

Remark 5. A torus which is a graph over $\mathbb{T}^\nu$, i.e. $\theta \mapsto j(\theta) = (\theta, I_1(\theta), z_1(\theta))$ is isotropic if and only if $I_1(\theta) = \eta + \nabla U(\theta) - \frac{1}{2}[Dz_1(\theta)]^T\bar{J}z_1(\theta)$ for some constant $\eta \in \mathbb{R}^\nu$ and $U : \mathbb{T}^\nu \to \mathbb{R}$. This follows from (27) and Corollary 1 of Sect. 4.

Since G is symplectic (note that Lemma 2 only requires i to be isotropic), the transformed Hamiltonian vector field

$$G^*X_H := (DG)^{-1} \circ X_H \circ G = X_K, \quad K := H \circ G,$$

is still Hamiltonian. By construction (see (23)) the torus $\{y = 0, w = 0\}$ is invariant and (16) implies $X_K(\psi, 0, 0) = (\omega, 0, 0)$ (see also Lemma 8). As a consequence, the Taylor expansion of the transformed Hamiltonian K in these new coordinates assumes the normal form

$$K = const + \omega \cdot y + \frac{1}{2}A(\psi)y \cdot y + \langle C(\psi)y, w\rangle + \frac{1}{2}\langle B(\psi)w, w\rangle + O_3(y, w) \tag{28}$$

where $O_3(y, w)$ collects all the terms at least cubic in (y, w), and the operators A and B are symmetric. If, furthermore, ω is Diophantine we can perform, by standard perturbation theory, a symplectic change of coordinates which conjugates K to another Hamiltonian of the form

$$K_1 := \omega \cdot y + \frac{1}{2}\bar{A}y \cdot y + \langle C_1(\psi)y, w\rangle + \frac{1}{2}\langle B_1(\psi)w, w\rangle + O_3(y, w) \tag{29}$$

where the constant matrix $\bar{A}$ is the average $\bar{A} := \int_{\mathbb{T}^\nu} A(\psi)d\psi$. This is the general normal form for a Hamiltonian near a Diophantine invariant torus.

Summarizing we have proved the following theorem:

Theorem 1. *Let $\mathcal{T} = i(\mathbb{T}^\nu)$ be an embedded torus, see* (12) *and* (13)*, which is a smooth graph over $\mathbb{T}^\nu$, see* (19) *and* (20)*, invariant for the Hamiltonian vector field X_H, and on which the flow is conjugate to the translation flow of vector ω, see* (14) *and* (15)*. Assume moreover that $\mathcal{T}$ is* ISOTROPIC*, a property which is automatically verified if ω is non-resonant.*

Then there exist symplectic coordinates (ψ, y, w) in which $\mathcal{T}$ is described by $\mathbb{T}^\nu \times \{0\} \times \{0\}$ and the Hamiltonian assumes the normal form (28)*, i.e. the torus $\mathcal{T} = G(\mathbb{T}^\nu \times \{0\} \times \{0\})$ where G is the symplectic diffeomorphism defined in* (23)*, and the Hamiltonian $H \circ G$ has the Taylor expansion* (28) *in a neighborhood of the invariant torus. If, moreover, ω is Diophantine, see* (18)*, there is a symplectic change of coordinates in which the Hamiltonian assumes the normal form* (29)*.*

Remark 6. If the torus $\mathcal{T}$ is isotropic, even if it is filled by periodic orbits (resonant torus), i.e. $\omega = 2\pi k/T$ for some $k \in \mathbb{Z}^\nu$, the previous theorem provides the normal form (28). For an application to Lagrangian tori see [1].

What is usually called a KAM normal form for isotropic Diophantine invariant tori is a Hamiltonian of the form

$$K_2 := \omega \cdot y + \frac{1}{2}\bar{A}y \cdot y + (\bar{C}y, w) + \frac{1}{2}(\bar{B}w, w) + O_3(y, w) \tag{30}$$

where also the matrices $\bar{B}, \bar{C}$ are constant, see e.g. [15, 20, 25, 28]. The possibility to obtain such a normal form is related to the verification of the so called "second order Melnikov" non resonance conditions. This may be a difficult task for PDEs in higher space dimension because of the possible multiplicity of the normal frequencies, see e.g. [16, 26] for NLS.

The normal form (28) is relevant also in view of a Nash-Moser approach, because it provides a control of the linearized equations in the normal bundle of the torus. The linearized Hamiltonian system associated to K at the trivial solution $(\psi, y, w)(t) = (\omega t, 0, 0)$ is

$$\begin{cases} \dot{\psi} - A(\omega t)y - [C(\omega t)]^T w = 0 \\ \dot{y} = 0 \\ \dot{w} - J\big(B(\omega t)w + C(\omega t)y\big) = 0. \end{cases} \tag{31}$$

For applying the Nash-Moser scheme (Sect. 4) we have to solve, at each step, the system of equations (31) with non-zero second members. Note that the second equation is decoupled from the others. Inserting its solution in the third equation we are reduced to solve a quasi-periodically forced linear equation in w. This may vary considerably for different PDEs. The difficulty is that $B(\omega t)$ is not constant. A way to solve it is to conjugate it to a constant coefficient equation (with second order Melnikov non resonance conditions), as for the normal form (30). For PDEs in higher space dimension this is not always possible and one proceeds with a multiscale analysis as in [4–8, 10, 11] which requires only the first order Melnikov non-resonance conditions. Finally one solves the first equation in (31) for the angle component.

3 A Nash-Moser Functional Approach to KAM

We now describe the strategy for proving an abstract normal form KAM theorem by using a Nash-Moser implicit function theorem. We choose the setting of a perturbation of a parameter dependent family of isochronous linear Hamiltonian systems.

Let $\mathcal{O} \subset \mathbb{R}^\nu$ be an open set of parameters. We consider a family of Hamiltonians $H : [0, \varepsilon_0) \times \mathcal{O} \times \mathcal{P} \to \mathbb{R}$ like

$$H = H(\varepsilon, \alpha, u) = \mathcal{N}(\alpha, \theta, I, z) + \varepsilon P(\varepsilon, \alpha, \theta, I, z), \tag{32}$$

which are perturbations of a parameter-dependent normal form

$$\mathcal{N}(\alpha, \theta, I, z) = \alpha \cdot I + \frac{1}{2}\langle N(\alpha, \theta)z, z\rangle \tag{33}$$

where $N(\alpha, \theta)$ is a symmetric operator. We suppose that, as in (10), (11), the Hamiltonian vector fields $z \mapsto JN(\alpha, \theta)z$, $J\nabla_z P(\alpha, \theta, I, z)$ are well defined and smooth maps from a dense subspace $E_1 \subset E$ into E. Note that $\mathcal{N}$ may depend on the angle variables θ (in the normal directions z).

Remark 7. In applications, the parameters α may vary with the "actions" of the unperturbed invariant tori (this approach was first used in [23]), or depend on the mass of a planet as in [17], or may be "external" parameters induced, for example, by the potential as in [21, 28], etc...

The normal form $\mathcal{N}$ possesses the invariant torus $\mathbb{T}^\nu \times \{0\} \times \{0\}$ on which the motion is endowed by the constant field α.

Remark 8. If the normal form $N(\alpha, \theta) = N(\alpha)$ is constant, i.e. it does not depend on the angles θ, the unperturbed torus $\mathbb{T}^\nu \times \{0\} \times \{0\}$ is said "reducible". In applications this is the common situation in order to start with perturbation theory.

The goal is then to prove that:

- for ε small enough, for "most" Diophantine vectors $\omega \in \mathcal{C} \subset \mathcal{O}$, there exists a value of the parameters $\alpha := \alpha_\infty(\omega, \varepsilon) = \omega + O(\varepsilon)$ and a ν-dimensional embedded torus $\mathcal{T} = i(\mathbb{T}^\nu)$ close to $\mathbb{T}^\nu \times \{0\} \times \{0\}$, invariant for the Hamiltonian vector field $X_{H(\varepsilon, \alpha_\infty(\omega,\varepsilon), \cdot)}$ and supporting quasi-periodic solutions with frequency ω. In view of (16), this is equivalent to looking for a solution $\varphi \mapsto i(\varphi) \in \mathbb{T}^\nu \times \mathbb{R}^\nu \times E$, close to $(\varphi, 0, 0)$, of the embedding equation

$$(\omega \cdot \partial_\varphi) i(\varphi) - X_{H(\varepsilon, \alpha_\infty(\omega,\varepsilon), \cdot)}(i(\varphi)) = 0\,, \tag{34}$$

for some value $\alpha := \alpha_\infty(\omega, \varepsilon)$ of the parameters to be determined. In order to have a unique determination of α_∞, one needs to impose another condition, such as : the mean value of the I component of the map i vanishes.

The set of frequencies $\omega \in \mathcal{C} \subset \mathcal{O}$ for which the invariant torus exists usually is a Cantor like set. The measure of the set $\mathcal{C}$ (in particular that $\mathcal{C} \neq \emptyset$) clearly depends on the properties of the Hamiltonian H, in particular for infinite dimensional Hamiltonian system. The parameter $\alpha := \alpha_\infty(\omega, \varepsilon)$ is adjusted along the iterative Nash-Moser proof in order control the average of the first component of the Hamilton's equation (36), in particular for solving the linearized equation (82).

The function $\omega \mapsto \alpha_\infty(\omega, \varepsilon)$ is invertible and it may be proved to be smooth in ω (if the Hamiltonian H is smooth). Then, in applications, one may ask if, *given* $\beta \in \mathbb{R}^\nu$, there exists a value of $\omega = \alpha_\infty^{-1}(\beta)$ in the Cantor set of parameters $\mathcal{C} \subset \mathcal{O}$ for which (34) has a solution. In such a case one has proved the existence of a quasi-periodic solution of the given Hamiltonian $\beta \cdot I + \frac{1}{2}\langle N(\beta, \theta)z, z\rangle + \varepsilon P$. This perspective is the spirit of the Théorème de conjugaison hypothétique of Herman presented in [17].

Remark 9. Variants are possible. For example we could develop a KAM theorem for Hamiltonians which are perturbations of a non-isocronous (or Kolmogorov) normal form

$$H = \alpha \cdot I + \frac{1}{2} L(\alpha, \theta) I \cdot I + \langle M(\alpha, \theta) I, z\rangle + \frac{1}{2}\langle N(\alpha, \theta) z, z\rangle + \varepsilon P\,.$$

This is the setting, for example, considered in [29]. Actually this case may be reduced to (32) by a rescaling $\mathcal{R}_\varepsilon : (I, z) \mapsto (\varepsilon^{2a} I, \varepsilon^a z)$. Note that the transformed symplectic 2-form is $\mathcal{R}_\varepsilon^* \Omega = \varepsilon^{2a}\Omega$. A technical advantage of dealing with the parameter dependent family of isochronous normal forms (33) is that the linearized equations are simpler.

In order to find solutions of (34) we look for zeros of the nonlinear operator

$$\mathcal{F}(\varepsilon, X) := (\omega \cdot \partial_\varphi) i(\varphi) - X_{H_\mu(\alpha, \cdot)}(i(\varphi)) \tag{35}$$

$$= \begin{pmatrix} \partial_\omega \theta_0(\varphi) - \partial_I H(\varepsilon, \alpha, i(\varphi)) \\ \partial_\omega I_0(\varphi) + \partial_\theta H(\varepsilon, \alpha, i(\varphi)) + \mu \\ \partial_\omega z_0(\varphi) - J\nabla_z H(\varepsilon, \alpha, i(\varphi)) \end{pmatrix}$$

$$= \begin{pmatrix} \partial_\omega \theta_0(\varphi) - \alpha - \varepsilon \partial_I P(\alpha, i(\varphi)) \\ \partial_\omega I_0(\varphi) + \frac{1}{2}\partial_\theta \langle N(\alpha, \theta_0(\varphi)) z_0(\varphi), z_0(\varphi)\rangle + \varepsilon \partial_\theta P(\alpha, i(\varphi)) + \mu \\ \partial_\omega z_0(\varphi) - JN(\alpha, \theta_0(\varphi)) z_0(\varphi) - \varepsilon J \nabla_z P(\alpha, i(\varphi)) \end{pmatrix} \tag{36}$$

in the unknowns

$$X := (\alpha, \mu, i(\varphi))$$

where the torus embedding

$$i(\varphi) := (\theta_0(\varphi), I_0(\varphi), z_0(\varphi)) := (\varphi, 0, 0) + (\Theta_0(\varphi), I_0(\varphi), z_0(\varphi)) \tag{37}$$

and we use the shorter notation

$$\partial_\omega := \omega \cdot \partial_\varphi \,.$$

Note that $\mathcal{F}(\varepsilon, X) = 0$ is the equation $\partial_\omega i(\varphi) - X_{H_\mu(\alpha,\cdot)}(i(\varphi)) = 0$ for an embedded invariant torus of the non-exact Hamiltonian system generated by the Hamiltonian

$$H_\mu := H_\mu(\alpha, \cdot) : \mathbb{R}^\nu \times \mathbb{R}^\nu \times E \to \mathbb{R}\,, \quad H_\mu := H + \mu \cdot \theta\,. \tag{38}$$

Remark that the Hamiltonian vector field X_{H_μ} is periodic in θ, unlike H_μ. Non-exact here means that $\Omega(X_{H_\mu}, \cdot) = -dH - \mu$ is a closed, non-exact 1-form on the phase space $\mathbb{T}^\nu \times \mathbb{R}^\nu \times E$. It is well-known that a non-exact Hamiltonian system does not possess invariant tori for $\mu \neq 0$. Actually, as proved in Lemma 3 below, if $\mathcal{F}(\varepsilon, X) = 0$ then $\mu = 0$ and so $\varphi \mapsto i(\varphi)$ is an invariant torus for X_H itself. The "counter-term" $\mu \in \mathbb{R}^\nu$ is introduced as a technical trick to control the average of the second component of the equation (36), in particular for solving the linearized equation (76).

Lemma 3. *Let us define the "error function"*

$$Z(\varphi) = (Z_1, Z_2, Z_3)(\varphi) := \mathcal{F}(\varepsilon, i, \alpha, \mu) = (\omega \cdot \partial_\varphi) i(\varphi) - X_{H_\mu(\alpha,\cdot)}(i(\varphi)) \tag{39}$$

where $\varphi \in \mathbb{T}^\nu$. Then

$$\mu = \frac{1}{(2\pi)^\nu} \int_{\mathbb{T}^\nu} -[DI_0(\varphi)]^T Z_1(\varphi) + [D\theta_0(\varphi)]^T Z_2(\varphi) + [Dz_0(\varphi)]^T \bar{J} Z_3(\varphi)\, d\varphi\,. \tag{40}$$

In particular, if $\partial_\omega i(\varphi) - X_{H_\mu}(i(\varphi)) = 0$ then $\mu = 0$ and so $\varphi \mapsto i(\varphi)$ is the embedding of an invariant torus of X_H.

Proof. Let $i_{\psi_0}(\varphi) := i(\varphi + \psi_0)$ be the translated torus embedding, for all $\psi_0 \in \mathbb{T}^\nu$. Since H is autonomous the "restricted" Hamiltonian action functional (recall (8))

$$\Phi(\psi_0) := \int_{\mathbb{T}^\nu} \lambda_{i_{\psi_0}(\varphi)}[\partial_\omega i_{\psi_0}(\varphi)] - H(i_{\psi_0}(\varphi))\, d\varphi = \Phi(0)$$

is constant. Differentiating Φ at $\psi_0 = 0$ and integrating by parts ∂_ω we get, for all $\zeta \in \mathbb{R}^\nu$, (see (9))

$$0 = D_{\psi_0}\Phi(0)[\zeta] = -\int_{\mathbb{T}^\nu} \Omega\big(\partial_\omega i(\varphi) - X_H(i(\varphi)), Di(\varphi)[\zeta]\big)d\varphi$$
$$= -\int_{\mathbb{T}^\nu} \Omega\big(Z(\varphi) - \mu \cdot \frac{\partial}{\partial I}, Di(\varphi)[\zeta]\big)d\varphi \tag{41}$$

by the definition of Z in (39), (38), and denoting the vector field $(0, \mu, 0) = \mu \cdot \frac{\partial}{\partial I}$. Recalling (6) and (7) the integral

$$\int_{\mathbb{T}^\nu} \Omega\Big(\mu \cdot \frac{\partial}{\partial I}, Di(\varphi)[\zeta]\Big)d\varphi = \int_{\mathbb{T}^\nu} \mu \cdot D\theta_0(\varphi)[\zeta]\, d\varphi = (2\pi)^\nu \mu \cdot \zeta$$

because the periodic function $D(\theta_0 - id) = D\Theta_0$ (see (37)) has zero average. Hence, by (41) we deduce

$$\mu \cdot \zeta = \frac{1}{(2\pi)^\nu}\int_{\mathbb{T}^\nu} \Omega\big(Z(\varphi), Di(\varphi)[\zeta]\big)d\varphi\,, \quad \forall \zeta \in \mathbb{R}^\nu\,,$$

which, recalling (5) and (6), gives (40). □

The optimal expected smallness condition for the KAM existence result, namely for finding solutions of the nonlinear equation $\mathcal{F}(X, \varepsilon) = 0$, is

$$\varepsilon\gamma^{-1} \ll 1 \tag{42}$$

where γ is the Diophantine constant in (18) of the frequency vector ω. This is certainly the case for finite dimensional Lagrangian tori (the optimality follows for example by a time rescaling argument). If ω has to satisfy other Diophantine conditions of first and second order Melnikov type the required smallness conditions may be stronger, see e.g. [3].

Remark 10. Other functional formulations are possible. We could look for zeros of

$$\mathcal{F}(\varepsilon, j, \alpha, c) = \begin{pmatrix} \partial_\omega\theta_0(\varphi) - \partial_I H(\varepsilon, \alpha, i(\varphi)) \\ H(\varepsilon, \alpha, j(\theta)) - c \\ \partial_\omega z_0(\varphi) - J\nabla_z H(\varepsilon, \alpha, i(\varphi)) \end{pmatrix}$$

where $j(\theta) = (\theta, I_1(\theta), z_1(\theta))$ defines an isotropic torus as described in remark 5 and $i = j \circ \theta_0$. The unknowns are the diffeomorphism θ_0 of $\mathbb{T}^\nu$, the component $z_0 = z_1 \circ \theta_0$ of the torus embedding, the constant $\eta \in \mathbb{R}^n$, the potential $U : \mathbb{T}^\nu \to \mathbb{R}$, and the value of the Hamiltonian $c \in \mathbb{R}$ and α. Actually, because of the presence of the parameter α, we may impose $\eta = 0$.

As already said, a solution of the nonlinear equation $\mathcal{F}(\varepsilon, X) = 0$ is obtained by a Nash-Moser iterative scheme. The first approximate solution is

$$X_0 = (\omega, 0, \varphi, 0, 0)$$

(namely $\alpha = \omega$, $\mu = 0$, $i(\varphi) = (\varphi, 0, 0)$) so that

$$\mathcal{F}(0, X_0) = O(\varepsilon).$$

Then the strategy is to obtain iteratively better and better approximate solutions of the equation $\mathcal{F}(\varepsilon, X) = 0$ by a quasi-quadratic scheme. Given an approximate solution X, we look for a better approximate solution $X' = X + h$ by a Taylor expansion (for simplicity we omit to write the dependence on ε)

$$\mathcal{F}(X') = \mathcal{F}(X + h) = \mathcal{F}(X) + d_X\mathcal{F}(X)[h] + O(|h|^2).$$

The idea of the classical Newton iterative scheme is to define h as the solution of $\mathcal{F}(X) + d_X\mathcal{F}(X)[h] = 0$. Since the invertibility of the linear operator $d_X\mathcal{F}(X)$ may be a quite difficult task, Zehnder [29] noted that it is sufficient to find only an approximate right inverse of $d_X\mathcal{F}(X)$, namely a linear operator $T(X)$ such that

$$d_X\mathcal{F}(X) \circ T(X) - Id = O(|\mathcal{F}(X)|). \tag{43}$$

Remark that, at a solution $\mathcal{F}(X) = 0$, the operator $T(X)$ is an exact right inverse of $d_X\mathcal{F}(X)$. Thus, defining the new approximate solution

$$X' = X + h, \quad h := -T(X)\mathcal{F}(X), \tag{44}$$

we get by (43) that

$$\mathcal{F}(X') = \mathcal{F}(X) - d_X\mathcal{F}(X)[T(X)\mathcal{F}(X)] + O(|h|^2) = O(|\mathcal{F}(X)|^2). \tag{45}$$

This scheme can be called a "quasi-Newton" scheme. In typical PDEs applications, the approximate right inverse $T(X)$ "loses derivatives" due to the small divisors. However, since the scheme (44) is quadratic by (45), it can nevertheless converge to a solution if $\mathcal{F}(X_0)$ is sufficiently small (depending also on the norm of T). The scheme (44) is usually implemented in Banach scales of analytic functions, as, for example,

$$A_\sigma := \Big\{ u(\varphi) = \sum_{k \in \mathbb{Z}^\nu} u_k\, e^{ik\cdot\varphi} \; : \; \|u\|_\sigma^2 := \sum_{k \in \mathbb{Z}^\nu} |u_k|^2\, e^{2|k|\sigma} (1 + |k|^{2s_0}) < +\infty \Big\} \tag{46}$$

for some $\sigma > 0$, $s_0 > \nu/2$. The approximate inverse operator T is usually "unbounded", satisfies Cauchy type estimates like

$$\|Tg\|_{\sigma'} \le \frac{C}{\gamma(\sigma - \sigma')^\tau} \|g\|_\sigma, \quad \forall \sigma' < \sigma, \tag{47}$$

and there is $\beta > 0$ such that, $\forall \sigma' < \sigma$, $\forall g \in A_\sigma$,

$$\left\| \left(d_X \mathcal{F}(X) \circ T(X) - Id\right) g \right\|_{\sigma'} \le C \frac{\|\mathcal{F}(\varepsilon, X)\|_\sigma}{\gamma(\sigma - \sigma')^\beta} \|g\|_\sigma . \tag{48}$$

The constants $\tau, \beta > 0$ are the "loss of derivatives".

On the other hand, in Banach spaces of functions with finite differentiability, as the Sobolev scale

$$H^s := \Big\{ u(\varphi) = \sum_{k \in \mathbb{Z}^\nu} u_k e^{ik\cdot\varphi} \ : \ \|u\|_s^2 := \sum_{k \in \mathbb{Z}^\nu} |u_k|^2 (1 + |k|^{2s}) < +\infty \Big\} , \tag{49}$$

the quasi-Newton scheme (44) does not converge because after finitely many steps the approximate solutions are no longer regular. Following Moser [22], it is necessary to insert a smoothing procedure at each step (Nash-Moser scheme). The approximate inverse usually satisfies estimates like: there are constants $p, \rho > 0$ ("loss of derivatives") such that, for all $s \in [s_0, S]$, $\forall g \in H^{s+\rho}$,

$$\|T(X)g\|_s \le C(s, \|X\|_{s_0+p}) \big(\|g\|_{s+\rho} + \|g\|_{s_0} \|X\|_{s+\rho} \big) , \tag{50}$$

and

$$\begin{aligned} \left\| \left(d_X \mathscr{F}(X) \circ T(X) - Id\right) g \right\|_s \le & C(s, \|X\|_{s_0+p}) \Big(\|\mathscr{F}(X)\|_{s_0} \|g\|_{s+\rho} + \\ & + \|\mathscr{F}(X)\|_{s+\rho} \|g\|_{s_0} + \|X\|_{s+\rho} \|\mathscr{F}(X)\|_{s_0} \|g\|_{s_0} \Big) . \end{aligned} \tag{51}$$

In this note we will not insist in the analytical aspects of the convergence, for which we refer to [7, 29], or [3, 6].

The linearized operator of (35) is

$$d_X \mathcal{F}(\varepsilon, X)[\hat{X}] = (\omega \cdot \partial_\varphi)\hat{\imath} - D_i X_{H_\mu(\alpha,\cdot)}(i)[\hat{\imath}] - D_\alpha X_{H_\mu(\alpha,\cdot)}(i)[\hat{\alpha}] + (0, \hat{\mu}, 0) .$$

It is rather difficult to invert it because all the components of the Hamiltonian vector field are coupled by $O(\varepsilon)$-non-constant coefficient terms. In the next section, following the ideas of Sect. 2, we present a symplectic change of variable which approximately decouples the tangential directions (i.e. $(\hat{\theta}, \hat{I})$) and the normal ones (i.e. $\hat{z}$), and thus enables to find an approximate right inverse of $d_X \mathcal{F}(\varepsilon, X)$.

4 Approximate Right Inverse

We first report a basic fact about 1-forms on a torus. We regard a 1-form $a = \sum_{i=1}^{\nu} a_i(\varphi) d\varphi_i$ equivalently as the vector field $\mathbf{a}(\varphi) = (a_1(\varphi), \dots, a_\nu(\varphi))$.

Given a function $g : \mathbb{T}^\nu \to \mathbb{R}$ with zero average, we denote by $u := \Delta^{-1} g$ the unique solution of $\Delta u = g$ with zero average.

Lemma 4 (Helmotz decomposition). *A smooth vector field* $\mathbf{a}$ *on* $\mathbb{T}^\nu$ *may be decomposed as the sum of a conservative and a divergence-free vector field:*

$$\mathbf{a} = \nabla U + \mathbf{c} + \boldsymbol{\rho}\,, \quad U : \mathbb{T}^\nu \to \mathbb{R}\,, \ \mathbf{c} \in \mathbb{R}^\nu\,, \ \mathrm{div}\boldsymbol{\rho} = 0\,, \ \int_{\mathbb{T}^\nu} \rho d\varphi = 0\,. \tag{52}$$

The above decomposition is unique if we impose that the mean value of U *vanishes. We have* $U = \Delta^{-1}(\mathrm{div}\, \mathbf{a})$, *the components of* ρ *are*

$$\rho_j(\varphi) = \Delta^{-1} \sum_{k=1}^{\nu} \partial_{\varphi_k} A_{kj}(\varphi)\,, \quad A_{kj} := \partial_{\varphi_k} a_j - \partial_{\varphi_j} a_k\,, \tag{53}$$

and $c_j = (2\pi)^{-\nu} \int_{\mathbb{T}^\nu} a_j(\varphi)\, d\varphi$, $j = 1, \dots, \nu$.

Proof. $\mathrm{div}(\mathbf{a} - \nabla U) = 0$ if and only if $\mathrm{div}\, \mathbf{a} = \Delta U$. This equation has the solution $U := \Delta^{-1}(\mathrm{div}\, \mathbf{a})$ (note that $\mathrm{div}\, \mathbf{a}$ has zero average). Hence (52) is achieved with $\rho := \mathbf{a} - \nabla U - \mathbf{c}$. By taking the φ-average we get that each $c_j = (2\pi)^{-\nu} \int_{\mathbb{T}^\nu} a_j(\varphi)\, d\varphi$. Let us now prove the expression (53) of ρ_j. We have $\partial_{\varphi_k} \rho_j - \partial_{\varphi_j} \rho_k = \partial_{\varphi_k} a_j - \partial_{\varphi_j} a_k =: A_{kj}$ because $\partial_{\varphi_j} \partial_{\varphi_k} U - \partial_{\varphi_k} \partial_{\varphi_j} U = 0$. For each $j = 1, \dots, \nu$ we differentiate $\partial_{\varphi_k} \rho_j - \partial_{\varphi_j} \rho_k = A_{kj}$ with respect to φ_k and we sum in k, obtaining

$$\Delta \rho_j - \sum_{k=1}^{\nu} \partial_{\varphi_k \varphi_j} \rho_k = \sum_{k=1}^{\nu} \partial_{\varphi_k} A_{kj}\,.$$

Since $\sum_{k=1}^{\nu} \partial_{\varphi_k \varphi_j} \rho_k = \partial_{\varphi_j} \mathrm{div} \boldsymbol{\rho} = 0$ then $\Delta \rho_j = \sum_{k=1}^{\nu} \partial_{\varphi_k} A_{kj}$ and (53) follows. □

Corollary 1. *Any closed* 1*-form on* $\mathbb{T}^\nu$ *has the form* $a(\varphi) = \mathbf{c} + dU$ *for some* $\mathbf{c} \in \mathbb{R}^\nu$.

Corollary 2. *Let* $a(\varphi)$ *be a* 1*-form on* $\mathbb{T}^\nu$, *and let* ρ *be defined by* (53). *Then* $a - \sum_{j=1}^{\nu} \rho_j(\varphi) d\varphi_j$ *is closed.*

We quantify how an embedded torus $i(\mathbb{T}^\nu)$ is approximately invariant for the Hamiltonian vector field X_{H_μ} in terms of the "error function" $Z(\varphi)$, defined in (39). A torus embedding $i(\varphi) = (\theta_0(\varphi), I_0(\varphi), z_0(\varphi))$ which is only approximately invariant may not be isotropic. Consider the pullback 1-form on $\mathbb{T}^\nu$ (see (8))

$$(i^* \lambda)(\varphi) = \sum_{k=1}^{\nu} a_k(\varphi) d\varphi_k\,, \tag{54}$$

where

$$\begin{aligned} a_k(\varphi) &:= \Big[[D\theta_0(\varphi)]^T I_0(\varphi) + \frac{1}{2} [Dz_0(\varphi)]^T \bar{J} z_0(\varphi) \Big]_k \\ &= I_0(\varphi) \cdot \frac{\partial \theta_0}{\partial \varphi_k}(\varphi) + \frac{1}{2} \langle \bar{J} z_0(\varphi), \frac{\partial z_0}{\partial \varphi_k}(\varphi) \rangle\,. \end{aligned} \tag{55}$$

The 1-form $i^*\lambda$ is only approximately closed, namely the 2-form (recall (7))

$$i^*\Omega = d(i^*\lambda) = \sum_{k<j} A_{kj}(\varphi) d\varphi_k \wedge d\varphi_j \,, \tag{56}$$

$$A_{kj}(\varphi) = \partial_{\varphi_k} a_j(\varphi) - \partial_{\varphi_j} a_k(\varphi) \,,$$

is small. We call the coefficients (A_{kj}) the "lack of isotropy" of the approximate torus embedding $\varphi \mapsto i(\varphi)$. In Lemma 5 below we quantify their size in terms of the error function Z defined in (39).

We first recall that the Lie derivative of a k-form β with respect to the vector field Y is $L_Y\beta := \frac{d}{dt}\big[(\Phi_Y^t)^*\beta\big]_{|t=0}$ where Φ_Y^t denotes the flow generated by Y.

Given a function $g(\varphi)$ with zero average, we denote by $u := \partial_\omega^{-1} g$ the unique solution of $\partial_\omega u = g$ with zero average.

Lemma 5. *The "lack of isotropy" coefficients A_{kj} satisfy, $\forall \varphi \in \mathbb{T}^\nu$,*

$$(\omega \cdot \partial_\varphi) A_{kj}(\varphi) = \Omega\big(DZ(\varphi)\underline{e}_k, Di(\varphi)\underline{e}_j\big) + \Omega\big(Di(\varphi)\underline{e}_k, DZ(\varphi)\underline{e}_j\big) \tag{57}$$

where $(\underline{e}_1, \ldots, \underline{e}_\nu)$ denotes the canonical basis of $\mathbb{R}^\nu$. Thus, since each A_{kj} has zero mean value, if the frequency vector $\omega \in \mathbb{R}^\nu$ is non-resonant,

$$A_{kj}(\varphi) = \partial_\omega^{-1}\Big(\Omega\big(DZ(\varphi)\underline{e}_k, Di(\varphi)\underline{e}_j\big) + \Omega\big(Di(\varphi)\underline{e}_k, DZ(\varphi)\underline{e}_j\big)\Big) \,. \tag{58}$$

Proof. We use Cartan's formula $L_\omega(i^*\Omega) = d\big((i^*\Omega)(\omega,\cdot)\big) + \big(d(i^*\Omega)\big)(\omega,\cdot)$. Since $d(i^*\Omega) = i^* d\Omega = 0$ by (7) we get

$$L_\omega(i^*\Omega) = d\big((i^*\Omega)(\omega,\cdot)\big) \,. \tag{59}$$

Now we compute, for $\hat\psi \in \mathbb{R}^\nu$ (denoting the vector field $(0,\mu,0) = \mu \cdot \frac{\partial}{\partial I}$)

$$\begin{aligned}(i^*\Omega)(\omega,\hat\psi) &= \Omega(Di(\varphi)\omega, Di(\varphi)\hat\psi) = \Omega\Big(X_H(i(\varphi)) + \mu \cdot \frac{\partial}{\partial I} + Z(\varphi), Di(\varphi)\hat\psi\Big) \\ &= -dH(i(\varphi))[Di(\varphi)\hat\psi] + \mu \cdot D\theta_0(\varphi)[\hat\psi] + \Omega(Z(\varphi), Di(\varphi)\hat\psi) \,.\end{aligned}$$

We obtain

$$(i^*\Omega)(\omega,\cdot) = \sum_{j=1}^{\nu} b_j(\varphi) d\varphi_j$$

$$b_j(\varphi) = (i^*\Omega)(\omega,\underline{e}_j) = -\frac{\partial(H \circ i)}{\partial\varphi_j}(\varphi) + \mu \cdot \frac{\partial\theta_0}{\partial\varphi_j}(\varphi) + \Omega(Z(\varphi), Di(\varphi)\underline{e}_j) \,.$$

Hence, by (59), the Lie derivative

$$L_\omega(i^*\Omega) = \sum\nolimits_{k<j} B_{kj}(\varphi)d\varphi_k \wedge d\varphi_j \tag{60}$$

with

$$\begin{aligned} B_{kj}(\varphi) &= \frac{\partial b_j}{\partial \varphi_k}(\varphi) - \frac{\partial b_k}{\partial \varphi_j}(\varphi) \\ &= \frac{\partial}{\partial \varphi_k}(\Omega(Z(\varphi), Di(\varphi)\underline{e}_j)) - \frac{\partial}{\partial \varphi_j}(\Omega(Z(\varphi), Di(\varphi)\underline{e}_k)) \\ &= \Omega(DZ(\varphi)\underline{e}_k, Di(\varphi)\underline{e}_j) + \Omega(Di(\varphi)\underline{e}_k, DZ(\varphi)\underline{e}_j)\,. \end{aligned} \tag{61}$$

Recalling (15) and (56) we have, $\forall \varphi \in \mathbb{T}^\nu$,

$$(\psi^t_\omega)^*(i^*\Omega)(\varphi) = i^*\Omega(\varphi + \omega t) = \sum\nolimits_{k<j} A_{kj}(\varphi + \omega t)d\varphi_k \wedge d\varphi_j\,.$$

Hence the Lie derivative

$$L_\omega(i^*\Omega)(\varphi) = \sum\nolimits_{k<j} (\omega \cdot \partial_\varphi)A_{kj}(\varphi)d\varphi_k \wedge d\varphi_j\,. \tag{62}$$

Comparing (60), (61) and (62) we deduce (57). □

The previous lemma provides another proof of Lemma 1. For an invariant torus embedding $i(\varphi)$ the "error function" $Z(\varphi) = 0$ (see (39)) and so each $A_{kj} = 0$. If ω is Diophantine (see (18)) then, by (58) the following size estimate holds

$$A_{kj} = O(Z\gamma^{-1})\,. \tag{63}$$

This estimate can be made quantitative once the norms are specified. For example, in scales of analytic functions as (46), it gives $\|A_{kj}\|_{\sigma'} \le \gamma^{-1}(\sigma - \sigma')^{-(\tau+1)}\|Z\|_\sigma$, for all $\sigma' < \sigma$. In the Sobolev spaces (49) it implies $\|A_{kj}\|_s \le \gamma^{-1}\|Z\|_{s+\tau+1}$. Since in the sequel of this note we will only focus on the algebraic aspect of the proof, we shall write only formal estimates as (63). We refer to [3, 6] for the analytic quantitative estimates.

We now prove that near an approximate isotropic torus there is an isotropic torus.

Lemma 6 (Isotropic torus).
The torus embedding $i_\delta(\varphi) = (\theta_0(\varphi), I_\delta(\varphi), z_0(\varphi))$ *defined by*

$$I_\delta(\varphi) = I_0(\varphi) - [D\theta_0(\varphi)]^{-T}\rho(\varphi)\,, \quad \rho_j := \Delta^{-1}\Big(\sum\nolimits_{k=1}^{\nu} \partial_{\varphi_j} A_{kj}(\varphi)\Big) \tag{64}$$

is isotropic. Thus $I_\delta - I_0 = O(\gamma^{-1}Z)$.

Proof. By Corollary 2 the 1-form $i^*\lambda - \rho$ is closed with ρ_j defined in (64), see also (53), (54). Actually $i^*\lambda - \rho = i_\delta^*\lambda$ is the pullback of the 1-form λ under the modified torus embedding i_δ defined in (64), see (55). Thus the torus $i_\delta(\mathbb{T}^\nu)$ is isotropic. □

Let

$$Z_\delta(\varphi) := \mathcal{F}(\varepsilon, i_\delta, \alpha, \mu) = \partial_\omega i_\delta(\varphi) - X_{H_\mu(\alpha,)}(i_\delta(\varphi)) \tag{65}$$

be the error function of the isotropic torus embedding i_δ. We now show that the isotropic torus embedding i_δ is a good approximate solution as i. This is needed for proving the convergence of the iterative Nash-Moser scheme under the minimal smallness condition $\varepsilon\gamma^{-1} \ll 1$, see (42).

Lemma 7. $Z_\delta = O(Z)$.

Proof. Let $Z_\delta(\varphi) = (Z_{1,\delta}, Z_{2,\delta}, Z_{3,\delta})(\varphi)$. Since the difference between the torus embeddings i_δ and i affects only the I-component (Lemma 6), and the normal form Hamiltonian vector field $X_{\mathcal{N}}$ is independent of I (see (35)), the components $Z_{1,\delta}, Z_{3,\delta}$ differ from Z_1, Z_3 for $O(\varepsilon|I_\delta - I_0|) = O(Z\gamma^{-1}\varepsilon) = O(Z)$. Moreover

$$Z_{2,\delta} - Z_2 = \partial_\omega(I_\delta - I_0) + \varepsilon\big(\partial_\theta P(i_\delta) - \partial_\theta P(i)\big) = -\partial_\omega v + O(\varepsilon Z\gamma^{-1})$$

where $v(\varphi) := [D\theta_0(\varphi)]^{-T}\rho(\varphi)$. We claim that $\partial_\omega v = O(Z)$ whence the lemma follows. We have $\partial_\omega v = (\partial_\omega[D\theta_0(\varphi)]^{-T})\rho + [D\theta_0(\varphi)]^{-T}\partial_\omega\rho$. The second term $\big[D\theta_0(\varphi)\big]^{-T}\partial_\omega\rho = O(Z)$ because (see (64)) each $\partial_\omega\rho_j = \Delta^{-1}\sum_{k=1}^{\nu}\partial_{\varphi_j}\partial_\omega A_{kj} = O(Z)$ by (57). We now prove that also the first term $(\partial_\omega[D\theta_0(\varphi)]^{-T})\rho = O(Z)$. Since $\rho = O(Z\gamma^{-1})$ (see (63), (64)) it is sufficient to prove that

$$\partial_\omega[D\theta_0(\varphi)]^{-T} = -\big[D\theta_0(\varphi)\big]^{-T}\big(\partial_\omega[D\theta_0(\varphi)]^T\big)\big[D\theta_0(\varphi)\big]^{-T} = O(\varepsilon)\,.$$

Differentiating in φ the first component $\partial_\omega\theta_0(\varphi) = \alpha + \varepsilon(\partial_I P)(\alpha, i(\varphi)) + Z_1(\varphi)$ of (36), we deduce

$$\begin{aligned}\partial_\omega D\theta_0(\varphi) &= \varepsilon D_i(\partial_I P)(i(\varphi))Di(\varphi) + DZ_1(\varphi)\\ &= \varepsilon(D_\theta(\partial_I P)D\theta_0 + D_I(\partial_I P)DI_0 + D_z(\partial_I P)Dz_0)(\varphi) + DZ_1(\varphi)\end{aligned}$$

and so its transposed $\partial_\omega\big[D\theta_0(\varphi)\big]^T = O(\varepsilon + Z) = O(\varepsilon)$. □

In analogy with Sect. 2 we now introduce a symplectic set of coordinates (ψ, y, w) near the isotropic torus $\mathcal{T}_\delta := i_\delta(\mathbb{T}^\nu)$ via the symplectic diffeomorphism

$$\begin{pmatrix}\theta\\ I\\ z\end{pmatrix} = G_\delta\begin{pmatrix}\psi\\ y\\ w\end{pmatrix} := \begin{pmatrix}\theta_0(\psi)\\ I_\delta(\psi) + [D\theta_0(\psi)]^{-T}y - \big[D\tilde{z}_0(\theta_0(\psi))\big]^T\bar{J}w\\ z_0(\psi) + w\end{pmatrix} \tag{66}$$

where $\tilde{z}_0 := z_0 \circ \theta_0^{-1}$. The map G_δ is symplectic by Lemma 2 because i_δ is isotropic (Lemma 6). In the new coordinates (ψ, y, w) the isotropic torus embedding i_δ is trivial, namely $i_\delta(\psi) = G_\delta(\psi, 0, 0)$.

Under the symplectic change of variable (66), the Hamiltonian vector field X_{H_μ} changes into

$$X_{K_\mu} = G_\delta^* X_{H_\mu} = (DG_\delta)^{-1} X_{H_\mu} \circ G_\delta \tag{67}$$

where (recall (38))

$$K_\mu := H_\mu \circ G_\delta = K + \theta_0(\psi) \cdot \mu \,, \qquad K := H \circ G_\delta \,. \tag{68}$$

In the above formula, θ_0 is the lift to $\mathbb{R}^\nu$ of the first component of i_δ (see Lemma 6). The Taylor expansion of the new Hamiltonian $K_\mu : \mathbb{R}^\nu \times \mathbb{R}^\nu \times E \to \mathbb{R}$ at the trivial torus $(\psi, 0, 0)$ is

$$\begin{aligned} K_\mu &= \theta_0(\psi) \cdot \mu + K_{00}(\alpha, \psi) + K_{10}(\alpha, \psi) \cdot y + \langle K_{01}(\alpha, \psi), w \rangle \\ &\quad + \frac{1}{2} K_{20}(\alpha, \psi) y \cdot y + \langle K_{11}(\alpha, \psi) y, w \rangle + \frac{1}{2} \langle K_{02}(\alpha, \psi) w, w \rangle + K_{\geq 3}(\alpha, \psi, y, w) \end{aligned} \tag{69}$$

where $K_{\geq 3}$ collects all the terms at least cubic in the variables (y, w). The Taylor coefficients of K (in the sequel we may omit to write their dependence on α) are $K_{00}(\psi) \in \mathbb{R}$, $K_{10}(\psi) \in \mathbb{R}^\nu$, $K_{01}(\psi) \in E$, $K_{20}(\psi) \in \mathrm{Mat}(\nu \times \nu)$ is a real symmetric matrix, $K_{02}(\psi)$ is a self-adjoint operator of E and $K_{11}(\psi) \in \mathcal{L}(\mathbb{R}^\nu, E)$.

The Hamiltonian system associated to K_μ then writes

$$\begin{cases} \dot{\psi} = K_{10}(\alpha, \psi) + K_{20}(\alpha, \psi) y + K_{11}^T(\alpha, \psi) w + \partial_y K_{\geq 3}(\psi, y, w) \\ \dot{y} = -[D\theta_0(\psi)]^T \mu - \partial_\psi K_{00}(\alpha, \psi) - [D_\psi K_{10}(\alpha, \psi)]^T y - [D_\psi K_{01}(\alpha, \psi)]^T w \\ \qquad -\partial_\psi \big(\tfrac{1}{2} K_{20}(\alpha, \psi) y \cdot y + \langle K_{11}(\alpha, \psi) y, w \rangle + \tfrac{1}{2} \langle K_{02}(\alpha, \psi) w, w \rangle + K_{\geq 3}(\psi, y, w) \big) \\ \dot{w} = J\big(K_{01}(\alpha, \psi) + K_{11}(\alpha, \psi) y + K_{02}(\alpha, \psi) w + \nabla_w K_{\geq 3}(\psi, y, w) \big). \end{cases} \tag{70}$$

As seen in Sect. 2, if i_δ were an invariant torus embedding, the coefficient $K_{00}(\psi) =$ const, $K_{10}(\psi) = \omega$ and $K_{01}(\psi) = 0$. Moreover also $\mu = 0$ by Lemma 3. We now express these coefficients in terms of the error function Z_δ of i_δ defined in (65) (equivalently Z, by Lemma 7).

Lemma 8. *The vector field*

$$\begin{aligned} X_{K_\mu}(\psi, 0, 0) &\overset{(70)}{=} \begin{pmatrix} K_{10}(\alpha, \psi) \\ -[D\theta_0(\psi)]^T \mu - \partial_\psi K_{00}(\alpha, \psi) \\ J K_{01}(\alpha, \psi) \end{pmatrix} \\ &= \begin{pmatrix} \omega \\ 0 \\ 0 \end{pmatrix} - \big(DG_\delta(\psi, 0, 0) \big)^{-1} Z_\delta(\psi) \,. \end{aligned} \tag{71}$$

Proof. By (67) and $i_\delta(\psi) = G_\delta(\psi, 0, 0)$, we have

$$X_{K_\mu}(\psi, 0, 0) = DG_\delta(\psi, 0, 0)^{-1} X_{H_\mu}(i_\delta(\psi)) = DG_\delta(\psi, 0, 0)^{-1}\big(\partial_\omega i_\delta(\psi) - Z_\delta(\psi)\big)$$

and (71) follows because $DG_\delta(\psi, 0, 0)^{-1} Di_\delta(\psi)[\omega] = (\omega, 0, 0)$. □

We now write the coefficient $K_{10}(\alpha, \psi)$ which describes in (69) and (70) how the tangential frequencies vary with respect to α, and the coefficients $K_{11}(\alpha, \psi)$, $K_{20}(\alpha, \psi)$ which are $O(\varepsilon)$.

Lemma 9. *The coefficients*

$$K_{10}(\alpha, \psi) = [D\theta_0(\psi)]^{-1}\alpha + \varepsilon[D\theta_0(\psi)]^{-1}(\partial_I P)(\varepsilon, \alpha, i_\delta(\psi))$$

$$K_{11}(\alpha, \psi) = \varepsilon D_I \nabla_z P(i_\delta(\psi))[D\theta_0(\psi)]^{-T} + \varepsilon \bar{J}(D\tilde{z}_0)(\theta_0(\psi))(D_I^2 P)(i_\delta(\psi))[D\theta_0(\psi)]^{-T}$$

$$K_{20}(\alpha, \psi) = \varepsilon[D\theta_0(\psi)]^{-1}(D_I^2 P)(i_\delta(\psi)))[D\theta_0(\psi)]^{-T}.$$

Proof. Differentiating $K = H \circ G_\delta$ we get $K_{10}(\psi) = [D\theta_0(\psi)]^{-1}(\partial_I H)(i_\delta(\psi))$ and the lemma follows by (32), (33). Similarly we deduce that

$$K_{11}(\psi) = D_I \nabla_z H(i_\delta(\psi))[D\theta_0(\psi)]^{-T} + \bar{J}(D_\theta \tilde{z}_0)(\theta_0(\psi))(D_I^2 H)(i_\delta(\psi))[D\theta_0(\psi)]^{-T}$$

$$K_{20}(\psi) = [D\theta_0(\psi)]^{-1}(D_I^2 H)(i_\delta(\psi)))[D\theta_0(\psi)]^{-T}$$

and by (32)–(33) the lemma follows. □

Under the linear change of variable (obtained linearizing (66) at $(\psi, y, w) = (\varphi, 0, 0)$)

$$\begin{pmatrix} \hat{\theta} \\ \hat{I} \\ \hat{z} \end{pmatrix} = DG_\delta(\varphi, 0, 0) \begin{pmatrix} \hat{\psi} \\ \hat{y} \\ \hat{w} \end{pmatrix} \tag{72}$$

the linearized operator $d_{i,\alpha,\mu}\mathcal{F}(\varepsilon, i_\delta, \alpha_0, \mu_0)$ is transformed approximately (see (88) for the precise expression of the error) into the one obtained when we linearize the Hamiltonian system (70) at $(\psi, y, w) = (\omega t, 0, 0)$ and differentiating also in α, μ at α_0, μ_0, namely

$$L(\hat{\psi}, \hat{y}, \hat{w}, \hat{\alpha}, \hat{\mu}) := \tag{73}$$

$$\begin{pmatrix} \partial_\omega \hat{\psi} - D_\alpha K_{10}(\alpha, \varphi)[\hat{\alpha}] - D_\psi K_{10}(\alpha, \varphi)[\hat{\psi}] - K_{20}(\alpha, \varphi)\hat{y} - K_{11}^T(\alpha, \varphi)\hat{w} \\ \partial_\omega \hat{y} + [D\theta_0(\varphi)]^T \hat{\mu} + [D^2\theta_0(\varphi)\hat{\psi}]^T[\mu_0] + \partial_\psi\big(D_\alpha K_{00}(\alpha, \varphi)[\hat{\alpha}]\big) \\ +D_{\psi\psi} K_{00}(\alpha, \varphi)\hat{\psi} + [D_\psi K_{10}(\alpha, \varphi)]^T \hat{y} + [D_\psi K_{01}(\alpha, \varphi)]^T \hat{w} \\ \partial_\omega \hat{w} - J\big(D_\alpha K_{01}(\alpha, \varphi)[\hat{\alpha}] + D_\psi K_{01}(\alpha, \varphi)[\hat{\psi}] + K_{11}(\alpha, \varphi)\hat{y} + K_{02}(\alpha, \varphi)\hat{w}\big) \end{pmatrix}.$$

For the convergence of the Nash Moser scheme, it is sufficient to invert the operator L defined in (73) only approximately, namely, in view of Lemmas 8 and 3, solve only the linear system

$$\mathbb{D}(\hat{\psi}, \hat{y}, \hat{w}, \hat{\alpha}, \hat{\mu}) := g(\varphi) = \begin{pmatrix} g_1(\varphi) \\ g_2(\varphi) \\ g_3(\varphi) \end{pmatrix} \tag{74}$$

with the "simpler" operator

$$\mathbb{D}(\hat{\psi}, \hat{y}, \hat{w}, \hat{\alpha}, \hat{\mu}) :=$$
$$\begin{pmatrix} \partial_\omega \hat{\psi} - D_\alpha K_{10}(\alpha, \varphi)[\hat{\alpha}] - K_{20}(\alpha, \varphi)\hat{y} - K_{11}^T(\alpha, \varphi)\hat{w} \\ \partial_\omega \hat{y} + [D\theta_0(\varphi)]^T \hat{\mu} + \partial_\psi \big(D_\alpha K_{00}(\alpha, \varphi)[\hat{\alpha}]\big) \\ \partial_\omega \hat{w} - J\big(D_\alpha K_{01}(\alpha, \varphi)[\hat{\alpha}] + K_{11}(\alpha, \varphi)\hat{y} + K_{02}(\alpha, \varphi)\hat{w}\big) \end{pmatrix} . \tag{75}$$

$\mathbb{D}$ is obtained from L in (73) neglecting the terms which are zero at an exact solution (α_0, μ_0, i_0) (with $\mu_0 = 0$) . System (74) may be solved in a triangular way. We first solve the second equation

$$\partial_\omega \hat{y} = -\partial_\psi \big(D_\alpha K_{00}(\alpha_0, \varphi)[\hat{\alpha}]\big) - [D\theta_0(\varphi)]^T \hat{\mu} + g_2 . \tag{76}$$

We choose $\hat{\mu}$ such that the φ-average of the right hand side

$$\langle -\partial_\psi (D_\alpha K_{00}(\alpha_0, \varphi)[\hat{\alpha}]) - [D\theta_0(\varphi)]^T \hat{\mu} + g_2 \rangle = 0 .$$

Note that the average of the total derivative $\partial_\psi\big(\partial_\alpha K_{00}(\alpha_0, \varphi)[\hat{\alpha}]\big)$ is zero, and the averaged matrix $\langle [D\theta_0(\varphi)]^T \rangle = Id + \langle [D\Theta_0(\varphi)]^T \rangle = Id$ because $\Theta_0(\varphi)$ is periodic in φ. Hence we find

$$\hat{\mu} := \langle g_2 \rangle , \tag{77}$$

and, by (76), we define

$$\hat{y} = -\partial_\omega^{-1}\big(\partial_\psi \big(D_\alpha K_{00}(\alpha_0, \varphi)[\hat{\alpha}]\big) + [D\theta_0(\varphi)]^T \langle g_2 \rangle - g_2\big) + c_1 \tag{78}$$

for some $c_1 \in \mathbb{R}^\nu$.

Next we consider the third equation

$$\partial_\omega \hat{w} - JK_{02}(\alpha_0, \varphi)\hat{w} = J(D_\alpha K_{01}(\alpha_0, \varphi)[\hat{\alpha}]) + JK_{11}(\alpha_0, \varphi)\hat{y} + g_3 . \tag{79}$$

Remark that (79) is a linear quasi-periodically forced PDE with a self adjoint operator K_{02} which is a perturbation of the normal form operator $N(\alpha, \theta)$ in (33). The solvability of (79) has to be checked case by case for a given PDE. We can say something when $N(\alpha, \theta) = N(\alpha)$ does not depend on θ, see remark 8.

What is relevant is the nature of spectrum of the Hamiltonian vector field $JN(\alpha)$: if their eigenvalues are real or purely imaginary, simple or multiple, their asymptotic expansions, etc... If, for example, $JN(\alpha)$ has real spectrum, bounded away from zero, then also the linear operator

$$\partial_\omega - JK_{02}(\alpha_0, \varphi) \tag{80}$$

is invertible with good bounds for the inverse. This is the case for the continuation of isotropic tori of hyperbolic type, as considered in [29] and in [18]. If $JN(\alpha)$ has purely imaginary discrete spectrum (elliptic tori) the main work is to prove that for "most" frequencies ω the quasi-periodic linear operator (80) is invertible, and its inverse satisfies good estimates in high norms. This may be hard work, see the forced PDEs [2, 4, 5, 8]. However, if it is solved, it is possible to define the solution $\hat{w}$ of the linear equation (79) by

$$\hat{w} := \big(\partial_\omega - JK_{02}(\alpha_0, \varphi)\big)^{-1}\Big(J(D_\alpha K_{01}(\alpha_0, \varphi)[\hat{\alpha}]) + JK_{11}(\alpha_0, \varphi)\hat{y} + g_3\Big). \tag{81}$$

Finally we solve also the first equation of (75), namely

$$\partial_\omega \hat{\psi} = D_\alpha K_{10}(\alpha_0, \varphi)[\hat{\alpha}] + K_{20}(\alpha_0, \varphi)\hat{y} + K_{11}^T(\alpha_0, \varphi)\hat{w} + g_1 . \tag{82}$$

We look for $\hat{\alpha}$ such that the right hand side in (82) has zero average, namely

$$\langle D_\alpha K_{10}(\alpha_0, \varphi)\rangle[\hat{\alpha}] + \langle K_{20}(\alpha_0, \varphi)\hat{y}\rangle + \langle K_{11}^T(\alpha_0, \varphi)\hat{w}\rangle + \langle g_1\rangle = 0 . \tag{83}$$

By Lemma 9, $D_\alpha K_{10}(\alpha_0, \varphi) = D\theta_0(\varphi)^{-1} + O(\varepsilon)$, hence

$$\langle D_\alpha K_{10}(\alpha_0, \varphi)\rangle = \langle D\theta_0(\varphi)^{-1}\rangle + O(\varepsilon) = Id + O(\varepsilon\gamma^{-1})$$

because $D\theta_0 = Id + O(\varepsilon\gamma^{-1})$. Note that $\hat{y}$ and $\hat{w}$ depend on $\hat{\alpha}$ (see (78), (81)) but, since K_{20}, K_{11}^T are $O(\varepsilon)$ by Lemma 9, the equation (83) takes the form

$$(Id + R_\varepsilon)[\hat{\alpha}] = \Gamma \qquad \text{with} \quad R_\varepsilon \to 0 \ \text{ as } \ \varepsilon \to 0 .$$

For ε small enough, $Id + R_\varepsilon$ is invertible and (83) has a unique solution $\hat{\alpha}$.

Remark 11. Above we suppose that, for example, the size of the inverse $(\partial_\omega - JK_{02}(\alpha_0, \varphi))^{-1} = O(\gamma^{-1})$ and $\varepsilon\gamma^{-1}$ is small. Variants are possible.

Next, from (82) we find

$$\hat{\psi} = \partial_\omega^{-1}\big(D_\alpha K_{10}(\alpha_0, \varphi)[\hat{\alpha}] + K_{20}(\alpha_0, \varphi)\hat{y} + K_{11}^T(\alpha_0, \varphi)\hat{w} + g_1\big) + c_2 \tag{84}$$

for some constant $c_2 \in \mathbb{R}^\nu$.

Remark 12. The constants $c_1, c_2 \in \mathbb{R}^\nu$ in the definition of $\hat{y}$ in (78) and $\hat{\psi}$ in (84) are free (we can set for instance $c_1 = c_2 = 0$). Thus the operator $d_{i,\alpha,\mu}\mathcal{F}(\varepsilon, i, \alpha, \mu)$ has only a right inverse. About c_1, the presence of the parameter α gives the freedom to impose an additional condition for I_0 (for instance $I_0(0) = 0$, or the mean value of I_0 vanishes). The presence of the constant c_2 is connected to the fact that if $i(\varphi)$ is a solution then all the translates $i(\varphi + c)$ are solutions too. It is usual to impose that the mean value of $\theta(\varphi) - \varphi$ is 0.

In conclusion, the solution of the linear system (74) is

$$\mathbb{D}^{-1} g := \big(\hat{\psi}, \hat{y}, \hat{w}, \hat{\alpha}, \hat{\mu}\big)$$

defined in (77), (78), (81), (83), (84). Recalling (72) we finally define the linear operator

$$T_{i,\alpha,\mu} := D\tilde{G}_\delta(\varphi, 0, 0) \circ \mathbb{D}^{-1} \circ DG_\delta(\varphi, 0, 0)^{-1}, \tag{85}$$

where we include in $\tilde{G}_\delta$ also the parameters components, namely

$$\tilde{G}_\delta(\psi, y, w, \alpha, \mu) := \big(G_\delta(\psi, y, w), \alpha, \mu\big).$$

By construction, the operator $T_{i,\alpha,\mu}$ is an approximate right inverse of $d_{i,\alpha,\mu}\mathcal{F}$, because it has been obtained neglecting terms which vanish at an exact solution: we first substituted the approximate torus embedding i with the isotropic one i_δ (which coincide at a solution by Lemma 6) and then we neglected the terms K_{00}, K_{10}, K_{01}, μ_0 which are naught at a solution (Lemmas 8, 3). Let us give a more formal proof.

Lemma 10. *The operator $T_{i,\alpha,\mu}$ is an approximate right inverse of $d_{i,\alpha,\mu}\mathcal{F}(\varepsilon, i, \alpha, \mu)$.*

Proof. By (35), since $X_\mathcal{N}$ does not depend on I, and i_δ differs from i only for the I component, we have

$$\begin{aligned} d_{i,\alpha,\mu}\mathcal{F}(i, \alpha_0) - d_{i,\alpha,\mu}\mathcal{F}(i_\delta, \alpha_0) &= \varepsilon\big(d_{i,\alpha,\mu}X_P(i, \alpha_0) - d_{i,\alpha,\mu}X_P(i_\delta, \alpha_0)\big) \\ &= \varepsilon \int_0^1 \partial_I d_{i,\alpha,\mu}X_P(i_\delta + s(i - i_\delta), \alpha_0)[I_0 - I_\delta]ds \\ &=: \mathcal{E}_0 \end{aligned} \tag{86}$$

which is $O(Z)$ by Lemma 6 and (42).

We denote by $\mathrm{u} := (\psi, y, w)$ the symplectic coordinates induced by G_δ in (66). Under the symplectic map G_δ, the nonlinear operator $\mathcal{F}$ in (35) is transformed into

$$\mathcal{F}(G_\delta(\mathrm{u}(\varphi)), \alpha, \mu) = DG_\delta(\mathrm{u}(\varphi))\big(\partial_\omega \mathrm{u}(\varphi) - X_{K_\mu}(\mathrm{u}(\varphi), \alpha)\big) \tag{87}$$

where $K_\mu = H_\mu \circ G_\delta$, see (68). Differentiating (87) at the trivial torus embedding $\mathrm{u}_\delta(\varphi) := G_\delta^{-1}(i_\delta(\varphi)) = (\varphi, 0, 0)$ for the values of the parameters $(\alpha, \mu) = (\alpha_0, \mu_0)$, we get

$$d_{i,\alpha,\mu}\mathcal{F}(i_\delta,\alpha_0,\mu_0) = DG_\delta(\mathrm{u}_\delta)\big(\partial_\omega - d_{\mathrm{u},\alpha,\mu}X_{K_\mu}(\mathrm{u}_\delta,\alpha_0,\mu_0)\big)D\tilde{G}_\delta(\mathrm{u}_\delta)^{-1} + \mathcal{E}_1\,,$$
$$\mathcal{E}_1 := D^2G_\delta(\mathrm{u}_\delta)\big[DG_\delta(\mathrm{u}_\delta)^{-1}\mathcal{F}(i_\delta,\alpha_0,\mu_0),\, DG_\delta(\mathrm{u}_\delta)^{-1}\Pi[\cdot]\big]\,, \qquad (88)$$

where Π is the projection $(\hat{\imath},\hat{\alpha},\hat{\mu}) \mapsto \hat{\imath}$. In expanded form $d_{\mathrm{u},\alpha,\mu}X_{K_\mu}(\mathrm{u}_\delta,\alpha_0,\mu_0)$ is provided in (73). We split $\partial_\omega - d_{\mathrm{u},\alpha,\mu}X_{K_\mu}(\mathrm{u}_\delta,\alpha_0,\mu_0) = \mathbb{D} + R_Z$ where $\mathbb{D}$ is defined in (75) and R_Z is the part which vanishes in Z. By (86) and (88)

$$d_{i,\alpha,\mu}\mathcal{F}(i,\alpha) = DG_\delta(\mathrm{u}_\delta) \circ \mathbb{D} \circ D\tilde{G}_\delta(\mathrm{u}_\delta)^{-1} + \mathcal{E}_0 + \mathcal{E}_1 + \mathcal{E}_2$$
$$\mathcal{E}_2 := DG_\delta(\mathrm{u}_\delta) \circ R_Z \circ D\tilde{G}_\delta(\mathrm{u}_\delta)^{-1}\,.$$

Applying T defined in (85) to the right, since $\mathbb{D} \circ \mathbb{D}^{-1} = Id$ we get

$$d_{i,\alpha,\mu}\mathcal{F}(i,\alpha_0,\mu_0) \circ T - Id = \mathcal{E} \circ T$$

where $\mathcal{E} := \mathcal{E}_0 + \mathcal{E}_1 + \mathcal{E}_2$ is $O(Z)$. □

Remark 13. In order to construct an approximate inverse for $d\mathcal{F}$, it is sufficient to have an approximate inverse of $\mathbb{D}$ in (74), i.e. we need in (81) only an approximate inverse for $\partial_\omega - JK_{02}(\alpha_0,\varphi)$.

The operator T usually satisfies estimates like (47) and (48) (in an analytic setting) or (50) and (51) (in a Sobolev scale) and the Nash-Moser iterative scheme with approximate right inverse converges.

References

1. Arcostanzo, M., Arnaud, M.C., Bolle, P., and Zavidovique, M.: Tonelli Hamiltonians without conjugate points and C^0 integrability. Math. Z. **280**, 165–194 (2015)
2. Baldi, P., Berti, M., Montalto, R.: KAM for quasi-linear and fully nonlinear forced perturbations of Airy equation. Math. Ann. **359**(1–2), 471–536 (2014)
3. Baldi, P., Berti, M., Montalto, R.: KAM for autonomous quasi-linear perturbations of KdV, preprint (2014)
4. Berti, M., Bolle, P.: Sobolev quasi periodic solutions of multidimensional wave equations with a multiplicative potential. Nonlinearity **25**, 2579–2613 (2012)
5. Berti, M., Bolle, P.: Quasi-periodic solutions with Sobolev regularity of NLS on $\mathbb{T}^d$ with a multiplicative potential. J. Eur. Math. Soc. **15**, 229–286 (2013)
6. Berti, M., Bolle, P.: Quasi-periodic solutions for autonomous NLW on $\mathbb{T}^d$ with a multiplicative potential, in preparation
7. Berti, M., Bolle, P., Procesi, M.: An abstract Nash-Moser theorem with parameters and applications to PDEs. Ann. I. H. Poincaré **1**, 377–399 (2010)
8. Berti, M., Corsi, L., Procesi, M.: An abstract Nash-Moser theorem and quasi-periodic solutions for NLW and NLS on compact Lie groups and homogeneous spaces. Comm. Math. Phys. **334**, 1413–1454 (2015)
9. Bourgain, J.: Construction of quasi-periodic solutions for Hamiltonian perturbations of linear equations and applications to nonlinear PDE. Int. Math. Res. Not. **11**, 475–497 (1994)

10. Bourgain, J.: Quasi-periodic solutions of Hamiltonian perturbations of 2*D* linear Schrödinger equations. Ann. Math. **148**, 363–439 (1998)
11. Bourgain, J.: Green's function estimates for lattice Schrödinger operators and applications. Annals of Mathematics Studies, vol. 158. Princeton University Press, Princeton (2005)
12. Craig, W.: Problèmes de petits diviseurs dans les équations aux dérivées partielles, Panoramas et Synthèses, vol. 9. Société Mathématique de France, Paris (2000)
13. Craig, W., Wayne, C.E.: Newton's method and periodic solutions of nonlinear wave equation. Comm. Pure Appl. Math. **46**, 1409–1498 (1993)
14. De la Llave, R., Gonzalez, A., Jorba, A., Villanueva, J.: KAM theory without action-angle variables. Nonlinearity **18**, 855–895 (2005)
15. Eliasson, L.H.: Perturbations of stable invariant tori for Hamiltonian systems. Annali della Scuola Normale Superiore di Pisa - Classe di Scienze (4) **15**(1), 115–147 (1988)
16. Eliasson, L.H., Kuksin, S.: KAM for non-linear Schrödinger equation. Ann. Math. **172**, 371–435 (2010)
17. Fejoz, J.: Démonstration du théorème d' Arnold sur la stabilité du système planétaire (d' après M. Herman). Ergodic Theory Dyn. Syst. **24**, 1–62 (2004)
18. Fontich, E., De La Llave, R., Sire, Y.: A method for the study of whiskered quasi-periodic and almost-periodic solutions in finite and infinite dimensional Hamiltonian systems. Electron. Res. Announc. Math. Sci. **16**, 9–22 (2009)
19. Herman, M.: Inégalités "a priori" pour des tores lagrangiens invariants par des difféomorphismes symplectiques. Publ. Math. Inst. Hautes Étud. Sci. **70**, 47–101 (1989)
20. Kuksin, S.: Hamiltonian perturbations of infinite-dimensional linear systems with imaginary spectrum. Funktsional. Anal. i Prilozhen. **21**(3), 22–37, 95 (1987)
21. Kuksin, S.: Analysis of Hamiltonian PDEs, Oxford University Press, Oxford (2000)
22. Moser, J.: A rapidly convergent iteration method and non-linear partial differential equations I-II. Ann. Scuola Norm. Sup. Pisa **3**(20), 265–315, 499–535 (1966)
23. Moser, J.: Convergent series expansions for quasi-periodic motions. Math. Ann. **169**, 136–176 (1967)
24. Moser, J., Zehnder, E.: Notes on Dynamical Systems, Courant Lecture Notes, vol. 12. Courant Institute of Mathematical Sciencce, New York (2005)
25. Pöschel, J.: A KAM-theorem for some nonlinear partial differential equations. Ann. Scuola Norm. Sup. Pisa Cl. Sci. (1) **4**(23), 119–148 (1996)
26. Procesi, M., Xu, X.: Quasi-Töplitz Functions in KAM Theorem. SIAM J. Math Anal. **45**(4), 2148–2181 (2013)
27. Salamon, D., Zehnder, E.: KAM theory in configuration space. Comm. Math. Helv. **64**, 84–132 (1989)
28. Wayne, E.: Periodic and quasi-periodic solutions of nonlinear wave equations via KAM theory. Comm. Math. Phys. **127**, 479–528 (1990)
29. Zehnder, E.: Generalized implicit function theorems with applications to some small divisors problems I-II. Comm. Pure Appl. Math. **28**, 91–140 (1975); **29**, 49–113 (1976)

On the Spectral and Orbital Stability of Spatially Periodic Stationary Solutions of Generalized Korteweg-de Vries Equations

Todd Kapitula and Bernard Deconinck

This paper is dedicated to Walter Craig on the occasion of his 60th birthday.

Abstract In this paper we generalize previous work on the spectral and orbital stability of waves for infinite-dimensional Hamiltonian systems to include those cases for which the skew-symmetric operator $\mathcal{J}$ is singular. We assume that $\mathcal{J}$ restricted to the orthogonal complement of its kernel has a bounded inverse. With this assumption and some further genericity conditions we (a) derive an unstable eigenvalue count for the appropriate linearized operator, and (b) show that the spectral stability of the wave implies its orbital (nonlinear) stability, provided there are no purely imaginary eigenvalues with negative Krein signature. We use our theory to investigate the (in)stability of spatially periodic waves to the generalized KdV equation for various power nonlinearities when the perturbation has the same period as that of the wave. Solutions of the integrable modified KdV equation are studied analytically in detail, as well as solutions with small amplitudes for higher-order pure power nonlinearities. We conclude by studying the transverse stability of these solutions when they are considered as planar solutions of the generalized KP-I equation.

1 Introduction

The study of the stability of spatially periodic stationary solutions of nonlinear wave equations has seen different advances the past few years. There are advances both in the numerical investigation of spectral stability [14, 43], as well as in the analytical study of spectral and orbital stability (see [5, 7, 18, 19, 24, 41] and the references

T. Kapitula
Department of Mathematics and Statistics, Calvin College, Grand Rapids, MI 49546, USA
e-mail: tmk5@calvin.edu

B. Deconinck (✉)
Department of Applied Mathematics, University of Washington, Seattle, WA 98195, USA
e-mail: bernard@amath.washington.edu

P. Guyenne et al. (eds.), *Hamiltonian Partial Differential Equations and Applications*, Fields Institute Communications 75,
DOI 10.1007/978-1-4939-2950-4_10

therein). We focus specifically on the study of the stability of periodic solutions of Hamiltonian partial differential equations, as in [24]. However, the results also apply to the study of localized waves for systems in which the appropriate linearized system has a compact resolvent, e.g., the Gross-Pitaevski equation (see [36] and the references therein). The equations of interest are written abstractly as

$$u_t = \mathscr{J}\mathscr{E}'(u), \qquad u(0) = u_0, \tag{1}$$

on a Hilbert space X, where $\mathscr{J} : X \to \text{range}(\mathscr{J}) \subset X$ is skew symmetric, and $\mathscr{E} : X \to \mathbb{R}$ is a C^2-functional. In previous works (see [24, 34] and the references therein) it was assumed that $\mathscr{J}$ is nonsingular with bounded inverse. We do not make that assumption here. We allow $\ker(\mathscr{J})$ to be nontrivial; however, we do assume that $\mathscr{J}|_{\ker(J)^\perp}$ has a *bounded* inverse.

We are interested in the spectral and orbital stability of spatially periodic waves to (1). The waves are realized as critical points of a constrained energy, and the stability of the waves is determined by closely examining the Hessian, say $\mathscr{L}$, of the constrained energy. It will henceforth be assumed that $\text{n}(\mathscr{L}) < \infty$, where the notation $\text{n}(\mathscr{S})$ is used to denote the number of negative eigenvalues (counting multiplicities) of the self-adjoint operator $\mathscr{S}$. If $\mathscr{J}$ is nonsingular with bounded inverse, then it was seen in [20] that there is a symmetric matrix $\boldsymbol{D}$ such that if $\text{n}(\mathscr{L}) - \text{n}(\boldsymbol{D}) = 0$, then the wave is orbitally stable. The matrix $\boldsymbol{D}$ is intimately related to the conserved quantities of (1) which are generated by its group invariances. It was shown in [24, 35] and the references therein that if this difference is positive, then there exists a close relationship between this difference and the structure of $\sigma(\mathscr{J}\mathscr{L})$, where $\mathscr{J}\mathscr{L}$ is the linearization of (1) about the critical point. As we demonstrate, this formula must be modified if $\mathscr{J}$ is singular. In particular, the formula must take into account the fact that the only nontrivial flow of (1) occurs on $\ker(\mathscr{J})^\perp$ (see Theorem 1 for a precise statement).

A concrete example to which the theory is applicable is the determination of the orbital (in)stability of spatially periodic stationary solutions of the generalized Korteweg-de Vries (gKdV) equations ($p \in \mathbb{N}_0$)

$$u_t = \partial_x\left(-u_{xx} \pm u^{p+1}\right). \tag{2}$$

Here $\mathscr{J} = \partial_x$, and $u \in \ker(\mathscr{J})^\perp$ if $\overline{u} = 0$, where $\overline{u}$ represents the spatial average. Note that the $\pm$ sign is irrelevant when p is odd, but not so when p is even. With $p = 1$ Eq. (2) is the integrable KdV equation, and with $p = 2$ it is the integrable modified KdV (mKdV) equation. For $p \geq 3$ the equation is not integrable by any definition. If $p \in \mathbb{R}^+$ is not an integer in the gKdV, then unless we know that solutions are always positive the nonlinear term must be replaced via $u^{p+1} \mapsto |u|^p u$. When we consider the integrable cases we can do all calculations explicitly for even large amplitude waves, providing examples that are far more robust than for the non-integrable cases. The stability theory we develop is applicable for *all* cases, as long as there is a local well-posedness theory for the initial value problem.

In this paper we restrict ourselves to proving the orbital stability of spatially periodic stationary solutions of gKdV for $p \geq 2$ with respect to perturbations of the same period. This corresponds to the case of zero Floquet exponent in [24], for which the determination of $\sigma(\mathcal{J}\mathcal{L})$ was left open there because it results in $\mathcal{J}$ being singular. For the mKdV equation the results depend on the period of the wave as well as on which sign is considered. A few non-integrable examples are discussed as well in the regime of small amplitude waves.

The paper is organized as follows. In Sect. 2 we discuss the spectral theory and the orbital stability theory for relative equilibria to (1) under the assumption that $\ker(\mathcal{J})$ is nontrivial. In Sect. 3 we apply the results of the theory to (2) in the case of periodic perturbations of the same period as the underlying solution. We consider the case of the mKdV equation ($p = 2$) in detail for three of its periodic solutions. We find that two of the solutions are orbitally stable, whereas the stability of the third solution depends upon its period. We finish the section by considering the case of small solutions for any $p \in \mathbb{N}$, and find that all of the solutions under consideration are orbitally stable. Finally, in Sect. 4 we consider the transverse stability of the gKdV solutions when they are considered as solutions of the generalized KP-I equation.

Addendum. This paper has a long history. The original work was completed in 2009, and a major revision was done in 2010. In its original form this paper was joined with the companion paper by Deconinck and Kapitula [13]. At the suggestion of an editor the two papers were severed, and the second was soon thereafter published. This paper was submitted elsewhere, where after some time and consideration the editors decided that it was not appropriate for the journal. We subsequently submitted this paper to another journal in the fall of 2011. As far as we can tell, the paper then fell through the cracks in the editor/referee system of that journal (we joked that it had fallen into a "Refereeing Purgatory"). While the paper as of Fall 2014 is still unpublished, the work has been noticed in the community. From the time of original submission to Fall 2014 the paper has been referenced to at least 31 times (according to Google Scholar). Some of the referencing papers are Benzoni-Gavage [3], Bottman et al. [6], Bronski et al. [8, 9], Chen et al. [11], Farah and Scialom [16], Hakkaev et al. [22], Johnson [27], Johnson and Zumbrun [28, 29], Johnson et al. [30], Kapitula and Promislow [31], Kapitula and Stefanov [33], Nivala and Deconinck [38], Pava and Natali [40], Pelinovsky [42], Stanislavova and Stefanov [45]. The results and ideas presented herein are also a highlighted example in the recent book on stability theory by Kapitula and Promislow [32, Chap. 6.1.2].

2 Theoretical Results

Much of the following discussion can be found in [34, Sect. 2]. It is included here for the sake of completeness. Let U, V, X denote three real Hilbert spaces with $U \subset X \subset V$ being dense and continuous embeddings. Throughout this paper, we use only the scalar product $\langle\cdot,\cdot\rangle$ on the space V. In particular, we have $U \subset X \subset V \subset X^*$ where X^* denotes the dual space of X. Adjoint operators are always taken with respect to the scalar product on V.

We are interested in relative equilibria of (1). These are solutions of (1) whose functional form is in some sense invariant under the dynamics. Typical examples include traveling wave solutions which are invariant under translation in x (as for (2), or which are invariant under multiplication by a unitary scalar (as for nonlinear Schrödinger-type equations). In order to make this notion of invariance precise so as to formally define what is meant by relative equilibria, we need a few elements from the theory of Lie groups and Lie algebras.

Let G be a finite-dimensional abelian Lie group with Lie algebra $\mathfrak{g}$. Denote by $\exp(\omega) = \mathrm{e}^{\omega}$ for $\omega \in \mathfrak{g}$ the exponential map from $\mathfrak{g}$ into G. Next, assume that $T : G \to L(V)$ is a unitary representation of G on V so that $T'(e)$ maps $\mathfrak{g}$ into the space of closed skew-symmetric operators on V with domain X. Since $\mathscr{J}$ is nonsingular with bounded inverse on $\ker(\mathscr{J})^{\perp}$, we make the additional assumption that:

Assumption 1. *The derivative of the group action $T(\cdot)$ satisfies $T_{\omega} : X \mapsto \ker(\mathscr{J})^{\perp}$.*

In Assumption 1 the notation $T_{\omega} := T'(e)\omega$ is used for the linear skew-symmetric operator which is the generator of the semigroup $T(\mathrm{e}^{\omega t})$. Note that T_{ω} is also linear in $\omega \in \mathfrak{g}$. We assume that U is contained in the domain of T_{ω}^2. The group orbit Gu of an element $u \in X$ is defined by $Gu := \{T(g)u;\ g \in G\}$.

2.1 Existence of Relative Equilibria

We need two compatibility assumptions. First, we assume that $\mathscr{E}$ is invariant under G so that

$$\mathscr{E}(T(g)u) = \mathscr{E}(u)$$

for all $u \in X$ and all $g \in G$. Second, there is a type of commutation between the group action and the skew operator,

$$T(g)\mathscr{J} = \mathscr{J}T(g^{-1})^*, \quad \text{all } g \in G. \tag{3}$$

As a consequence of Assumption 1 we can define the bounded functional $M_{\omega} : X \mapsto \mathbb{R}$ by

$$M_{\omega}(u) := \frac{1}{2}\langle \mathscr{J}^{-1}T_{\omega}u, u\rangle, \quad \omega \in \mathfrak{g}.$$

Its second derivative $M''_\omega(u) = \mathcal{J}^{-1}T_\omega : X \mapsto \ker(\mathcal{J})^\perp$ is a bounded symmetric linear operator by (3). Note that $M_\omega(T(g)u) = M_\omega(u)$, i.e., that M_ω is a conserved functional under the group action.

We are now positioned to define what we mean by relative equilibria of (1). These are solutions $u(t)$ whose time orbit is contained in the group orbit Gu_0 so that $u(t) \in Gu(0)$ for all t. Thus, $\phi \in X$ is a relative equilibrium of (1) if and only if there is an $\omega \in \mathfrak{g}$ so that $u(t) = T(\mathrm{e}^{\omega t})\phi$ satisfies (1). Substituting the ansatz $u(t) = T(\mathrm{e}^{\omega t})\phi$ into (1) we get

$$T_\omega\phi = \mathcal{J}\mathcal{E}'(\phi), \quad \omega \in \mathfrak{g}. \tag{4}$$

As a consequence of Assumption 1 both sides are in $\ker(\mathcal{J})^\perp$. Let the operator $P_\mathcal{J} : X \mapsto \ker(\mathcal{J})^\perp$ be the orthogonal projection onto the range of $\mathcal{J}$. Since $P_\mathcal{J} T_\omega = T_\omega P_\mathcal{J}$, (4) is equivalent to

$$P_\mathcal{J}\left[\mathcal{E}'(\phi) - \mathcal{J}^{-1}T_\omega\phi\right] = 0. \tag{5}$$

Note that (5) implies that

$$\mathcal{E}'(\phi) - \mathcal{J}^{-1}T_\omega\phi \in \ker(\mathcal{J}).$$

In conclusion, we see from (5) that $\phi \in X$ is a relative equilibrium if and only if $P_\mathcal{J}\mathcal{H}'_\omega(\phi) = 0$, where

$$\mathcal{H}_\omega := \mathcal{E} - M_\omega, \quad \omega \in \mathfrak{g}.$$

Note that this does not necessarily imply that ϕ is a critical point of $\mathcal{H}_\omega$. We assume throughout that there exists a smooth family of bound states:

Assumption 2 (Relative Equilibria). *There exists a non-empty open set $\Omega \subset \mathfrak{g}$ and a $\mathcal{C}^1$ function $\phi : \Omega \to U, \omega \mapsto \phi_\omega$ such that ϕ_ω is a relative equilibrium of* (1), *i.e., $P_\mathcal{J}\mathcal{H}'_\omega(\phi_\omega) = 0$ for each $\omega \in \Omega$. We assume that the isotropy subgroups $\{g \in G;\ T(g)\phi_\omega = \phi_\omega\}$ are discrete for all ω so that the group orbits $G\phi_\omega$ satisfy* $\dim(G\phi_\omega) = \dim(G)$.

Remark 1. Since G is abelian, the entire group orbit $T(g)\phi_\omega$ with $g \in G$ consists of relative equilibria with time evolution $T(\mathrm{e}^{\omega t})$.

2.2 Formulation of the Evolution Equation

Without loss of generality we will henceforth assume that the relative equilibrium is a critical point of $\mathcal{H}_\omega$. Indeed, if the relative equilibrium satisfies

$$\mathcal{H}'_\omega(\phi_\omega) = z, \quad z \in \ker(\mathcal{J}),$$

then the mapping

$$\widetilde{\mathcal{H}_\omega}(u) = \mathcal{H}_\omega(u) - \langle z, u \rangle$$

yields that the relative equilibrium is a critical point of $\widetilde{\mathcal{H}_\omega}(u)$. The self-adjoint Hessian of the energy at the relative equilibrium ϕ_ω is defined as

$$\mathcal{L} := \mathcal{H}_\omega''(\phi_\omega) : U \mapsto V. \tag{6}$$

Note that the linearization of (1) around the relative equilibrium ϕ_ω in the "co-moving" frame is given by $\mathcal{J}\mathcal{L}$. Due to the invariance of H_ω under the abelian group G, one has that the tangent space of the group orbit $G\phi_\omega$ at ϕ_ω is contained in $\ker(\mathcal{L})$. As a consequence of Assumption 2 it follows from [20, p. 314] that

$$\ker(\mathcal{L}) := \mathrm{span}\{T_\alpha \phi_\omega;\ \alpha \in \mathfrak{g}\}. \tag{7}$$

Upon writing $X = \ker(\mathcal{J}) \oplus H_1$, where $H_1 := \ker(\mathcal{J})^\perp$, let $Q_\mathcal{J} := \mathbb{1} - P_\mathcal{J} : X \mapsto \ker(\mathcal{J})$ be the orthogonal projection onto $\ker(\mathcal{J})$, where $P_\mathcal{J}$ was defined in the previous subsection. One may rewrite the system

$$u_t = \mathcal{J}\mathcal{H}_\omega'(\phi_\omega + u), \quad u(0) = u_0, \tag{8}$$

for which the relative equilibrium ϕ_ω satisfies $P_\mathcal{J}\mathcal{H}_\omega'(\phi_\omega) = 0$, as the system,

$$\begin{aligned} \partial_t P_\mathcal{J} u &= \mathcal{J} P_\mathcal{J} \mathcal{H}_\omega'(\phi_\omega + P_\mathcal{J} u + Q_\mathcal{J} u), \quad P_\mathcal{J} u(0) = P_\mathcal{J} u_0 \\ \partial_t Q_\mathcal{J} u &= 0, \quad Q_\mathcal{J} u(0) = Q_\mathcal{J} u_0. \end{aligned} \tag{9}$$

From (9) it is seen that $Q_\mathcal{J} u(t) = Q_\mathcal{J} u(0)$ for all $t \geq 0$; in other words, nontrivial evolution of the initial data only occurs in H_1. Consider an initial condition for (9) which satisfies $Q_\mathcal{J} u_0 = 0$. One sees from (9) that $Q_\mathcal{J} u(t) = 0$ for all $t > 0$. Using $P_\mathcal{J} u = u$ for all $t \geq 0$, the evolution equation of interest is given by

$$u_t = \mathcal{J} P_\mathcal{J} \mathcal{H}_\omega'(\phi_\omega + u), \quad u(0) = u_0. \tag{10}$$

Since the evolution occurs on H_1, $\mathcal{J}$ is now skew symmetric with *bounded* inverse.

2.3 The Eigenvalue Count

The spectral problem associated with the stability problem for relative equilibria of (10) is

$$\mathcal{J}\mathcal{L}|_{H_1} u = \lambda u, \quad \mathcal{L}|_{H_1} := P_\mathcal{J} \mathcal{L} P_\mathcal{J}. \tag{11}$$

By assumption $\mathscr{J}$ is skew-symmetric, and it has bounded inverse on H_1. It is clear that $\mathscr{L}$, and hence $\mathscr{L}|_{H_1}$, are self-adjoint. In what follows, the following assumptions are used:

Assumption 3. *It is assumed that:*

1. *$\mathscr{L}$ has a compact resolvent, and $\sigma(\mathscr{L}) \cap \mathbb{R}^-$ is a finite set,*
2. *There is a self-adjoint operator $\mathscr{L}_0$ with compact resolvent such that:*
   a. *$\mathscr{L} = \mathscr{L}_0 + \mathscr{A}$, where $\mathscr{A}$ is $\mathscr{L}_0$-compact and satisfies*

$$\|\mathscr{A}u\| \leq a\|u\| + b\| \, |\mathscr{L}_0|^r u\|,$$

   for some positive constants a, b and $r \in [0, 1)$,

   b. *The increasing sequence of nonzero eigenvalues ω_j of $\mathscr{L}_0$ satisfies*

$$\sum_{j=1}^{\infty} |\omega_j|^{-s} < \infty,$$

   for some $s \geq 1$,

   c. *There exists a subsequence of eigenvalues $\{\omega_{n_k}\}_{k \in \mathbb{N}}$ and constants $c > 0$ and $r' > r$ such that*

$$\omega_{n_k+1} - \omega_{n_k} \geq c\, \omega_{n_k+1}^{r'}.$$

3. *$\mathrm{Im}(\mathscr{J}) = \mathrm{Im}(\mathscr{L}|_{H_1}) = 0$, where Im denotes the imaginary part.*

Remark 2. Assumption 3(a)–(b) are also assumed in [24]. It is known that these assumptions are not absolutely necessary (see [34] where the assumptions are removed), but they are satisfied for the applications we have in mind. It is clear that Assumption 3(a)–(b) for $\mathscr{L}$ imply that $\mathscr{L}|_{H_1}$ has the same properties. As seen in [24], Assumption 3(c) is not necessary, and it is assumed here only for the sake of simplicity. As a consequence of this assumption, eigenvalues for (11) come in quartets $\{\pm\lambda, \pm\lambda^*\}$.

In contrast to [24] we do not assume that $\mathscr{L}|_{H_1}$ is nonsingular. In fact, as a consequence of Assumption 1 one has $\ker(\mathscr{L}) \subset H_1$; consequently, $\ker(\mathscr{L}) \subset \ker(\mathscr{L}|_{H_1})$. One has that $\ker(\mathscr{L}) \subset H_1$ and $\ker(\mathscr{J}) \subset \ker(\mathscr{L})^{\perp}$. Upon defining

$$\ker_{\mathrm{a}}(\mathscr{L}) := \{z \in \ker(\mathscr{J}) : \mathscr{L}^{-1}z \in H_1\},$$

one has

$$\ker(\mathscr{L}|_{H_1}) = \ker(\mathscr{L}) \oplus \ker_{\mathrm{a}}(\mathscr{L}).$$

The definition of $\ker_a(\mathscr{L})$ makes sense because $\ker(\mathscr{J}) \subset \ker(\mathscr{L})^\perp$. Let $\ker(\mathscr{J}) = \mathrm{Span}(z_1, \dots, z_m)$, where $\{z_1, \dots, z_m\}$ are orthonormal. Let $\boldsymbol{J} \in \mathbb{C}^{m\times m}$ be the Hermitian matrix whose entries are given by

$$\boldsymbol{J}_{ij} = \langle z_i, \mathscr{L}^{-1} z_j\rangle, \tag{12}$$

i.e., $\boldsymbol{J}$ is a matrix representation for the quadratic form $\langle u, \mathscr{L}^{-1}|_{\ker(\mathscr{J})} u\rangle$. Note that

$$\dim[\ker(\boldsymbol{J})] = \dim[\ker_a(\mathscr{L})];$$

hence, $\ker(\mathscr{L}) = \ker(\mathscr{L}|_{H_1})$ if and only if $\boldsymbol{J}$ is nonsingular. This is henceforth assumed.

With $\boldsymbol{J}$ nonsingular, one has

$$\ker(\mathscr{L}|_{H_1})^\perp = \ker(\mathscr{L})^\perp \cap H_1.$$

Using the notation

$$\mathscr{A}(S) := \{y : y = \mathscr{A}s \text{ for some } s \in S\},$$

one has that as a consequence of [34, Eq. (3.2)],

$$\mathscr{J}^{-1}(\ker(\mathscr{L}|_{H_1})) \subset \ker(\mathscr{L}|_{H_1})^\perp. \tag{13}$$

Since $\ker(\mathscr{L}|_{H_1}) \subset H_1$ one can refine (13) to say that $\mathscr{J}^{-1}(\ker(\mathscr{L}|_{H_1})) \subset \ker(\mathscr{L}|_{H_1})^\perp \cap H_1$; hence, there is a generalized eigenspace $X_{\mathscr{L}} \subset H_1$ such that

$$\mathscr{L}|_{H_1} X_{\mathscr{L}} = \mathscr{J}^{-1}(\ker(\mathscr{L}|_{H_1})).$$

Define

$$\boldsymbol{D}_{ij} := \langle y_i, \mathscr{L}|_{H_1} y_j\rangle, \tag{14}$$

where $\{y_i\} \subset X_{\mathscr{L}}$ is any basis for $X_{\mathscr{L}}$. If $\boldsymbol{D}$ is nonsingular, then by the Fredholm alternative

$$\mathrm{gker}(\mathscr{J}\mathscr{L}|_{H_1}) = \ker(\mathscr{L}|_{H_1}) \oplus X_{\mathscr{L}}, \tag{15}$$

with $\mathrm{m_g}(0) = \dim(\ker(\mathscr{L}|_{H_1}))$ and $\mathrm{m_a}(0) = 2\mathrm{m_g}(0)$. Here $\mathrm{m_g}(\lambda)$ is the geometric multiplicity of the eigenvalue λ, $\mathrm{m_a}(\lambda) \geq \mathrm{m_g}(\lambda)$ is the algebraic multiplicity, and $\mathrm{gker}(\mathscr{A})$ refers to the generalized kernel of the operator $\mathscr{A}$.

In order to apply the results of [24] we must recast the eigenvalue problem in the appropriate subspace so that the operator $\mathscr{L}|_{H_1}$ no longer has a nontrivial kernel. Let $P_1 : H_1 \mapsto \ker(\mathscr{L}|_{H_1})^\perp$ be the orthogonal projection, and set $Q_1 := \mathbb{1} - P_1$. For $\lambda \neq 0$ rewrite (11) as

$$\mathcal{L}|_{H_1} u = \lambda \mathcal{J}^{-1} u. \tag{16}$$

Upon using the projections P_1, Q_1 one sees that (16) is equivalent to the system,

$$\begin{aligned} P_1 \mathcal{L}|_{H_1} P_1 \cdot P_1 u &= \lambda P_1 \mathcal{J}^{-1} P_1 u + \lambda P_1 \mathcal{J}^{-1} Q_1 u \\ 0 &= \lambda Q_1 \mathcal{J}^{-1} P_1 u + \lambda Q_1 \mathcal{J}^{-1} Q_1 u. \end{aligned} \tag{17}$$

As a consequence of (13) one has that $Q_1 \mathcal{J}^{-1} Q_1 u = 0$; thus, from the second line of (17) one has for $\lambda \neq 0$ the identities,

$$\begin{aligned} P_1 \mathcal{J}^{-1} P_1 u &= (P_1 + Q_1) \mathcal{J}^{-1} P_1 u = \mathcal{J}^{-1} P_1 u \\ P_1 \mathcal{J}^{-1} Q_1 u &= (P_1 + Q_1) \mathcal{J}^{-1} Q_1 u = \mathcal{J}^{-1} Q_1 u. \end{aligned}$$

The first line of (17) can be rewritten as

$$\mathcal{J} P_1 \mathcal{L}|_{H_1} P_1 \cdot P_1 u = \lambda P_1 u + \lambda Q_1 u, \tag{18}$$

which, upon using $P_1 Q_1 = 0$, becomes

$$P_1 \mathcal{J} P_1 \cdot P_1 \mathcal{L}|_{H_1} P_1 \cdot P_1 u = \lambda P_1 u. \tag{19}$$

In other words, when looking for nonzero eigenvalues (11) is equivalent to (19). Once (19) is solved, then $Q_1 u$ is uniquely determined by (18). It is an exercise to show that the same conclusion holds when considering generalized eigenfunctions.

By the Fredholm alternative, the solvability of (19) requires that

$$P_1 u \in \ker(P_1 \mathcal{L}|_{H_1} P_1 \cdot P_1 \mathcal{J} P_1)^{\perp}.$$

Since $P_1 \mathcal{L}|_{H_1} P_1$ is nonsingular on $\ker(\mathcal{L}|_{H_1})^{\perp}$, the above is equivalent to requiring that

$$P_1 u \in \ker(P_1 \mathcal{J} P_1)^{\perp} = [\mathcal{J}^{-1}(\ker(\mathcal{L}|_{H_1}))]^{\perp}.$$

Next, define the orthogonal projection $P_2 : H_1 \mapsto [\mathcal{J}^{-1}(\ker(\mathcal{L}|_{H_1}))]^{\perp}$ and set $Q_2 = \mathbb{1} - P_2$. Note that as a consequence of (13), $P_1 P_2 = P_2 P_1$, and that $P_2(P_1 \mathcal{J} P_1) = (P_1 \mathcal{J} P_1) P_2$. Applying the operator P_2 to (19) yields

$$\Pi \mathcal{J} \Pi \cdot \Pi \mathcal{L}|_{H_1} \Pi \cdot \Pi u + \Pi \mathcal{J} \Pi \cdot \Pi \mathcal{L}|_{H_1} \Pi \cdot Q_2 P_1 u = \lambda \Pi u, \quad \Pi := P_1 P_2. \tag{20}$$

Applying the operator Q_2 to (19) and using the fact that $Q_2 : H_1 \mapsto \ker(P_1 \mathcal{J} P_1)$ one gets

$$0 = \lambda Q_2 P_1 u. \tag{21}$$

Thus, by (21) for nonzero λ (20) becomes

$$\Pi \mathscr{J} \Pi \cdot \Pi \mathscr{L}|_{H_1} \Pi \cdot \Pi u = \lambda \Pi u, \tag{22}$$

which is equivalent to (19) and hence (11).

Set

$$R_{\mathscr{L}} := [\ker(\mathscr{L}) \oplus \mathscr{J}^{-1}(\ker(\mathscr{L}))]^{\perp},$$

and note that $\Pi : R_{\mathscr{L}} \mapsto R_{\mathscr{L}}$ and $\Pi \mathscr{J} \Pi \cdot \Pi \mathscr{L}|_{H_1} \Pi : R_{\mathscr{L}} \mapsto R_{\mathscr{L}}$. By construction it is clear that $\Pi \mathscr{J} \Pi : R_{\mathscr{L}} \mapsto R_{\mathscr{L}}$ is nonsingular. In addition,

$$\ker(\Pi \mathscr{L}|_{H_1} \Pi) = \ker(\mathscr{L}|_{H_1}) \oplus X_{\mathscr{L}},$$

and since $\boldsymbol{D}$ is nonsingular one has that

$$\ker(\Pi \mathscr{L}|_{H_1} \Pi) \cap R_{\mathscr{L}} = \{0\}.$$

This follows from $P_1 X_{\mathscr{L}} = X_{\mathscr{L}}$ and the fact that for any $y \in X_{\mathscr{L}}$,

$$P_2 y = y - \sum \langle y, \mathscr{J}^{-1} \ell_i \rangle \mathscr{J}^{-1} \ell_i = y - \sum \langle y, \mathscr{L}|_{H_1} y_i \rangle \mathscr{J}^{-1} \ell_i,$$

where $\{\ell_i\} \subset \ker(\mathscr{L}|_{H_1})$ are chosen so that $\{\mathscr{J}^{-1}\ell_i\}$ is an orthonormal basis for $\mathscr{J}^{-1}(\ker(\mathscr{L}|_{H_1}))$, and $\mathscr{J} \mathscr{L}|_{H_1} y_i = \ell_i$. In conclusion, both of the operators $\Pi \mathscr{J} \Pi$ and $\Pi \mathscr{L}|_{H_1} \Pi$ are nonsingular when acting on $R_{\mathscr{L}}$. Furthermore, these operators satisfy Assumption 3.

Before continuing our study of the eigenvalue problem in (22), we need to briefly discuss the notion of the Krein signature of purely imaginary eigenvalues. For a nonzero purely imaginary eigenvalue λ let $E_\lambda \subset R_{\mathscr{L}}$ be its associated eigenspace. The Krein signature of λ is determined via the nonsingular quadratic form $\langle w, (\mathscr{L}|_{H_1})|_{E_\lambda} w \rangle$. For a self-adjoint operator $\mathscr{A}$, let $\mathrm{n}(\langle w, \mathscr{A} w \rangle)$ denote the dimension of the maximal subspace for which $\langle w, \mathscr{A} w \rangle < 0$. The eigenvalue is said to have negative Krein signature if

$$k_{\mathrm{i}}^{-}(\lambda) := \mathrm{n}(\langle w, (\mathscr{L}|_{H_1})|_{E_\lambda} w \rangle) \geq 1;$$

otherwise, if $k_{\mathrm{i}}^{-}(\lambda) = 0$, then the eigenvalue is said to have positive Krein signature. If the eigenvalue λ is geometrically and algebraically simple with eigenfunction u_λ, then

$$k_{\mathrm{i}}^{-}(\lambda) = \begin{cases} 0, & \langle u_\lambda, \mathscr{L}|_{H_1} u_\lambda \rangle > 0 \\ 1, & \langle u_\lambda, \mathscr{L}|_{H_1} u_\lambda \rangle < 0 \end{cases}.$$

We set the total Krein signature to be

$$k_{\mathrm{i}}^{-} := \sum_{\lambda \in \mathrm{i}\mathbb{R}\backslash\{0\}} k_{\mathrm{i}}^{-}(\lambda).$$

Since Assumption 3(c) implies that $k_{\mathrm{i}}^{-}(\lambda) = k_{\mathrm{i}}^{-}(\overline{\lambda})$, one has that k_{i}^{-} is necessarily even.

The theoretical ideas and results in [24] can now be applied to (22). For a given $w \in H_1$ one has

$$w = w_{\mathscr{L}} + w_X + w_R; \quad w_{\mathscr{L}} \in \ker(\mathscr{L}),\ w_X \in \mathscr{J}^{-1}(\ker(\mathscr{L}|_{H_1})),\ w_R \in R_{\mathscr{L}}.$$

Since $\mathscr{L}|_{H_1} : X_{\mathscr{L}} \mapsto \mathscr{J}^{-1}(\ker(\mathscr{L}|_{H_1}))$, and since $\boldsymbol{D}$ is nonsingular, one may write alternatively

$$w = w_{\mathscr{L}} + y + w_R; \quad w_{\mathscr{L}} \in \ker(\mathscr{L}),\ y \in X_{\mathscr{L}},\ w_R \in R_{\mathscr{L}}. \tag{23}$$

A simple modification of the proof leading to [24, Proposition 2.8] yields

$$\begin{aligned} \langle w, \mathscr{L}|_{H_1} w\rangle &= \langle w_{\mathscr{L}}, \mathscr{L}|_{H_1} w_{\mathscr{L}}\rangle + \langle y, \mathscr{L}|_{H_1} y\rangle + \langle w_R, \mathscr{L}|_{H_1} w_R\rangle \\ &= \langle y, \mathscr{L}|_{H_1} y\rangle + \langle w_R, \mathscr{L}|_{H_1} w_R\rangle. \end{aligned} \tag{24}$$

By [24, Theorem 2.13] applied to (22) one has that

$$k_{\mathrm{r}} + k_{\mathrm{c}} + k_{\mathrm{i}}^{-} = \mathrm{n}(\langle w_R, \mathscr{L}|_{H_1} w_R\rangle). \tag{25}$$

Here k_{r} refers to the number of real eigenvalues of $\mathscr{J}\mathscr{L}|_{H_1}$ in the open right-half plane, k_{c} (even) is its number of complex-valued eigenvalues in the open right-half plane, and k_{i}^{-} (even) is the total negative Krein signature. In conclusion, upon noting that $\mathrm{n}(\mathscr{A}) = \mathrm{n}(\langle w, \mathscr{A} w\rangle)$ for self-adjoint operators $\mathscr{A}$, one has from (24) and (25) that

$$\mathrm{n}(\mathscr{L}|_{H_1}) = \mathrm{n}(\boldsymbol{D}) + k_{\mathrm{r}} + k_{\mathrm{c}} + k_{\mathrm{i}}^{-}. \tag{26}$$

We wish to further refine (26). For $x \in X$ write $x = z_1 + w_1$, where $z_1 \in \ker(\mathscr{J})$ and $w_1 \in H_1$. Since $\boldsymbol{J}$ is nonsingular one can alternatively write $x = \mathscr{L}^{-1} z + w$, where $z \in \ker(\mathscr{J})$ and $w \in H_1$. Since

$$\langle x, \mathscr{L} x\rangle = \langle z, \mathscr{L}^{-1} z\rangle + \langle w, \mathscr{L} w\rangle,$$

and since $\langle w, \mathscr{L} w\rangle = \langle w, \mathscr{L}|_{H_1} w\rangle$, one has from the above that

$$\mathrm{n}(\mathscr{L}) = \mathrm{n}(\boldsymbol{J}) + \mathrm{n}(\mathscr{L}|_{H_1}). \tag{27}$$

Combining (26) and (27) leads to the following theorem:

Theorem 1. *Suppose that Assumption 1 and Assumption 3 hold, and that* $\boldsymbol{J}$ *given in* (12) *and* $\boldsymbol{D}$ *given in* (14) *are nonsingular. Then for the eigenvalue problem of* (11)

$$k_{\mathrm{r}} + k_{\mathrm{c}} + k_{\mathrm{i}}^{-} = \mathrm{n}(\mathscr{L}) - \mathrm{n}(\boldsymbol{J}) - \mathrm{n}(\boldsymbol{D}).$$

2.4 Orbital Stability

Assume that there is a local well-posedness theory for (10); in other words, the initial value problem has a unique solution for at least some time. The key condition that must be verified in order to demonstrate orbital stability is that $\mathrm{n}(\mathscr{L}|_{H_1}) = \mathrm{n}(\boldsymbol{D})$ [20, Theorem 4.1]. By (27) one has that

$$\mathrm{n}(\mathscr{L}|_{H_1}) = \mathrm{n}(\mathscr{L}) - \mathrm{n}(\boldsymbol{J});$$

thus, by applying Theorem 1 one has:

Theorem 2. *Under the assumptions of Theorem 1, if* $k_{\mathrm{r}} = k_{\mathrm{c}} = k_{\mathrm{i}}^{-} = 0$*, then the relative equilibria of* (4) *are orbitally stable. In other words, when considering solutions of* (8) *it is true that for each* $\varepsilon > 0$ *there is a* $\delta > 0$ *such that*

$$Q_{\mathscr{J}} u_0 = 0,\ \|u_0 - \phi_\omega\| < \delta \quad \Longrightarrow \quad \inf_{g \in G} \|u(t) - T(g)\phi_\omega\| < \varepsilon.$$

Here $Q_{\mathscr{J}}$ *is the orthogonal projection onto* $\ker(\mathscr{J})$.

Remark 3. Theorem 2 can be considered as a natural generalization of the results of [4], as well as the related works of [2, 41]. Unfortunately, the result cannot be used to furnish an alternate proof of the results in [4], for in this work the operator $P_{\mathscr{J}} \mathscr{J} P_{\mathscr{J}}$ does not have a bounded inverse.

3 Application: Generalized KdV

Consider the generalized KdV equation with power nonlinearity,

$$u_t + (u_{xx} \pm u^{p+1})_x = 0, \quad p \geq 1. \tag{28}$$

The equation with the plus sign is referred to as the focusing gKdV, whereas that with the minus sign is the defocusing gKdV. As discussed in [12], if p is odd then the sign is irrelevant, but if p is even then the two cases are genuinely distinct. Our interest in this section is in the orbital stability of $2L$-periodic solutions of (28) with respect to perturbations of period $2L$, i.e., harmonic perturbations, using the terminology of [5]. It is seen in [12] that global solutions exist to (28) for integer $1 \leq p \leq 3$ for initial data of any size in the appropriate space, whereas solutions are known to exist globally for integer $p \geq 4$ for sufficiently small initial data in the appropriate space. For explicit calculations we focus most of our attention on the cases $p = 1$ (KdV) and $p = 2$ (mKdV). As stated in Sect. 1, both of these cases are completely integrable.

Following the notation of the previous section one has that $\mathcal{J} = \partial_x$. The space H_1 is given by

$$H_1 = \{u \in L^2_{\mathrm{per}}[-L, L] : \overline{u} = 0\}, \quad \overline{u} := \frac{1}{2L}\int_{-L}^{L} u(x)\,\mathrm{d}x.$$

The projection operator is $P_{\mathcal{J}}u = u - \overline{u}$, so $Q_{\mathcal{J}}u$ has zero mean. The relevant group action is $T(\omega)u(x,t) = u(x+\omega, t)$. Since $T_\omega = \omega\partial_x = \omega\mathcal{J}$, Assumption 1 holds. The Hamiltonian associated with (28) is

$$\mathcal{E}(u) = \int_{-L}^{L}\left(\frac{1}{2}u_x^2 \mp \frac{1}{p+2}u^{p+2}\right)\mathrm{d}x, \tag{29}$$

while the functional M_ω is given by

$$M_\omega = \frac{\omega}{2}\int_{-L}^{L} u^2\,\mathrm{d}x,$$

leading to

$$\mathcal{H}_\omega(u) = \mathcal{E}(u) - M_\omega(u) = \int_{-L}^{L}\left(\frac{1}{2}u_x^2 \mp \frac{1}{p+2}u^{p+2} - \frac{\omega}{2}u^2\right)\mathrm{d}x.$$

We find that the relative equilibria of interest satisfy

$$u_{xx} = cu \mp u^{p+1}, \tag{30}$$

with $c = -\omega$. This familiar result is usually obtained by writing (28) in a moving frame via $x \mapsto x - ct$, and looking for stationary solutions. The point of introducing the functional M_ω is that our approach works equally well for other equations where different symmetry reductions lead to the relative equilibria, such as the nonlinear Schrödinger equation.

Remark 4. Note that the solutions we use by no means exhaust the stationary periodic solutions of the gKdV equation; in particular, the full class of solutions can be found only by considering the ODE

$$u_{xx} = cu \mp u^{p+1} + a_0, \quad a_0 \in \mathbb{R} \tag{31}$$

(for instance, see [25], where a larger class of stationary solutions is constructed). We restrict ourselves to the case $a_0 = 0$ because the functional form of the solutions is the simplest, enabling explicit calculations. The case of $a_0 \neq 0$ was carried out in [8, 26].

Let a $2L$-periodic solution to (30) be denoted by $U(x)$. Upon linearizing, we obtain a linear eigenvalue problem of the form (11) with

$$\mathscr{J} = \partial_x, \quad \mathscr{L} = -\partial_x^2 + c \mp (p+1)U^p(x). \tag{32}$$

Note that $\mathscr{L}$ is a Hill operator [37]. In the space $L^2_{\text{per}}[-L, L]$ it is known that the countable set of eigenvalues for $\mathscr{L}$ can be ordered as $\lambda_0 < \lambda_1 \leq \lambda_2 \leq \cdots$ with $\lim_{n\to\infty} \lambda_n = +\infty$, and that the associated normalized eigenfunctions ϕ_j form an orthonormal basis.

Before continuing, we need to verify Assumption 3. Since $\mathscr{L}$ is a Hill operator, it has compact resolvent, and the number of negative eigenvalues is finite. Moreover, using $\mathscr{L}_0 = -\partial_x^2$ the Hill operator satisfies the compactness condition of (b)-(1) with $b = r = 0$. Since the eigenvalues are explicitly given by $(j\pi/L)^2$ (double eigenvalues for $j \geq 1$), the growth conditions on the eigenvalues, (b)-(2) and (b)-(3), are also satisfied. The final condition (c) clearly holds.

As a consequence of the spatial translation invariance associated with (28) one knows that $\mathscr{L}U_x = 0$. Using the notation from Sect. 2,

$$\begin{aligned} &\ker(\mathscr{J}) = \text{Span}\{1\}, \quad \ker(\mathscr{L}) = \text{Span}\{U_x\} \\ &\mathscr{J}^{-1}(\ker(\mathscr{L})) = \text{Span}\{U - \overline{U}\}, \quad \mathscr{L}^{-1}(\ker(\mathscr{J})) = \text{Span}\{\mathscr{L}^{-1}(1)\}. \end{aligned}$$

Note that the assumption that $\boldsymbol{J}$ is nonsingular in Theorem 1 is equivalent to $\langle \mathscr{L}^{-1}(1), 1\rangle \neq 0$.

In order to construct the one-by-one dimensional matrix $\boldsymbol{D}$ one must first find a basis for the one-dimensional generalized eigenspace $X_{\mathscr{L}}$. Assume that $\boldsymbol{J}$ is nonsingular. Let

$$u_C = \mathscr{L}^{-1}(U) - C\mathscr{L}^{-1}(1), \quad C = \frac{\langle \mathscr{L}^{-1}(U), 1\rangle}{\langle \mathscr{L}^{-1}(1), 1\rangle}. \tag{33}$$

It is clear that

$$\mathscr{L}u_C = U - C,$$

and since $u_C \in H_1$ one has that

$$\mathscr{L}|_{H_1} u_C = U - \overline{U}.$$

It follows that u_C provides a basis for $X_{\mathscr{L}}$, where C is given in (33). Since $u_C \in H_1$ one has that

$$\langle u_C, \mathscr{L}|_{H_1} u_C\rangle = \langle u_C, \mathscr{L} u_C\rangle,$$

which upon using (33) and simplifying finally yields

$$\boldsymbol{D} = \frac{\begin{vmatrix} \langle \mathscr{L}^{-1}(U), U\rangle & \langle \mathscr{L}^{-1}(U), 1\rangle \\ \langle \mathscr{L}^{-1}(U), 1\rangle & \langle \mathscr{L}^{-1}(1), 1\rangle \end{vmatrix}}{\langle \mathscr{L}^{-1}(1), 1\rangle}, \tag{34}$$

where we have used that $\mathscr{L}^{-1}$ is self adjoint.

Remark 5. Assuming that $\boldsymbol{J}$ is nonsingular, one sees from (34) that $\boldsymbol{D} = 0$ if and only u_C satisfies $\langle u_C, U\rangle = 0$.

The expression in (34) is difficult to calculate in general, hence we wish to simplify it. Assume that $U(x)$ is even, which in particular implies that $\mathscr{L}$ maps even (odd) functions to even (odd) functions. As stated, one solution of $\mathscr{L}\phi = 0$ is $\phi = U'(x)$. Reduction of order provides a second solution Ψ, which satisfies

$$\mathscr{L}\Psi = 0, \qquad \begin{vmatrix} U' & \Psi \\ U'' & \Psi' \end{vmatrix} = 1. \tag{35}$$

Formally,

$$\Psi(x) = U'(x) \int^x \frac{1}{[U'(s)]^2}\,\mathrm{d}s.$$

The above formulation is problematic because U' will generally have at least one zero. For any even f which is $2L$-periodic one has by variation of parameters that

$$\mathscr{L}^{-1}(f) = U'(x) \int_0^x \Psi(s)f(s)\,\mathrm{d}s - \Psi(x) \int_0^x U'(s)f(s)\,\mathrm{d}s + c_f \Psi(x), \tag{36}$$

where

$$c_f := \int_0^L U'(s)f(s)\,\mathrm{d}s - \frac{1}{2} U''(L) \frac{\langle \Psi, f\rangle}{\Psi'(L)}$$

is chosen so that

$$\frac{\mathrm{d}}{\mathrm{d}x}\mathscr{L}^{-1}(f)|_{x=L} = 0.$$

This final condition guarantees that $\mathscr{L}^{-1}(f)$ is $2L$-periodic. Using (36) it is straightforward to check that

$$\begin{aligned}
\langle \mathscr{L}^{-1}(1), 1\rangle &= \left(2U(L) - \frac{1}{2}U''(L)\frac{\langle \Psi, 1\rangle}{\Psi'(L)}\right)\langle \Psi, 1\rangle - 2\langle \Psi, U\rangle \\
\langle \mathscr{L}^{-1}(U), 1\rangle &= \frac{1}{2}U^2(L)\langle \Psi, 1\rangle - \frac{3}{2}\langle \Psi, U^2\rangle + \left(U(L) - \frac{1}{2}U''(L)\frac{\langle \Psi, 1\rangle}{\Psi'(L)}\right)\langle \Psi, U\rangle \\
\langle \mathscr{L}^{-1}(U), U\rangle &= -\langle \Psi, U^3\rangle + \left(U^2(L) - \frac{1}{2}U''(L)\frac{\langle \Psi, U\rangle}{\Psi'(L)}\right)\langle \Psi, U\rangle. \qquad (37)
\end{aligned}$$

Further simplification of these expressions is possible for specific values of p, and for a choice of focusing or defocusing. An example of such simplifications is found below, where we consider specific instances of the generalized KdV equation.

3.1 Modified KdV: $p = 2$

With $p = 2$ and for the focusing case, we simplify the expressions in (37) even more. Integration by parts and use of the fact that U and Ψ are even yields

$$\langle \Psi'', U\rangle = 2U(L)\Psi'(L) + \langle \Psi, U''\rangle.$$

Consequently,

$$0 = \langle \mathscr{L}\Psi, U\rangle = -2U(L)\Psi'(L) + \langle \Psi, \mathscr{L}U\rangle,$$

and since $\mathscr{L}U = -2U^3$

$$0 = -2U(L)\Psi'(L) - 2\langle \Psi, U^3\rangle \quad \Longrightarrow \quad \langle \Psi, U^3\rangle = -U(L)\Psi'(L).$$

Similarly,

$$0 = \langle \mathscr{L}\Psi, 1\rangle \quad \Longrightarrow \quad \langle \Psi, U^2\rangle = \frac{1}{3}c\langle \Psi, 1\rangle - \frac{2}{3}\Psi'(L).$$

Substitution of the above into (37) gives

$$\begin{aligned}
\langle \mathscr{L}^{-1}(1), 1\rangle &= \left(2U(L) - \frac{1}{2}U''(L)\frac{\langle \Psi, 1\rangle}{\Psi'(L)}\right)\langle \Psi, 1\rangle - 2\langle \Psi, U\rangle \\
\langle \mathscr{L}^{-1}(U), 1\rangle &= \frac{1}{2}(U^2(L) - c)\langle \Psi, 1\rangle + \Psi'(L) + \left(U(L) - \frac{1}{2}U''(L)\frac{\langle \Psi, 1\rangle}{\Psi'(L)}\right)\langle \Psi, U\rangle \\
\langle \mathscr{L}^{-1}(U), U\rangle &= U(L)\Psi'(L) + \left(U^2(L) - \frac{1}{2}U''(L)\frac{\langle \Psi, U\rangle}{\Psi'(L)}\right)\langle \Psi, U\rangle. \qquad (38)
\end{aligned}$$

In order to evaluate (34) one must compute the expressions in (38). This is done in subsequent sections for two explicit cases.

For the focusing case with $p = 2$ we work with the two families of stationary periodic solutions given by

$$U = \sqrt{2}\,\mu\,\mathrm{dn}(\mu x, k), \quad c = \mu^2(2 - k^2),$$
$$U = \sqrt{2}\,\mu k\,\mathrm{cn}(\mu x, k), \quad c = \mu^2(-1 + 2k^2).$$

Here $\mu > 0$, $0 \le k < 1$, and $\mathrm{dn}(y, k)$ and $\mathrm{cn}(y, k)$ are Jacobi elliptic functions. For the defocusing case we use

$$U = \sqrt{2}\,\mu k\,\mathrm{sn}(\mu x, k), \quad c = -\mu^2(1 + k^2).$$

The waves proportional to $\mathrm{cn}(y, k)$ and $\mathrm{sn}(y, k)$ have period $4K(k)/\mu$, where $K(k)$ is the complete elliptic integral of the first kind. The wave proportional to $\mathrm{dn}(y, k)$ has period $2K(k)/\mu$. The function $K(k)$ is smooth, strictly increasing, and satisfies the limits

$$\lim_{k\to 0+} K(k) = \frac{\pi}{2}, \quad \lim_{k\to 1-} K(k) = +\infty.$$

Thus, the inclusion of the parameter μ allows us to consider the entire family of cnoidal solutions for fixed L. In all that follows we set $\mu = 1$, as the effect of including μ is simply an overall eigenvalue scaling in all of the calculations listed below. In particular, the parameter μ is needed only to consider the entire family of cnoidal waves for a fixed period.

Define

$$\mathscr{L}_0 := -\frac{\mathrm{d}^2}{\mathrm{d}x^2} + 6k^2\,\mathrm{sn}^2(x, k).$$

Since $\mathscr{L}_0$ has a two-gap potential, when considering $\sigma(\mathscr{L}_0)$ on $L^2_{\mathrm{per}}([-2K(k), 2K(k)]; \mathbb{C})$ it is known that the first five eigenvalues are simple, and that all other eigenvalues have multiplicity two. In particular, the first five eigenvalues, as well as the associated eigenfunctions, are given by

$$\begin{aligned}
&\lambda_0 = 2(1 + k^2 - a(k));\ \phi_0(x) = k^2\,\mathrm{sn}^2(x, k) - \frac{1}{3}(1 + k^2 + a(k)),\\
&\lambda_1 = 1 + k^2;\ \phi_1(x) = \partial_x\,\mathrm{sn}(x, k),\\
&\lambda_2 = 1 + 4k^2;\ \phi_2(x) = \partial_x\,\mathrm{cn}(x, k),\\
&\lambda_3 = 4 + k^2;\ \phi_3(x) = \partial_x\,\mathrm{dn}(x, k),\\
&\lambda_4 = 2(1 + k^2 + a(k));\ \phi_4(x) = k^2\,\mathrm{sn}^2(x, k) - \frac{1}{3}(1 + k^2 - a(k)),
\end{aligned} \tag{39}$$

where $a(k) := \sqrt{1-k^2+k^4}$ [17]. The following proposition is useful for all subsequent calculations. As a consequence of Proposition 1 the evaluation of $\langle \mathscr{L}^{-1}(1), 1\rangle$ will be straightforward.

Proposition 1. *For* $j \notin \{0, 4\}$, $\langle \phi_j, 1\rangle = 0$.

Proof. An observation of (39) reveals that the result holds for $j = 1, 2, 3$. Now consider $j \geq 5$. Upon using the fact that $\mathscr{L}_0\phi_j = \lambda_j\phi_j$ and integrating both sides over one period one sees that

$$\lambda_j\langle \phi_j, 1\rangle = 6\langle \phi_j, k^2 \operatorname{sn}^2(x,k)\rangle.$$

Set $b(k) := (1+k^2-a(k))/3$. Using the representation of $\phi_4(x)$ given in (39) allows one to rewrite the above as

$$(\lambda_j - 6b(k))\langle \phi_j, 1\rangle = 6\langle \phi_j, \phi_4\rangle = 0,$$

where the second inequality follows from the orthogonality of the eigenfunctions. Since $\lambda_j - 6b(k) > 0$ for $j \geq 5$, the desired conclusion follows. □

3.1.1 Focusing mKdV: Solution $\sqrt{2}\,\mathrm{dn}(x,k)$

The wave is given by

$$U(x;k) := \sqrt{2}\,\mathrm{dn}(x,k), \quad c(k) := 2-k^2. \tag{40}$$

The fundamental period of $U(x;k)$ is $2K(k)$, so that $L = K(k)$. Upon using the identity $1 - k^2\,\mathrm{dn}^2(x,k) = k^2\,\mathrm{sn}^2(x,k)$ one finds that the linearization around U yields the operator

$$\mathscr{L} := -\frac{\mathrm{d}^2}{\mathrm{d}x^2} + c - 3U^2 = \mathscr{L}_0 - (4+k^2). \tag{41}$$

The spectrum of $\mathscr{L}$ is derived from that of $\mathscr{L}_0$ via $\lambda_j \mapsto \lambda_j - (4+k^2)$ in (39). Set $\tilde{\lambda}_j := \lambda_j - (4+k^2)$. Since the fundamental period is $2K(k)$, the eigenvalues associated with λ_1, λ_2 are not relevant, as the associated eigenfunctions have fundamental period $4K(k)$. Consequently, $\mathrm{n}(\mathscr{L}) = 1$ for all $k \in [0, 1)$.

As a consequence of Proposition 1 one has that for any $0 \leq k < 1$,

$$\langle \mathscr{L}^{-1}(1), 1\rangle = \frac{\langle \phi_0, 1\rangle^2}{\tilde{\lambda}_0} + \frac{\langle \phi_4, 1\rangle^2}{\tilde{\lambda}_4}. \tag{42}$$

Using the explicit expressions given in (39), and using the identities

$$\int_{-K(k)}^{K(k)} k^2 \operatorname{sn}^2(x,k)\,\mathrm{d}x = 2(K(k)-E(k))$$
$$\int_{-K(k)}^{K(k)} k^4 \operatorname{sn}^4(x,k)\,\mathrm{d}x = \frac{2}{3}[(2+k^2)K(k)-2(1+k^2)E(k)], \tag{43}$$

we have explicitly

$$\langle \mathscr{L}^{-1}(1),1\rangle = \frac{2K(k)-(2+k^2)E(k)}{k^4} > 0, \tag{44}$$

for $k > 0$. This inequality is easily verified using the series expansions of $K(k)$ and $E(k)$ [39]. As a consequence, it is now known that $\mathrm{n}(\boldsymbol{J}) = 0$.

In order to complete the calculation, we compute $\boldsymbol{D}$. Recall that formally

$$\Psi(x) = U'(x)\int^x \frac{1}{[U'(s)]^2}\,\mathrm{d}s.$$

Upon using the identities

$$\frac{1}{[\partial_x \operatorname{dn}(x;k)]^2} = \frac{1}{k^4}\left(\frac{1}{\operatorname{sn}^2(x;k)}+\frac{1}{\operatorname{cn}^2(x;k)}\right),$$
$$\frac{\partial}{\partial x}\frac{\operatorname{sn}(x;k)}{\operatorname{cn}(x;k)} = \frac{\operatorname{dn}(x;k)}{\operatorname{cn}^2(x;k)},\quad \frac{\partial}{\partial x}\frac{\operatorname{cn}(x;k)}{\operatorname{sn}(x;k)} = -\frac{\operatorname{dn}(x;k)}{\operatorname{sn}^2(x;k)},$$

integrating by parts, and normalizing with (35) to get $\Psi(0) = 1/(\sqrt{2}k^2)$, one eventually gets

$$\Psi(x) = \frac{1}{\sqrt{2}k^2}\left(\frac{1-2\operatorname{sn}^2(x;k)}{\operatorname{dn}(x;k)} - k^2\operatorname{sn}(x;k)\operatorname{cn}(x;k)\int_0^x \frac{1-2\operatorname{sn}^2(t;k)}{\operatorname{dn}^2(t;k)}\,\mathrm{d}t\right).$$

Differentiating and evaluating at $x = L = K(k)$ yields

$$\Psi'(L) = \frac{1}{2}U(L)\int_0^L \frac{1-2\operatorname{sn}^2(t;k)}{\operatorname{dn}^2(t;k)}\,\mathrm{d}t.$$

Upon integrating by parts one has that

$$\langle\Psi,1\rangle = \frac{1}{k^2}U(L)\int_0^L \frac{1-2\operatorname{sn}^2(t;k)}{\operatorname{dn}^2(t;k)}\,\mathrm{d}t \quad\Longrightarrow\quad \Psi'(L) = \frac{1}{2}k^2\langle\Psi,1\rangle.$$

Since $U''(L) = k^2 U(L)$, one can now rewrite (38) as

$$\begin{aligned}\langle \mathscr{L}^{-1}(1), 1\rangle &= U(L)\langle \Psi, 1\rangle - 2\langle \Psi, U\rangle,\\ \langle \mathscr{L}^{-1}(U), 1\rangle &= \frac{1}{2}(U^2(L) - c + k^2)\langle \Psi, 1\rangle = 0,\\ \langle \mathscr{L}^{-1}(U), U\rangle &= U^2(L)\langle \Psi, U\rangle + \frac{1}{2}k^2 U(L)\langle \Psi, 1\rangle - U(L)\frac{\langle \Psi, U\rangle^2}{\langle \Psi, 1\rangle}.\end{aligned} \tag{45}$$

Using the results of (45) in (34) one sees that

$$\boldsymbol{D} = \langle \mathscr{L}^{-1}(U), U\rangle.$$

An expression for $\langle \Psi, 1\rangle$ is given above. Upon integrating by parts and simplifying with (43) one sees that

$$\langle \Psi, U\rangle = \frac{1}{k^4}\left(2E(k) - (2-k^2)K(k) + k^2(1-k^2)\int_0^L \frac{1 - 2\,\mathrm{sn}^2(t;k)}{\mathrm{dn}^2(t;k)}\,\mathrm{d}t\right).$$

In order to complete the calculation the integral must be computed. Using the fact that the integrand is even in t and $2L$-periodic one has that

$$\int_0^L \frac{1 - 2\,\mathrm{sn}^2(t;k)}{\mathrm{dn}^2(t;k)}\,\mathrm{d}t = \int_L^{2L} \frac{1 - 2\,\mathrm{sn}^2(t;k)}{\mathrm{dn}^2(t;k)}\,\mathrm{d}t.$$

Since

$$\mathrm{sn}(t+L;k) = \frac{\mathrm{sn}(t;k)}{\mathrm{dn}(t;k)}, \quad \mathrm{dn}(t+L;k) = \frac{\sqrt{1-k^2}}{\mathrm{dn}(t;k)},$$

one has

$$\int_L^{2L} \frac{1 - 2\,\mathrm{sn}^2(t;k)}{\mathrm{dn}^2(t;k)}\,\mathrm{d}t = \frac{1}{1-k^2}\int_0^L [-1 + (2-k^2)\,\mathrm{sn}^2(t;k)]\,\mathrm{d}t.$$

The latter integral can be computed with (43) to finally get

$$\int_0^L \frac{1 - 2\,\mathrm{sn}^2(t;k)}{\mathrm{dn}^2(t;k)}\,\mathrm{d}t = \frac{1}{k^2(1-k^2)}(2(1-k^2)K(k) - (2-k^2)E(k)).$$

Consequently, one concludes that

$$\langle \Psi, 1\rangle = \frac{U(L)}{k^4(1-k^2)}(2(1-k^2)K(k) - (2-k^2)E(k)), \quad \langle \Psi, U\rangle = -\frac{K(k) - E(k)}{k^2}.$$

Plugging these expressions into (45) and evaluating the resulting expression yields

$$\langle \mathcal{L}^{-1}(U), U\rangle = -\frac{(1-k^2)K^2(k) - E^2(k)}{2(1-k^2)K(k) - (2-k^2)E(k)} < 0,$$

for $k > 0$. Here the inequality follows as before, using the series expansions of $E(k)$ and $K(k)$ [39] to establish that the denominator has a definite sign. The definite sign of the numerator follows similarly from $E(k) > \sqrt{1-k^2}K(k)$, for $k > 0$.

In conclusion, from Theorem 1 it is seen that $k_\mathrm{r} = k_\mathrm{c} = k_\mathrm{i}^- = 0$. Recalling that the scaling μ was unimportant in the above calculations, and applying Theorem 2, the following result is obtained.

Theorem 3. *Consider the solution* $U_\mu(x) = \sqrt{2}\,\mu\,\mathrm{dn}(\mu x, k)$ *on* $L^2_{\mathrm{per}}([-K(k)/\mu, K(k)/\mu];\mathbb{R})$ *endowed with the natural inner-product. For a given* $\varepsilon > 0$ *sufficiently small there is a* $\delta > 0$ *such that if* $\|u(0) - U_\mu\| < \delta$ *with* $\overline{u(0)} = \overline{U_\mu}$, *then*

$$\inf_{\omega\in\mathbb{R}} \|u(t) - U_\mu(\cdot + \omega)\| < \varepsilon.$$

Thus the dn solution of the focusing mKdV equation is orbitally stable with respect to periodic perturbations of the same period for all values of the elliptic modulus.

Remark 6. The result of Theorem 3 was recently established in [1]; however, the proof there is different than that presented here. In particular, the proof in [1] fails when $\mathrm{n}(\mathcal{L}) = 2$, which will be the case in next problem.

3.1.2 Focusing mKdV: Solution $\sqrt{2}k\,\mathrm{cn}(x,k)$

Next set

$$U(x;k) := \sqrt{2}\,k\,\mathrm{cn}(x,k), \quad c(k) := -1 + 2k^2. \tag{46}$$

For this second case, the fundamental period of $U(x;k)$ is $4K(k)$, so that now $L = 2K(k)$. The linearization around U gives the operator

$$\mathcal{L} := -\frac{\mathrm{d}^2}{\mathrm{d}x^2} + c - 3U^2 = \mathcal{L}_0 - (1+4k^2). \tag{47}$$

The spectrum of $\mathcal{L}$ is derived from that of $\mathcal{L}_0$ via $\lambda_j \mapsto \lambda_j - (1+4k^2)$ in (39). Set $\tilde{\lambda}_j := \lambda_j - (1+4k^2)$. Since the fundamental period is $4K(k)$, all of the eigenvalues in (39) are relevant. Consequently, $\mathrm{n}(\mathcal{L}) = 2$ for all k.

The result of Proposition 1 still holds. Thus, $\langle \mathscr{L}^{-1}(1), 1\rangle$ is still given by (42) with the appropriate substitution. Using the explicit expressions given in (39), and using (43), allows one to explicitly compute

$$\langle \mathscr{L}^{-1}(1), 1\rangle = -4(2E(k) - K(k)) = -4\frac{\mathrm{d}}{\mathrm{d}k}kE(k). \tag{48}$$

Since

$$\lim_{k\to 0^+} \frac{\mathrm{d}}{\mathrm{d}k}kE(k) = E(0) > 0, \quad \lim_{k\to 1^-} \frac{\mathrm{d}}{\mathrm{d}k}kE(k) = -\infty, \quad \frac{\mathrm{d}^2}{\mathrm{d}k^2}kE(k) < 0,$$

there is a unique $k^* \sim 0.909$ such that $\langle \mathscr{L}^{-1}(1), 1\rangle = 0$ for $0 \le k < k^*$. In conclusion,

$$\mathrm{n}(\langle \mathscr{L}^{-1}(1), 1\rangle) = \begin{cases} 1, & 0 \le k < k^*, \\ 0, & k^* < k < 1. \end{cases} \tag{49}$$

Now we compute $\boldsymbol{D}$ in order to complete the calculation. The calculation is similar to that presented in the previous subsection, and hence only the highlights will be given. One has

$$\Psi(x) = \frac{1}{\sqrt{2}\,k}\left(\mathrm{cn}(x;k) - k^2\,\mathrm{sn}(x;k)\,\mathrm{dn}(x;k)\int_0^x \frac{2-\mathrm{sn}^2(t;k)}{\mathrm{dn}^2(t;k)}\,\mathrm{d}t\right),$$

from which one gets that

$$\Psi'(L) = -\frac{1}{2}U(L)\int_0^L \frac{2-\mathrm{sn}^2(t;k)}{\mathrm{dn}^2(t;k)}\,\mathrm{d}t = -\frac{1}{2}\langle \Psi, 1\rangle.$$

The second equality is again found by integrating by parts. The analogue of (45) with $U''(L) = -U(L)$ is now

$$\begin{aligned} &\langle \mathscr{L}^{-1}(1), 1\rangle = U(L)\langle \Psi, 1\rangle - 2\langle \Psi, U\rangle, \\ &\langle \mathscr{L}^{-1}(U), 1\rangle = 0, \\ &\langle \mathscr{L}^{-1}(U), U\rangle = U^2(L)\langle \Psi, U\rangle - \frac{1}{2}U(L)\langle \Psi, 1\rangle - U(L)\frac{\langle \Psi, U\rangle^2}{\langle \Psi, 1\rangle}; \end{aligned} \tag{50}$$

hence, as in the previous subsection we conclude with

$$\boldsymbol{D} = \langle \mathscr{L}^{-1}(U), U\rangle.$$

Note that the potential singularity at $k = k^*$ has been removed.

Calculating as before one has that

$$\int_0^L \frac{2-\mathrm{sn}^2(t;k)}{\mathrm{dn}^2(t;k)}\,\mathrm{d}t = 2\int_{L/2}^L \frac{2-\mathrm{sn}^2(t;k)}{\mathrm{dn}^2(t;k)}\,\mathrm{d}t$$
$$= \frac{2}{k^2(1-k^2)}((1-k^2)K(k)-(1-2k^2)E(k)),$$

from which one eventually gets that

$$\langle \Psi, 1\rangle = \frac{2U(L)}{k^2(1-k^2)}((1-k^2)K(k)-(1-2k^2)E(k)), \quad \langle \Psi, U\rangle = \frac{2}{1-k^2}E(k).$$

The second equality requires the use of (43) and the identity

$$\frac{\partial}{\partial x}\frac{\mathrm{cn}(x;k)}{\mathrm{dn}(x;k)} = -(1-k^2)\frac{\mathrm{sn}(x;k)}{\mathrm{dn}^2(x;k)}.$$

Substituting the above into the expression for $\langle \mathscr{L}^{-1}(U), U\rangle$ yields a negative sign for $k > 0$, as before. Thus $\mathrm{n}(\boldsymbol{D}) = 1$.

Upon using the result of Theorem 1 one has that

$$k_\mathrm{r} + k_\mathrm{c} + k_\mathrm{i}^- = \begin{cases} 0, & 0 \le k < k^*, \\ 1, & k^* < k < 1. \end{cases}$$

Since k_i^- and k_c are even, it then follows that $k_\mathrm{c} = k_\mathrm{i}^- = 0$, but

$$k_\mathrm{r} = \begin{cases} 0, & 0 \le k < k^*, \\ 1, & k^* < k < 1. \end{cases} \tag{51}$$

Hence, for $k < k^*$ the cn wave of period $2L$ is a constrained minimizer and thus stable with respect to periodic perturbations of period $2L$. On the other hand, the cn wave is unstable for $k > k^*$.

Theorem 4. *Consider the solution* $U_\mu(x) = \sqrt{2}\,\mu k\,\mathrm{cn}(\mu x, k)$ *on* $L^2_{\mathrm{per}}([-2K(k)/\mu, 2K(k)/\mu];\mathbb{R})$ *endowed with the natural inner-product. Define* k^* *as the unique value satisfying* $K(k^*) = 2E(k*)$. *If* $k < k^* \sim 0.909$, *then for a given* $\varepsilon > 0$ *sufficiently small there is a* $\delta > 0$ *such that if* $\|u(0) - U_\mu\| < \delta$ *with* $\overline{u(0)} = \overline{U_\mu}$, *then*

$$\inf_{\omega\in\mathbb{R}} \|u(t) - U_\mu(\cdot + \omega)\| < \varepsilon.$$

Thus, the cn solution of the focusing mKdV equation is orbitally stable with respect to periodic perturbations of the same period, provided the elliptic modulus $k < k^*$. If $k > k^*$, then the wave is unstable.

Remark 7. Since $\mathrm{n}(\boldsymbol{D}) = 1$ for all k, the spectral structure at $k = k^*$ is such that the origin is an eigenvalue of $\mathscr{J}\mathscr{L}|_{H_1}$ of algebraic multiplicity four and geometric multiplicity two; furthermore, there are two nontrivial Jordan blocks.

Remark 8. If the operator $\mathscr{J}$ were nonsingular, then $\mathrm{n}(\mathscr{L}) = 2$ with $\mathrm{n}(\boldsymbol{D}) = 1$ would imply that $k_{\mathrm{r}} = 1$ for all values of k. This example shows that the modification of the result of [35] presented in Theorem 1 is indeed necessary, and is not simply a technical detail.

Remark 9. Without loss of generality assume that $\mu = 1$. A numerical calculation of $\sigma(\mathscr{J}\mathscr{L})$ shows that the cnoidal wave for focusing mKdV is unstable for any value of k in the space $L^2_{\mathrm{per}}([-2nK(k), 2nK(k)]; \mathbb{R})$ for any integral $n \geq 2$. This is illustrated in Fig. 1. The spectra illustrated there were computed using SpectrUW2.0, using 20 Fourier modes and 400 equally-spaced Floquet exponents, with $P = 1$ [10]. Note the different scalings of the different figures. Since the density of the eigenvalues computed is not uniform, there are some parts of the spectrum where we have less information than elsewhere. Nevertheless, between what is known theoretically about the spectra and what we observe numerically, we feel the statements made below are safe inferences based on the numerical results plotted. For all plots in Fig. 1, the entire imaginary axis is part of the spectrum, whereas the real axis is not (except for the origin, and perhaps two other points, see below). It should also be pointed out that all plots are consistent with the results of Bronski and Johnson [7]: at the origin, the spectrum should generically consist of either the imaginary axis (with multiplicity three), or of three distinct components, all intersecting at the origin. In our case, the second scenario unfolds for all but one value of k, see below. From the numerical results, it appears that all self-intersection points of the spectrum occur at eigenvalues corresponding to eigenfunctions with period $4K(k)$.

- For $k < k^*$, in addition to the imaginary axis, the spectrum consists of the boundary of two lobes, each cut in half by the imaginary axis. Of course, the lobes are symmetric with respect to the real and imaginary axes. The lobes touch at the origin. The boundary of the upper lobe has a second intersection point with the imaginary axis which approaches the origin as $k \to k^{*-}$. One of the eigenfunctions corresponding to this point is periodic with period $4K(k)$. It is the first (in terms of distance to the origin) non-zero eigenvalue on the imaginary axis corresponding to a period $4K(k)$ eigenfunction. This case is illustrated in Fig. 1a–c. For the last panel $k = 0.908$, very close to $k^* \sim 0.909$. For $k < k^*$, the origin is the only point on the real axis that is in the spectrum.
- It appears that three eigenvalues with periodic eigenfunctions collide at the origin for $k = k^*$. For $k < k^*$ all three eigenvalues are on the imaginary axis. For $k > k^*$ all three are on the real axis, giving rise to unstable and stable directions. For $k = k^*$, it appears the spectrum near the origin appears to consists of more

than three components (counting multiplicities), leading to the conclusion that for $k = k^*$ the discriminant of Bronski and Johnson [7] is zero. The case of a zero discriminant is not discussed in [7].

- For $k > k^*$, in addition to the imaginary axis, the spectrum consists of additional curves, bounding a total of six regions in $\mathbb{C}$, three in the right-half plane, three in the left-half plane. We describe the ones in the right-half plane. Using the left-right symmetry of the spectrum completes the picture. There is a region touching the origin, which has a point furthest from the origin, which is a self-intersection point of the spectrum. This point corresponds to an eigenfunction with period $4K(k)$. It is the unstable eigenvalue given by (48). To the right of this point are two more lobes, one above the real axis, one below. For k greater than but close to k^*, the region touching the origin is small, and the two remaining lobes are relatively large. This is illustrated in Fig. 1d–e. Figure 1e is a zoom-in of Fig. 1d near the origin. Due to the low density of computed eigenvalues the left (right) side of the right (left) upper and lower lobes is not visible in Fig. 1e. For Fig. 1f, $k = 0.95$, and the outer lobes have decreased in size, whereas the regions touching the origin have grown.

In summary, the numerical results show that the cn solution of the mKdV equation is unstable with respect to perturbations of period $4nK(k)$ for any $n \geq 2$, even if it is stable with respect to perturbations of period $4K(k)$.

3.1.3 Defocusing mKdV: Solution $\sqrt{2}k\,\mathrm{sn}(x,k)$

In the defocusing regime there exists a branch of solutions

$$U(x;k) := \sqrt{2}k\,\mathrm{sn}(x,k), \quad c(k) := -(1+k^2). \tag{52}$$

The linearization around U yields the operator

$$\mathscr{L} := -\frac{\mathrm{d}^2}{\mathrm{d}x^2} + c + 3U^2 = \mathscr{L}_0 - (1+k^2). \tag{53}$$

The spectrum of $\mathscr{L}$ is derived from that of $\mathscr{L}_0$ via $\lambda_j \mapsto \lambda_j - (1+k^2)$ in (39). Setting $\tilde{\lambda}_j := \lambda_j - (1+k^2)$, one sees that $\mathrm{n}(\mathscr{L}) = 1$. Arguing as in the previous cases gives that $\langle \mathscr{L}^{-1}(1), 1\rangle < 0$ (the explicit calculations are left for the interested reader). Since $\mathrm{n}(\mathscr{L}) = \mathrm{n}(\boldsymbol{J}) = 1$, by applying Theorem 1 one has the following:

Theorem 5. *Consider the solution* $U_\mu(x) = \sqrt{2}\,\mu k\,\mathrm{sn}(\mu x, k)$ *on* $L^2_{\mathrm{per}}([-2K(k)/\mu, 2K(k)/\mu];\mathbb{R})$ *endowed with the natural inner-product. For a given* $\varepsilon > 0$ *sufficiently small there is a* $\delta > 0$ *such that if* $\|u(0) - U_\mu\| < \delta$ *with* $\overline{u(0)} = \overline{U_\mu}$, *then*

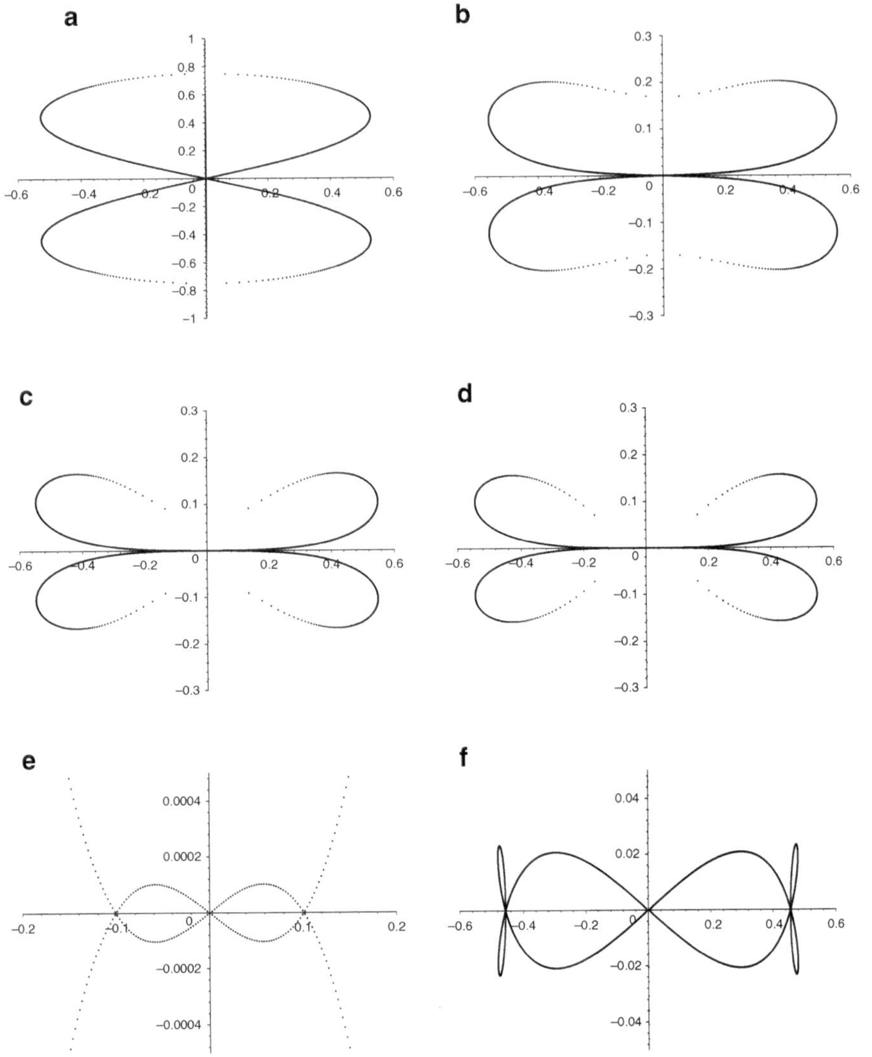

Fig. 1 Numerical computations of $\sigma(\mathcal{J}\mathcal{L})$ for the cnoidal wave of focusing mKdV using the space $L^2_{\text{per}}([-2nK(k)/\mu, 2nK(k)/\mu]; \mathbb{R})$ with $n = 400$. For a detailed explanation, see the main text. (**a**) $k = 0.7 < k^*$. (**b**) $k = 0.9 < k^*$. (**c**) $k = 0.908 < k^*$. (**d**) $k = 0.91 > k^*$. (**e**) $k = 0.91 > k^*$ (zoom). (**f**) $k = 0.95 > k^*$

$$\inf_{\omega \in \mathbb{R}} \|u(t) - U_\mu(\cdot + \omega)\| < \varepsilon.$$

Thus the sn solution of the defocusing mKdV equation is orbitally stable with respect to perturbations of the same period.

Remark 10. The comments of Remark 8 also apply here, except now $\mathrm{n}(\mathscr{L}) = 1$ with $\mathrm{n}(\boldsymbol{D}) = 0$.

Remark 11. Using the integrability of the mKdV equation extensively, more explicit statements are possible, see [15], where the stability and instability of all stationary solutions of the mKdV equation is discussed. For instance, it is shown there that the stationary solutions of the defocusing equation are orbitally stable with respect to subharmonic perturbations. For the focusing equation, the stability with respect to these perturbations depends on whether the solutions have higher (dn-like solutions; stable) or lower (cn-like solutions; unstable) energy than the soliton solutions which act as a separatrix between these two classes in the phase plane of the stationary equation (31).

3.2 Perturbative Results: $p \geq 3$

As a consequence of [26, Theorem 5.6] it is known that for any $p \neq 2$ the periodic waves which are analogous to the dn-wave when $p = 2$ are orbitally stable if they are sufficiently small perturbations of the appropriate nonzero constant state. For this reason, these waves are not discussed here. Instead, we focus on the orbital stability of the small waves analogous to the cn-wave (focusing) and sn-wave (defocusing) when $p \in \mathbb{N}_0$.

After rescaling x, t, u (28) can be rewritten as

$$u_t + (\omega u_{xx} + u + \varepsilon\delta u^{p+1})_x = 0, \quad \delta \in \{-1, +1\}, \tag{54}$$

where the interest will be on the orbital stability of 2π-periodic solutions for $0 < \varepsilon \ll 1$. Note that the wave speed has been fixed. The free parameter ω allows us to perturbatively construct the desired steady-state solutions via the Poincaré-Lindstedt method.

First consider the existence problem for the steady-state solutions, i.e.,

$$\omega u'' + u + \varepsilon\delta u^{p+1} = a_0, \quad a_0 \in \mathbb{R}. \tag{55}$$

Looking for even solutions yields the expansions

$$u(x) = U(x) \sim \underbrace{a_0 + b\cos x}_{U_0} + \varepsilon U_1(x), \quad \omega \sim 1 + \varepsilon\omega_1.$$

Setting $\mathscr{L}_0 := -(\partial_x^2 + 1)$, one sees that at $\mathscr{O}(\varepsilon)$,

$$\mathscr{L}_0 U_1 = 2\omega_1 \partial_x^2 U_0 + \delta U_0^{p+1}. \tag{56}$$

The standard solvability condition for (56) that removes the secular terms is

$$\omega_1 = \frac{\delta}{2\pi b^2}\langle U_0^{p+2} - a_0 U_0^{p+1}, 1\rangle. \tag{57}$$

Furthermore, using a finite cosine-series representation,

$$U_1(x) = -\frac{\delta}{2\pi}\langle U_0^{p+1}, 1\rangle + \sum_{j=2}^{p+1} c_j \cos(jx), \tag{58}$$

for suitably chosen constants c_j.

The linearization about the wave U gives the linear operator

$$\mathcal{L} \sim \mathcal{L}_0 + \varepsilon\mathcal{L}_\varepsilon; \quad \mathcal{L}_\varepsilon := -\omega_1\frac{\mathrm{d}^2}{\mathrm{d}x^2} - \delta(p+1)U_0^p(x). \tag{59}$$

The principal eigenvalue is given by $\lambda_0 = -1 + \mathcal{O}(\varepsilon)$, and the associated eigenfunction $\phi_0 = 1/\sqrt{2\pi} + \mathcal{O}(\varepsilon)$ satisfies $\langle \phi_0, 1\rangle = 1 + \mathcal{O}(\varepsilon)$. The next nonzero eigenvalue is given by $\lambda_1 = \varepsilon\lambda_\varepsilon + \mathcal{O}(\varepsilon^2)$, and the associated eigenfunction is

$$\phi_1 = \phi_1^0 + \varepsilon\phi_\varepsilon + \mathcal{O}(\varepsilon^2), \quad \phi_1^0 := \frac{1}{\sqrt{\pi}}\cos x.$$

Using regular perturbation theory results in

$$\begin{aligned}\lambda_\varepsilon &= \frac{1}{\pi}\langle \cos x, \mathcal{L}_\varepsilon(\cos x)\rangle \\ &= -\frac{\delta}{\pi b^2}\left(p\langle U_0^{p+2} - a_0U_0^{p+1}, 1\rangle - a_0(p+1)\langle U_0^{p+1} - a_0U_0^p, 1\rangle\right).\end{aligned} \tag{60}$$

A Maple-assisted calculation reveals that when $p = 1$, $\lambda_\varepsilon = 0$; otherwise, for p odd one has $\lambda_\varepsilon = -\delta a_0 f(a_0, b, p)$, where $f > 0$, while for p even $\lambda_\varepsilon = -\delta g(a_0, b, p)$, where again $g > 0$. The above calculations characterize $\sigma(\mathcal{L})$ in the following manner: if $\lambda_1 < 0$, then a left band edge of $\sigma(\mathcal{L})$ is at $\lambda = 0$; otherwise, the right band edge is at the origin (see Fig. 2). In conclusion, for $0 < \varepsilon \ll 1$,

$$\mathrm{n}(\mathcal{L}) = \begin{cases} 1, & \lambda_1 > 0 \\ 2, & \lambda_1 < 0. \end{cases} \tag{61}$$

Remark 12. For $p = 1$ one finds by continuing the perturbation expansion that

$$\lambda_1 \sim \left(2a_0^2 + \frac{11}{6}b^2 + b^3\right)\varepsilon^2 > 0.$$

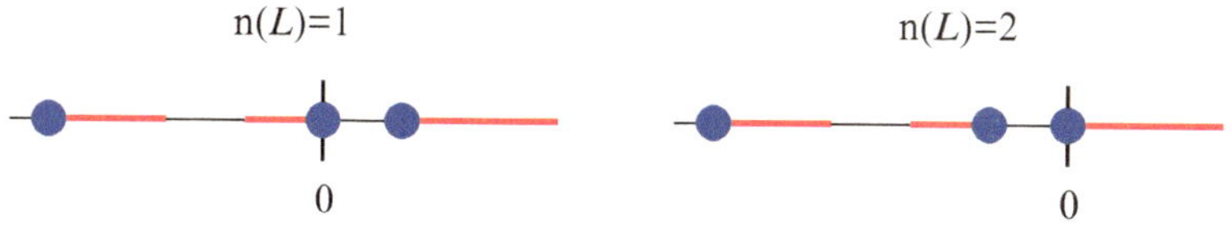

Fig. 2 The spectrum of the operator $\mathscr{L}$ for the perturbative solutions is marked with thick (red) lines. The eigenvalues associated with eigenfunctions which are 2π-periodic are marked with filled (blue) circles

Furthermore, note that $\lambda_\varepsilon \neq 0$ for all (a_0, b) if p is even, and for p odd it is true as long as $a_0 \neq 0$.

Next, we compute $\boldsymbol{J}$ and $\boldsymbol{D}$. Using the expansion of (59),

$$\mathscr{L}^{-1}(1) = -1 + \mathscr{O}(\varepsilon);$$

hence,

$$\langle \mathscr{L}^{-1}(1), 1\rangle = -2\pi + \mathscr{O}(\varepsilon) \quad \Longrightarrow \quad \mathrm{n}(\boldsymbol{J}) = 1.$$

Since

$$\langle \mathscr{L}^{-1}(U_0), 1\rangle = \langle \mathscr{L}^{-1}(1), U_0\rangle = -2\pi a_0 + \mathscr{O}(\varepsilon),$$

an examination of (34) reveals that the dominant term for $\boldsymbol{D}$ follows from the calculation of $\langle \mathscr{L}^{-1}(U_0), U_0\rangle$. This in turn requires an expansion for Ψ. First write

$$\tilde{\Psi} = \cos x + \varepsilon \Psi_\varepsilon + \mathscr{O}(\varepsilon^2).$$

Then

$$\mathscr{L}_0 \Psi_\varepsilon = -\mathscr{L}_\varepsilon \cos x,$$

which upon solving yields the existence of constants e_j such that

$$\Psi_\varepsilon = \frac{1}{2}\lambda_\varepsilon x \sin x + e_0 + \sum_{j=2}^{p+1} e_j \cos(jx).$$

Setting

$$\Psi = \frac{1}{b}\left(\cos x + \varepsilon\left([U_1''(0) - \Psi_\varepsilon(0)]\cos x + \Psi_\varepsilon\right) + \mathscr{O}(\varepsilon^2)\right)$$

yields the normalization

$$U_0''(0)\Psi(0) = -1 + \mathcal{O}(\varepsilon^2).$$

Since

$$\Psi'(\pi) = -\frac{\pi}{2b}\lambda_\varepsilon\varepsilon + \mathcal{O}(\varepsilon^2),$$

an examination of the last line of (37) yields that

$$\langle \mathcal{L}^{-1}(U_0), U_0\rangle = \frac{b^2\pi}{\lambda_1} + \mathcal{O}(1).$$

In conclusion, for $p \geq 2$ and $\varepsilon > 0$ sufficiently small one has

$$\mathrm{n}(\boldsymbol{D}) = \begin{cases} 0, & \lambda_1 > 0, \\ 1, & \lambda_1 < 0. \end{cases} \tag{62}$$

Upon using (61) and (62), along with the fact that $\mathrm{n}(\boldsymbol{J}) = 1$, one can conclude via Theorem 1 that $k_{\mathrm{r}} = k_{\mathrm{c}} = k_{\mathrm{i}}^- = 0$. By Theorem 2 this yields:

Theorem 6. *Let $p \geq 2$, and let (a_0, b) be such that λ_ε given in (60) is nonzero. Consider the solution $U_0(x) = a_0 + b\cos x + \mathcal{O}(\varepsilon)$ to (54) on the space $L^2_{\mathrm{per}}([-\pi,\pi];\mathbb{R})$ endowed with the natural inner product. If $\varepsilon > 0$ is sufficiently small, then U_0 is orbitally stable.*

4 Transverse Instabilities of gKP-I

Recently Rousset and Tzvetkov [44] considered the parameter-dependent eigenvalue problem

$$\mathcal{J}(\ell)\mathcal{L}(\ell)u = \lambda u, \quad \ell \in [0, +\infty). \tag{63}$$

They considered the situation where $\mathcal{J}(\ell)$ is skew symmetric and invertible for all ℓ, and where the self-adjoint operator $\mathcal{L}(\ell)$ satisfies the assumptions:

1. there is an $L > 0$ and $\alpha > 0$ such that $\mathcal{L}(\ell) \geq \alpha\mathbb{1}$ for $\ell \geq L$,
2. if $\ell_1 > \ell_2$, then $\mathcal{L}(\ell_1) > \mathcal{L}(\ell_2)$; furthermore, if for some $\ell > 0$ the operator $\mathcal{L}(\ell)$ has a nontrivial kernel, then $\langle \mathcal{L}'(\ell)\phi, \phi\rangle > 0$ for any $\phi \in \ker(\mathcal{L}(\ell))$ (here $\mathcal{L}'(\ell)$ is the derivative of $\mathcal{L}(\ell)$ with respect to ℓ),
3. $\mathrm{n}(\mathcal{L}(0)) = 1$.

There is also an assumption on the essential spectrum; however, we do not state it here, since it is not relevant for our considerations. Under these assumptions Rousset and Tzvetkov [44] showed there is an $\ell > 0$ such that (63) has a bounded solution for $\operatorname{Re}\lambda > 0$. In other words, they developed an instability criterion. The goal in this section is to restate the instability result in terms of the index theory from our previous results. The theoretical result will be applied to the spectral stability study of periodic waves of the generalized Kadomtsev-Petviashvili equation with strong surface tension (gKP-I).

Consider the theory and arguments leading to the result of Theorem 1. They are clearly independent of any parameter dependence of the operators. For the parameter-dependent problem it is often the case that the operator $\mathscr{L}(\ell)$ is invertible except for a finite number of values ℓ; hence, without loss of generality, we assume that $\mathscr{L}(\ell)$ is nonsingular. We further assume that $\mathscr{J}(\ell)$ is nonsingular with bounded inverse. The theorem can be restated to say:

Theorem 7. *Consider the eigenvalue problem* (63). *Suppose that* $\mathscr{L}(\ell)$ *satisfies Assumption 3 for each* $\ell \geq 0$, *and further assume that* $\mathscr{J}(\ell)$ *has bounded inverse for each* $\ell \geq 0$. *If* $\mathscr{L}(\ell)$ *is nonsingular, then for each* $\ell \geq 0$,

$$k_{\mathrm{r}} + k_{\mathrm{c}} + k_{\mathrm{i}}^{-} = \mathrm{n}(\mathscr{L}(\ell)).$$

In particular, if $\mathrm{n}(\mathscr{L})$ *is odd, then* $k_{\mathrm{r}} \geq 1$.

Remark 13. The instability proof used in [44] required that $\mathrm{n}(\mathscr{L}(0)) = 1$. The other assumptions are clearly not necessary for the statement of Theorem 7; however, if it is the case that $\mathrm{n}(\mathscr{L}(\ell)) = 0$ for $\ell \geq L$, then there will be no unstable spectrum for $\ell \geq L$.

Let us apply the result of Theorem 7 to the study of spatially periodic waves of the generalized KP-I equation (gKP-I), which is a two-dimensional version of the gKdV given by

$$u_t = \partial_x(-u_{xx} + cu \mp u^{p+1}) + \partial_x^{-1} u_{yy}. \tag{64}$$

This problem was recently studied in [21] as an application of the results of [44] (also see [29] for an Evans function analysis). We consider solutions to (64) that are spatially periodic in both x and y; in particular, we assume that

$$u(x, y + 2\pi/\ell) = u(x, y), \quad \ell > 0. \tag{65}$$

The period in x, say $2L_x$, will be determined by one of the y-independent solutions considered in the previous sections. In order for the operator ∂_x^{-1} to make sense, it must be assumed that

$$\int_{-L_x}^{L_x} u_{yy}(x, y)\,\mathrm{d}x = 0, \tag{66}$$

i.e., in the language of the previous sections, it must be the case that

$$u_{yy} \in H_{1,x},$$

where the notation $H_{1,x}$ denotes the fact that the spatial average must be zero in the x-direction only.

Let $U(x)$ represent a y-independent spatially periodic steady-state solution, i.e.,

$$-U'' + cU \mp U^{p+1} = 0, \quad U(x + 2L_x) = U(x),$$

and set the linearization about this wave to be

$$\mathcal{L}_0 = -\partial_x^2 + c \mp (p + 1)U^p.$$

Writing the perturbation as

$$u(x, y) = U(x) + v(x, y),$$

the linearized eigenvalue problem is

$$\partial_x \mathcal{L}_0 v + \partial_x^{-1} v_{yy} = \lambda v, \quad v(x + 2L_x, y) = v(x, y). \tag{67}$$

Using the fact that the perturbation is $2\pi/\ell$-periodic in y implies that we can write

$$v(x, y) = \sum_{\ell=-\infty}^{+\infty} v_\ell(x) \mathrm{e}^{\mathrm{i}\ell y}.$$

Plugging this expansion into (67) yields the parameter dependent eigenvalue problem

$$\partial_x \mathcal{L}_0 v_\ell - \ell^2 \partial_x^{-1} v_\ell = \lambda v_\ell, \quad v_\ell(x + 2L_x) = v_\ell(x). \tag{68}$$

The system (68) is not yet in the desired form of (63). Write

$$v_\ell = \partial_x u, \tag{69}$$

so that (68) becomes

$$-\partial_x \mathcal{L}_0 \partial_x u + \ell^2 u = -\lambda \partial_x u, \quad u(x + 2L_x) = u(x). \tag{70}$$

Equating $\mathcal{J} = -\partial_x^{-1}$ yields

$$\mathcal{J} \underbrace{\left(-\partial_x \mathcal{L}_0 \partial_x + \ell^2\right)}_{\mathcal{L}(\ell)} u = \lambda u, \quad u(x + 2L_x) = u(x). \tag{71}$$

This is the desired Hamiltonian form for the eigenvalue equation.

As for (71), some care must be taken regarding the space in which it is being considered. The transformation (69), and the periodicity condition of (68), require that $u \in H_{1,x}$. As we have already seen, in this space ∂_x has bounded inverse; hence, for (71) $\mathcal{J}$ is a skew-symmetric operator with bounded inverse. The operator $\mathcal{L}(\ell)$ on the space of $2L_x$-periodic functions is self adjoint and has a compact resolvent. Hence, the same is true for the operator $\mathcal{L}(\ell)|_{H_{1,x}}$. Assuming that $\mathcal{L}(\ell)|_{H_{1,x}}$ is nonsingular, by Theorem 7 we have for the eigenvalue problem (71)

$$k_{\mathrm{r}} + k_{\mathrm{c}} + k_{\mathrm{i}}^{-} = \mathrm{n}\left(\mathcal{L}(\ell)|_{H_{1,x}}\right). \tag{72}$$

We wish to make (72) more definitive. Let us first consider the problem (71) for $\ell \ll 1$. With $\ell = 0$ the problem is

$$\mathcal{L}_0\partial_x u = \lambda u \quad \Rightarrow \quad (\partial_x\mathcal{L}_0)^{\mathrm{a}} u = (-\lambda)u,$$

where $\mathcal{T}^{\mathrm{a}}$ is used to denote the adjoint of the operator $\mathcal{T}$. This is precisely the adjoint problem for the problem studied in the previous sections; hence, the spectrum is completely known for the worked examples. If $\langle \mathcal{L}_0^{-1}(1), 1\rangle \neq 0$, then $\lambda = 0$ is an eigenvalue with $\mathrm{m_g}(0) = 1$ and $\mathrm{m_a}(0) = 2$. For $\ell > 0$ but small the double eigenvalue at zero splits into a pair of eigenvalues, each of which is $\mathcal{O}(\ell)$ (see [34, Theorem 4.1]). Since eigenvalues come in quartets, it must be the case that the pair is either purely real, and thus contributes to a linear instability, or purely imaginary. Furthermore, if there are any eigenvalues with $\mathrm{Re}\,\lambda > 0$ when $\ell = 0$, these continue to have positive real part for small ℓ.

In order to compute the right-hand side of (72) for $\ell = \mathcal{O}(1)$, we need to consider $\sigma\left(\mathcal{L}(\ell)|_{H_{1,x}}\right)$. The eigenvalue problem

$$\mathcal{L}(\ell)u = \lambda u$$

is equivalent to

$$-\partial_x\mathcal{L}_0\partial_x u = \gamma u, \quad \gamma = \lambda - \ell^2. \tag{73}$$

Thus, if one can compute $\mathrm{n}(-\partial_x\mathcal{L}_0\partial_x|_{H_{1,x}})$, one can readily compute the desired quantity $\mathrm{n}(\mathcal{L}(\ell)|_{H_{1,x}})$. In particular, if the negative eigenvalues of $-\partial_x\mathcal{L}_0\partial_x|_{H_{1,x}}$ are ordered as $\lambda_0 < \lambda_1 \leq \cdots \leq \lambda_N < 0$, then the (potential) negative eigenvalues of $\mathcal{L}(\ell)|_{H_{1,x}}$ are given by $\lambda_0 + \ell^2 < \cdots < \lambda_N + \ell^2$. It is clearly the case that if $\ell^2 > -\lambda_0$, then $\mathcal{L}(\ell)|_{H_{1,x}}$ is a positive operator; otherwise, the operator will have a finite number of negative directions, and this number will decrease as increasing ℓ moves a negative eigenvalue λ_j across the origin.

In conclusion, in order to refine (72) we must compute $\mathrm{n}(-\partial_x\mathcal{L}_0\partial_x|_{H_{1,x}})$. We will show that it will be enough to consider $\mathrm{n}(\mathcal{L}_0|_{H_{1,x}})$. Integrating (73) yields that for $\gamma \neq 0$,

$$0 = \langle(-\partial_x\mathcal{L}_0\partial_x)u, 1\rangle = \gamma\langle u, 1\rangle.$$

This implies that all eigenfunctions associated with nonzero eigenvalues are in $H_{1,x}$, so that

$$\mathrm{n}\left(-\partial_x \mathscr{L}_0 \partial_x|_{H_{1,x}}\right) = \mathrm{n}(-\partial_x \mathscr{L}_0 \partial_x).$$

Regarding the computation of $\mathrm{n}(-\partial_x \mathscr{L}_0 \partial_x)$, let us consider the quadratic form

$$\langle -\partial_x \mathscr{L}_0 \partial_x v, v\rangle = \langle \mathscr{L}_0 u, u\rangle, \quad u = \partial_x v.$$

In order for the two quadratic forms to be equivalent it must be the case that both u and v are $2L_x$-periodic, which in turn implies that $u \in H_{1,x}$. In other words, we have

$$\langle -\partial_x \mathscr{L}_0 \partial_x v, v\rangle = \langle \mathscr{L}_0|_{H_{1,x}} u, u\rangle \quad \Rightarrow \quad \mathrm{n}(-\partial_x \mathscr{L}_0 \partial_x) = \mathrm{n}\left(\mathscr{L}_0|_{H_{1,x}}\right).$$

But, from (27) we have

$$\mathrm{n}\left(\mathscr{L}_0|_{H_{1,x}}\right) = \mathrm{n}(\mathscr{L}_0) - \mathrm{n}(\langle \mathscr{L}_0^{-1}(1), 1\rangle);$$

thus, we conclude that

$$\mathrm{n}\left(-\partial_x \mathscr{L}_0 \partial_x|_{H_{1,x}}\right) = \mathrm{n}(\mathscr{L}_0) - \mathrm{n}(\langle \mathscr{L}_0^{-1}(1), 1\rangle). \tag{74}$$

The computation of the right-hand side of (74) has been done in the previous sections for specific examples. Summarizing, we have that for $p = 2$,

$$U(x) = \sqrt{2}\,\mathrm{dn}(x,k): \quad \mathrm{n}(\mathscr{L}_0) - \mathrm{n}(\langle \mathscr{L}_0^{-1}(1), 1\rangle) = 1,$$

$$U(x) = \sqrt{2}k\,\mathrm{cn}(x,k): \quad \mathrm{n}(\mathscr{L}_0) - \mathrm{n}(\langle \mathscr{L}_0^{-1}(1), 1\rangle) = \begin{cases} 1, & 0 \le k < k^*, \\ 2, & k^* < k < 1, \end{cases}$$

$$U(x) = \sqrt{2}k\,\mathrm{sn}(x,k): \quad \mathrm{n}(\mathscr{L}_0) - \mathrm{n}(\langle \mathscr{L}_0^{-1}(1), 1\rangle) = 0.$$

While we do not prove it here, it is not difficult to show in the second case, i.e., $U(x) \propto \mathrm{cn}(x,k)$, that the two negative eigenvalues are simple. For the perturbative results proved for $p \ge 3$, we have that

$$\mathrm{n}(\mathscr{L}_0) - \mathrm{n}(\langle \mathscr{L}_0^{-1}(1), 1\rangle) = \begin{cases} 0, & \lambda_1 > 0, \\ 1, & \lambda_1 < 0. \end{cases}$$

In other words, the count is zero when the right band edge is at the origin (left panel of Fig. 2), and is one when the left band is at the origin (right panel of Fig. 2). In all of the above cases except when $U(x) \propto \mathrm{cn}(x,k)$ for $k > k^*$ we saw that the waves were orbitally stable; otherwise, we saw that $k_r = 1$, and that the unstable eigenvalue is $\mathcal{O}(1)$. We can now apply the result of Theorem 7 to say the following about transverse instabilities of these waves:

Theorem 8. *Suppose that for a gKdV spatially periodic wave* $\mathrm{n}(\mathcal{L}_0) = 1$ *(left panel of Fig. 2). If* $\langle \mathcal{L}_0^{-1}(1), 1\rangle < 0$*, then the wave is spectrally transversely stable. Otherwise, the wave is spectrally unstable transversely with* $k_\mathrm{r} = 1$ *to perturbations of period* $2\pi/\ell$ *for* $0 < \ell < \sqrt{-\lambda_0}$*, where* $0 > \lambda_0 \in \sigma(\mathcal{L}_0|_{H_{1,x}})$ *is the ground state eigenvalue. All waves are spectrally stable for* $\ell > \sqrt{-\lambda_0}$*. If* $\mathrm{n}(\mathcal{L}_0) = 2$ *(right panel of Fig. 2), the wave is spectrally transversely unstable.*

Remark 14. For all of the unstable waves considered in this paper we have that $k_\mathrm{r} = 1$ expect when $p = 2$ with $U(x) \propto \mathrm{cn}(x, k)$. In this case, if $0 < k < k^*$, then $k_\mathrm{r} = 1$, whereas if $k^* < k < 1$, we have that $k_\mathrm{r} = 1$ for $\sqrt{-\lambda_0} < \ell < \sqrt{-\lambda_1}$, and $k_\mathrm{r} = 2$ for $0 < \ell < \sqrt{-\lambda_1}$. Here $0 > \lambda_1 > \lambda_0 \in \sigma(\mathcal{L}_0|_{H_{1,x}})$. It should be noted that these cnoidal waves were not studied in [21]; furthermore, the methods of [21] fail for $k > k^*$.

Remark 15. If $p = 1$, then we saw in Remark 12 that for small waves, $\mathrm{n}(\mathcal{L}_0) = 1$. The following calculation which led to $\langle \mathcal{L}_0^{-1}(1), 1\rangle = -2\pi + \mathcal{O}(\varepsilon)$ was independent of p; hence, by Theorem 8 we can conclude the wave is spectrally transversely stable. This is precisely the result of Hărăguş [23], which was derived by a careful perturbation calculation for the entire spectrum for the problem (68).

Remark 16. When $p = 2$, the wave proportional to $\mathrm{sn}(x, k)$ is spectrally stable for all $\ell > 0$, as $\sigma(\mathcal{L}_0) \subset \mathbb{R}^+$. This answers an open question posed in [21, Sect. 5].

Remark 17. If $p \geq 3$, then if the right band edge is at the origin, the wave is spectrally stable for all $\ell > 0$, while if the left band edge is at the origin, then $k_\mathrm{r} = 1$ for $0 < \ell < \sqrt{-\lambda_0} = \mathcal{O}(1)$.

Remark 18. While stated in a different way, the result of theorem 8 regarding transverse instability to long wavelengths ($0 < \ell \ll 1$) is precisely that given by Johnson and Zumbrun [29, Theorem 1]. In that paper the transverse instability criteria for gKP-I required a calculation of the quantity $\{T, M\}_{E,a}$; in particular,

$$\{T, M\}_{E,a} > 0 \quad \Rightarrow \quad \text{transversely unstable.}$$

Now, Bronski et al. [8] show that

$$\{T, M\}_{E,a} = T_E \langle \mathcal{L}_0^{-1}(1), 1\rangle,$$

where for the problems at hand,

$$T_E \begin{cases} < 0, & \mathrm{n}(\mathcal{L}_0) = 2, \\ > 0, & \mathrm{n}(\mathcal{L}_0) = 1. \end{cases}$$

Thus, if $T_E > 0\,(\mathrm{n}(\mathscr{L}_0) = 1)$, the instability criteria becomes $\langle \mathscr{L}_0^{-1}(1), 1\rangle > 0$, which is precisely what we have. On the other hand, if $T_E < 0\,(\mathrm{n}(\mathscr{L}_0) = 2)$, then [29] provide only a partial instability result. In particular, they show that the wave is transversely unstable if $\langle \mathscr{L}_0^{-1}(1), 1\rangle < 0$, which implies $k_\mathrm{r} = 1$. The reason for the lack of a complete description on their part is that their results depend upon the calculation of a parity (orientation) index, which yields definitive results only if k_r is odd for small ℓ.

Acknowledgements BD acknowledges support from the National Science Foundation through grant NSF-DMS-0604546. TK gratefully acknowledges the support of the Jack and Lois Kuipers Applied Mathematics Endowment, a Calvin Research Fellowship, and the National Science Foundation under grant DMS-0806636. Any opinions, findings, and conclusions or recommendations expressed in this material are those of the authors and do not necessarily reflect the views of the funding sources.

References

1. Angulo, J.: Non-linear stability of periodic travelling wave solutions to the Schrödinger and the modified Korteweg-de Vries equations. J. Differ. Equ. **235**, 1–30 (2007)
2. Angulo, J., Quintero, J.: Existence and orbital stability of cnoidal waves for a 1D Boussinesq equation. Int. J. Math. Math. Sci. **2007**, 52020 (2007)
3. Benzoni-Gavage, S.: Planar traveling waves in capillary fluids. Differ. Integr. Equ. **26**(3/4), 439–485 (2013)
4. Bona, J., Souganidis, P., Strauss, W.: Stability and instability of solitary waves of Korteweg-de Vries type. Proc. R. Soc. Lond. A **411**, 395–412 (1987)
5. Bottman, N., Deconinck, B.: KdV cnoidal waves are linearly stable. Disc. Cont. Dyn. Sys. A **25**, 1163–1180 (2009)
6. Bottman, N., Nivala, M., Deconinck, B.: Elliptic solutions of the defocusing NLS equation are stable. J. Phys. A **44**, 285201 (2011)
7. Bronski, J., Johnson, M.: The modulational instability for a generalized KdV equation. Arch. Ration. Mech. Anal. **197** (2), 357–400 (2010)
8. Bronski, J., Johnson, M., Kapitula, T.: An index theorem for the stability of periodic traveling waves of KdV type. Proc. Roy. Soc. Edinburgh Sect. A **141**(6), 1141–1173 (2011)
9. Bronski, J., Johnson, M., Kapitula, T.: An instability index theory for quadratic pencils and applications. Comm. Math. Phys. **327**(2), 521–550 (2014)
10. Carter, J., Deconinck, B., Kiyak, F., Kutz, J.N.: SpectrUW: a laboratory for the numerical exploration of spectra of linear operators. Math. Comp. Sim. **74**, 370–379 (2007)
11. Chen, M., Curtis, C., Deconinck, B., Lee, C., Nguyen, N.: Spectral stability of stationary solutions of a boussinesq system describing long waves in dispersive media. SIAM J. Appl. Dyn. Sys. **9**, 999–1018 (2010)
12. Colliander, J., Keel, M., Staffilani, G., Takaoka, H., Tao, T.: Sharp global well-posedness for KdV and modified KdV on $\mathbb{R}$ and $\mathbb{T}$. J. Am. Math. Soc. **16**(3), 705–749 (2003)
13. Deconinck, B., Kapitula, T.: The orbital stability of the cnoidal waves of the Korteweg-de Vries equation. Phys. Lett. A **374**, 4018–4022 (2010)
14. Deconinck, B., Kutz, J.N.: Computing spectra of linear operators using Hill's method. J. Comput. Phys. **219**, 296–321 (2006)
15. Deconinck, B., Nivala, B.: The stability analysis of the periodic traveling wave solutions of the mkdv equation. Stud. Appl. Math. **126**, 17–48 (2010)

16. Farah, L., Scialom, M.: On the periodic "good" Boussinesq equation. Proc. Am. Math. Soc. **138**(3), 953–964 (2010)
17. Finkel, F., González-López, A., Rodríguez, M.: A new algebraization of the Lamé equation. J. Phys. A Math. Gen. **33**, 1519–1542 (2000)
18. Gallay, T., Hǎrǎguş, M.: Orbital stability of periodic waves for the nonlinear Schrödinger equation. J. Dyn. Differ. Equ. **19**, 825–865 (2007)
19. Gallay, T., Hǎrǎguş, M.: Stability of small periodic waves for the nonlinear Schrödinger equation. J. Differ. Equ. **234**, 544–581 (2007)
20. Grillakis, M., Shatah, J., Strauss, W.: Stability theory of solitary waves in the presence of symmetry, II. J. Funct. Anal. **94**, 308–348 (1990)
21. Hakkaev, S., Stanislavova, M., Stefanov, A.: Transverse instability for periodic waves of KP-I and Schrödinger equations. Indiana Univ. Math. J. **61**(2), 461–492 (2012)
22. Hakkaev, S., Stanislavova, M., Stefanov, A.: Linear stability analysis for periodic traveling waves of the Boussinesq equation and the Klein-Gordon-Zakharov system. Proc. Roy. Soc. Edinburgh A **144**(3), 455–489 (2014)
23. Hǎrǎguş, M.: Transverse spectral stability of small periodic traveling waves for the KP equation. Stud. Appl. Math. **126**, 157–185 (2011)
24. Hǎrǎguş, M., Kapitula, T.: On the spectra of periodic waves for infinite-dimensional Hamiltonian systems. Phys. D **237**(20), 2649–2671 (2008)
25. Jian-Ping, Y., Yong-Li, S.: Weierstrass elliptic function solutions to nonlinear evolution equations. Commun. Theor. Phys. **50**(2), 295–298 (2008)
26. Johnson, M.: Nonlinear stability of periodic traveling wave solutions of the generalized Korteweg-de Vries equation. SIAM J. Math. Anal. **41**(5), 1921–1947 (2009)
27. Johnson, M.: The transverse instability of periodic waves in Zakharov–Kuznetsov type equations. Stud. Appl. Math. **124**(4), 323–345 (2010)
28. Johnson, M., Zumbrun, K.: Rigorous justification of the Whitham Modulation Equations for the generalized Korteweg-de Vries equation. Stud. Appl. Math. **125**(1), 69–89 (2010)
29. Johnson, M., Zumbrun, K.: Transverse instability of periodic traveling waves in the generalized Kadomtsev-Petviashvili equation. SIAM J. Math. Anal. **42**(6), 2681–2702 (2010)
30. Johnson, M., Zumbrun, K., Bronski, J.: On the modulation equations and stability of periodic generalized Korteweg–de Vries waves via Bloch decompositions. Phys. D **239**(23–24), 2057–2065 (2010)
31. Kapitula, T., Promislow, K.: Stability indices for constrained self-adjoint operators. Proc. Am. Math. Soc. **140**(3), 865–880 (2012)
32. Kapitula, T., Promislow, K.: Spectral and Dynamical Stability of Nonlinear Waves. Springer, New York (2013)
33. Kapitula, T., Stefanov, A.: A Hamiltonian-Krein (instability) index theory for KdV-like eigenvalue problems. Stud. Appl. Math. **132**(3), 183–211 (2014)
34. Kapitula, T., Kevrekidis, P., Sandstede, B.: Counting eigenvalues via the Krein signature in infinite-dimensional Hamiltonian systems. Phys. D **195**(3&4), 263–282 (2004)
35. Kapitula, T., Kevrekidis, P., Sandstede, B.: Addendum: Counting eigenvalues via the Krein signature in infinite-dimensional Hamiltonian systems. Phys. D **201**(1&2), 199–201 (2005)
36. Kapitula, T., Law, K., Kevrekidis, P.: Interaction of excited states in two-species Bose-Einstein condensates: a case study. SIAM J. Appl. Dyn. Sys. **9**(1), 34–61 (2010)
37. Magnus, W., Winkler, S.: Hill's Equation, volume 20 of Interscience Tracts in Pure and Applied Mathematics. Interscience Publishers, New York (1966)
38. Nivala, M., Deconinck, B.: Periodic finite-genus solutions of the KdV equation are orbitally stable. Phys. D **239**(13), 1147–1158 (2010)
39. Olver, F., Lozier, D., Boisvert, R., Clark, C. (eds.): NIST handbook of mathematical functions. US Department of Commerce National Institute of Standards, Washington, DC (2010)
40. Pava, J., Natali, F.: (non)linear instability of periodic traveling waves: Klein–Gordon and KdV type equations. Adv. Nonlinear Anal. **3**(2), 95–123 (2014)
41. Pava, J., Bona, J., Scialom, M.: Stability of cnoidal waves. Adv. Differ. Equ. **11**(12), 1321–1374 (2006)

42. Pelinovsky, D.: Spectral stability of nonlinear waves in KdV-type evolution equations. In: Kirillov, O., Pelinovsky, D. (eds.) Nonlinear Physical Systems: Spectral Analysis, Stability, and Bifurcations, pp. 377–400. Wiley-ISTE, Hoboken (2014)
43. Rademacher, J., Sandstede, B., Scheel, A.: Computing absolute and essential spectra using continuation. Phys. D **229**(1&2), 166–183 (2007)
44. Rousset, F., Tzvetkov, N.: A simple criterion of transverse linear instability for solitary waves. Math. Res. Lett. **17**(1) 157–169 (2010)
45. Stanislavova, M., Stefanov, A.: Stability analysis for traveling waves of second order in time PDE's. Nonlinearity **25**, 2625–2654 (2012)

Time-Averaging for Weakly Nonlinear CGL Equations with Arbitrary Potentials

Guan Huang, Sergei Kuksin, and Alberto Maiocchi

Dedicated to Walter Craig on his 60th birthday

Abstract Consider weakly nonlinear complex Ginzburg–Landau (CGL) equation of the form:

$$u_t + i(-\triangle u + V(x)u) = \varepsilon\mu\triangle u + \varepsilon\mathscr{P}(\nabla u, u), \quad x \in \mathbb{R}^d, \qquad (*)$$

under the periodic boundary conditions, where $\mu \geq 0$ and $\mathscr{P}$ is a smooth function. Let $\{\zeta_1(x), \zeta_2(x), \dots\}$ be the L_2-basis formed by eigenfunctions of the operator $-\triangle + V(x)$. For a complex function $u(x)$, write it as $u(x) = \sum_{k\geq 1} v_k\zeta_k(x)$ and set $I_k(u) = \frac{1}{2}|v_k|^2$. Then for any solution $u(t, x)$ of the linear equation $(*)_{\varepsilon=0}$ we have $I(u(t, \cdot)) = const$. In this work it is proved that if equation $(*)$ with a sufficiently smooth real potential $V(x)$ is well posed on time-intervals $t \lesssim \varepsilon^{-1}$, then for any its solution $u^\varepsilon(t, x)$, the limiting behavior of the curve $I(u^\varepsilon(t, \cdot))$ on time intervals of order ε^{-1}, as $\varepsilon \to 0$, can be uniquely characterized by a solution of a certain well-posed effective equation:

$$u_t = \varepsilon\mu\triangle u + \varepsilon F(u),$$

where $F(u)$ is a resonant averaging of the nonlinearity $\mathscr{P}(\nabla u, u)$. We also prove similar results for the stochastically perturbed equation, when a white in time and

G. Huang
CMLS, Ecole Polytechnique, Palaiseau, France
e-mail: huangguan.at.nju@gmail.com

S. Kuksin (✉)
IMJ, Université Paris Diderot-Paris 7, 5 rue Thomas Mann, 75205 Paris Cedex 13, France
e-mail: kuksin@math.jussieu.fr

A. Maiocchi
Université de Cergy-Pontoise, Cergy-Pontoise, France
e-mail: alberto.maiocchi@unimi.it

© Springer Science+Business Media New York 2015
P. Guyenne et al. (eds.), *Hamiltonian Partial Differential Equations and Applications*, Fields Institute Communications 75,
DOI 10.1007/978-1-4939-2950-4_11

smooth in x random force of order $\sqrt{\varepsilon}$ is added to the right-hand side of the equation. The approach of this work is rather general. In particular, it applies to equations in bounded domains in $\mathbb{R}^d$ under Dirichlet boundary conditions.

1 Introduction

Equations. We consider a weakly nonlinear CGL equation on a rectangular d-torus $T^d = \mathbb{R}/(L_1\mathbb{Z}) \times \mathbb{R}/(L_2\mathbb{Z}) \times \cdots \times \mathbb{R}/(L_d\mathbb{Z})$, $L_1, \ldots, L_d > 0$,

$$u_t + i(-\Delta + V(x))u = \varepsilon\mu\Delta u + \varepsilon\mathscr{P}(\nabla u, u), \quad u = u(t,x),\ x \in T^d, \tag{1}$$

where $\mu \geq 0$, $\mathscr{P} : \mathbb{C}^{d+1} \to \mathbb{C}$ is a C^∞-smooth function, ε is a small parameter and $V(\cdot) \in C^n(T^d)$ is a sufficiently smooth real-valued function on T^d (we will assume that n is large enough). If $\mu = 0$, then the nonlinearity $\mathscr{P}$ should be independent of the derivatives of the unknown function u. For simplicity, we assume that $\mu > 0$. The case $\mu = 0$ can be treated exactly in the same way (even simpler).

For any $s \in \mathbb{R}$ we denote by H^s the Sobolev space of complex-valued functions on T^d, provided with the norm $\|\cdot\|_s$,

$$\|u\|_s^2 = \langle(-\Delta)^s u, u\rangle + \langle u, u\rangle, \quad \text{if} \quad s \geq 0\,,$$

where $\langle\cdot,\cdot\rangle$ is the real scalar product in $L^2(T^d)$,

$$\langle u, v\rangle = \Re\int_{T^d} u\bar{v}\mathrm{d}x, \quad u, v \in L^2(T^d).$$

For any $s > d/2+1$, it is known that the mapping $\mathscr{P} : H^s \to H^{s-1},\ u \mapsto \mathscr{P}(\nabla u, u)$, is smooth and locally Lipschitz, see below Lemma 3.

Our goal is to study the dynamics of Eq. (1) on time intervals of order ε^{-1} when $0 < \varepsilon \ll 1$. Introducing the slow time $\tau = \varepsilon t$, we rewrite the equation as

$$\dot{u} + \varepsilon^{-1} i(-\Delta + V(x))u = \mu\Delta u + \mathscr{P}(\nabla u, u), \tag{2}$$

where $u = u(\tau, x)$, $x \in T^d$, and the upper dot $\dot{}$ stands for $\frac{\mathrm{d}}{\mathrm{d}\tau}$. We assume

Assumption A. *There exists a number $s_* \in (d/2+1, n]$ and for every $M_0 > 0$ there exists $T = T(s_*, M_0) > 0$ such that if $u_0 \in H^{s_*}$ and $\|u_0\|_{s_*} \leq M_0$, then Eq. (2) has a unique solution $u(\tau, x) \in C([0,T], H^{s_*})$ with the initial datum u_0, and $\|u(\tau,x)\|_{s_*} \leq C(s_*, M_0, T)$ for $\tau \in [0,T]$.*

This assumption can be verified for Eq. (1) with various nonlinearities $\mathscr{P}$. For example when $\mu = 0$ and $V(x) \equiv 0$ it holds if $\mathscr{P}(u)$ is any smooth function. Indeed, taking the scalar product in the space H^{s_*} of eq. (2) with $u(t)$ and using the Gronwall lemma we get the Assumption A with suitable positive constants $T(s_*, M_0)$ and

$C(s_*, M_0, T)$.

$$p \in \mathbb{N},\ p < \infty, \quad \text{if} \quad d = 1, 2, \quad \text{and} \quad p = 1, 2 \quad \text{if} \quad d = 3.$$

When $\mu > 0$, the assumption with any $T > 0$ is satisfied by Eq. (1) with nonlinearity $\mathscr{P}(u) = -\gamma_R f_p(|u|^2)u - i\gamma_I f_q(|u|^2)u$, where $\gamma_R, \gamma_I > 0$, the functions $f_p(r)$ and $f_q(r)$ are the monomials $|r|^p$ and $|r|^q$, smoothed out near zero, and

$$0 \le p, q < \infty \quad \text{if} \quad d = 1, 2 \quad \text{and} \quad 0 \le p, q < \min\left\{\frac{d}{2}, \frac{2}{d-2}\right\} \quad \text{if} \quad d \ge 3,$$

see, e.g. [8].

We denote by A_V the Schrödinger operator

$$A_V u := -\Delta u + V(x)u.$$

Let $\{\lambda_k\}_{k\ge 1}$ be its eigenvalues, ordered in such a way that

$$\lambda_1 \le \lambda_2 \le \lambda_3 \le \cdots,$$

and let $\{\zeta_k,\ k \ge 1\}$ of $L^2(T^d)$ be an orthonormal basis, formed by the corresponding eigenfunctions. We denote $\Lambda = (\lambda_1, \lambda_2, \dots)$ and call Λ the *frequency vector* of Eq. (2). For a complex-valued function $u \in H^s$, we denote by

$$\Psi(u) := v = (v_1, v_2, \dots), \quad v_k \in \mathbb{C}, \tag{3}$$

the vector of its Fourier coefficients with respect to the basis $\{\zeta_k\}_{k\ge 1}$: $u = \sum_{k\ge 1} v_k \zeta_k$. Note that Ψ is a real operator: it maps real functions $u(x)$ to real vectors v. In the space of complex sequences $v = (v_1, v_2, \dots)$, we introduce the norms

$$|v|_s^2 = \sum_{k=1}^{+\infty} (|\lambda_k|^s + 1)\,|v_k|^2, \quad s \in \mathbb{R},$$

and denote $h^s = \{v : |v|_s < \infty\}$. Clearly Ψ defines an isomorphism between the spaces H^s and h^s.

Now we write Eq. (2) in the v-variables:

$$\dot{v}_k + \varepsilon^{-1} i\lambda_k v_k = -\mu\lambda_k v_k + P_k(v), \quad k \in \mathbb{N}, \tag{4}$$

where

$$P(v) := (P_k(v),\ k \in \mathbb{N}) = \Psi\Big(\mu V(x)u + \mathscr{P}(\nabla u, u)\Big), \quad u = \Psi^{-1}v. \tag{5}$$

For every $k \in \mathbb{N}$ we set

$$I_k(v) = \frac{1}{2} v_k \bar{v}_k, \text{ and } \varphi_k(v) = \text{Arg } v_k \in \mathbb{T}^1 = \mathbb{R}/(2\pi\mathbb{Z}) \text{ if } v_k \neq 0, \text{ else } \varphi_k = 0. \tag{6}$$

Then $v_k = \sqrt{2I_k} e^{i\varphi_k}$. Notice that the quantities I_k are conservation laws of the linear equation $(1)_{\varepsilon=0}$, and that the variables $(I, \varphi) \in \mathbb{R}_+^\infty \times \mathbb{T}^\infty$ are its action-angles. For any $(I, \varphi) \in \mathbb{R}_+^\infty \times \mathbb{T}^\infty$ we denote

$$v = v(I, \varphi) \quad \text{if} \quad v_k = \sqrt{2I_k} e^{i\varphi_k}, \quad \forall\, k. \tag{7}$$

If this relation holds, we will write $v \sim (I, \varphi)$. We introduce the weighted l^1-space h_I^s:

$$h_I^s := \{I = (I_k,\ k \in \mathbb{N}) \in \mathbb{R}^\infty : |I|_s^\sim = \sum_{k=1}^{+\infty} 2(|\lambda_k|^s + 1)|I_k| < \infty\}.$$

Then $|v|_s^2 = |I(v)|_s^\sim$, for each $v \in h^s$. Using the action-angle variables (I, φ), we write Eq. (4) as a slow-fast system:

$$\dot{I}_k = v_k \cdot \big(-\mu\lambda_k v_k + P_k(v)\big), \quad \dot{\varphi}_k = -\varepsilon^{-1}\lambda_k + |v_k|^{-2} \cdots, \quad k \in \mathbb{N}.$$

Here $a \cdot b$ denotes $\Re(a\bar{b})$, for $a, b \in \mathbb{C}$, and the dots stand for a factor of order 1 (as $\varepsilon \to 0$).

Effective Equations. Our task is to study the evolution of the actions I_k when $\varepsilon \ll 1$ and $0 \leq \tau \lesssim 1$. An efficient way to deal with this problem is through the so-called *interaction representation*. Let us define

$$a_k(\tau) = e^{i\varepsilon^{-1}\lambda_k \tau} v_k(\tau)\,. \tag{8}$$

Then

$$|a_k|^2 = |v_k|^2 = 2I_k\,, \tag{9}$$

so to study the evolution of the actions we can use the a-variables instead of the v-variables. Using Eq. (4), we obtain for $a = (a_1, a_2, \dots)$ the system of equations

$$\dot{a}_k(\tau) = -\mu\lambda_k a_k + e^{i\varepsilon^{-1}\lambda_k \tau} P_k(\Phi_{-\varepsilon^{-1}\Lambda\tau} a), \quad k \in \mathbb{N}\,, \tag{10}$$

where for each $\theta = (\theta_k,\ k \in \mathbb{N}) \in \mathbb{R}^\infty$, Φ_θ stands for the linear operator in h^s defined by

$$\Phi_\theta v = v', \quad v'_k = e^{i\theta_k} v_k \quad \forall\, k.$$

Clearly Φ_θ defines isometries of all Hilbert spaces h^s, and in the action-angle variables it reads $\Phi_\theta(I, \varphi) = (I, \varphi + \theta)$.

To approximately describe the dynamics of Eq. (10) with $\varepsilon \ll 1$ we introduce an effective equation:

$$\dot{\tilde{a}}_k = -\mu\lambda_k\tilde{a}_k + R_k(\tilde{a}), \quad k \in \mathbb{N}, \tag{11}$$

where $R(\tilde{a}) := (R_k(\tilde{a}),\ k \in \mathbb{N})$ and

$$R(\tilde{a}) = \lim_{T\to\infty} \frac{1}{T}\int_0^T \Phi_{\Lambda t}P(\Phi_{-\Lambda t}\tilde{a})dt. \tag{12}$$

We will see in Sects. 2 and 3 that the limit in (12) is well defined and that Eq. (11) is well posed, at least locally in time.
Results. In Sect. 4 we prove that the actions of solutions for the effective equation approximate well the actions $I_k(v(\tau))$ of solutions v for (4). Let us fix any $M_0 > 0$.

Theorem 1. *Let $u(\tau, x)$, $0 \le \tau \le T = T(s_*, M_0)$, be a solution of (2), such that $u(0, x) = u_0(x)$, $\|u_0\|_{s_*} \le M_0$, existing by Assumption A. Denote $v(\tau) = \Psi(u(\tau, \cdot))$, $0 \le \tau \le T$. Then a solution $\tilde{a}(\tau)$ of (11), such that $\tilde{a}(0) = v(0)$, exists for $0 \le \tau \le T$, and for any $s_1 < s_*$ we have*

$$\sup_{0\le\tau\le T} |I(v(\tau)) - I(\tilde{a}(\tau))|^{\sim}_{s_1} \to 0, \quad as \quad \varepsilon \to 0\,.$$

The rate of the convergence does not depend on u_0, if $\|u_0\|_{s_*} \le M_0$.

This theorem may be regarded as a PDE-version of the Bogolyubov averaging principle, see [3] and [1], Sect. 6.1. The result and its proof may be easily recasted to a theorem on perturbations of linear Hamiltonian systems with discrete spectrum. Instead of doing this, below we briefly discuss its generalisations to other nonlinear PDE problems.

In the second part of the paper (Sects. 5–7) we consider the CGL equation (1) with added small random force:

$$u_t + i(-\Delta + V(x))u = \varepsilon\mu\Delta u + \varepsilon\mathscr{P}(\nabla u, u) + \sqrt{\varepsilon}\frac{d}{dt}\sum_{l\ge1} b_l\beta_l(t)e_l(x), \tag{13}$$

where $u = u(t, x)$, $x \in T^d$, the coefficients b_l decay fast enough with $|l|$, $\{\beta_l(t)\}$ are standard independent complex Wiener processes and $\{e_l(x)\}$ is the usual trigonometric basis of the space $L_2(T^d)$, parametrized by natural numbers. It turns out that the effective equation for (13) is the equation (11), perturbed by a suitable stochastic forcing, see Sect. 5. Assuming that the function $\mathscr{P}$ has at most a polynomial growth and that the equation satisfies a suitable stochastic analogy of the Assumption A we prove a natural stochastic version of Theorem 1 (see Theorem 2). Next, supposing that the stochastic effective equation is mixing and has a unique stationary measure μ_0, we prove in Theorem 3 that if μ_ε is a stationary measure for Eq. (13), then $\Psi \circ \mu_\varepsilon$ converge to μ_0 as $\varepsilon \to 0$. So if the stochastic effective equation is mixing, then it comprises asymptotical properties of solutions for Eq. (13) as $t \to \infty$ and $\varepsilon \to 0$.

The proof of the theorems in this work follows the Anosov approach to averaging in finite-dimensional systems (see in [1, 18]), its version for averaging in resonant systems (see in [1]) and its stochastic version due to Khasminski [12]. The crucial idea that for averaging in PDEs the averaged equations for actions (which are equations with singularities) should be considered jointly with suitable effective equations (which are regular equations) was suggested in [13] for averaging in stochastic PDEs, and later was used in [14] and [8, 9, 15, 16]. It was realised in the second group of publications that for perturbations of linear systems the method may be well combined with the interaction representation of solutions, well known and popular in nonlinear physics (see [3, 19]), and which already was used for purposes of completely resonant averaging, corresponding to constant coefficient PDEs with small nonlinearities on the square torus (see [5, 7]).

For the case when the spectrum of the unperturbed linear system is non-resonant (see below Example 1), the results of this paper were obtained in [8, 14], while for the case when the spectrum is completely resonant—in [9, 15]. The novelty of this work is a version of the Anosov method of averaging, applicable to nonlinear PDEs with small nonlinearities, which does not impose restrictions on the spectrum of the unperturbed equation.

Alternatively, the averaging for weakly nonlinear PDEs may be studied, using the normal form techniques, e.g. see [2] and references therein. Compared to the Anosov approach, exploited in this work, the method of normal form is much more demanding to the spectrum of the unperturbed equation, and more sensitive to its perturbations. So usually it applies only in small vicinities of equilibriums. Its advantage is that it may imply stability on longer time intervals, while the method of this work is restricted to the first-order averaging. So in the deterministic setting it allows to control solutions of ε-perturbed equations only on time-intervals of order ε^{-1} (still, in the stochastic setting it also allows to control the stationary measure, which describes the asymptotic behaviour of solutions as $t \to \infty$).

Generalizations. The Anosov-like method of resonant averaging, presented in this work, is very flexible. With some slight changes, it easily generalizes to weakly nonlinear CGL equations, involving high order derivatives,

$$u_t + i(-\Delta u + V(x)u) = \varepsilon \mathscr{P}(\nabla^2 u, \nabla u, u, x), \quad x \in T^d, \tag{14}$$

provided that the Assumption A holds and the corresponding effective equation is well posed locally in time. See in Appendix (also see [8], where a similar result is proven for the case of non-resonant spectra).

The method applies to Eqs. (1) and (13) in a bounded domain $\mathscr{O} \subset \mathbb{R}^d$ under Dirichlet boundary conditions. Indeed, if $d \leq 3$, then to treat the corresponding boundary-value problem we can literally repeat the argument of this work, replacing there the space H^s with the Hilbert space $H_0^2(\mathscr{O}) = \{u \in H^2(\mathscr{O}) : u\,|_{\partial\mathscr{O}} = 0\}$. If $d \geq 4$, then H^s should be replaced with an L_p-based Banach space $W_0^{2,p}(\mathscr{O})$, where $p > d/2$.

Obviously the method applies to weakly nonlinear equations of other types; e.g. to weakly nonlinear wave equations. In [16] the method in its stochastic form was applied to the Hasegawa-Mima equation, regarded as a perturbation of the Rossby equation $(-\Delta + K)\psi_t(t,x,y) - \psi_x = 0$, while in [4] it is applied to systems of non-equilibrium statistical physics, where each particle is perturbed by an ε-small Langevin thermostat, and is studied the limit $\varepsilon \to 0$ (similar to the same limit in Eq. (13)).

The averaging for perturbations of nonlinear integrable PDEs is more complicated. Due to the lack in the functional phase-spaces of an analogy of the Lebesgue measure (required by the Anosov approach to the finite-dimensional deterministic averaging), in this case the results for stochastic perturbations are significantly stronger than the deterministic results. See in [10].

2 Resonant Averaging in Hilbert Spaces

The goal of this section is to show that the limit in (12) is well-defined in some suitable settings and study its properties. Below for an infinite-vector $v = (v_1, v_2, \dots)$ and any $m \in \mathbb{N}$ we denote

$$v^m = (v_1, \dots, v_m), \quad \text{or} \quad v^m = (v_1, \dots, v_m, 0, \dots),$$

depending on the context. This agreement also applies to elements $\varphi = (\varphi_1, \varphi_2, \dots)$ of the torus $\mathbb{T}^\infty$. For m-vectors I^m, φ^m, v^m we write $v^m \sim (I^m, \varphi^m)$ if (7) holds for $k = 1, \dots, m$. By Π^m, $m \geq 1$, we denote the Galerkin projection

$$\Pi^m : h^0 \to h^0, (v_1, v_2, \dots) \mapsto v^m = (v_1, \dots, v_m, 0, \dots).$$

For a continuous complex function f on a Hilbert space H, we say that f is locally Lipschitz and write $f \in Lip_{loc}(H)$ if

$$|f(v) - f(v')| \leq \mathscr{C}(R)\|v - v'\|, \quad \text{if} \quad \|v\|, \|v'\| \leq R, \tag{15}$$

for some continuous non-decreasing function $\mathscr{C} : \mathbb{R}^+ \to \mathbb{R}^+$ which depends on f. We write

$$f \in Lip_{\mathscr{C}}(H) \ \text{ if (15) holds and } |f(v)| \leq \mathscr{C}(R) \text{ if } \|v\| \leq R. \tag{16}$$

If $f \in Lip_{\mathscr{C}}(H)$, where $\mathscr{C}(\cdot) =$ Const, then f is a bounded (globally) Lipschitz function. If B is a Banach space, then the space $Lip_{loc}(H, B)$ of locally Lipschitz mappings $H \to B$ and its subsets $Lip_{\mathscr{C}}(H, B)$ are defined similarly.

For any vector $W = (w_1, w_2, \dots) \in \mathbb{R}^\infty$ we set

$$\langle f \rangle_{W,l}^T(v) = \frac{1}{T}\int_0^T e^{iw_l t} f(\Phi_{-Wt} v)\,dt, \tag{17}$$

and if the limit of $\langle f \rangle_{W,l}^T(v)$ when $T \to \infty$ exists, we denote

$$\langle f \rangle_{W,l} = \lim_{T\to\infty} \langle f \rangle_{W,l}^T(v).$$

Concerning this definition we have the following lemma. Denote

$$B(M, h^s) = \{v \in h^s : |v|_s \le M\}, \quad M > 0.$$

Lemma 1. *Let $f \in Lip_{\mathscr{C}}(h^{s_0})$ for some $s_0 \ge 0$ and some function $\mathscr{C}$ as above. Then*

(i) For every $T \neq 0$, $\langle f \rangle_{W,l}^T \in Lip_{\mathscr{C}}(h^{s_0})$.
(ii) The limit $\langle f \rangle_{W,l}(v)$ exists for $v \in h^{s_0}$ and this function also belongs to $Lip_{\mathscr{C}}(h^{s_0})$.
(iii) For $s > s_0$ and any $M > 0$, the functions $\langle f \rangle_{W,l}^T(v)$ converge, as $T \to \infty$, to $\langle f \rangle_{W,l}(v)$ uniformly for $v \in B(M, h^s)$.
(iv) The convergence is uniform for $f \in Lip_{\mathscr{C}}(h^{s_0})$ with a fixed function $\mathscr{C}$.

Proof. (i) It is obvious since the transformations Φ_θ are isometries of h^{s_0}.

(ii) To prove this, consider the restriction of f to $B(M, h^{s_0})$, for any fixed $M > 0$. Let us take some $v \in B(M, h^{s_0})$ and fix any $\rho > 0$. Below in this proof by $O(v)$, $O_1(v)$, etc, we denote various functions $g(v) = g(I, \varphi)$, defined for $|v|_{s_0} \le M$ and bounded by 1.

Let us choose any $m = m(\rho, M, v, \mathscr{C})$ such that

$$\mathscr{C}(M)\,|v - \Pi^m v|_{s_0} \le \rho\,.$$

Then $|f(v) - f(\Pi^m v)| < \rho$, and by (i)

$$\left|\langle f \rangle_{W,l}^T(v) - \langle f \rangle_{W,l}^T(\Pi^m v)\right| < \rho\,,$$

for every $T > 0$.

Let us set

$$\mathscr{F}^m(I^m, \varphi^m) = \mathscr{F}^m(v^m) = f\big(v^m\big), \qquad \forall\, v^m \sim (I^m, \varphi^m) \in \mathbb{C}^m\,,$$

where in the r.h.s. v^m is regarded as the vector $(v^m, 0, \dots)$. Clearly, the function $\varphi^m \mapsto \mathscr{F}^m(I^m, \varphi^m)$ is Lipschitz-continuous on $\mathbb{T}^m$. So its Fejer polynomials

$$\sigma_K(\mathscr{F}^m) = \sum_{k\in\mathbb{Z}^m,\, |k|_\infty \le K} a_k^K e^{ik\cdot(\varphi^m)}, \qquad K \ge 1\,,$$

where $a_k^K = a_k^K(m, I^m)$, converges to $\mathscr{F}^m(I^m, \varphi^m)$ uniformly on $\mathbb{T}^m$. Moreover, the rate of convergence depends only on its Lipschitzian norm and the dimension m (see e.g. Theorem 1.20, Chap. XVII of [22]). Therefore, there exists $K = K(\mathscr{C}, M, \rho, m) > 0$ such that

$$\mathscr{F}^m(I^m, \varphi^m) = \sum_{k\in\mathbb{Z}^m, |k|_\infty \le K} a_k^K e^{ik\cdot\varphi^m} + \rho O_1(I^m, \varphi^m)\,. \tag{18}$$

Now we define

$$\mathscr{F}_K^{res}(I^m, \varphi^m) = \sum_{k\in S(K)} a_k^K e^{ik\cdot\varphi^m}, \quad S(K) = \{k \in \mathbb{Z}^m : \; |k|_\infty \le K, w_l - \sum_{j=1}^m k_i w_i = 0\}.$$

Since

$$\mathscr{F}^m\big(\Phi_{-W^m t}(\Pi^m v)\big) = \mathscr{F}^m(I^m, \varphi^m - Wt)\,,$$

then

$$\langle e^{ik\cdot\varphi^m}\rangle_{W,l}^T = e^{ik\cdot\varphi^m} \quad \text{if} \quad k \in S(K)\,,$$

$$\left|\langle e^{ik\cdot\varphi^m}\rangle_{W,l}^T\right| \le \frac{2T^{-1}}{|w_l - k\cdot W^m|} \quad \text{if} \quad |k|_\infty \le K,\; k \notin S(K)\,,$$

where we regard $e^{ik\cdot\varphi^m}$ as a function of v. Accordingly,

$$\begin{aligned}\langle f\rangle_{W,l}^T(v) &= \langle \mathscr{F}^m(I^m, \varphi^m)\rangle_{W,l}^T + \rho O_2(v) \\ &= \mathscr{F}_K^{res}(I^m, \varphi^m) + C(\rho, M, W, f, l)T^{-1}O_3(v) + \rho O_4(v)\,.\end{aligned}$$

So there exists $\bar{T} = T(\rho, M, W, f, l) > 0$ such that if $T \ge \bar{T}$, then

$$\left|\langle f\rangle_{W,l}^T - \mathscr{F}_K^{res}(I^m, \varphi^m)\right| < 2\rho\,,$$

and for any $T' \ge T'' \ge \bar{T}$, we have

$$\left|\langle f\rangle_{W,l}^{T'}(v) - \langle f\rangle_{W,l}^{T''}(v)\right| < 4\rho.$$

This implies that the limit $\langle f\rangle_{W,l}(v)$ exists for every $v \in B(M, h^{s_0})$. Using (i) we obtain that $\langle f\rangle_{W,l}(\cdot) \in Lip_{\mathscr{C}}(h^{s_0})$.

(iii) This statement follows directly from (ii) since the family of functions $\{\langle f\rangle_{W,l}^T(v)\}$ is uniformly continuous on balls $B(R, h^{s_0})$ by (i) and each ball $B(M, h^s)$, $s > s_0$, is compact in h^{s_0}.

(iv) From the proof of (ii) we see that for any $\rho > 0$ and $v \in h^{s_0}$, there exists $T = T(W, \rho, v, \mathscr{C})$ such that if $T' \geq T$, then $|\langle f\rangle_{W,l}^{T'}(v) - \langle f\rangle_{W,l}(v)| \leq \rho$. This implies the assertion. □

We now give some examples of the limits $\langle f\rangle_{W,l}$.

Example 1. If the vector W is non-resonant, i.e., non-trivial finite linear combinations of w_j's with integer coefficients do not vanish (this property holds for typical potentials $V(x)$, see [14]), then the set $S(K)$ reduces to one trivial resonance $e_l = (0, \dots, 0, 1, 0, \dots 0)$, where 1 stands on the l-th place (if $m < l$, then $S(K) = \emptyset$). Let $f(v)$ be any finite polynomial of v. We write it in the form $\sum_{k,l \in \mathbb{N}^\infty, |k|,|l|<\infty} f_{k,l}(I) v^k \bar{v}^l$, where $f_{k,l}$ are polynomials of I and finite vectors k, l are such that if $k_j \neq 0$, then $l_j = 0$, and vice versa. Then $\langle f\rangle_{W,l} = f_{e_l,I}(I) v_l$.

Example 2. If f is a linear functional, $f = \sum_{i=1}^\infty b_i v_i$, then for any $l \in \mathbb{N}$,

$$\langle f\rangle_{W,l} = \sum_{i \in \mathscr{A}_l^1} b_i v_i, \quad \mathscr{A}_l^1 = \{i \in \mathbb{N} : w_i - w_l = 0\}.$$

If f is polynomial of v, e.g. $f = \sum_{i+j+m=k} a_{i,j,k} v_i v_j v_k$, then

$$\langle f\rangle_{W,l} = \sum_{(i,j,m) \in \mathscr{A}_l^3} a_{i,j,k} v_i v_j v_k, \quad \mathscr{A}_l^3 = \{(i,j,m) \in \mathbb{N}^3 : w_l - w_i - w_j - w_m = 0\}.$$

We may also consider the averaging

$$\langle\langle f\rangle\rangle_W^T(v) = \frac{1}{T}\int_0^T f(\Phi_{-Wt} v)\, dt, \quad \langle\langle f\rangle\rangle_W(v) = \lim_{T\to\infty} \langle\langle f\rangle\rangle_W^T(v). \tag{19}$$

Lemma 2. *Let $f \in Lip_{\mathscr{C}}(h^{s_0})$. Then*

(a) for the averaging $\langle\langle\cdot\rangle\rangle_W$ hold natural analogies of all assertions of Lemma 1.
(b) The function $\langle\langle f\rangle\rangle_W$ commutes with the transformations Φ_{Wt}, $t \in \mathbb{R}$.

Proof. To prove (a) we repeat for the averaging $\langle\langle\cdot\rangle\rangle_W$ the proof of Lemma 1, replacing there w_l by 0. Assertion (b) immediately follows from the formula for $\langle\langle f\rangle\rangle_W^T$ in (19). □

3 The Effective Equation

Let $V(x) \in C^n(T^d)$. As in the introduction, A_V is the operator $-\Delta + V$ and $\{\lambda_k, k \in \mathbb{N}\}$ are its eigenvalues.

The following result is well known, see Sect. 5.5.3 in [20].

Lemma 3. *If $f(x) : \mathbb{C} \to \mathbb{C}$ is C^∞, then the mapping*

$$M_f : H^s \to H^s, \quad u \mapsto f(u),$$

is C^∞-smooth for $s > d/2$. Moreover, $M_f \in Lip_{\mathscr{C}_s}(H^s, H^s)$ for a suitable function $\mathscr{C}_s$.

Consider the map $P(v)$ defined in (5). From Lemma 3, we have

$$P(\cdot) \in Lip_{\mathscr{C}_s}(h^s, h^{s-1}), \quad \forall\, s \in (d/2 + 1, n]\,, \tag{20}$$

for some $\mathscr{C}_s$. We recall that Λ is the frequency vector of Eq. (2). For any $T \in \mathbb{R}$, we denote

$$\langle P\rangle_\Lambda^T(v) := (\langle P_k\rangle_{\Lambda,k}^T(v), k \in \mathbb{N}) = \frac{1}{T}\int_0^T \Phi_{\Lambda t} P(\Phi_{-\Lambda t} v)dt\,,$$

and

$$R(v) = \langle P\rangle_\Lambda(v) := (\langle P_k\rangle_{\Lambda,k}(v), k \in \mathbb{N})\,.$$

Example 3. If P is a diagonal operator, $P_k(v) = \gamma_k v_k$ for each k, where γ_k's are complex numbers, then in view of Example 2, $\langle P\rangle_\Lambda = P$.

We have the following lemma:

Lemma 4. *(i) For every $d/2 < s_1 < s - 1 \le n - 1$ and $M > 0$, we have*

$$\left|\langle P\rangle_\Lambda^T(v) - R(v)\right|_{s_1} \to 0, \quad as \quad T \to \infty, \tag{21}$$

uniformly for $v \in B(M, h^s)$;
(ii) $R(\cdot) \in Lip_{\mathscr{C}_s}(h^s, h^{s-1})$, $s \in (d/2 + 1, n]$;
(iii) R commutes with $\Phi_{\Lambda t}$, for each $t \in \mathbb{R}$.

Proof. (i) There exists $M_1 > 0$, independent from v and T, such that

$$\left|\langle P\rangle_\Lambda^T(v) - R(v)\right|_{s-1} \le M_1, \quad v \in B(M, h^s).$$

So for any $\rho > 0$ we can find $m_\rho > 0$ such that

$$\left|(\mathrm{Id} - \Pi^{m_\rho})\big[\langle P\rangle_\Lambda^T(v) - R(v)\big]\right|_{s_1} < \rho/2, \quad v \in B(M, h^s).$$

By Lemma 1(iii), there exists T_ρ such that for $T > T_\rho$,

$$\left|\Pi^{m_\rho}\big[\langle P\rangle_\Lambda^T(v) - R(v)\big]\right|_{s_1} < \rho/2, \quad v \in B(M, h^s).$$

Therefore if $T > T_\rho$, then

$$\left|\langle P\rangle_\Lambda^T(v) - R(v)\right|_{s_1} < \rho, \quad v \in B(M, h^s).$$

This implies the first assertion.

(ii) Using the fact that the linear maps $\Phi_{\Lambda t}$, $t \in \mathbb{R}$ are isometries in h^s, we obtain that for $T \in \mathbb{R}$ and $v', v'' \in B(M, h^s)$,

$$\left|\langle P\rangle_\Lambda^T(v') - \langle P\rangle_\Lambda^T(v'')\right|_{s-1} \le \mathscr{C}_s(M)\left|v' - v''\right|_s .$$

Therefore

$$\left|R(v') - R(v'')\right|_{s-1} \le \mathscr{C}_s(M)\left|v' - v''\right|_s, \quad v', v'' \in B(M, h^s).$$

This estimate, the convergence (21) and the Fatou lemma imply that R is a locally Lipschitz mapping with a required estimate for the Lipschitz constant. A bound on its norm may be obtained in a similar way, so the second assertion follows.

(iii) We easily verify that

$$\left|\langle P\rangle_\Lambda^{T+t}(v) - \Phi_{\Lambda t}\langle P\rangle_\Lambda^T(\Phi_{-\Lambda t}v)\right|_{s-1} \le 2\,\mathscr{C}_s(|v|_s)\frac{|t|}{|T+t|} .$$

Passing to the limit as $T \to \infty$ we recover (iii).

Corollary 1. *For $d/2 < s_1 < s - 1 \le n - 1$ and any $v \in h^s$,*

$$\langle P\rangle_\Lambda^T(v) = R(v) + \varkappa(T; v),$$

where $|\varkappa(T; v)|_{s_1} \le \overline{\varkappa}(T; |v|_s)$. Here for each T, $\overline{\varkappa}(T; r)$ is an increasing function of r, and for each $r \ge 0$, $\overline{\varkappa}(T; r) \to 0$ as $T \to \infty$.

Example 4. In the completely resonant case, when

$$L_1 = \cdots = L_d = 2\pi \quad \text{and} \quad V = 0\,, \tag{22}$$

the frequency vector is $\Lambda = (|\mathbf{k}|^2, \mathbf{k} \in \mathbb{Z}^d)$. If $\mathscr{P}(u) = i|u|^2u$, then

$$P(v) = (P_{\mathbf{k}}(v), \mathbf{k} \in \mathbb{Z}^d), \quad v = (v_{\mathbf{k}}, \mathbf{k} \in \mathbb{Z}^d), \quad u = \sum_{\mathbf{k}\in\mathbb{Z}^d} v_{\mathbf{k}} e^{i\mathbf{k}\cdot x},$$

with

$$P_{\mathbf{k}}(v) = \sum_{\mathbf{k}_1 - \mathbf{k}_2 + \mathbf{k}_3 = \mathbf{k}} i v_{\mathbf{k}_1}\bar{v}_{\mathbf{k}_2}v_{\mathbf{k}_3}, \quad \mathbf{k} \in \mathbb{Z}^d.$$

Therefore $\langle P\rangle_\Lambda = (\langle P_{\mathbf{k}}\rangle_{\Lambda,\mathbf{k}}, \mathbf{k} \in \mathbb{Z}^d)$, with

$$\langle P_{\mathbf{k}}\rangle_{\Lambda,\mathbf{k}} = \sum_{(\mathbf{k}_1,\mathbf{k}_2,\mathbf{k}_3)\in Res(\mathbf{k})} i v_{\mathbf{k}_1} \bar{v}_{\mathbf{k}_2} v_{\mathbf{k}},$$

where $Res(\mathbf{k}) = \{(\mathbf{k}_1, \mathbf{k}_2, \mathbf{k}_3) : |\mathbf{k}_1|^2 - |\mathbf{k}_2|^2 + |\mathbf{k}_3|^2 - |\mathbf{k}|^2 = 0\}$.

Lemma 4 implies that the effective equation (11) is a quasi-linear heat equation. So it is locally well-posed in the spaces h^s, $s \in (d/2+1, n]$.

4 Proof of the Averaging Theorem

In this section we will prove Theorem 1. We recall that $d/2 + 1 < s_* \le n$ and $s_1 < s_*$, where s_* is the number from Assumption A and n is a sufficiently big integer (the smoothness of the potential $V(x)$). Without loss of generality we assume that

$$s_1 > d/2 + 1 \quad \text{and} \quad s_1 > s_* - 2\,,$$

and that Assumption A holds with $T = 1$.

Let $u^\varepsilon(\tau, x)$ be the solution of Eq. (2) from Theorem 1,

$$\|u^\varepsilon(0,x)\|_{s*} \le M_0\,,$$

and $v^\varepsilon(\tau) = \Psi(u^\varepsilon(\tau,\cdot))$. Then there exists $M_1 \ge M_0$ such that

$$v^\varepsilon(\tau) \in B(M_1, h^{s_*}), \quad \tau \in [0,1]\,,$$

for each $\varepsilon > 0$. The constants in estimates below in this section may depend on M_1, and this dependence may be non-indicated.

Let

$$a^\varepsilon(\tau) = \Phi_{\tau\varepsilon^{-1}\Lambda}(v^\varepsilon(\tau))$$

be the interaction representation of $v^\varepsilon(\tau)$ (see Introduction),

$$a^\varepsilon(0) = v(0) =: v_0\,.$$

For every $v = (v_k, k \in \mathbb{N})$, denote

$$\hat{A}_V(v) = (\lambda_k v_k, k \in \mathbb{N}) = \Psi(A_V u)\,, \quad u = \Psi^{-1} v\,.$$

Then

$$\dot{a}^\varepsilon(\tau) = -\mu \hat{A}_V(a^\varepsilon(\tau)) + Y\big(a^\varepsilon(\tau), \varepsilon^{-1}\tau\big)\,, \tag{23}$$

where

$$Y(a,t) = \Phi_{t\Lambda}\Big(P(\Phi_{-t\Lambda}(a))\Big). \tag{24}$$

Let $r \in (d/2+1, n]$. Since the operators $\Phi_{t\Lambda}$, $t \in \mathbb{R}$, define isometries of h^r, then, in view of (20), for any $t \in \mathbb{R}$ we have

$$Y(\cdot,t) \in Lip_{\mathscr{C}_r}(H^r, H^{r-1}). \tag{25}$$

For any $s \geq 0$ we denote by X^s the space

$$X^s = C([0,T], h^s),$$

given the supremum-norm. Then

$$|a^\varepsilon|_{X^{s*}} \leq M_1, \qquad |\dot{a}^\varepsilon|_{X^{s*-2}} \leq C(M_1). \tag{26}$$

Since for $0 \leq \gamma \leq 1$ we have

$$|v|_{\gamma(s_*-2)+(1-\gamma)s_*} \leq |v|^{\gamma}_{s_*-2}|v|^{1-\gamma}_{s_*}$$

by the interpolation inequality, then in view of (26) for any $s_* - 2 < \bar{s} < s_*$ and $0 \leq \tau_1 \leq \tau_2 \leq 1$ we have

$$|a^\varepsilon(\tau_2) - a^\varepsilon(\tau_1)|_{\bar{s}} \leq C(M_1)^\gamma(\tau_2-\tau_1)^\gamma(2M_1)^{1-\gamma}, \tag{27}$$

for a suitable $\gamma = \gamma(\bar{s}, s_*) > 0$, uniformly in ε.

Denote

$$\mathscr{Y}(v,t) = Y(v,t) - R(v).$$

Then by Lemma 4 relation (25) also holds for the map $v \mapsto \mathscr{Y}(v,t)$, for any t.

The following lemma is the main step of the proof.

Lemma 5. *For every $s' > d/2+1$, $s_* - 2 < s' < s_*$ we have*

$$\Big|\int_0^{\tilde{\tau}} \mathscr{Y}(a^\varepsilon(\tau), \varepsilon^{-1}\tau)d\tau\Big|_{s'} \leq \delta(\varepsilon, M_1), \quad \forall\, \tilde{\tau} \in [0,1], \tag{28}$$

where $\delta(\varepsilon, M_1) \to 0$ as $\varepsilon \to 0$.

Proof. Below in this proof we write $a^\varepsilon(\tau)$ as $a(\tau)$. We divide the time interval $[0, 1]$ into subintervals $[b_{l-1}, b_l]$, $l = 1, \cdots, N$ of length $L = \varepsilon^{1/2}$:

$$b_k = Lk \quad \text{for} \quad k = 0, \ldots, N-1, \quad b_N = 1, b_N - b_{N-1} \leq L,$$

where $N \leq 1/L + 1 \leq 2/L$.

In virtue of (25) and Lemma 4(ii),

$$\Big|\int_{b_{N-1}}^{b_N} \mathscr{Y}(a(\tau),\varepsilon^{-1}\tau)\,d\tau\Big|_{s'} \leq LC(s',s,M_1)\,. \tag{29}$$

Similar, if $\bar{\tau}\in[b_r,b_{r+1})$ for some $0\leq r<N$, then $|\int_{b_r}^{\bar{\tau}}\mathscr{Y}\,d\tau|_{s'}$ is bounded by the r.h.s. of (29).

Now we estimate the integral of $\mathscr{Y}$ over any segment $[b_l,b_{l+1}]$, where $l\leq N-2$. To do this we write it as

$$\begin{aligned}\int_{b_l}^{b_{l+1}} \mathscr{Y}(a(\tau),\varepsilon^{-1}\tau)\,d\tau &= \int_{b_l}^{b_{l+1}} \big(Y(a(b_l),\varepsilon^{-1}\tau)-R(a(b_l))\big)\,d\tau\\ &+\int_{b_l}^{b_{l+1}} \big(Y(a(\tau),\varepsilon^{-1}\tau)-Y(a(b_l),\varepsilon^{-1}\tau)\big)\,d\tau\\ &+\int_{b_l}^{b_{l+1}} \big(R(a(b_l))-R(a(\tau))\big)\,d\tau\,.\end{aligned}$$

In view of Lemma 4 and (27) the $h^{s'}$-norm of the second and third terms in the r.h.s. are bounded by $C(s',s,M_1)L^{1+\gamma}$. Since

$$\begin{aligned}&\varepsilon\int_0^{\varepsilon^{-1}L} Y(a(b_l),\varepsilon^{-1}b_l+s)\,ds\\ &= L\Phi_{\Lambda\varepsilon^{-1}b_l}\frac{1}{L^{-1}}\int_0^{L^{-1}} \Phi_{\Lambda s}P(\Phi_{-\Lambda s}(\Phi_{-\Lambda\varepsilon^{-1}b_l}a(b_l)))ds\,,\end{aligned}$$

then using Corollary 1 and Lemma 4(iii) we see that this equals $LR(a(b_l))+\varkappa_1(L^{-1})$, where $|\varkappa_1(L^{-1})|_{s'}\leq\overline{\varkappa}(L^{-1};M_1)$ and $\overline{\varkappa}\to 0$ when $L^{-1}\to\infty$. We have arrived at the estimate

$$\Big|\int_{b_l}^{b_{l+1}} \mathscr{Y}(a(\tau),\varepsilon^{-1}\tau)\,d\tau\Big|_{s'} \leq L\Big(\overline{\varkappa}(L^{-1};M_1)+CL^{\gamma}\Big)\,. \tag{30}$$

Since $N\leq 2/L$ and $L=\varepsilon^{1/2}$, then by (30) and (29) the l.h.s. of (28) is bounded by $2\overline{\varkappa}(\varepsilon^{-1/2};M_1)+C\varepsilon^{\gamma/2}+C\varepsilon^{1/2}$. It implies the assertion of the lemma. □

Consider the effective equation (11). By Lemma 4 this is the linear parabolic equation $\dot{u}-\Delta u+V(x)u=0$, written in the v-variables, perturbed by a locally Lipschitz operator of order one. So its solution $\tilde{a}(\tau)$ such that $\tilde{a}(0)=v_0$ exists (at least) locally in time. Denote by $\tilde{T}$ the stopping time

$$\tilde{T}=\min\{\tau\in[0,1]:\ |\tilde{a}(\tau)|_{s*}\geq M_1+1\}\,,$$

where, by definition, $\min\emptyset=1$.

Now consider the family of curves $a^{\varepsilon}(\cdot) \in X^{s*}$. In view of (26), (27) and the Arzelà-Ascoli theorem (e.g. see in [11]) this family is pre-compact in each space X^{s_1}, $s_1 < s_*$. Hence, for any sequence $\varepsilon_j' \to 0$ there exists a subsequence $\varepsilon_j \to 0$ such that

$$a^{\varepsilon_j}(\cdot) \underset{\varepsilon_j \to 0}{\longrightarrow} a^0(\cdot) \quad \text{in} \quad X^{s_1}.$$

By this convergence, (26) and the Fatou lemma,

$$|a^0(\tau)|_{s*} \le M_1 \qquad \forall\, 0 \le \tau \le 1. \tag{31}$$

In view of Lemma 5, the curve $a^0(\tau)$ is a mild solution of Eq. (11) in the space h^{s_1}, i.e,

$$a(\tau) - a(0) = \int_0^{\tau} \big(-\mu \hat{A}_V a(s) + R(a(s)) \, ds\,, \quad \forall\, 0 \le \tau \le 1$$

(the equality holds in the space h^{s_1-2}). So $a^0(\tau) = \tilde{a}(\tau)$ for $0 \le \tau \le \tilde{T}$. In view of (31) and the definition of the stopping time $\tilde{T}$ we see that $\tilde{T} = 1$. That is, $\tilde{a} \in X^{s*}$ and

$$a^{\varepsilon}(\cdot) \longrightarrow \tilde{a}(\cdot) \quad \text{in} \quad X^{s_1}, \tag{32}$$

where $\varepsilon = \varepsilon_j \to 0$. Since the limit $\tilde{a}$ does not depend on the sequence $\varepsilon_j \to 0$, then the convergence holds as $\varepsilon \to 0$.

Now we show that the convergence (32) holds uniformly for $v_0 \in B(M_0, h^{s*})$. Assume the opposite. Then there exists $\delta > 0$, sequences $\tau_j \in [0,1], a_0^j \in B(M_0, h^{s*})$, and $\varepsilon_j \to 0$ such that if $a^{\varepsilon_j}(\cdot)$ is a solution of (23) with initial data a_0^j and $\varepsilon = \varepsilon_j$, and $\tilde{a}^j(\cdot)$ is a solution of the effective equation (11) with the same initial data, then

$$|a^{\varepsilon_j}(\tau_j) - \tilde{a}^j(\tau_j)|_{s_1} \ge \delta. \tag{33}$$

Using again the Arzelà-Ascoli theorem and (27), replacing the subsequence $\varepsilon_j \to 0$ by a suitable subsequence, we have that

$$\begin{aligned}
&\tau_j \to \tau_0 \in [0,1],\\
&a_0^j \to a_0 \quad \text{in} \quad h^{s_1}, \quad \text{where} \quad a_0 \in h^{s*},\\
&a^{\varepsilon_j}(\cdot) \to a^0(\cdot) \quad \text{in} \quad X^{s_1},\\
&\tilde{a}^j(\cdot) \to \tilde{a}^0(\cdot) \quad \text{in} \quad X^{s_1}.
\end{aligned}$$

Clearly, $\tilde{a}^0(\cdot)$ is a solution of Eq. (11) with the initial datum a_0. Due to Lemma 5, $a^0(\cdot)$ is a mild solution of Eq. (11) with $a^0(0) = a_0$. Hence we have $a^0(\tau) = \tilde{a}^0(\tau)$, $\tau \in [0, 1]$, particularly, $a^0(\tau_0) = \tilde{a}^0(\tau_0)$. This contradicts with (33), so the convergence (32) is uniform in $v_0 \in B(M_0, h^{s_*})$.

Since

$$|I(a) - I(\tilde{a})|_{s_1}^{\sim} \le 4|a - \tilde{a}|_{s_1}(|a|_{s_1} + |\tilde{a}|_{s_1}),$$

then the convergence (32) implies the statement of Theorem 1.

5 The Randomly Forced Case

We study here the effect of the addition a random forcing to Eq. (1). Namely, we consider equation (13). We suppose that

$$B_s = 2\sum_{j=1}^{\infty} \lambda_j^{2s} b_j^2 < \infty \quad \text{for } s = s_* \in (d/2 + 1, n],$$

and impose a restriction on the nonlinearity $\mathscr{P}$ by assuming that there exists $\bar{N} \in \mathbb{N}$ and for each $s \in (d/2 + 1, n]$ there exists C_s such that

$$\|\mathscr{P}(\nabla u, u)\|_{s-1} \le C_s(1 + \|u\|_s)^{\bar{N}}, \quad \forall\, u \in H^s \tag{34}$$

(this assumption holds e.g. if $\mathscr{P}(\nabla u, u)$ is a polynomial in $(u, \nabla u)$).

Passing to the slow time $\tau = \varepsilon t$, Eq. (13) becomes (cf. (2))

$$\dot{u} + \varepsilon^{-1} i(-\Delta + V(x))u = \mu\Delta u + \mathscr{P}(\nabla u, u) + \frac{d}{d\tau}\sum_{k=1}^{\infty} b_k \boldsymbol{\beta}_k e_k(x), \qquad u = u(\tau, x), \tag{35}$$

which, in the v-variables, takes the form (cf. (4))

$$dv_k + \varepsilon^{-1} i\lambda_k v_k\, d\tau = (-\mu\lambda_k v_k + P_k(v))\, d\tau + \sum_{l=1}^{\infty} \Psi_{kl} b_l d\boldsymbol{\beta}_l\,, \quad k \in \mathbb{N}\,, \tag{36}$$

where we have denoted by $\{\Psi_{kl}, k, l \ge 1\}$ the matrix of the operator Ψ (see (3)) with respect to the basis $\{e_k\}$ in H^0 and $\{\zeta_k\}$ in h^0. We assume

Assumption A′. *There exist $s_* \in (d/2 + 1, n]$ and an ε-independent $T > 0$ such that for any $u_0 \in H^{s_*}$, Eq. (35) has a unique strong solution $u(\tau, x)$, $0 \le \tau \le T$, equal to u_0 at $\tau = 0$. Furthermore, for each p there exists a $C = C_p(\|u_0\|_{s_*}, B_{s_*}, T)$ such that*

$$\mathbf{E} \sup_{0 \le \tau \le T} \|u(\tau)\|_{s_*}^p \le C\,. \tag{37}$$

Remark 1. The Assumption A′ is not too restrictive. In particular, in [14] it is verified for Eq. (13) if $\mu > 0$ and $\mathscr{P}(u) = -u + zf_p(|u|^2)u$, where $f_p(r)$ is a smooth function, equal $|r|^p$ for $|r| \geq 1$, and $\Im z \leq 0, \Re z \leq 0$. The degree p is any real number if $d = 1, 2$ and $p < 2/(d-2)$ if $d \geq 3$.

Under this assumption, a result analogous to Theorem 1 holds. Namely, the limiting behaviour of the action variables I_k (see (6)) is described by the stochastically forced effective equation (cf. (11))

$$d\tilde{a}_k = (-\mu\lambda_k\tilde{a}_k + R_k(\tilde{a}))\,d\tau + \sum_{l=1}^{\infty} B_{kl}d\boldsymbol{\beta}_l\,, \quad k \in \mathbb{N}\,, \tag{38}$$

where we have defined $\{B_{kr}, k, r \geq 1\}$ as the principal square root of the real matrix

$$A_{kr} = \begin{cases} \sum_l b_l^2 \Psi_{kl}\Psi_{rl} \text{ if } \lambda_k = \lambda_r\,, \\ 0 \qquad\qquad\qquad \text{else}\,, \end{cases} \tag{39}$$

which defines a nonnegative selfadjoint compact operator in h^0. Note that since R is locally Lipschitz by Lemma 4, then strong solutions for (38) exist and are unique till the stopping time $\tau_K = \inf\{\tau \geq 0 : |\tilde{a}(\tau)|_{s*} = K\}$, where K is any positive number.

In the theorem below $v^\varepsilon(\tau)$ denotes a solution of (36) with the initial value $v_0 \in h^{s*}$.

Theorem 2. *If Assumption A′ holds, there exists a unique strong solution $\tilde{a}(\tau)$, $0 \leq \tau \leq T$, of Eq. (38) such that $\tilde{a}(0) = v_0 = \Psi(u_0) \in h^{s*}$, and*

$$\mathscr{D}(I(v^\varepsilon(\tau))) \rightharpoonup \mathscr{D}(I(\tilde{a}(\tau))) \quad \text{as } \varepsilon \to 0\,,$$

in $C([0,T], h_I^{s_1})$, for any $s_1 < s_*$.

In the theorem's assertion and below the arrow $\rightharpoonup$ stands for the weak convergence of measures. Let us assume further:

Assumption B′. *(i) Equation (13) has a unique strong solution $u(\tau)$, $u(0) = u_0 \in H^{s*}$, defined for $\tau \geq 0$, and*

$$\mathbf{E} \sup_{\theta \leq \tau \leq \theta+1} \|u(\tau)\|_{s*}^p \leq C \quad \text{for any } \theta \geq 0\,, \tag{40}$$

where $C = C(\|u_0\|_{s*}, B_{s*})$.
(ii) Equation (38) has a unique stationary measure μ^0 and is mixing.

Remark 2. The assumption (i) is fulfilled, for example, for equations, discussed in Remark 1. Assumption (ii) holds trivially if for a.e. realisation of the random force any two solutions of Eq. (38) converge exponentially fast.[1] For less trivial examples, corresponding to perturbations of linear systems with non-resonant or completely resonant spectra, see [14, 15].

Assumption B′ (i) and the Bogolyubov-Krylov argument, applies for solutions, starting from 0, imply that Eq. (13) has a stationary measure μ^ε, supported by the space H^{s_*}, and inheriting estimates (40).

Theorem 3. *Let us suppose that Assumptions A′ and B′ hold. Then*

$$\lim_{\varepsilon\to 0} \mu^\varepsilon = \mu^0 , \tag{41}$$

weakly in h^{s_1}, for any $s_1 < s_*$. The measure μ^0 is invariant with respect to transformations $\Phi_{t\Lambda}$, $t \in \mathbb{R}$. If, in addition, (13) is mixing and μ^ε is its unique stationary measure, then for any solution $u^\varepsilon(t)$ of (13) with ε-independent initial data $u_0 \in H^{s_*}$, we have

$$\lim_{\varepsilon\to 0}\lim_{t\to\infty} \mathscr{D}(v^\varepsilon(t)) = \mu^0 ,$$

where $v^\varepsilon(t) = \Psi(u^\varepsilon(t))$.

For examples of mixing Eq. (13) see [14] and references in that work. In particular, (13) is mixing if $\mathscr{P}(u, \nabla u) = \mathscr{P}(u)$ is a smooth function such that all its derivatives are bounded uniformly in u, cf. Remark 2.

For the case when the spectrum Λ is non-resonant (see Example 1) or is completely resonant, i.e. (22) holds, the theorem was proved in [14, 15].

The proofs of Theorem 2 and 3 closely follow the arguments in [14–16]. Proof of Theorem 3, in addition, uses some technical ideas from [4] (see there Corollary 4.2). The proofs are given, respectively, in Sects. 6 and 7.

6 Proof of Theorem 2

As in the proof of Theorem 1, let us assume, without loss of generality, that $T = 1$, $s_1 > d/2 + 1$ and $s_1 > s_* - 2$ (recall that $s_1 < s_*$ and $s_* \in (d/2 + 1, n]$).

Following the suite of [15] (see also [16]) we pass once again to the a-variables, defined in (8)). In view of (36), they satisfy the system (cf. (10))

$$da_k = \left(-\mu\lambda_k a_k + Y_k(a, \varepsilon^{-1}\tau)\right) d\tau + e^{i\varepsilon^{-1}\lambda_k\tau} \sum_l \Psi_{kl} b_l d\boldsymbol{\beta}_l, \quad k \in \mathbb{N} , \tag{42}$$

[1] This is fulfilled, for example, if (i) holds and $\mathscr{P}(u) = -u + \mathscr{P}_0(u)$, where the Lipschitz constant of $\mathscr{P}_0$ is less than one.

where Y is defined in (24). For any p we denote

$$X^p = C([0,1], h^p), \qquad X_I^p = C([0,1], h_I^p).$$

Let a^ε be a solution of (42) such that $a^\varepsilon(0) = v_0 = \Psi(u_0) \in h^{s*}$; we will often write a for a^ε to shorten notation. Denote the white noise in (42) as $\dot{\zeta}(t,x)$ and denote $U_1(\tau) = Y(a(\tau), \varepsilon^{-1}\tau)$, $U_2(\tau) = -\hat{A}_V a(\tau)$. Then

$$\dot{a} - \dot{\zeta} = U_1 + U_2 .$$

In view of (34), $\|U_1\|_{s*-1} = |P(v)|_{s*-1} \le C(1 + \|u(\tau)\|_{s*}^{\bar{N}})$. So, by (37),

$$\mathbf{E}\int_\tau^{(\tau+\tau')\wedge 1} \|U_1\|_{s*-1}\, dt \le C\int_\tau^{(\tau+\tau')\wedge 1} \mathbf{E}C(1 + \|u(t)\|_{s*}^{\bar{N}})\, dt \le C(\|u_0\|_{s*}, B_{s*})\tau' ,$$

for any $\tau \in [0,1]$ and $\tau' > 0$. Similar,

$$\mathbf{E}\int_\tau^{(\tau+\tau')\wedge 1} \|U_2\|_{s*-2}\, dt \le \mu C\mathbf{E}\int_\tau^{(\tau+\tau')\wedge 1} \|u\|_{s*} \le \mu C(\|u_0\|_{s*}, B_{s*})\tau' .$$

Hence, there exists $\gamma > 0$ such that

$$\mathbf{E}\left\|(a-\zeta)((\tau+\tau')\wedge 1) - (a-\zeta)(\tau)\right\|_{s_1} \le C(\|u_0\|_{s*}, B_{s*})\tau'^{\gamma} ,$$

in virtue of the interpolation and Hölder inequalities (cf. (27)). It is classical that

$$\mathbf{P}\{\|\zeta\|_{C^{1/3}([0,1],h^{s_1})} \le R_3\} \to 1 \quad \text{as} \quad R_3 \to \infty .$$

In view of what was said, for any $\delta > 0$ there is a set $Q_\delta^1 \subset X^{s_1}$, formed by equicontinuous functions, such that

$$\mathbf{P}\{a^\varepsilon \in Q_\delta^1\} \ge 1 - \delta ,$$

for each ε. By (37),

$$\mathbf{P}\{\|a^\varepsilon\|_{X^{s*}} \ge C\delta^{-1}\} \le \delta ,$$

for a suitable C, uniformly in ε. Consider the set

$$Q_\delta = \left\{a^\varepsilon \in Q_\delta^1 : \|a\|_{X^{s*}} \le C\delta^{-1}\right\} .$$

Then $\mathbf{P}\{a^\varepsilon \in Q_\delta\} \ge 1 - 2\delta$, for each ε. By this relation and the Arzelà-Ascoli theorem (e.g., see [11], ¯8), the set of laws $\{\mathscr{D}(a^\varepsilon(\cdot)),\ 0 < \varepsilon \le 1\}$, is tight in X^{s_1}.

So by the Prokhorov theorem there is a sequence $\varepsilon_l \to 0$ and a Borel measure $\mathcal{Q}^0$ on X^{s_1} such that

$$\mathcal{D}(a^{\varepsilon_l}(\cdot)) \rightharpoonup \mathcal{Q}^0 \quad \text{as} \quad \varepsilon_l \to 0\,. \tag{43}$$

Accordingly, due to (9), for actions of solutions v^ε we have the convergence

$$\mathcal{D}\left(I\left(v^{\varepsilon_l}(\cdot)\right)\right) \rightharpoonup I \circ \mathcal{Q}^0 \quad \text{as} \quad \varepsilon_l \to 0\,, \tag{44}$$

in $X_I^{s_1}$.

Theorem 2 follows then as a simple corollary from

Proposition 1. *There exists a unique weak solution $a(\tau)$ of the effective equation* (38) *such that $\mathcal{D}(a) = \mathcal{Q}^0$, $a(0) = v^0$ a.s.; and the convergences* (43) *and* (44) *hold as $\varepsilon \to 0$.*

Proof. The proof follows the Khasminski scheme (see [6, 12]), as expounded in [15]. Namely, we show that the limiting measure $\mathcal{Q}^0$ is a martingale solution of the limiting equation, which turns out to be exactly the Eq. (38). Since the equation has a unique solution, then the convergences (43), (44) hold as $\varepsilon \to 0$.

For $\tau \in [0, 1]$ consider the processes

$$N_k^{\varepsilon_l} = a_k^{\varepsilon_l}(\tau) - \int_0^\tau \left(-\mu\lambda_k a_k^{\varepsilon_l}(s) + R_k(a^{\varepsilon_l}(s))\right)\, ds\,, \quad k \geq 1$$

(cf. Eq. (38)). Due to (42) we write $N_k^{\varepsilon_l}$ as

$$N_k^{\varepsilon_l}(\tau) = \tilde{N}_k^{\varepsilon_l}(\tau) + \overline{N}_k^{\varepsilon_l}(\tau)\,,$$

where $\tilde{N}_k^{\varepsilon_l}(\tau) = a^{\varepsilon_l}(\tau) - \int_0^\tau (-\mu\lambda_k a^{\varepsilon_l}(s) + Y_k(a^{\varepsilon_l}(s), \varepsilon_l^{-1}s))ds$ is a $\mathcal{Q}^0$ martingale and the disparity $\overline{N}_k^{\varepsilon_l}$ is

$$\overline{N}_k^{\varepsilon_l}(\tau) = \int_0^\tau \mathcal{Y}_k(a^{\varepsilon_l}(s), \varepsilon_l^{-1}s)ds$$

(as before, $\mathcal{Y}(a,t) = Y(a,t) - R(a)$).

The key point is then a stochastic counterpart of Lemma 5, which is proved below:

Lemma 6. *For every $k \in \mathbb{N}$, $\mathbf{E}\,\mathfrak{A}_k^\varepsilon \to 0$ as $\varepsilon \to 0$, where*

$$\mathfrak{A}_k^\varepsilon = \max_{0 \leq \tilde{\tau} \leq 1} \left| \int_0^{\tilde{\tau}} \mathcal{Y}_k(a^\varepsilon(\tau), \varepsilon^{-1}\tau)d\tau \right|\,.$$

This lemma and the convergence (43) imply that the processes

$$N_k(\tau) = a_k(\tau) - \int_0^\tau (-\mu\lambda_k a_k + R_k(a))\, ds\,, \quad k \geq 1\,,$$

are $\mathscr{Q}^0$ martingales, considered on the probability space ($\Omega = X^{s_1}, \mathscr{F}, Q^0$) ($\mathscr{F}$ is the Borel sigma-algebra), given the natural filtration ($\mathscr{F}_\tau, 0 \leq \tau \leq 1$). For details see [17], Proposition 6.3).

Consider then the diffusion matrix $\{\mathscr{A}_{kr}, k, r \geq 1\}$ for the system (42), i.e.,

$$\mathscr{A}_{kr} = \exp(i\varepsilon^{-1}\tau(\lambda_k - \lambda_r)) \sum_{l=1}^{\infty} b_l^2 \Psi_{kl} \bar{\Psi}_{rl}\,.$$

Clearly, $\int_0^{\tilde{\tau}} \mathscr{A}_{kr} d\tau \to A_{kr}\tilde{\tau}$, as $\varepsilon \to 0$, where A denotes the diffusion matrix for the system (38) (cf. (39)). Similar to Lemma 6, we also find that

$$\mathbf{E} \max_{0 \leq \tilde{\tau} \leq 1} \left| \int_0^{\tilde{\tau}} \mathscr{Y}_k(a^\varepsilon(\tau), \varepsilon^{-1}\tau) d\tau \right|^2 \to 0 \quad \text{as } \varepsilon \to 0\,.$$

Then, using the same argument as before, we see that the processes

$$\begin{aligned} N_k(\tau)N_r(\tau) - A_{kr}\tau &= \left(\tilde{N}_k \tilde{N}_r - \int_0^\tau \mathscr{A}_{kr} ds \right) \\ &\quad + \left(\overline{N}_k \overline{N}_r + \overline{N}_k \tilde{N}_r + \tilde{N}_k \overline{N}_l - \int_0^\tau (\mathscr{A}_{kr} - A_{kr})\, ds \right) \end{aligned}$$

are $\mathscr{Q}^0$ martingales. That is, $\mathscr{Q}^0$ is a solution of the martingale problem with the drift R and the diffusion A (see [21]), so the assertion follows. □

Proof of Lemma 6. We adopt a convenient notation from our previous publications. Namely, we denote by $\varkappa(r)$ various functions of r such that $\varkappa \to 0$ as $r \to \infty$. We write $\varkappa(r; M)$ to indicate that $\varkappa(r)$ depends on a parameter M. Besides for events Q and O and a random variable f we write $\mathbf{P}_O(Q) = \mathbf{P}(O \cap Q)$ and $\mathbf{E}_O(f) = \mathbf{E}(\chi_O f)$.

The constants below may depend on k, but this dependence is not indicated since k is fixed through the proof. By $M \geq 1$ we denote a constant which will be specified later. Denote by $\Omega_M = \Omega_M^\varepsilon$ the event

$$\Omega_M = \left\{ \sup_{0 \leq \tau \leq 1} |a^\varepsilon(\tau)|_{s_*} \leq M \right\}\,.$$

Then, by (37),

$$\mathbf{P}(\Omega_M^c) \leq \varkappa(M). \tag{45}$$

In view of Lemma 4(ii) and (34), for any $t \in [0, \varepsilon^{-1}]$ and any $a \in h^{s_*}$ the difference $\mathscr{Y} = Y - R$ satisfies

$$|\mathscr{Y}_k(a,t)| \le |Y_k(a,t)| + |R_k(a)| \le |P_k(v)| + |R_k(a)| \le C(1 + |a|_{s_*})^{\bar N}. \tag{46}$$

Using this and (45) we get

$$\begin{aligned} \mathbf{E}_{\Omega_M^c} \mathfrak{A}_k^\varepsilon &\le \int_0^1 \mathbf{E}_{\Omega_M^c} |\mathscr{Y}_k(a(\tau), \varepsilon^{-1}\tau)| d\tau \\ &\le C\left(\mathbf{P}(\Omega_M^c)\right)^{1/2} \int_0^1 \left(\mathbf{E}(1 + |a|_{s_*})^{2\bar N}\right)^{1/2} d\tau \le \varkappa(M)\,. \end{aligned} \tag{47}$$

To estimate $\mathbf{E}_{\Omega_M} \mathfrak{A}_k^\varepsilon$, as in Lemma 5 we consider a partition of [0, 1] by the points

$$b_n = nL, \quad 0 \le n \le N-1, \;\; b_{N-1} \ge 1 - L, \; b_N = 1\,, \qquad L = \varepsilon^{1/2}\,,$$

$N \sim 1/L$. Let us denote

$$\eta_l = \int_{b_l}^{b_{l+1}} \mathscr{Y}_k(a(\tau), \varepsilon^{-1}\tau) d\tau\,, \quad 0 \le l \le N-1\,.$$

Since for $\omega \in \Omega_M$ and any $\tau' < \tau''$ such that $\tau'' - \tau' \le L$, in view of (46) we have $\left|\int_{\tau'}^{\tau''} \mathscr{Y}_k(a(\tau), \varepsilon^{-1}\tau) d\tau\right| \le LC(M)$, then

$$\mathbf{E}_{\Omega_M} \mathfrak{A}_k^\varepsilon \le LC(M) + \mathbf{E}_{\Omega_M} \sum_{l=0}^{N-1} |\eta_l|\,. \tag{48}$$

Let us fix any $\bar s > d/2 + 1$, $s_* - 2 < \bar s < s_*$, sufficiently small $\gamma > 0$, and consider the event

$$\mathscr{F}_l = \left\{ \sup_{b_l \le \tau \le b_{l+1}} |a^\varepsilon(\tau) - a^\varepsilon(b_l)|_{\bar s} \ge L^\gamma \right\}\,.$$

By the equicontinuity of the processes $\{a^\varepsilon(\tau)\}$ on suitable events with arbitrarily close to one ε-independent probability (as shown above), the probability of $\mathbf{P}(\mathscr{F}_l)$ goes to zero with L, uniformly in l and ε. Since $|\eta_l| \le C(M)L$ for $\omega \in \Omega_M$ and each l, then

$$\sum_{l=0}^{N-1} \left| \mathbf{E}_{\Omega_M} |\eta_l| - \mathbf{E}_{\Omega_M \setminus \mathscr{F}_l} |\eta_l| \right| \le C(M)L \sum_{l=0}^{N-1} \mathbf{P}_{\Omega_M}(\mathscr{F}_l) \le C(M)\varkappa(L^{-1})\,, \tag{49}$$

and it remains to estimate $\sum_l \mathbf{E}_{\Omega_M \setminus \mathscr{F}_l} |\eta_l|$.

We have

$$|\eta_l| \le \left|\int_{b_l}^{b_{l+1}} \left(\mathscr{Y}_k(a(\tau),\varepsilon^{-1}\tau) - \mathscr{Y}_k(a(b_l),\varepsilon^{-1}\tau)\right) d\tau\right| + \left|\int_{b_l}^{b_{l+1}} \left(\mathscr{Y}_k(a(b_l),\varepsilon^{-1}\tau)\right) d\tau\right| =: \Upsilon_l^1 + \Upsilon_l^2 .$$

By (20) and Lemma 4(ii), in Ω_M the following inequality hold:

$$\left|\mathscr{Y}_k(a(\tau),\varepsilon^{-1}\tau) - \mathscr{Y}_k(a(b_l),\varepsilon^{-1}\tau)\right| \le C(M)\,|a(\tau) - a(b_l)|_{\tilde{s}} .$$

So that, by the definition of $\mathscr{F}_l$,

$$\sum_l \mathbf{E}_{\Omega_M\setminus\mathscr{F}_l}\Upsilon_l^1 \le L^{\gamma} C(M) = \varkappa(\varepsilon^{-1};M) . \tag{50}$$

It remains to estimate the expectation of $\sum \Upsilon_l^2$. In view of (30) (with $M_1 = M$) we have

$$\sum_l \mathbf{E}_{\Omega_M\setminus\mathscr{F}_l}\Upsilon_l^2 \le NL\varkappa_1(\varepsilon^{-1};M) = \varkappa(\varepsilon^{-1};M). \tag{51}$$

Now the inequalities (47)–(51) jointly imply that

$$\mathbf{E}\,\mathfrak{A}_k^{\varepsilon} \le \varkappa(M) + \varkappa(\varepsilon^{-1};M) .$$

Choosing first M large and then ε small we make the r.h.s. arbitrarily small. This proves the lemma.

Lemma 6 estimates integrals of the differences

$$e^{i\varepsilon^{-1}\tau\lambda_k}P_k\left(\Phi_{-\varepsilon^{-1}\tau\lambda_k}(a^{\varepsilon}(\tau)\right) - \langle P\rangle_{\Lambda,k}(a^{\varepsilon}(\tau)) .$$

Similar result holds if we replace the averaging $\langle\cdot\rangle_{\Lambda,k}$ by $\langle\langle\cdot\rangle\rangle_{\Lambda}$ and the function P_k by any Lipschitz function:

Lemma 7. *Let $f \in Lip_1(h^{s_1}) =: Lip_1$ (i.e., f is a bounded Lipschitz function on h^{s_1}). Then*

(i) $\mathbf{E}\int_0^1 \left(f(\Phi_{-\tau\varepsilon^{-1}\Lambda}a^{\varepsilon}(\tau)) - \langle\langle f\rangle\rangle_{\Lambda}(a^{\varepsilon}(\tau))\right) d\tau \to 0$ *as* $\varepsilon \to 0$;

(ii) if in (i) f is replaced by $f^{\theta} = f\circ\Phi_{\theta}$, $\theta \in \mathbb{T}^{\infty}$, then the rate of convergence does not depend on θ.

Proof. To get (i) we literally repeat the proof of Lemma 6, using Lemma 2 instead of Lemma 1. The assertion (ii) follows from Lemma 2 and item (iv) of Lemma 1. □

7 Proof of Theorem 3

Let $v^\varepsilon(\tau)$, $0 \le \tau \le 1$, be a stationary solution for Eq. (13) such that $\mathscr{D}(v^\varepsilon(\tau)) \equiv \mu^\varepsilon$, and let $a^\varepsilon(\tau) = \Phi_{\varepsilon^{-1}\Lambda\tau} v^\varepsilon(\tau)$ be its interaction representation. Since v inherits the a-priori estimate (40) (with $u_0 = 0$), then an analogy of the convergence (43) holds for a suitable sequence $\varepsilon_l \to 0$. The argument from the proof of Proposition 1 applies and imply that

$$\mathscr{D}(a^{\varepsilon_l}(\cdot)) \rightharpoonup \mathscr{D}(a^0(\cdot)) \quad \text{in} \quad X^{s_1} \quad \text{as} \quad \varepsilon_l \to 0\,, \tag{52}$$

where a^0 is a weak solution of (38). We may also assume that

$$\mu^{\varepsilon_l} \rightharpoonup \bar{\mu}^0 \quad \text{in} \quad h^{s_1}\,, \tag{53}$$

for some measure $\bar{\mu}^0$.

Let us take any $f \in Lip_1(h^{s_1})$. Then

$$\mathbf{E}\int_0^1 f(v^\varepsilon(\tau))\,d\tau = \mathbf{E}\int_0^1 f(\Phi_{-\varepsilon^{-1}\Lambda\tau} a^\varepsilon(\tau))\,d\tau\,.$$

Applying to the second integral Lemma 7 we find that

$$\int_0^1 \mathbf{E}f(v^\varepsilon(\tau))\,d\tau = \int_0^1 \mathbf{E}\langle\langle f\rangle\rangle_\Lambda(a^\varepsilon(\tau))\,d\tau + \varkappa(\varepsilon^{-1})\,. \tag{54}$$

Since the function $\langle\langle f\rangle\rangle_\Lambda$ is invariant with respect to transformations $\Phi_{\Lambda t}$, $t \in \mathbb{R}$ (see item b) of Lemma 2), then $\langle\langle f\rangle\rangle_\Lambda(a^\varepsilon(\tau)) = \langle\langle f\rangle\rangle_\Lambda(v^\varepsilon(\tau))$. So both integrands in (54) are independent from τ, and

$$\mathbf{E}f(v^\varepsilon(\tau)) = \mathbf{E}\langle\langle f\rangle\rangle_\Lambda(a^\varepsilon(\tau)) + \varkappa(\varepsilon^{-1}) \quad \forall\,\tau\,. \tag{55}$$

Now let us take for f the function $\tilde{f} = \tilde{f}_{\varepsilon^{-1}\tau} = f \circ \Phi_{\varepsilon^{-1}\Lambda\tau}$ (which also belongs to $Lip_1(h^{s_1})$). Then

$$\mathbf{E}f(a^\varepsilon(\tau)) = \mathbf{E}\tilde{f}(v^\varepsilon(\tau)) = \mathbf{E}\langle\langle \tilde{f}\rangle\rangle_\Lambda(a^\varepsilon(\tau)) + \varkappa(\varepsilon^{-1}) = \mathbf{E}\langle\langle f\rangle\rangle_\Lambda(a^\varepsilon(\tau)) + \varkappa(\varepsilon^{-1})\,,$$

where $\varkappa$ may be chosen the same for all functions $\tilde{f}$ in view of Lemma 7(i). Comparing this with (55) and using (53) we find that

$$\mathbf{E}f(a^{\varepsilon_l}(\tau)) \rightharpoonup \langle f, \bar{\mu}^0\rangle \quad \text{as} \quad \varepsilon_l \to 0,$$

for each τ. Therefore, in virtue of (52), $\mathscr{D}(a^0(\tau)) \equiv \bar{\mu}^0$. So $a^0(\tau)$ is a stationary solution for (38), and $\bar{\mu}^0$ is a stationary measure for this equation. Since the latter is unique, $\bar{\mu}^0 \equiv \mu^0$, and (53) implies the convergence (41).

Replacing in (55) f by $\tilde{f}_t$ and using Lemma 2 b) we see that

$$\langle f, \Phi_{\Lambda t} \circ \mu^{\varepsilon} \rangle = \langle \tilde{f}_t, \mu^{\varepsilon} \rangle = \langle f, \mu^{\varepsilon} \rangle + \varkappa(\varepsilon^{-1}).$$

Passing to the limit as $\varepsilon \to 0$ we get the claimed invariance of the measure μ^0. Finally, the last assertion immediately follows from (41). □

Acknowledgements We are thankful to Anatoli Neishtadt for discussing the finite-dimensional averaging. This work was supported by l'Agence Nationale de la Recherche through the grant STOSYMAP (ANR 2011BS0101501).

Appendix

Consider the CGL equation (14), where $\mathscr{P} : \mathbb{C}^{d(d+1)/2+d+1} \times T^d \to \mathbb{C}$ is a C^{∞}-smooth function. We write it in the v-variables and slow time $\tau = \varepsilon t$:

$$\dot{v}_k + \varepsilon^{-1} i\lambda_k v_k = P_k(v), \quad k \in \mathbb{N},$$

where

$$P(v) := (P_k(v), k \in \mathbb{N}) = \Psi(\mathscr{P}(\nabla^2 u, \nabla u, u, x)), \quad u = \Psi^{-1} v,$$

and introduce the effective equation

$$\dot{\tilde{a}} = \langle P \rangle_{\Lambda}(\tilde{a}). \tag{56}$$

By Lemma 3 P defines smooth locally Lipschitz mappings $h^s \to h^{s-2}$ for $s > 2 + d/2$. So by a version of Lemma 4, $\langle P \rangle_{\Lambda} \in Lip_{loc}(h^s; h^{s-2})$ for $s > 2 + d/2$. Assume that

Assumption E. *There exists $s_0 \in (d/2, n]$ such that the effective equation (56) is locally well posed in the Hilbert spaces h^s, with $s \in [s_0, n] \cap \mathbb{N}$.*

Let $u^{\varepsilon}(t,x)$ be a solution of Eq. (14) with initial datum $u_0 \in H^s$, $v^{\varepsilon}(\tau) = \Psi(u(\varepsilon^{-1}\tau, x))$, and $\tilde{a}(\tau)$ be a solution of Eq. (56) with initial datum $\Psi(u_0)$. Then we have the following result:

Theorem 4. *If Assumptions A and E hold and $s > \max\{s_0 + 2, d/2 + 4\}$, then the solution of the effective equation exists for $0 \le \tau \le T$, and for any $s_1 < s$ we have*

$$I(v^{\varepsilon}(\cdot)) \underset{\varepsilon \to 0}{\longrightarrow} I(\tilde{a}(\cdot)) \quad in \quad C([0,T], h_I^{s_1}).$$

The proof of this theorem follows that of Theorem 1, with slight modifications. Cf. [8], where the result is proven for the non-resonant case.

References

1. Arnold, V., Kozlov, V.V., Neistadt, A.I.: *Mathematical Aspects of Classical and Celestial Mechanics*, 3rd edn. Springer, Berlin (2006)
2. Bambusi, D.: Galerkin averaging method and Poincaré normal form for some quasilnear PDEs. Ann. Scuola Norm. Sup. Pisa C1. Sci. **4**, 669–702 (2005)
3. Bogoljubov, N.N., Mitropol'skij, J.A.: *Asymptotic Methods in the Theory of Non-linear Oscillations*. Gordon & Breach, New York (1961)
4. Dymov, A.: Nonequilibrium statistical mechanics of weakly stochastically perturbed system of oscillators. Preprint (2015) [arXiv: 1501.04238]
5. Faou, E., Germain, P., Hani, Z.: The weakly nonlinear large box limit of the 2D cubic nonlinear Schrödinger equation. Preprint (2013) [arXiv1308.6267]
6. Freidlin, M.I., Wentzell, A.D.: Averaging principle for stochastic perturbations of multifrequency systems. Stoch. Dyn. **3**, 393–408 (2003)
7. Gérard, P., Grellier, S.: Effective integrable dynamics for a certain nonlinear wave equation. Anal. PDE **5**, 1139–1154 (2012)
8. Huang, G.: An averaging theorem for nonlinear Schrödinger equations with small nonlinearities. DCDS-A **34**(9), 3555–3574 (2014)
9. Huang, G.: Long-time dynamics of resonant weakly nonlinear CGL equations. J. Dyn. Diff. Equat. 1–13 (2014). doi:10.1007/s10884-014-9391-0
10. Huang, G., Kuksin, S.B.: KdV equation under periodic boundary conditions and its perturbations. Nonlinearity **27**, 1–28 (2014)
11. Kelley, J.L., Namioka, I.: *Linear Topological Spaces*. Springer, New York/Heidelberg (1976)
12. Khasminski, R.: On the averaging principle for Ito stochastic differential equations. Kybernetika **4**, 260–279 (1968) (in Russian)
13. Kuksin, S.: Damped-driven KdV and effective equations for long-time behavior of its solutions. Geom. Funct. Anal. **20**, 1431–1463 (2010)
14. Kuksin, S.: Weakly nonlinear stochastic CGL equations. Ann. Inst. H. Poincaré B **49**(4), 1033–1056 (2013)
15. Kuksin, S., Maiocchi, A.: Resonant averaging for weakly nonlinear stochastic Schrödinger equations. Nonlinearity **28**, 2319–2341 (2015)
16. Kuksin, S., Maiocchi, A.: The limit of small Rossby numbers for randomly forced quasi-geostrophic equation on β-plane. Nonlinearity **28**, 2319–2341 (2015)
17. Kuksin, S., Piatnitski, A.: Khasminskii-Whitham averaging for randomly perturbed KdV equation. J. Math. Pures Appl. **89**, 400–428 (2008)
18. Lochak, P., Meunier, C.: *Multiphase Averaging for Classical Systems*. Springer, New York/Berlin/Heidelberg (1988)
19. Nazarenko, S.: *Wave Turbulence*. Springer, Berlin (2011)
20. Runst, T., Sickel, W.: *Sobolev Spaces of Fractional Order, Nemytskij Operators, and Nonlinear Partial Differential Equations*, vol. 3. de Gruyter, Berlin (1996)
21. Stroock, D., Varadhan, S.R.S.: *Multidimensional Diffusion Processes*. Springer, Berlin (1979)
22. Zygmund, A.: *Trigonometric Series*, vol. II, 3th edn. Cambridge University Press, Cambridge (2002)

Partial Differential Equations with Random Noise in Inflationary Cosmology

Robert H. Brandenberger

Abstract Random noise arises in many physical problems in which the observer is not tracking the full system. A case in point is inflationary cosmology, the current paradigm for describing the very early universe, where one is often interested only in the time-dependence of a subsystem. In inflationary cosmology it is assumed that a slowly rolling scalar field leads to an exponential increase in the size of space. At the end of this phase, the scalar field begins to oscillate and transfers its energy to regular matter. This transfer typically involves a parametric resonance instability. This article reviews work which the author has done in collaboration with Walter Craig studying the role which random noise can play in the parametric resonance instability of matter fields in the presence of the oscillatory inflation field. We find that the particular idealized form of the noise studied here renders the instability more effective. As a corollary, we obtain a new proof of finiteness of the localization length in the theory of Anderson localization.

1 Background

This article reviews work done in collaboration with Walter Craig applying rigorous results from the theory of random matrix differential equations to problems motivated by early Universe cosmology [1, 2]. As a corollary, we obtain a new proof of the positivity of the Lyapunov exponent, corresponding to the finiteness of the localization length in the theory of Anderson localization [3].

Over the past two decades, cosmology has developed into a data-driven field. Thanks to new telescopes we are obtaining high precision data about the structure of the universe on large scales. Optical telescopes are probing the distribution of stellar matter to greater depths, microwave telescopes have allowed us to make detailed maps of anisotropies in the cosmic microwave background radiation at fractions of 10^{-5} of the mean temperature. In the coming years microwave telescopes outfitted with polarimeters will allow us to produce polarization maps of the microwave

R.H. Brandenberger (✉)
Physics Department, McGill University, Montreal, QC, Canada H3A 2T8
e-mail: rhb@physics.mcgill.ca

P. Guyenne et al. (eds.), *Hamiltonian Partial Differential Equations and Applications*, Fields Institute Communications 75,
DOI 10.1007/978-1-4939-2950-4_12

background, and prototype telescopes are being developed which will allow us to measure the three-dimensional distribution of all baryonic matter (not just the stellar component): this is by measuring the redshifted 21 cm radiation.

The data from optical telescopes yield three dimensional maps of the density distribution of stellar matter in space. This data can be quantified by taking a Fourier transform of the data and determining the density power spectrum, the square of the amplitude of the Fourier modes, as a function of wavenumber k. Similarly, the sky maps of the temperature of the cosmic microwave background can be quantified by expanding the maps in spherical harmonics and determining the square of the amplitudes of the coefficients as a function of the angular quantum number l. One of the goals of modern cosmology is to find a causal mechanism which can explain the origin of these temperature and density fluctuations.

The data are being interpreted in a theoretical framework in which space-time is a four dimensional pseudo-Riemannian manifold $\mathcal{M}$ with a metric $g_{\mu\nu}$ with signature $(+,-,-,-)$, and evolves in the presence of matter as determined by the Einstein field equations

$$G_{\mu\nu} = 8\pi G T_{\mu\nu}\,, \tag{1}$$

where $G_{\mu\nu}$ is the Einstein tensor constructed from the metric and its first and second derivatives, G is Newton's gravitational constant, and $T_{\mu\nu}$ is the energy-momentum tensor of matter.

In physics, it is believed that all fundamental equations of motion follow from an action principle. The physical trajectories extremize the action when considering fluctuations of the fields. The total action for space-time and matter is

$$S = \int d^4x\sqrt{-g}\Big[\frac{R}{8\pi G} + \mathcal{L}_m(\varphi_i)\Big]\,, \tag{2}$$

where φ_i are matter fields (functions of space-time which represent matter), $\mathcal{L}_M$ is the Lagrangian for the matter fields (which is obtained by covariantizing the matter action in Special Relativity), g is the determinant of the metric tensor, and R is the Ricci scalar. For simplicity, cosmologists usually consider scalar matter fields (and not the fermionic and gauge fields which represent most of the matter particles which are known to exist in Nature—the only scalar field known to exist is the Higgs field).

The gravitational field Equations (1) follow from varying the joint gravitational and matter action S with respect to the metric, and the equations for matter follow from varying S with respect to each of the matter fields, leading to

$$\mathcal{D}_g^2(\varphi_i) = -\frac{\partial V}{\partial \varphi_i}\,, \tag{3}$$

where $\mathcal{D}_g^2$ is the covariant d'Alembertian operator in the metric g, and $V(\varphi)$ is the total potential energy density of the matter fields. We have assumed above that the kinetic terms of the matter fields are independent of each other and of canonical form (the reader not familiar with this physics jargon can simply take (3) to define what the form of the matter Lagrangian is).

Cosmologists are lucky since observations show that the metric of space-time is to a first order homogeneous and isotropic on large length scales, and hence describable by the metric

$$ds^2 = dt^2 - a(t)^2\left(dx^2 + dy^2 + dz^2\right). \tag{4}$$

In the above, t is physical time, and x, y, and z are Cartesian coordinates on the three-dimensional constant time hypersurfaces. For simplicity (and because current observations show that this is an excellent approximation) we have assumed that the spatial hypersurfaces are spatially flat as opposed to positively curved three spheres or negatively curved hyperspheres (the three possibilities for the spatial hypersurfaces consistent with homogeneity and isotropy).

The function $a(t)$ is called the "cosmic scale factor". In the presence of matter, space-time cannot be static. In the absence of external forces, matter follows geodesics, and matter initially at rest remains at constant values of x, y, and z. Hence, these coordinates are called "comoving". The function $a(t)$ thus represents the spatial radius of a ball of matter locally at rest. Currently, the Universe is expanding and hence $a(t)$ is an increasing function of time. The Einstein equation (1) yield the following equations for the scale factor:

$$H^2 = \frac{8\pi G}{3}\rho, \tag{5}$$

$$\dot{\rho} = -3H(\rho + p), \tag{6}$$

where ρ and p are the energy density and pressure density of matter, respectively, and

$$H(t) \equiv \frac{\dot{a}}{a} \tag{7}$$

is the Hubble expansion rate.

In Standard Big Bang cosmology matter is given as a superposition of pressure-less "cold matter" with $p = 0$ and relativistic radiation with $p = \rho/3$. At late times, the cold matter dominates and it then follows from (5) and (6) that

$$a(t) \sim \left(\frac{t}{t_0}\right)^{2/3}, \tag{8}$$

where t_0 is the normalization time (often taken to be the present time). With and without radiation, Standard Big Bang cosmology suffers from a singularity at $t = 0$. At that time, the curvature of space-time as well as temperature and density of matter blow up. This is clearly unphysical: no physical detector can ever measure an infinite

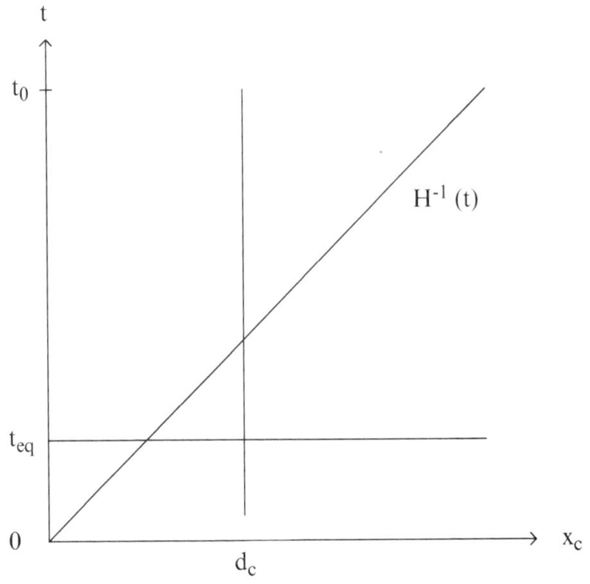

Fig. 1 Space-time sketch depicting the "structure formation problem" of Standard Big Bang cosmology. The *vertical axis* is time, the *horizontal* one denotes comoving spatial coordinates. The *solid vertical line* denotes the wavelength of a mode for which fluctuations are observed, the diagonal curve through the origin is the cosmological horizon which denotes the limit of causal influence. As depicted, at early times (in particular at the time t_{eq} when structures can start to grow) the wavelength of the mode is larger than the horizon and hence no causal structure formation scenario is possible

result, and in addition the assumption that matter can be treated as an ideal classical fluid breaks down at the high energy densities when quantum and particle physics effects become important.

In addition, Standard Big Bang cosmology cannot explain the observed homogeneity and isotropy of the universe, and it cannot provide a causal mechanism for the generation of the structure in the universe which current data reveal. The last point is illustrated in the space-time sketch of Fig. 1. The vertical axis is time, the horizontal axis gives the comoving dimension of space. The region of causal influence of a point at the initial time is bounded by the "horizon", the forward light cone of the initial point. In Standard Big Bang cosmology the horizon increases as t. In contrast, the physical length of a particular structure in the universe (which is not gravitationally bound) grows in proportion to $a(t)$ which in Standard cosmology grows much more slowly than t. Hence, if we trace back the wavelength $\lambda(t)$ of structures seen at the present time on large cosmological scales, we see that $\lambda(t) > t$ at early times. Hence, it is impossible to explain the origin of the seeds which develop into the structures observed today in a causal way (since the seeds have to be present in the very early universe). These problems of Standard Big Bang cosmology motivated the development of the "Inflationary Universe" scenario.

2 The Inflationary Universe

The idea behind the Inflationary Universe scenario is very simple [4]: it is postulated that there is an epoch in the very early stage of cosmology during which the scale factor expands exponentially, i.e.

$$a(t) \sim e^{Ht}, \tag{9}$$

where here H is a constant. This period lasts from some initial time t_i to a final moment t_R (see Fig. 2).

In inflationary cosmology the time evolution of the horizon and of $\lambda(t)$ are modified compared to what happens in Standard Cosmology: the horizon expands exponentially in the interval between t_i and t_R, and so does $\lambda(t)$. In contrast, the Hubble radius $l_H(t)$ defined as the inverse Hubble expansion rate

$$l_H(t) = H^{-1}(t) \tag{10}$$

is constant. Provided that the period of inflation is sufficiently long, then the horizon will at all times be larger than $\lambda(t)$ for any wavelength which can currently be observed (see Fig. 3). Thus, there is no causality problem to have homogeneity and isotropy on scales currently observed. As follows from the study of linearized fluctuations about the background (4), the Hubble radius is the upper limit on the length scales on which fluctuations can be created. From Fig. 3 it can be seen that in inflationary cosmology perturbation modes originate with a length smaller than the Hubble radius. Thus, it is possible that inflation could provide a causal mechanism for the formation of the structures which are currently observed. In fact, it turns out the quantum vacuum fluctuations in the exponentially expanding phase yield such a mechanism [5], but this is not the focus of this article (see [6] for a review of this topic).

In order to obtain inflationary expansion in the context of Einstein's theory of space and time, it follows from Eqs. (5) and (6) that a form of matter with

$$p = -\rho \tag{11}$$

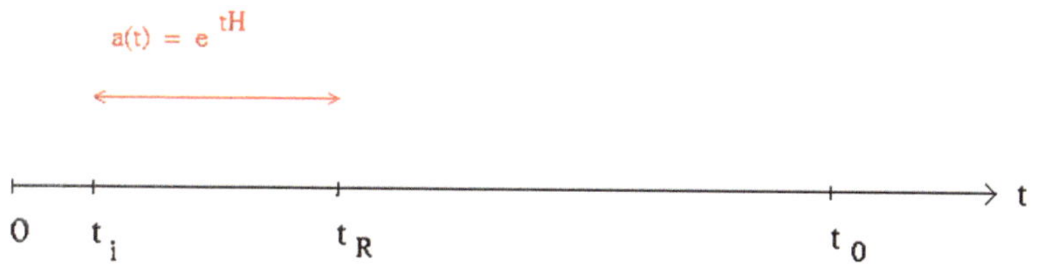

Fig. 2 Time line of inflationary cosmology. The period of inflation begins at the time t_i and ends at t_R

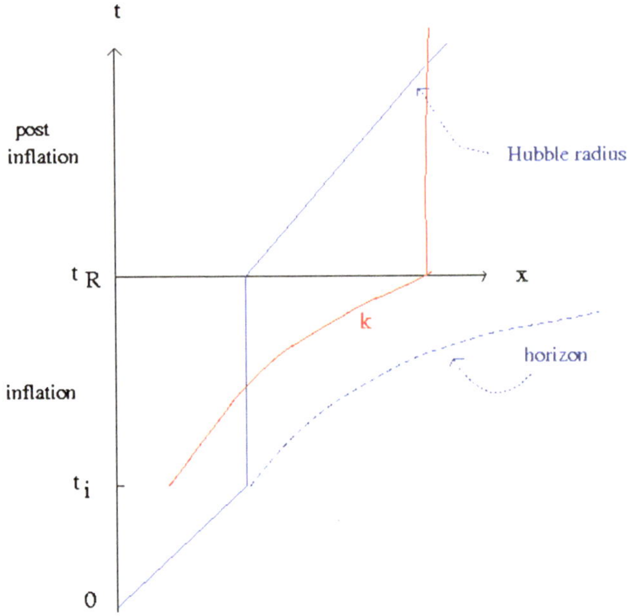

Fig. 3 Space-time sketch of inflationary cosmology. The *vertical axis* denotes time, the *horizontal axis* is physical distance. During the period of inflation, the horizon expands exponentially. Similarly, the physical wavelength of a perturbation mode (*the curve* denoted by *k*) grows exponentially, and it is smaller than the Hubble radius at the beginning of the period of inflation, provided the period of inflation lasts sufficiently long

is required. No such equation of state can be obtained using classical fluids, nor can it be obtained from fields representing the usual fermionic and gauge degrees of matter, at least in the context of renormalizable matter theories. Hence, a scalar field is required to obtain inflation. Even with scalar fields, it is not easy to obtain inflation since we must ensure that the potential energy density dominates over kinetic and spatial gradient energies over a long period of time. This follows from the following expressions for the energy density and pressure of scalar field matter

$$\rho = \frac{1}{2}(\dot{\varphi})^2 + \frac{1}{2}(\nabla\varphi)^2 + V(\varphi)\,, \tag{12}$$

$$p = \frac{1}{2}(\dot{\varphi})^2 - \frac{1}{6}(\nabla\varphi)^2 - V(\varphi)\,. \tag{13}$$

A typical potential for a scalar field which can lead to inflation is [7]

$$V(\varphi) = \frac{1}{2}m^2\varphi^2\,. \tag{14}$$

The equation of motion for φ which follows from (3) is

$$\ddot{\varphi} + 3H\dot{\varphi} - \frac{\nabla^2}{a^2}\varphi = -V'(\varphi), \tag{15}$$

where the prime indicates the derivative with respect to φ. Inflation can arise if φ is slowly rolling, i.e. $\ddot{\varphi} \ll H\dot{\varphi}$ and $\dot{\varphi}^2 \ll V(\varphi)$. Slow rolling is possible for field values $|\varphi| > m_{pl}$, where m_{pl} is the Planck mass which is given by

$$m_{pl}^2 = \left(\frac{8\pi}{3}G\right)^{-1}. \tag{16}$$

The slow roll trajectory is given by

$$\varphi(t) = -\left(\frac{2}{9}\right)^{1/2} m m_{pl} t, \tag{17}$$

and it is in fact a local attractor in initial condition space for large field values [8].

Once $|\varphi|$ drops below m_{pl}, the slow-roll approximation breaks down, the inflationary period ends, and φ begins to oscillate about the minimum of its potential at $\varphi = 0$. The amplitude of the oscillations is damped by the expansion of space, i.e. by the second term on the left hand side of (15).

3 The Reheating Challenge

The field φ which leads to inflation cannot be any of the fields whose particles we have observed (except possibly the Higgs field if the latter is non-minimally coupled to gravity [9]). Any regular matter (matter which is not φ) which might have been present at the beginning of the period of inflation is exponentially diluted during inflation. Thus, at the end of inflation we have a state in which no regular matter is present (the regular matter fields are in their vacuum state), and all energy is locked up in the φ field. Thus, to make inflation into a viable model of the early universe, a mechanism is needed to convert the energy density in φ at the end of inflation into Standard Model particles.

As was discovered in [10, 11] and worked out later in more detail in [12–14] (see [15] for a recent review) the initial energy transfer proceeds via a parametric resonance instability which is described in more generality by Floquet theory. Let us represent regular matter by a scalar field χ which is weakly coupled to φ by an interaction term of the form

$$\mathcal{L}_I = \frac{1}{2} g \chi^2 \varphi, \tag{18}$$

where g is a constant which has dimensions of mass. The free action for χ is assumed to be that of a canonical massless scalar field with no bare potential (i.e. no self-interactions). In this case, the equation of motion for χ becomes

$$\ddot{\chi} + 3H\dot{\chi} - \left(\frac{\nabla^2}{a^2} + g\varphi\right)\chi = 0. \quad (19)$$

Since this is a linear differential equation, each Fourier mode χ_k of χ will evolve independently according to

$$\ddot{\chi}_k + 3H\dot{\chi}_k + \left(\frac{k^2}{a^2} - g\varphi\right)\chi_k = 0. \quad (20)$$

After the end of the period of inflation, φ undergoes damped oscillations. This leads to a periodic variation of the mass term in (20). This, in turn, leads to a resonant instability and to energy transfer from φ to χ.

Let us for a moment neglect the expansion of space. In this case $H = 0$ and $a = 1$ and then the basic matter equation (20) becomes

$$\ddot{\chi}_k + \big(k^2 - g\mathscr{A}\cos(\omega t)\big)\chi_k = 0, \quad (21)$$

where $\mathscr{A}$ is the amplitude of the oscillations of φ (constant if the expansion of space is neglected), and ω is the frequency of the oscillations (which equals m in our case). Readers will recognize this equation as the Mathieu equation [16], an equation which has exponentially growing solutions in resonance bands for k which are centered around half integer multiples of ω. Because of this instability, there will be conversion of energy between the φ field driving the resonance and the matter fields, as first pointed out in [10]. This instability was later given the name "preheating" [12]. Since the instability is exponential, we expect that the time scale of the energy conversion is small compared to the expansion time H^{-1}, and that hence the approximation of neglecting the expansion of space is self-consistent.

As discussed in [12–14], the expansion of space can be included in an elegant way. In terms of a rescaled field $X_k = a^{3/2}\chi_k$, the equation of motion (20) becomes an equation of the form

$$\ddot{X}_k + \Omega_k^2(t)X_k = 0, \quad (22)$$

with an effective frequency Ω which contains a periodically oscillating term. An exponential instability persists in this setup, and this justifies the simplified approach which we focus on in this article, where we neglect the expansion of space. In the following, we will extract the periodic term from the effective frequency, i.e.

$$\Omega_k^2(t) = \omega_k^2 + p(\omega t). \quad (23)$$

Our starting equation will be

$$\ddot{\chi}_k + \big(\omega_k^2 + p(\omega t)\big)\chi_k = 0, \quad (24)$$

where ω_k^2 generalizes the previous setup to the case in which the χ field has a non-vanishing mass m_χ

$$\omega_k^2 = k^2 + m_\chi^2 , \tag{25}$$

and p is a function with period 2π. Let us denote the amplitude of p by P. Previous work (see [15] for a review) has shown that if $P \ll \omega^2$ there is "narrow-band resonance" (only k modes within narrow resonance bands experience the instability), whereas if $P \gg \omega^2$ then there is "broad-band resonance" in which all modes with $k \ll \omega$ undergo exponential instability. In inflationary universe modes with broad-band resonance the reheating process is very rapid on Hubble time scale.

The Mathieu equation (24) is a special case of a Floquet type equation. According to Floquet theory [16, 17], the solutions of (24) scale as

$$\chi_k(t) \sim e^{\mu_k t + i\alpha_k t} , \tag{26}$$

where the constant $\mu_k + i\alpha_k$ is called the Floquet exponent, and the real part of it, μ_k, is the Lyapunov exponent. The constant α_k is the rotation number of the solution, which in the context of the theory of Schrödinger operators is the "integrated density of states".

The above setup is, however, too idealized for the purposes of real cosmology. The field φ which yields the inflationary expansion and the field χ are only two of many fields. All of them are excited in the early universe, and they are directly or indirectly coupled to χ, and they will hence give correction terms to the basic equation (24). The extra fields are called the "environment" in which the system under consideration lives. The environment is typically describable by random noise, which is uncorrelated in time with the time-dependence of $\varphi(t)$. We will now consider an idealized equation which includes effects of the noise:

$$\ddot{\chi} + \left(-\nabla^2 + p(\omega t) + q(x,t)\right)\chi = 0 , \tag{27}$$

where $q(x,t)$ is a stochastic variable whose time-dependence is uncorrelated with that of $\varphi(t)$.

4 Homogeneous Noise

Our basic equation (27) is a second order partial differential equation with random coefficients. In phase space, we obtain a random matrix equation which is first order in time.

To simplify the analysis, we will first consider the case of homogeneous noise, i.e. we will assume that the noise function q depends only on time. In this case, each Fourier mode of χ continues to evolve independently and satisfies the equation

$$\ddot{\chi}_k + \left(k^2 + p(\omega t) + q(t)\right)\chi = 0 . \tag{28}$$

This dramatically reduces the mathematical complexity of the problem: we now have a second order ordinary differential equation rather than a PDE.

To re-write this equation in the form of a random matrix equation we introduce the transfer matrix $\Phi_q(t,0)$ made up of two independent solutions $\phi_1(t;q)$ and $\phi_2(t;q)$ of (28) and their time derivatives:

$$\Phi_q(t,0) = \begin{pmatrix} \phi_1(t;q) & \phi_2(t;q) \\ \dot{\phi}_1(t;q) & \dot{\phi}_2(t;q) \end{pmatrix},$$

which satisfies the first order matrix equation

$$\dot{\Phi}_q = M(q(t),t)\Phi_q , \tag{29}$$

where the matrix M is given by

$$M = \begin{pmatrix} 0 & 1 \\ -(\omega_k^2 + p + q) & 0 \end{pmatrix}.$$

The transfer matrix describes the evolution of the system from initial time $t = 0$ to final time t.

Let us denote the transfer matrix in the absence of noise by $\Phi_0(t,0)$. According to Floquet theory (see e.g [16, 17]) for mathematical background), this matrix takes the form

$$\Phi_0(t,0) = P_0(t)e^{Ct} , \tag{30}$$

where $P_0(t)$ is a periodic matrix function with period ω^{-1}, and C is a constant matrix whose spectrum is

$$\text{spec}(C) = \{\pm\mu(0)\} , \tag{31}$$

where $\mu(0)$ is called the Lyapunov exponent in the absence of noise.

To study the effects of noise, we re-write the full transfer matrix by extracting the transfer matrix in the absence of noise:

$$\Phi_q(t,0) = \Phi_0(t,0)\Psi_q(t,0) , \tag{32}$$

where the non-triviality of the reduced matrix Ψ_q describes the effects of the noise. The reduced transfer matrix satisfies the equation

$$\dot{\Psi}_q = S\Psi_q , \tag{33}$$

where S is the following matrix:

$$S = \varphi_0^{-1} \begin{pmatrix} 0 & 0 \\ -q & 0 \end{pmatrix} \varphi_0 .$$

Let T be the period of the oscillation of ϕ. We can now write the transfer matrix as a product of transfer matrices over individual oscillation times:

$$\Phi_q(NT, 0) = \prod_{j=1}^{N} \Phi_q(jT, (j-1)T) , \tag{34}$$

where N is an integer.

Let us assume that the noise is uncorrelated in time when considered in different oscillation periods. In addition, let us assume that the noise is drawn from some probability measure on $\mathscr{C}(\mathscr{R})$ such that q restricted to a period fills a neighborhood of $\mathscr{C}(\mathscr{R})$. In this case, the overall Lyapunov exponent is well defined and can be extracted using the limit [1]

$$\mu(q) = \lim_{N\to\infty} \frac{1}{NT} \log || \prod_{j=1}^{N} \Phi_q(JT, (j-1)T) || , \tag{35}$$

where $\| \cdot \|$ indicates a matrix norm. Note that the dependence on the particular matrix norm vanishes in the limit $N \to \infty$.

The key result of [1] is that noise which obeys the above-mentioned conditions renders the instability stronger. More specifically, we have the following theorem:

Theorem 1. *Given a random noise function $q(t)$ which is uncorrelated on the time scale T and which is drawn from a probability measure on $\mathscr{C}(\mathscr{R})$ such that q restricted to a period fills a neighborhood of $\mathscr{C}(\mathscr{R})$, then*

$$\mu(q) > \mu(0) . \tag{36}$$

Note the strict inequality in the above theorem. At first sight, this result could be surprising since one might expect that noise could cut off an instability which occurs in the absence of noise. However, a physical way to understand the result of the above theorem is to realize that the noise we have introduced can only add energy to the system rather than drain energy. Thus, it is consistent to find that noise renders the resonant instability more effective.

The above theorem follows from the Furstenberg Theorem [18] on random matrices. This theorem takes the following form:

Theorem 2. *Given a probability distribution dA on $\Psi \in SL(2n, \mathscr{R})$ and defining G_A as the smallest subgroup of $SL(2n, \mathscr{R})$ containing the support of dA, then if G_A is not compact, and G_A restricted to lines has no invariant measure, then for almost* *all independent random sequences* $\{\Psi_j\}_{j=1}^{\infty}$ *distributed according to dA we have*

$$\lim_{N\to\infty} \frac{1}{N} \log || \prod_{j=1}^{N} \Psi_j || = \lambda > 0 . \tag{37}$$

In addition, for almost all vectors v_1 *and* v_2 *in* $\mathscr{R}^{2n}$ *the exponent* λ *can be extracted via*

$$\lambda = \lim_{N\to\infty} \frac{1}{N} \log < v_1, \prod_{j=1}^{N} \Psi_j v_2 > . \tag{38}$$

Note once again the strict inequality in the above theorem. Applied to our reheating problem, then for any mode k, the above theorems in the case of $n = 1$ can be used, and they imply that, in the presence of noise, the Floquet exponent increases for each value of k. In particular, if for a particular value of k there is no instability in the absence of noise, an instability will develop in the presence of noise.

The way that our result (36) follows[1] from Furstenberg's Theorem is the following. Let us take v_1 to be an eigenvector of $\Phi_0(T,0)^t$, the transverse of the noiseless transfer matrix with eigenvalue $exp(\mu(0)T)$. Then,

$$\begin{aligned}
\mu(q) &= \lim_{N\to\infty} \frac{1}{NT} \log\big(< v_1, \Phi_q(NT,0) v_2 >\big) \\
&= \lim_{N\to\infty} \frac{1}{NT} \log\big(e^{\mu_0 NT} < v_1, \prod_{j=1}^{N} \Psi_j(jT,(j-1)T) v_2 >\big) \\
&= \mu(0) + \lambda > \mu(0) ,
\end{aligned} \tag{39}$$

where in the last step we have used Furstenberg's Theorem.

From the point of view of physics, the restriction to homogeneous noise is not realistic. We must allow for inhomogeneous noise functions $q(x,t)$. This is the topic we turn to in the following section.

[1]There is, in fact, a small hole in our proof of Theorem 1: in the case of values of k in the resonance band of the noiseless system, the Ψ_j are not necessarily identically distributed on $SL(2)$ because of the exponential factor which enters. We still obtain the rigorous result $\mu(q) > 0$ for all values of k, and numerical evidence confirms the validity of the statement $\mu(q) > \mu(0)$ even for values of k which are in the resonance band. The application of our result to Anderson localization involves values of k which are in the stability bands of the noiseless system and is hence robust - I thank Walter Craig for pointing out this issue.

5 Inhomogeneous Noise

In the case of inhomogeneous noise we must return to the original partial differential equation (27) with random noise. By going to phase space we obtain a first order matrix operator differential equation

$$\dot{\Phi}_q = M(q(t,\cdot),t)\Phi_q \tag{40}$$

for the fundamental solution matrix operator Φ_q. The Floquet exponent for the inhomogeneous system is defined by

$$\mu_q = \lim_{N\to\infty}\log||\Phi_q(NT,0)||\,, \tag{41}$$

where, as before, $\|\cdot\|$ indicates a norm on the matrix operator space, and T is the period of the unperturbed system.

As in the previous section, we will separate out the effects of the noise by defining

$$\Phi_q(t) = \Phi_{q=0}(t)\Psi_q(t)\,, \tag{42}$$

where $\Phi_{q=0}$ is the fundamental solution matrix in the absence of noise, and $\Psi_q(t)$ is the matrix which encodes the effects of the random noise. We wish to compare the value of the Floquet exponent in the presence of noise with that of the noiseless system.

To dramatically reduce the complexity of the problem we apply a trick which is commonly used in physics. First, we introduce an infrared cutoff by replacing the infinite spatial sections $\mathscr{R}^3$ by a three-dimensional torus of side length L. This renders Fourier space discrete. Secondly, we impose an ultraviolet cutoff, namely we eliminate high "energy" modes with $k < \Lambda$, where Λ is the cutoff scale. The fundamental solution matrix of the cutoff problem is denoted by $\Phi_q^{L,\Lambda}(t)$, and the corresponding Floquet exponents are also denoted by superscripts.

After the above steps, our problem can be written in Fourier space as a ordinary matrix differential equation in $\mathscr{R}^{2n}$, where n is the number of Fourier modes which are left. The first pair of coordinates corresponds to the phase space coordinates of the first Fourier mode and so forth. In the absence of noise, the fundamental solution matrix is block diagonal—there is no mixing between different Fourier modes. In each block, $\Phi_{q=0}^{L,\Lambda}$ reduces to the transfer matrix of the corresponding Fourier mode discussed in the previous section. The noise term Ψ_q introduces mixing between the different blocks.

Since Furstenberg's Theorem is valid on $\mathscr{R}^{2n}$, the results of the previous section immediately apply and we have

$$\mu_q^{L,\Lambda} > \mu_0^{L,\Lambda}\,. \tag{43}$$

Note the fact that we have a strict inequality. Note also that the Floquet exponent in the noiseless case is the maximum of the Floquet exponents over all values of k:

$$\mu_0^{L,\Lambda} = \max_k(\mu_{k,0}), \tag{44}$$

where $\mu_{k,0}$ is the Floquet exponent for Fourier mode k in the absence of noise.

Let us now consider removing the limits, i.e. taking $L \to \infty$ and $\Lambda \to \infty$. Since in the absence of noise, there is no resonance for large k modes, the limit of the right hand side of (43) is well defined (in fact, the right hand side is independent of the cutoffs). For any finite value of the cutoffs, the result (43) is true. Hence, the result persists in the limit when the cutoffs are taken to infinity. However, one loses the strict inequality sign. Hence, assuming that the limit of the left hand side of (43) in fact exists, we obtain our final result

$$\mu_q \geq \max_k(\mu_{k,0}). \tag{45}$$

We in fact expect a stronger result. Let us denote by $\mu_q(k)$ the Floquet exponent of the dynamical system restricted to the k'th Fourier mode (the restriction made at the end of the evolution). Then we expect that due to the mode mixing the maximal growth rate over all Fourier modes of the noiseless system will influence all Fourier modes of the system with noise, and that hence

$$\mu_q(k) \geq \max_k(\mu_{k,0}). \tag{46}$$

Although we have numerical evidence [2] for the validity of this result, we have not been able to provide a proof.

6 New Proof of Anderson Localization

It is well known that there is a correspondence between classical time-dependent problems and a time-independent Schrödinger equation. Let us start from the second order differential equation (28) in the case of homogeneous noise. Let us now make the following substitutions:

$$\begin{aligned} \chi &\to \psi \\ t &\to x \\ \omega_k^2 &\to 2mE \\ p(\omega t) &\to -2mV_p(\omega x) \\ q(t) &\to -2mV_R(x). \end{aligned} \tag{47}$$

Then, the Eq. (28) becomes

$$H\psi = E\psi, \tag{48}$$

with the operator H given by

$$H = -\frac{1}{2m}\frac{\partial^2}{\partial x^2} + V_p(\omega x) + V_q(x), \tag{49}$$

which is the time-independent Schrödinger equation for the wavefunction ψ of an electron of mass m in a periodic potential $V_p(\omega x)$ of period ω^{-1} in the presence of a random noise term V_R in the potential.

In the absence of noise there are bands of values of E where there is no instability, and where hence the wave functions are oscillatory. In condensed matter physics, the corresponding solutions for ψ are known as Bloch wave states. Theorem 1 now implies that if a random potential is added, then the solutions for ψ become unstable. This means that there is one exponentially growing mode and one exponentially decaying mode. In quantum mechanics the growing mode is unphysical since it is not normalizable. Hence, the decaying mode is the only physical mode. This solution corresponds to a localized wave function. Thus, we have obtained a new proof of the finiteness of the localization length in the theory of "Anderson localization" [19] (for reviews see e.g. [20]).

Theorem 3. *Consider the time-independent Schrödinger equation for a particle in a periodic potential $V_p(\omega x)$, and consider a random noise contribution $V_q(x)$ which is uncorrelated on the length scale of the period of V_p and which is drawn from a probability measure on $\mathscr{C}(\mathscr{R})$ such that q restricted to a period fills a neighborhood of $\mathscr{C}(\mathscr{R})$. Then the presence of the noise localizes the wavefunction, and the localization strength is exponential, i.e. the wavefunction ψ_q in the presence of noise scales as*

$$\psi_q(x) \sim exp(-\mu(q)x), \tag{50}$$

where $\mu(q)$ is strictly positive on the basis of Theorem 1.

Note that our method can only be applied to study Anderson localization in one spatial dimension.

7 Conclusions

We have applied rigorous results from random matrix theory to study the effects of noise on reheating in inflationary cosmology. We have found that the type of noise studied here, namely a random noise contribution to the mass term in the Klein-Gordon equation for a scalar field representing Standard Model matter, renders the parametric resonance instability of matter production in the presence of an oscillating inflation field more effective. After the standard duality mapping between a time-dependent classical field theory problem and a time-independent quantum mechanical Schrödinger problem, we obtain a new proof of the finiteness of the localization length in the theory of "Anderson localization", a famous result in condensed matter physics. Our work is an example of how the same rigorous mathematics result can find interesting applications to diverse physics problems.

Acknowledgements I wish to thank P. Guyenne, D. Nicholls and C. Sulem for organizing this conference in honor of Walter Craig, and for inviting me to contribute. Walter Craig deserves special thanks for collaborating with me on the topics discussed here, for his friendship over many years, and for comments on this paper. The author is supported in part by an NSERC Discovery Grant and by funds from the Canada Research Chair program.

References

1. Zanchin, V., Maia Jr. A., Craig, W., Brandenberger, R.: Reheating in the presence of noise. Phys. Rev. D **57**, 4651 (1998). [arXiv:hep-ph/9709273]
2. Zanchin, V., Maia Jr. A., Craig, W., Brandenberger, R.: Reheating in the presence of inhomogeneous noise. Phys. Rev. D **60**, 023505 (1999). [arXiv:hep-ph/9901207]
3. Brandenberger, R., Craig, W.: Towards a new proof of Anderson localization. Eur. Phys. J. C **72**, 1881 (2012). [arXiv:0805.4217 [hep-th]]
4. Guth, A.H.: The inflationary universe: a possible solution to the Horizon and Flatness Problems. Phys. Rev. D **23**, 347–356 (1981)
5. Mukhanov, V.F., Chibisov, G.V.: Quantum fluctuation and nonsingular universe. JETP Lett. **33**, 532 (1981) [Pisma Zh. Eksp. Teor. Fiz. **33**, 549 (1981) (In Russian)]
6. Brandenberger, R.H.: Lectures on the theory of cosmological perturbations. Lect. Notes Phys. **646**, 127–167 (2004). [hep-th/0306071]
7. Linde, A.D.: Chaotic inflation. Phys. Lett. B **129**, 177–181 (1983)
8. Brandenberger, R.H., Kung, J.H.: Chaotic inflation as an attractor in initial condition space. Phys. Rev. D **42**(4), 1008–1015 (1990);
   Brandenberger, R.H., Feldman, H., Kung, J.: Initial conditions for chaotic inflation. Phys. Scripta T **36**, 64–69 (1991)
9. Bezrukov, F.L., Shaposhnikov, M.: The standard model Higgs boson as the inflaton. Phys. Lett. B **659**, 703–706 (2008). [arXiv:0710.3755 [hep-th]]
10. Traschen, J.H., Brandenberger, R.H.: Particle production during out-of-equilibrium phase transitions. Phys. Rev. D **42**, 2491–2504 (1990)
11. Dolgov, A.D., Kirilova, D.P.: On particle creation by a time dependent scalar field. Sov. J. Nucl. Phys. **51**, 172–177 (1990) [Yad. Fiz. **51**, 273 (1990)]
12. Kofman, L., Linde, A.D., Starobinsky, A.A.: Reheating after inflation. Phys. Rev. Lett. **73**, 3195–3198 (1994). [hep-th/9405187]
13. Shtanov, Y., Traschen, J.H., Brandenberger, R.H.: Universe reheating after inflation. Phys. Rev. D **51**, 5438–5455 (1995). [hep-ph/9407247]
14. Kofman, L., Linde, A.D., Starobinsky, A.A.: Towards the theory of reheating after inflation. Phys. Rev. D **56**, 3258–3295 (1997). [hep-ph/9704452]
15. Allahverdi, R., Brandenberger, R., Cyr-Racine, F.-Y., Mazumdar, A.: Reheating in inflationary cosmology: theory and applications. Ann. Rev. Nucl. Part. Sci. **60**, 27–51 (2010). [arXiv:1001.2600 [hep-th]]
16. McLachlan, N.: Theory and Applications of Mathieu Functions. Oxford University Press, Clarendon (1947)

17. Carmona, R., Lacroix, J.: Spectral Theory of Random Schrödinger Operators. Birkhauser, Boston (1990) [See p. 198 for a the proof of existence of the Floquet exponents]
18. Pastur, L., Figotin, A.: Spectra of Random and Almost-Periodic Operators. Springer, Berlin (1991) [See p. 256 for the theory of generalized Floquet (Lyapunov) exponents and p. 344 for Furstenberg's theorem]
19. Anderson, P.W.: Absence of diffusion in certain random lattices. Phys. Rev. **109**, 1492–1505 (1958);
Mott, N.F., Twose, W.D.: The theory of impurity conduction. Adv. Phys. **10**, 107–163 (1961);
Abrahams, E., Anderson, P.W., Licciardello, D.C., Ramakrishnan, T.V.: Scaling theory of localization: absence of quantum diffusion in two dimensions. Phys. Rev. Lett. **42**, 673–676 (1970)
20. Thouless, D.J.: Electrons in disordered systems and the theory of localization. Rep. Prog. Phys. **13**, 93–142 (1974);
Lee, P.A., Ramakrishnan, T.V.: Disordered electronic systems. Rep. Mod. Phys. **57**, 287–337 (1985);
Kramer, B., MacKimmon, A.: Localization: theory and experiment. Rep. Prog. Phys. **56**, 1469–1564 (1993)

Local Isometric Immersions of Pseudo-Spherical Surfaces and Evolution Equations

Nabil Kahouadji, Niky Kamran, and Keti Tenenblat

To Walter Craig, with friendship and admiration.

Abstract The class of differential equations describing pseudo-spherical surfaces, first introduced by Chern and Tenenblat (Stud. Appl. Math. **74**, 55–83, 1986), is characterized by the property that to each solution of a differential equation within the class, there corresponds a two-dimensional Riemannian metric of curvature equal to -1. The class of differential equations describing pseudo-spherical surfaces carries close ties to the property of complete integrability, as manifested by the existence of infinite hierarchies of conservation laws and associated linear problems. As such, it contains many important known examples of integrable equations, like the sine-Gordon, Liouville and KdV equations. It also gives rise to many new families of integrable equations. The question we address in this paper concerns the local isometric immersion of pseudo-spherical surfaces in $\mathbf{E}^3$ from the perspective of the differential equations that give rise to the metrics. Indeed, a classical theorem in the differential geometry of surfaces states that any pseudo-spherical surface can be locally isometrically immersed in $\mathbf{E}^3$. In the case of the sine-Gordon equation, one can derive an expression for the second fundamental form of the immersion that depends only on a jet of finite order of the solution of the pde. A natural question is to know if this remarkable property extends to equations other than the sine-Gordon equation within the class of differential equations describing pseudo-spherical surfaces. In an earlier paper (Kahouadji et al., Second-order equations and local isometric immersions of pseudo-spherical surfaces, 25 pp. [arXiv:1308.6545], to appear in Comm. Analysis and Geometry (2015), we have shown that this

N. Kahouadji
Department of Mathematics, Northwestern University, Evanston, IL 60208, USA
e-mail: nabil@math.northwestern.edu

N. Kamran (✉)
Department of Mathematics and Statistics, McGill University, Montreal, QC, Canada
e-mail: nkamran@math.mcgill.ca

K. Tenenblat
Departamento de Matemàtica, Universidade de Brasilia, Brasflia, Brazil
e-mail: keti@mat.unb.br

P. Guyenne et al. (eds.), *Hamiltonian Partial Differential Equations and Applications*, Fields Institute Communications 75,
DOI 10.1007/978-1-4939-2950-4_13

property fails to hold for all other second order equations, except for those belonging to a very special class of evolution equations. In the present paper, we consider a class of evolution equations for $u(x,t)$ of order $k \geq 3$ describing pseudo-spherical surfaces. We show that whenever an isometric immersion in $\mathbf{E}^3$ exists, depending on a jet of finite order of u, then the coefficients of the second fundamental form are universal, that is they are functions of the independent variables x and t only.

1 Introduction and Statement of Results

The notion of a partial differential equation describing pseudo-spherical surfaces was defined and studied extensively in a paper by Chern and Tenenblat [3]. The class of these equations is of particular interest because it enjoys a remarkable set of integrability properties in the case when a parameter playing the role of a spectral parameter is present in the 1-forms associated to the pseudo-spherical structure. Indeed, one obtains in that case an infinite sequence of conservation laws and an associated linear problem whose integrability condition is the given partial differential equation.[1] We begin by recalling some basic definitions. A partial differential equation

$$\Delta(t, x, u, u_t, u_x, \ldots, u_{t^l x^{k-l}}) = 0, \tag{1}$$

is said to describe pseudo-spherical surfaces if there exist 1-forms

$$\omega^i = f_{i1}dx + f_{i2}dt, \quad 1 \leq i \leq 3, \tag{2}$$

where the coefficients f_{ij}, $1 \leq i \leq 3$, $1 \leq j \leq 2$, are smooth functions of t, x, u and finitely many derivatives of u, such that the structure equations for a surface of Gaussian curvature equal to -1,

$$d\omega^1 = \omega^3 \wedge \omega^2, \quad d\omega^2 = \omega^1 \wedge \omega^3, \quad d\omega^3 = \omega^1 \wedge \omega^2 \tag{3}$$

are satisfied if and only if u is a solution of (1) for which $\omega^1 \wedge \omega^2 \neq 0$. In other words, for every smooth solution of (1) such that ω^1 and ω^2 are linearly independent, we obtain a Riemannian metric

$$ds^2 = (\omega^1)^2 + (\omega^2)^2, \tag{4}$$

[1]It is worth pointing out at this stage that the conservation laws arising from the geometry of pseudo-spherical surfaces may be non-local. We refer to [18] and the references therein for an explicit treatment of the relationship between the conservation laws obtained in [2] and the standard series of conservation laws obtained via the classical Wahlquist-Estabrook construction. We also refer to [9, 19, 20] for the study of the integrability properties of some specific families of equations describing pseudo-spherical surfaces.

of constant Gaussian curvature equal to -1, with ω^3 being the Levi-Civita connection 1-form. This condition is equivalent to the integrability condition for the linear problem given by

$$dv^1 = \frac{1}{2}(\omega^2 v^1 + (\omega^1 - \omega^3)v^2), \quad dv^2 = \frac{1}{2}((\omega^1 + \omega^3)v^1 - \omega^2 v^2). \tag{5}$$

For the purposes of this paper, the motivating example of a partial differential equation describing pseudo-spherical surfaces is the sine-Gordon equation

$$u_{tx} = \sin u, \tag{6}$$

for which a choice of 1-forms (2) satisfying the structure equations (3) is given by

$$\omega^1 = \frac{1}{\eta}\sin u\, dt, \quad \omega^2 = \eta\, dx + \frac{1}{\eta}\cos u\, dt, \quad \omega^3 = u_x\, dx, \tag{7}$$

where η is a non-vanishing real parameter. This continuous parameter is closely related to the parameter appearing in the classical Bäcklund transformation for the sine-Gordon equation and is central to the existence of infinitely many conservation laws for the sine-Gordon equation. It is noteworthy that there may be different choices of 1-forms satisfying the structure equations (3) for a given differential equation. For example, for the sine-Gordon equation (6), a choice different from the one given in (7) is given by

$$\omega^1 = \cos\frac{u}{2}(dx + dt), \quad \omega^2 = \sin\frac{u}{2}(dx - dt), \quad \omega^3 = \frac{u_x}{2}dx - \frac{u_t}{2}dt. \tag{8}$$

Partial differential equations (1) which describe pseudo-spherical surfaces and for which one of the components f_{ij} can be chosen to be a continuous parameter will be said to describe η pseudo-spherical surfaces. One important feature of the differential equations describing η-pseudo-spherical surfaces is that each such differential equation is the integrability condition of a linear system of the form (5), which may be used as a starting point in the inverse scattering method and lead to solutions of the differential equation (see for example [1]).

It is therefore an interesting problem to characterize the class of differential equations describing η-pseudo-spherical surfaces, and this is precisely what Chern and Tenenblat [3] did for k-th order evolution equations

$$u_t = F(u, u_x, \dots, u_{x^k}). \tag{9}$$

These results were extended to more general classes of differential equations in [13–15], and [16]. One can also remove the assumption that $f_{21} = \eta$ and perform a complete characterization of evolution equations of the form (9) which describe pseudo-spherical surfaces, as opposed to η pseudo-spherical surfaces [12]. It is also noteworthy that the classification results obtained by Chern and Tenenblat [3] for η-pseudo-spherical surfaces were extended in [17] to differential equations of the form $u_t = F(x, t, u, u_x, \ldots, \partial u/\partial x^k)$. Finally we mention that the concept of a differential equation that describes pseudo-spherical surfaces has a spherical counterpart [4], where similar classification results have been obtained. Further developments can be found in [2, 5–8, 10, 17, 18].

The property of a surface being pseudo-spherical is by definition intrinsic since it only depends on its first fundamental form. It is only recently [11] that the problem has been considered of locally isometrically immersing in $\mathbf{E}^3$ the pseudo-spherical surfaces arising from the solutions of partial differential equations describing pseudo-spherical surfaces. Let us first recall that any pseudo-spherical surface can be locally isometrically immersed into three-dimensional Euclidean space $\mathbf{E}^3$. This means that to any solution u of a partial differential equation (1) describing pseudo-spherical surfaces (for which $\omega^1 \wedge \omega^2 \neq 0$), there corresponds a local isometric immersion into $\mathbf{E}^3$ for the corresponding metric of constant Gaussian curvature equal to -1. The problem investigated in [11] was to determine to what extent the *second fundamental form* of the immersion could be expressed in terms of the solution u of a second-order equation and *finitely many* of its derivatives. The motivation for this question came from a remarkable property of the sine-Gordon equation, which we now explain. Let us first derive a set of necessary and sufficient conditions that the components a, b, c of the second fundamental form of any local isometric immersion into $\mathbf{E}^3$ of a metric of constant curvature equal to -1 must satisfy. Recall that a, b, c are defined by the relations

$$\omega_1^3 = a\omega^1 + b\omega^2, \quad \omega_2^3 = b\omega^1 + c\omega^2, \tag{10}$$

where the 1-forms ω_1^3, ω_2^3 satisfy the structure equations

$$d\omega_1^3 = -\omega_2^3 \wedge \omega_1^2, \quad d\omega_2^3 = -\omega_1^3 \wedge \omega_2^1, \tag{11}$$

equivalent to the Codazzi equations, and the Gauss equation

$$ac - b^2 = -1. \tag{12}$$

For the sine-Gordon equation, with the choice of 1-forms ω^1, ω^2 and $\omega^3 = \omega_1^2$ given by (8), it is easily verified that the 1-forms ω_1^3, ω_2^3 are given by

$$\omega_1^3 = \sin\frac{u}{2}(dx + dt) = \tan\frac{u}{2}\omega^1,$$
$$\omega_2^3 = -\cos\frac{u}{2}(dx - dt) = -\cot\frac{u}{2}\omega^2.$$

It is a most remarkable fact that the components a, b, c of the second fundamental form that we have just obtained depend only on the solution u of the sine-Gordon equation. Our main goal in [11] was to investigate to what extent this property was true for all second-order equations describing pseudo-spherical surfaces, in the sense that a, b, c should only depend on u and at most *finitely many derivatives* of u. We showed that this is an extremely rare event, essentially confined to the sine-Gordon equation. Indeed, what we proved was that except for the equation

$$u_{xt} = F(u), \quad F''(u) + \alpha u = 0, \tag{13}$$

where α is a positive constant, every second-order partial differential equation describing η-pseudo-spherical surfaces is such that either a, b, c are universal functions of t, x, meaning that they are independent of u, or they do not factor through any finite-order jet of u. The starting point of the proof of this result is a set of necessary and sufficient conditions, given in terms of the coefficients f_{ij} of the 1-forms (2), for a, b and c to be the components of the second fundamental form of a local isometric immersion corresponding to a solution of (1). These are equivalent to the Gauss and Codazzi equations, and are easily derived. We consider the pair of vector fields (e_1, e_2) dual to the coframe (ω^1, ω^2). It is given by

$$\begin{vmatrix} f_{11} & f_{21} \\ f_{12} & f_{22} \end{vmatrix} e_1 = f_{22}\partial_x - f_{21}\partial_t, \qquad \begin{vmatrix} f_{11} & f_{21} \\ f_{12} & f_{22} \end{vmatrix} e_2 = -f_{12}\partial_x + f_{11}\partial_t. \tag{14}$$

By feeding these expressions into the structure equations (11), we obtain

$$d\omega_1^3 = \Big(db(e_1) - da(e_2)\Big)\omega^1 \wedge \omega^2 - a\,\omega^2 \wedge \omega^3 + b\,\omega^1 \wedge \omega^3, \tag{15}$$

$$d\omega_2^3 = \Big(dc(e_1) - db(e_2)\Big)\omega^1 \wedge \omega^2 - b\,\omega^2 \wedge \omega^3 + c\,\omega^1 \wedge \omega^3. \tag{16}$$

Denoting by D_t and D_x the total derivative operators, these are equivalent to

$$f_{11}D_t a + f_{21}D_t b - f_{12}D_x a - f_{22}D_x b - 2b \begin{vmatrix} f_{11} & f_{31} \\ f_{12} & f_{32} \end{vmatrix} + (a-c) \begin{vmatrix} f_{21} & f_{31} \\ f_{22} & f_{32} \end{vmatrix} = 0, \tag{17}$$

$$f_{11}D_t b + f_{21}D_t c - f_{12}D_x b - f_{22}D_x c + (a-c) \begin{vmatrix} f_{11} & f_{31} \\ f_{12} & f_{32} \end{vmatrix} + 2b \begin{vmatrix} f_{21} & f_{31} \\ f_{22} & f_{32} \end{vmatrix} = 0. \tag{18}$$

These differential constraints, which amount to the Codazzi equations, have to be augmented by the Gauss equation

$$ac - b^2 = -1. \tag{19}$$

The proof of the main result of [11] consists in a detailed case-by-case analysis of the Eqs. (17), (18) and (19), where use is made of the expressions and constraints on the f_{ij}'s that result from the classification results of [3, 15], and where we assume that a, b and c depend on t, x, u and only finitely many derivatives of u.

Our goal in the present paper is to extend the results of [11] concerning the components a, b, c of the second fundamental form to the case of k-th order evolution equations. Our main result is given by:

Theorem 1. *Let*

$$u_t = F(u, u_x, \ldots, u_{x^k}) \tag{20}$$

be an evolution equation of order k describing η-pseudo-spherical surfaces. If there exists a local isometric immersion of a surface determined by a solution u for which the coefficients of the second fundamental form depend on a jet of finite order of u, i.e., a, b and c depend on $x, t, u, \ldots, u_{x^l}$, where l is finite, then a, b and c are universal, that is $l = 0$ and a, b and c depend at most on x and t only.

In Sect. 2, we give a proof of Theorem 1 based on a careful order-by-order analysis of the Codazzi equations (17), (18) and the Gauss equation (19). In Sect. 3, we show by means of an example that the class of evolution equations of order $k \geq 3$ for which the components a, b, c are universal in the sense of Theorem 1, that is independent of u and its derivatives, is non-empty.

2 Proof of the Main Result

In the case of a differential equation describing η-pseudo-spherical surfaces, the structure equations (3) are equivalent to

$$D_t f_{11} - D_x f_{12} = \Delta_{23} \tag{21}$$

$$D_x f_{22} = \Delta_{13} \tag{22}$$

$$D_t f_{31} - D_x f_{32} = -\Delta_{12} \tag{23}$$

where D_t and D_x are the total derivative operators and

$$\Delta_{12} := f_{11} f_{22} - \eta f_{12}; \quad \Delta_{13} := f_{11} f_{32} - f_{31} f_{12}; \quad \Delta_{23} = \eta f_{32} - f_{31} f_{22}. \tag{24}$$

We shall use the notation

$$z_i = u_{x^i} = \frac{\partial^i u}{\partial x^i}, \quad 0 \leq i \leq k, \tag{25}$$

introduced in [3] to denote the derivatives of u with respect to x and write the evolution equation (20) as

$$z_{0,t} = F(z_0, z_1, \ldots, z_k). \tag{26}$$

We will thus think of $(t, x, z_0, \ldots, z_k)$ as local coordinates on an open set of the submanifold of the jet space $J^k(\mathbf{R}^2, \mathbf{R})$ defined by the differential equation (20). We first recall the following lemma from [3]:

Lemma 1. *Let (26) be a k-th order evolution equation describing η-pseudo-spherical surfaces, with associated 1-forms (2) such that $f_{21} = \eta$. Then necessary conditions for the structure equations (3) to hold are given by*

$$f_{11,z_k} = \cdots = f_{11,z_0} = 0 \tag{27}$$

$$f_{21} = \eta \tag{28}$$

$$f_{31,z_k} = \cdots = f_{31,z_0} = 0 \tag{29}$$

$$f_{12,z_k} = 0 \tag{30}$$

$$f_{22,z_k} = f_{22,z_{k-1}} = 0 \tag{31}$$

$$f_{32,z_k} = 0 \tag{32}$$

$$f_{11,z_0}^2 + f_{31,z_0}^2 \neq 0. \tag{33}$$

We now proceed with the proof of Theorem 1. If a, b, c depend of a jet of finite order, that is a, b, c are functions of $x, t, z_0, \ldots, z_l$ for some finite l, then (17) and (18) become

$$f_{11}a_t + \eta b_t - f_{12}a_x - f_{22}b_x - 2b\Delta_{13} + (a - c)\Delta_{23} - \sum_{i=0}^{l}(f_{12}a_{z_i} + f_{22}b_{z_i})z_{i+1} + \sum_{i=0}^{l}(f_{11}a_{z_i} + \eta b_{z_i})z_{i,t} = 0,$$

and

$$f_{11}b_t + \eta c_t - f_{12}b_x - f_{22}c_x + (a - c)\Delta_{13} + 2b\Delta_{23} - \sum_{i=0}^{l}(f_{12}b_{z_i} + f_{22}c_{z_i})z_{i+1} + \sum_{i=0}^{l}(f_{11}b_{z_i} + \eta c_{z_i})z_{i,t} = 0.$$

Differentiating (17) and (18) with respect to z_{l+k}, and using the fact that $F_{z_k} \neq 0$ and $\eta \neq 0$, it follows that

$$b_{z_l} = -\frac{f_{11}}{\eta}a_{z_l}, \qquad c_{z_l} = \left(\frac{f_{11}}{\eta}\right)^2 a_{z_l}. \tag{34}$$

Differentiating the Gauss equation (19) with respect to z_l leads to $ca_{z_l}+ac_{z_l}-2bb_{z_l}=0$, and substituting (34) in the latter leads to

$$\left[c+\left(\frac{f_{11}}{\eta}\right)^2 a+2\frac{f_{11}}{\eta}b\right]a_{z_l}=0. \tag{35}$$

We therefore have two cases to deal with. The first case corresponds to

$$c+\left(\frac{f_{11}}{\eta}\right)^2 a+2\frac{f_{11}}{\eta}b\neq 0. \tag{36}$$

It follows then by (35) that $a_{z_l}=0$, and hence by (34) that $b_{z_l}=c_{z_l}=0$. Successive differentiating leads to $a_{z_i}=b_{z_i}=c_{z_i}=0$ for all $i=0,\ldots,l$. Finally, if the functions a,b and c depend on a jet of finite order, then there are universal, i.e., they are functions of x and t only. We now turn to the second case, defined by the condition

$$c+\left(\frac{f_{11}}{\eta}\right)^2 a+2\frac{f_{11}}{\eta}b=0, \tag{37}$$

on an open set, for which the analysis is far more elaborate. Substituting the expression of c in the Gauss equation $-ac+b^2=1$ leads to $(f_{11}a/\eta+b)^2=1$ so that

$$b=\pm 1-\frac{f_{11}}{\eta}a,\qquad c=\left(\frac{f_{11}}{\eta}\right)^2 a\mp 2\frac{f_{11}}{\eta}. \tag{38}$$

We have then

$$D_t b=-\tfrac{f_{11}}{\eta}D_t a-\tfrac{a}{\eta}f_{11,z_0}F,\quad D_t c=\left(\frac{f_{11}}{\eta}\right)^2 D_t a+\frac{2}{\eta}\left(\frac{f_{11}}{\eta}a\mp 1\right)f_{11,z_0}F,$$

$$D_x b=-\tfrac{f_{11}}{\eta}D_x a-\tfrac{a}{\eta}f_{11,z_0}z_1,\quad D_x c=\left(\frac{f_{11}}{\eta}\right)^2 D_x a+\frac{2}{\eta}\left(\frac{f_{11}}{\eta}a\mp 1\right)f_{11,z_0}z_1,$$

and hence

$$f_{11}D_t a+\eta D_t b=-af_{11,z_0}F, \tag{39}$$

$$f_{11}D_t b+\eta D_t c=\left(\frac{f_{11}}{\eta}a\mp 2\right)f_{11,z_0}F, \tag{40}$$

$$f_{12}D_x a+f_{22}D_x b=-\frac{\Delta_{12}}{\eta}D_x a-\frac{af_{22}}{\eta}f_{11,z_0}z_1, \tag{41}$$

$$f_{12}D_xb + f_{22}D_xc = \frac{f_{11}}{\eta}\frac{\Delta_{12}}{\eta}D_xa + \frac{\Delta_{12}}{\eta^2}af_{11,z_0}z_1 + \frac{f_{22}}{\eta}\left(\frac{f_{11}}{\eta}a \mp 2\right)f_{11,z_0}z_1. \quad (42)$$

Substituting the latter four equalities in (17) and (18) leads to

$$-af_{11,z_0}F + \frac{\Delta_{12}}{\eta}D_xa + \frac{af_{22}}{\eta}f_{11,z_0}z_1 - 2b\Delta_{13} + (a-c)\Delta_{23} = 0 \quad (43)$$

and

$$\left(\frac{f_{11}}{\eta}a \mp 2\right)f_{11,z_0}F - \frac{f_{11}\Delta_{12}}{\eta^2}D_xa - \frac{\Delta_{12}}{\eta^2}af_{11,z_0}z_1 -$$
$$\frac{f_{22}}{\eta}\left(\frac{f_{11}}{\eta}a \mp 2\right)f_{11,z_0}z_1 + (a-c)\Delta_{13} + 2b\Delta_{23} = 0$$

which are equivalent to

$$-af_{11,z_0}F + \frac{\Delta_{12}}{\eta}a_x + \frac{\Delta_{12}}{\eta}\sum_{i=0}^{l}a_{z_i}z_{i+1} + \frac{af_{22}}{\eta}f_{11,z_0}z_1 - 2b\Delta_{13} + (a-c)\Delta_{23} = 0 \quad (44)$$

and

$$\left(\frac{f_{11}}{\eta}a \mp 2\right)\left(F - \frac{f_{22}}{\eta}z_1\right)f_{11,z_0} - \frac{f_{11}\Delta_{12}}{\eta^2}a_x - \quad (45)$$
$$\frac{f_{11}\Delta_{12}}{\eta^2}\sum_{i=0}^{l}a_{z_i}z_{i+1} - \frac{\Delta_{12}}{\eta^2}af_{11,z_0}z_1 + (a-c)\Delta_{13} + 2b\Delta_{23} = 0.$$

We are now led to several cases depending on the value of l.

- If $l \geq k$, then differentiating (44) with respect to z_{l+1} leads to $\Delta_{12}a_{z_l}/\eta = 0$. Thus $a_{z_l} = 0$ and also $b_{z_l} = c_{z_l} = 0$ for $l \geq k$ since $\Delta_{12} \neq 0$.
- If $l = k-1$, then differentiating (44) and (45) with respect to z_k leads to

$$-af_{11,z_0}F_{z_k} + \frac{\Delta_{12}}{\eta}a_{z_{k-1}} = 0, \quad (46)$$

$$\left(\frac{f_{11}}{\eta}a \mp 2\right)f_{11,z_0}F_{z_k} - \frac{f_{11}}{\eta}\frac{\Delta_{12}}{\eta}a_{z_{k-1}} = 0. \quad (47)$$

  Taking into account (46), Eq. (47) becomes $\mp 2f_{11,z_0}F_{z_k} = 0$. It follows then from (46) that $a_{z_{k-1}} = 0$, and therefore that $b_{z_{k-1}} = c_{z_{k-1}} = 0$.
- If $l \leq k-2$, then differentiating (44) and (45) with respect to z_k leads to

$$-af_{11,z_0}F_{z_k} = 0, \quad (48)$$

$$\left(\frac{f_{11}}{\eta}a \mp 2\right)f_{11,z_0}F_{z_k} = 0, \tag{49}$$

which imply that

$$f_{11} = \mu, \tag{50}$$

for some real constant μ. Equations (44) and (45) then become

$$\frac{\Delta_{12}}{\eta}a_x + \frac{(\mu f_{22} - \eta f_{12})}{\eta}\sum_{i=0}^{l} a_{z_i}z_{i+1} - 2b(\mu f_{32} - f_{31}f_{12}) + (a-c)(\eta f_{32} - f_{31}f_{22}) = 0 \tag{51}$$

and

$$-\frac{f_{11}\Delta_{12}}{\eta^2}a_x - \frac{\mu(\mu f_{22} - \eta f_{12})}{\eta^2}\sum_{i=0}^{l} a_{z_i}z_{i+1} + (a-c)(\mu f_{32} - f_{31}f_{12}) + 2b(\eta f_{32} - f_{31}f_{22}) = 0, \tag{52}$$

where

$$\Delta_{12} = \mu f_{22} - \eta f_{12}. \tag{53}$$

Note that when $f_{11,z_0} = 0$, the structure equation (21) becomes $D_x f_{12} = -\Delta_{23}$, or equivalently

$$f_{12,z_{k-1}}z_k + \cdots + f_{12,z_0}z_1 = f_{31}f_{22} - \eta f_{32}. \tag{54}$$

Differentiating (54) with respect to z_k leads then to $f_{12,z_{k-1}} = 0$. If $l = k-2$, then taking into account the latter, and differentiating (51) and (52) with respect to z_{k-1} lead to

$$\frac{\mu f_{22} - \eta f_{12}}{\eta}a_{z_{k-2}} - 2b\mu f_{32,z_{k-1}} + (a-c)\eta f_{32,z_{k-1}} = 0, \tag{55}$$

$$-\frac{\mu(\mu f_{22} - \eta f_{12})}{\eta^2}a_{z_{k-2}} + (a-c)\mu f_{32,z_{k-1}} + 2b\eta f_{32,z_{k-1}} = 0. \tag{56}$$

Note that $f_{11} = \mu \neq 0$, otherwise we would have $b = \pm 1$ and therefore (56) would become $\pm 2\eta f_{32,z_{k-1}} = 0$ which would lead to a contradiction. Indeed, differentiating the structure equation (23) with respect to z_k, we obtain $f_{31,z_0}F_{z_k} = f_{32,z_{k-1}}$. The vanishing of $f_{32,z_{k-1}}$ would then imply the vanishing of f_{31,z_0}, but this is not possible because $f_{11,z_0}^2 + f_{31,z_0}^2 \neq 0$. Therefore, $f_{11} = \mu \neq 0$ and $f_{32,z_{k-1}} \neq 0$. Now, multiplying (55) by μ/η and adding (56), we now obtain that

$$\mu(a-c) + 2\eta b = 2b\mu^2/\eta - (a-c)\mu, \tag{57}$$

which is equivalent to

$$\eta\mu(a-c) = (\mu^2 - \eta^2)b. \tag{58}$$

If $\mu^2 = \eta^2$, then (58) leads to $a - c = 0$ which runs into a contradiction because it follows from (37), (50) and the hypothesis $\mu^2 = \eta^2$ that $a - c = \mp 2\mu/\eta \neq 0$. We have then $\mu^2 - \eta^2 \neq 0$. Substituting (38) in (58) leads to

$$\eta\mu\left[\left(1 - \frac{\mu^2}{\eta^2}\right)a \pm 2\frac{\mu}{\eta}\right] = (\mu^2 - \eta^2)\left[\pm 1 - \frac{\mu}{\eta}a\right], \tag{59}$$

which simplifies to $\mu^2 + \eta^2 = 0$ which runs into a contradiction. Finally, if $l < k - 2$, where $k \geq 3$, then differentiating (51) and (52) with respect to z_{k-1} and using the non-vanishing of $f_{32,z_{k-1}}$ leads to

$$\eta(a-c) - 2\mu b = 0 \tag{60}$$

$$\mu(a-c) + 2\eta b = 0. \tag{61}$$

Since $\eta^2 + \mu^2 \neq 0$, we have $a = c$ and $b = 0$ which runs into a contradiction with the Gauss equation.

Therefore, for all l, (17), (18) and the Gauss equation form an inconsistent system. Hence, if the immersion exists then (36) holds and a, b and c are functions of x and t only. This completes the proof of our theorem.

3 An Example

We now show by displaying an example that the class of evolution equations of order $k \geq 3$ for which the components a, b, c are universal in the sense of Theorem 1, that is independent of u and its derivatives, is non-empty. Consider the following fourth-order evolution equation obtained in [5]

$$u_t = u_{xxxx} + m_1 u_{xxx} + m_2 u_{xx} - u u_x + m_0 u^2, \tag{62}$$

where m_0, m_1, m_2 are arbitrary real constants. Letting

$$\phi = (m_1 + 2m_0)u_{xx} + Bu_x - \frac{u^2}{2} + 2m_0 B, \quad r_0 = -4m_0^2 B, \tag{63}$$

where

$$B = 4m_0^2 + 2m_0 m_1 + m_2, \tag{64}$$

it is straightforward to check that the 1-forms

$$\omega^1 = udx + (u_{xxx} + \phi)dt, \tag{65}$$

$$\omega^2 = -2m_0 dx + r_0 dt, \tag{66}$$

$$\omega^3 = udx + (u_{xxx} + \phi)dt, \tag{67}$$

satisfy the structure equations (3) whenever u is a solution of (62). Let now

$$h = e^{2(-2m_0 x + r_0 t)}, \tag{68}$$

and let γ and l be real constants such that $l > 0$ and $l^2 > 4\gamma^2$. The functions a, b, c defined by

$$a = \sqrt{lh - \gamma^2 h^2 - 1},\ b = \gamma h,\ c = \frac{\gamma^2 h^2 - 1}{a}, \tag{69}$$

satisfy the Gauss equation (19) and the Codazzi equations (17), (18) whenever u is a solution of (62).

Acknowledgements Research partially supported by NSERC Grant RGPIN 105490-2011 and by the Ministério de Ciência e Tecnologia, Brazil, CNPq Proc. No. 303774/2009-6.

References

1. Beals, R., Rabelo, M., Tenenblat, K.: Bäcklund transformations and inverse scattering for some pseudospherical surface equations. Stud. Appl. Math. **81**, 121–151 (1989)
2. Cavalcante, D., Tenenblat, K.: Conservation laws for nonlinear evolution equations. J. Math. Phys. **29**, 1044–1049 (1988)
3. Chern, S.-S., Tenenblat, K.: Pseudospherical surfaces and evolution equations. Stud. Appl. Math. **74**, 55–83 (1986)
4. Ding, Q., Tenenblat, K.: On differential equations describing surfaces of constant curvature. J. Diff. Equ. **184**, 185–214 (2002)
5. Ferraioli, D.C., Tenenblat, K.: Fourth order evolution equations which describe pseudospherical surfaces. J. Differ. Equ. **257**, 3165–3199 (2014)
6. Foursov, V., Olver, P.J., Reyes, E.: On formal integrability of evolution equations and local geometry of surfaces. Differ. Geom. Appl. **15**, 183–199 (2001)

7. Gomes Neto, V.P.: Fifth-order evolution equations describing pseudospherical surfaces. J. Differ. Equ. **249**, 2822–2865 (2010)
8. Gorka, P., Reyes, E.: The modified Hunter-Saxton equation. J. Geom. Phys. **62**, 1793–1809 (2012)
9. Huber, A.: The Cavalcante–Tenenblat equation—does the equation admit physical significance? Appl. Math. Comput. **212**, 14–22 (2009)
10. Jorge, L., Tenenblat, K.: Linear problems associated to evolution equations of type $u_{tt} = F(u, u_x, \ldots, u_{x^k})$. Stud. Appl. Math. **77**, 103–117 (1987)
11. Kahouadji, N., Kamran, N., Tenenblat, K.: Second-order equations and local isometric immersions of pseudo-spherical surfaces, 25 pp. [arXiv:1308.6545], to appear in Comm. Analysis and Geometry (2015)
12. Kamran, N., Tenenblat, K.: On differential equations describing pseudo-spherical surfaces. J. Differ. Equ. **115**, 75–98 (1995)
13. Rabelo, M.: A characterization of differential equations of type $u_{xt} = F(u, u_x, \ldots, u_{x^k})$ which describe pseudo-spherical surfaces. An. Acad. Bras. Cienc. **60**, 119–126 (1988)
14. Rabelo, M.: On equations which describe pseudo-spherical surfaces. Stud. Appl. Math. **81**, 221–148 (1989)
15. Rabelo, M., Tenenblat, K.: On equations of the type $u_{xt} = F(u, u_x)$ which describe pseudo-spherical surfaces. J. Math. Phys. **29**, 1400–1407 (1990)
16. Rabelo, M., Tenenblat, K.: A classification of equations of the type $u_t = u_{xxx} + G(u, u_x, u_{xx})$ which describe pseudo-spherical surfaces. J. Math. Phys. **33**, 1044–149 (1992)
17. Reyes, E.: Pseudospherical surfaces and integrability of evolution equations. J. Differ. Equ. **147**, 195–230 (1998)
18. Reyes, E.: Pseudopotentials, nonlocal symmetries and integrability of some shallow water wave equations. Sel. Math. N. Ser. **12**, 241–270 (2006)
19. Sakovich, A., Sakovich, S.: Solitary wave solutions of the short pulse equation. J. Phys. A **39**, L361-L367 (2006)
20. Sakovich, A., Sakovich, S.: On transformations of the Rabelo equations. SIGMA Symmetry Integrability Geom. Methods Appl. **3**, 8 p., paper 086 (2007)

IST Versus PDE: A Comparative Study

Christian Klein and Jean-Claude Saut

To Walter Craig with friendship and admiration

Abstract We survey and compare, mainly in the two-dimensional case, various results obtained by IST and PDE techniques for integrable equations. We also comment on what can be predicted from integrable equations on non integrable ones.

1 Introduction

The theory of nonlinear dispersive equations has been flourishing during the last thirty years. Partial differential equations (PDE) techniques (in the large) have led to striking results concerning the resolution of the Cauchy problem, blow-up issues, stability analysis of various "localized" solutions. On the other hand, a few nonlinear dispersive equations or systems are integrable by Inverse Scattering Transform (IST) techniques. This allows, as already pointed out in [209], to have a deep understanding of the equation dynamic and also to make relevant conjectures on close, non integrable equations. The best example is the Korteweg-de Vries (KdV) equation for which IST allows to prove that any solution to the Cauchy problem with sufficiently smooth and decaying initial data decomposes into a finite train of solitons traveling to the right and a dispersive tail traveling to the left (see [207] and the references therein and Sect. 1).

The aim of the present paper is to survey and compare, on specific examples, the advantages and shortcomings of PDE and IST techniques and also how they can

C. Klein
Institut de Mathématiques de Bourgogne, UMR 5584, Université de Bourgogne,
9 avenue Alain Savary, 21078 Dijon Cedex, France
e-mail: Christian.Klein@u-bourgogne.fr

J.-C. Saut (✉)
Laboratoire de Mathématiques, UMR 8628, Université Paris-Sud et CNRS,
91405 Orsay, France
e-mail: jean-claude.saut@math.u-psud.fr

P. Guyenne et al. (eds.), *Hamiltonian Partial Differential Equations and Applications*, Fields Institute Communications 75,
DOI 10.1007/978-1-4939-2950-4_14

benefit from each other. We also hope to bring closer two communities that work on very similar objects by quite different tools. We will restrain to the Cauchy problem posed on the whole space $\mathbb{R}$ or $\mathbb{R}^2$ since the corresponding *periodic* problems lead to rather different issues, both by the two approaches.

The paper will be organized as follows. The first section is devoted to one-dimensional (spatial) problems. After recalling the KdV case for which the IST techniques yield the more complete results and which can be seen as a paradigm of what is expected for "close", possibly not integrable equations, we then consider two nonlocal integrable equations, the Benjamin-Ono (BO) and Intermediate Long Wave (ILW) equations that have a much less complete IST theory than the KdV equation. We will in this section present some numerical simulations from Klein and Saut [127] showing that the long time dynamics of KdV solutions seems to be inherited by those of some non integrable equations, such as the fractional KdV or BBM equations. We close this Section by the one-dimensional Gross-Pitaevskii equation, a defocusing nonlinear Schrödinger equation for which non trivial boundary conditions at infinity provide some *focusing* behavior. At this example, one can compare, for the specific problem of the stability of the *black soliton*, the differences between the two methods. We will provide here some details since this example might be less known than for instance the KdV equation.

We then turn to two-dimensional equations. Section two is devoted to the Kadomtsev-Petviashvili equations (KP). Finally, the last section deals with the family of Davey-Stewartson systems, two members of which are integrable (the so-called DSI and DS II system). An important issue is whether or not some of the remarkable properties of the integrable DS systems persist in the non integrable case.

We conclude by a short mention of two other integrable two-dimensional systems, the Ishimori and the Novikov-Veselov systems.

2 Notations

The following notations will be used throughout this article. The partial derivative will be denoted by $u_x, \dots$ or $\partial_x \phi, \dots$ For any $s \in \mathbb{R}$, $D^s = (-\Delta)^{\frac{s}{2}}$ and $J^s = (I - \Delta)^{\frac{s}{2}}$ denote the Riesz and Bessel potentials of order $-s$, respectively.

The Fourier transform of a function f is denoted by $\hat{f}$ or $\mathscr{F}(f)$ and the dual variable of $x \in \mathbb{R}^d$ is denoted ξ. For $1 \leq p \leq \infty$, $L^p(\mathbb{R})$ is the usual Lebesgue space with the norm $|\cdot|_p$, and for $s \in \mathbb{R}$, the Sobolev spaces $H^s(\mathbb{R}^2)$ are defined via the usual norm $\|\phi\|_s = |J^s \phi|_2$.

$\mathscr{S}(\mathbb{R}^d)$ will denote the Schwartz space of smooth rapidly decaying functions in $\mathbb{R}^d$, and $\mathscr{S}'(\mathbb{R}^d)$ the space of tempered distributions.

3 The One-Dimensional Case

3.1 The KdV Equation

The KdV equation is historically the first nonlinear dispersive equation which has been written down. It was in fact derived formally by Boussinesq (1877). We refer to Darrigol [53] for a complete historical account. The full rigorous derivation from the water wave system is due to Craig [51]. We refer to Darrigol [53] for historical aspects and to Lannes [142] for the systematic rigorous derivation of water waves models in various regimes. The KdV equation is (as in fact most of the classical nonlinear dispersive equations or systems) a "universal" asymptotic equation describing a specific dynamic (in the long wave, weakly nonlinear regime) of a large class of complex nonlinear dispersive systems.[1]

The KdV equation

$$u_t - 6uu_x + u_{xxx} = 0, \tag{1}$$

is also the first nonlinear PDE for which the Inverse scattering technique was successively applied (see for instance [73, 145]). For the sake of simplicity we will summarize the traditional approach to IST through the Gelfand-Levitan-Marchenko equation. Much recent progress in IST has been made using the Riemann-Hilbert (RH) formulation of the problem. We refer to the survey article [55] and to the key papers [56–60] where the RH method is used to obtain asymptotic behavior of NLS, mKdV and KdV equations.

The KdV equation is associated to the spectral problem for the Schrödinger operator

$$L(t) = -\frac{d^2\psi}{dx^2} + u(\cdot, t)\psi$$

considered as an unbounded operator in $L^2(\mathbb{R})$.

We thus consider the spectral problem

$$\psi_{xx} + (k^2 - u(x,t))\psi = 0, \quad -\infty < x < +\infty.$$

Given $u_0 = u(\cdot, 0)$ sufficiently smooth and decaying at $\pm\infty$, say in the Schwartz space $\mathscr{S}(\mathbb{R})$, one associates to $L(0)$ its spectral data, that is a finite (possibly empty) set of negative eigenvalues $-\kappa_1^2 < -\kappa_2^2 \cdots < -\kappa_N^2$, together with right normalization coefficients c_j^r and right reflection coefficients $b_r(k)$ (see [207] for precise definitions and properties of those objects).

[1]Note however that the KdV equation is a one-way model. For waves traveling in both directions, the same asymptotic regime would lead to the class of Boussinesq systems (see eg [32] in the context of water waves) which are not integrable.

The spectral data consist thus in the collection of $\{b_r(k), \kappa_j, c_j^r\}$. It turns out that if $u(x,t)$ evolves according to the KdV equation, the scattering data evolves in a very simple way:

$$\kappa_j(t) = \kappa_j,$$
$$c_j^r(t) = c_j^r \exp(4\kappa_j^3 t), \quad j = 1, 2, \cdots, N,$$
$$b_r(k,t) = b_r(k) \exp(8ik^3 t), \quad -\infty < k < +\infty.$$

The potential $u(x,t)$ is recovered as follows. Let

$$\Omega(\xi; t) = 2\sum_{j=1}^{N} [c_j^r(t)]^2 e^{-2\kappa_j \xi} + \frac{1}{\pi}\int_{-\infty}^{\infty} b_r(k,t) e^{2ik\xi} dk.$$

One then solves the linear integral equation (Gelfand-Levitan-Marchenko equation):

$$\beta(y; x, t) + \Omega(x+y; t) + \int_0^\infty \Omega(x+y+z; t)\beta(z; x, t) dz = 0, \quad y > 0,\ x \in \mathbb{R},\ t > 0. \tag{2}$$

The solution of the Cauchy problem (1) is then given by

$$u(x,t) = -\frac{\partial}{\partial x}\beta(0^+; x, t), \quad x \in \mathbb{R}, \quad t > 0.$$

One obtains explicit solutions when $b_r = 0$. A striking case is obtained when the scattering data are $\{0, \kappa_j, c_j^r(t)\}$. This corresponds to the so-called $N-$ *soliton* solution $u_d(x,t)$ according to its asymptotic behavior obtained by Tanaka [218]:

$$\lim_{t\to\infty} \sup_{x\in\mathbb{R}} \left| u_d(x,t) - \sum_{p=1}^{N} \left(-2\kappa_p^2 \operatorname{sech}^2[\kappa_p(x - x_p^+ - 4\kappa_p^2 t)]\right) \right| = 0, \tag{3}$$

where

$$x_p^+ = \frac{1}{2\kappa_p} \log\left\{ \frac{[c_p^r]^2}{2\kappa_p} \prod_{l=1}^{p-1} \left(\frac{\kappa_l - \kappa_p}{\kappa_l + \kappa_p}\right)^2 \right\}.$$

In other words, $u_d(x,t)$ appears for large positive time as a sequence of N solitons, with the largest one in the front, uniformly with respect to $x \in \mathbb{R}$.

For $u_0 \in \mathscr{S}(\mathbb{R})$,[2] the solution of (1) has the following asymptotics

[2]This condition can be weakened, but a decay property is always needed.

$$\sup_{x \geq -t^{1/3}} |u(x,t)| = O(t^{-2/3}), \quad \text{as } t \to \infty, \tag{4}$$

in the absence of solitons (that is when $L(0)$ has no negative eigenvalues) and

$$\sup_{x \geq -t^{1/3}} |u(x,t) - u_d(x,t)| = O(t^{-1/3}), \quad \text{as } t \to \infty \tag{5}$$

in the general case, the N in u_d being the number of negative eigenvalues of $L(0)$. One has moreover the convergence result

$$\lim_{t \to +\infty} \sup_{x \geq -t^{1/3}} \left| u(x,t) - \sum_{p=1}^{N} \left(-2\kappa_p^2 \text{sech}^2[\kappa_p(x - x_p^+ - 4\kappa_p^2 t)]\right) \right| = 0. \tag{6}$$

In both cases, a "dispersive tail" propagates to the left.

Remark 1. The shortcoming of those remarkable results is of course that they apply only to the integrable KdV equation and also to the modified KdV equation

$$u_t + 6u^2 u_x + u_{xxx} = 0.$$

However, though they are out of reach of "classical" PDE methods, they give hints on the behavior of other, non integrable, equations whose dynamics could be in some sense similar.

Remark 2. As previously noticed, the results obtained by IST methods necessitate a decay property of the initial data, the minimal condition being

$$I(0) = \int_{-\infty}^{\infty} (1 + |x|)|u_0(x)|dx < \infty. \tag{7}$$

This condition ensures in particular [156] that $L(0)$ has a finite number of discrete eigenvalues, more precisely [41], the number N of eigenvalues of $L(0)$ is bounded by $1 + I(0)$.

This excludes for instance initial data in the energy space $H^1(\mathbb{R})$ in which the Cauchy problem is globally well-posed [115]. The global behavior of the flow might thus be different from the aforementioned results for such initial data.

Remark 3.

1. Stemming from the seminal work of Bourgain [39], PDE techniques allow to prove the well-posedness of the Cauchy problem for the KdV equation with initial data in very large spaces, namely $H^s(\mathbb{R})$, $s > -\frac{3}{4}$, see [116] which includes in particular measures, for instance the Dirac distribution.

2. PDE techniques yield also the asymptotic stability of the solitary waves for subcritical KdV equations with a rather general nonlinearity [10, 157, 158] which can be seen as a first step towards the *soliton resolution conjecture*, see eg [220].

In order to see to what extent the long time dynamics of the KdV equation is in some sense generic, we will consider as a toy model the fractional KdV (fKdV) equation

$$u_t + uu_x - D^\alpha u_x = 0, \quad u(.,0) = u_0, \tag{8}$$

where $\widehat{D^\alpha f}(\xi) = |\xi|^\alpha \hat{f}(\xi)$, $\alpha > -1$. Using the Fourier multiplier operator notation

$$\widehat{Lu}(\xi) = p(\xi)\hat{u}(\xi), \quad p(\xi) = |\xi|^\alpha,$$

it can be rewritten as

$$u_t + uu_x - Lu_x = 0, \quad u(.,0) = u_0. \tag{9}$$

When $\alpha = 2$ (resp. 1) (8) reduces to the KdV (resp. Benjamin-Ono) equation. If the symbol $|\xi|^\alpha$ is replaced by

$$p(\xi) = \left(\frac{\tanh \xi}{\xi}\right)^{1/2},$$

one gets the so-called Whitham equation [224] that models surface gravity waves in an appropriate regime. This symbol behaves like $|\xi|^{-1/2}$ for large frequencies.

When surface tension is included in the Whitham equation, one gets

$$p(\xi) = (1 + \beta|\xi|^2)^{1/2}\left(\frac{\tanh \xi}{\xi}\right)^{1/2}, \quad \beta \geq 0$$

which behaves like $|\xi|^{1/2}$ for large $|\xi|$.

The following quantities are formally conserved by the flow associated to (8),

$$M(u) = \int_{\mathbb{R}} u^2(x,t)dx, \tag{10}$$

and the Hamiltonian

$$H(u) = \int_{\mathbb{R}} \Big(\frac{1}{2}|D^{\frac{\alpha}{2}} u(x,t)|^2 - \frac{1}{6}u^3(x,t)\Big)dx. \tag{11}$$

One notices that the values $\alpha = 1/3$ and $\alpha = 1/2$ correspond respectively to the so-called energy critical and to the L^2 critical cases. Actually, Eq. (8) is invariant under the scaling transformation

$$u_\lambda(x,t) = \lambda^\alpha u(\lambda x, \lambda^{\alpha+1} t), \tag{12}$$

for any positive number λ. A straightforward computation shows that $\|u_\lambda\|_{\dot{H}^s} = \lambda^{s+\alpha-\frac{1}{2}} \|u_\lambda\|_{\dot{H}^s}$, and thus the critical index corresponding to (8) is $s_\alpha = \frac{1}{2} - \alpha$. Thus, Eq. (8) is L^2-critical for $\alpha = \frac{1}{2}$. On the other hand the Hamiltonian does not make sense in the *energy space* $H^{\alpha/2}(\mathbb{R})$ when $\alpha < \frac{1}{3}$. The numerical simulations in [127] suggest that the Cauchy problem (8) has global solutions (for arbitrary large suitably localized and smooth initial data) if and only if $\alpha > 1/2$. This has been rigorously proven when $\alpha \geq 1$ see [69, 70] but is an open problem when $1/2 < \alpha < 1$. On the other hand, the local Cauchy problem is for $\alpha > 0$ locally well-posed in $H^s(\mathbb{R}), s > \frac{3}{2} - \frac{3\alpha}{8}$, [150].

More surprising is the fact that the resolution into solitary waves plus dispersion seems to be still valid when $\alpha > 1/2$ as also suggested from the numerical simulations in [127] from which we extract the following figures. In Fig. 1, one can see the solution for the fKdV equation in the mass subcritical case $\alpha = 0.6$ for the initial data $5\text{sech}^2 x$.

In Fig. 2 we have fitted the humps with the computed solitary waves. This is an evidence for the above mentioned *soliton resolution conjecture*.

A similar behavior seems to occur for the fractional BBM equation (fBBM)

$$u_t + u_x + uu_x + D^\alpha u_t = 0 \tag{13}$$

in the subcritical case $\alpha > \frac{1}{3}$. Again the simulations in [127] suggest that the soliton resolution also holds (see Fig. 3 below). See [159, 160] for the inelastic collision of solitons when $\alpha = 2$.

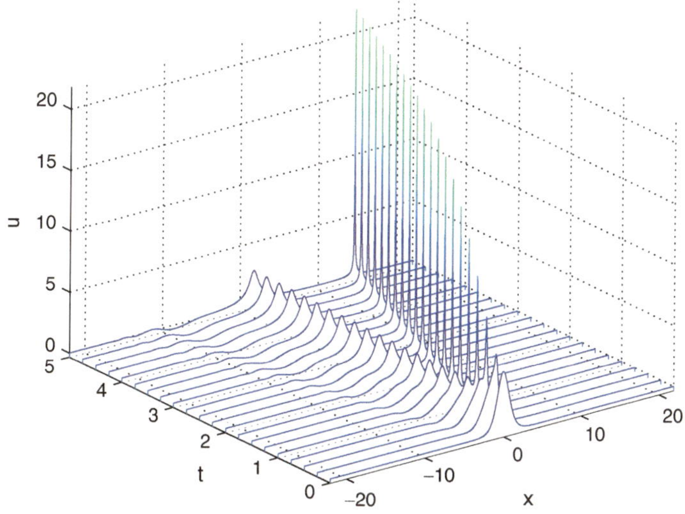

Fig. 1 Solution to the fKdV equation (8) for $\alpha = 0.6$, for the initial data $u_0 = 5\text{sech}^2 x$

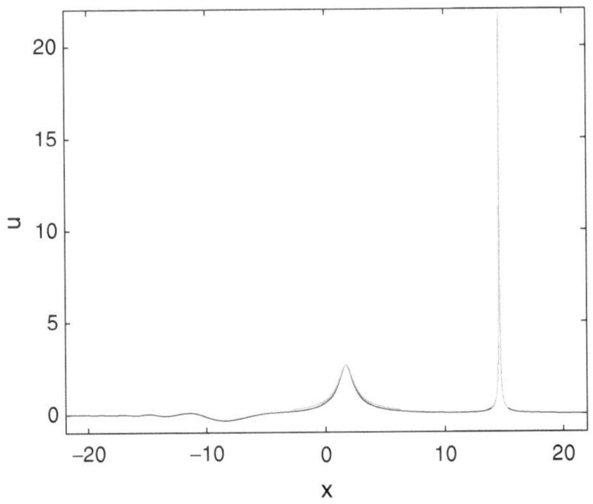

Fig. 2 Solution to the fKdV equation (8) for $\alpha = 0.6$, and the initial data $u_0 = 5\text{sech}^2 x$ for $t = 5$ in *blue*, fitted solitons at the humps in *green*

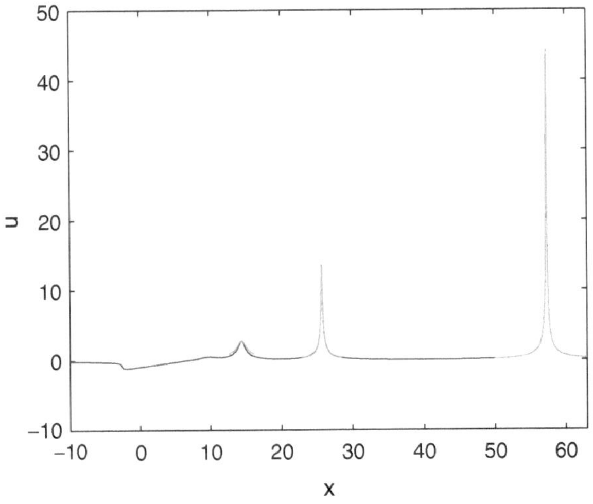

Fig. 3 Solution to the fBBM equation (13) for $\alpha = 0.5$, and the initial data $u_0 = 10\text{sech}^2 x$ for $t = 10$ in *blue*, fitted solitons at the humps in *green*

3.2 The Benjamin-Ono and Intermediate Long Wave Equations

The Intermediate Long Wave Equation (ILW) and the Benjamin-Ono equation (BO) are asymptotic models in an appropriate regime for a two-fluid system when the depth δ of the bottom layer is very large with respect to the upper one (ILW) or infinite (BO) see [33].

The ILW corresponds (in the notations of [1]) to

$$p(\xi) = 2\pi\xi \coth(2\pi\delta\xi) - \frac{1}{\delta}$$

and the BO equation to

$$p(\xi) = 2\pi|\xi|.$$

Alternatively they can be respectively written in convolution form

$$u_t + uu_x + \frac{1}{\delta}u_x + \mathscr{T}_\delta(u_{xx}) = 0, \tag{14}$$

$$\mathscr{T}_\delta(u)(x) = -\frac{1}{2\delta}PV\int_{-\infty}^{\infty} \coth\left(\frac{\pi(x-y)}{2\delta}\right)u(y)dy$$

and

$$u_t + uu_x + \mathscr{H}u_{xx} = 0 \tag{15}$$

where $\mathscr{H}$ is the Hilbert transform

$$\mathscr{H}u(x) = \frac{1}{\pi}PV\int_{-\infty}^{\infty}\frac{u(y)}{x-y}dy.$$

3.2.1 The Benjamin-Ono Equation

A striking difference between KdV and BO equations is that the latter is *quasilinear* rather that *semilinear*. This means that the Cauchy problem for BO cannot be solved by a Picard iterative scheme implemented on the integral Duhamel formulation, for initial data in any Sobolev spaces $H^s(\mathbb{R})$, $s \in \mathbb{R}$. Alternatively, this implies that the flow map $u_0 \mapsto u(t)$ cannot be *smooth* in the same spaces [167], and actually not even locally Lipschitz [133]. We will give a precise statements of those facts later on for the KP I equation which also is quasilinear in this sense.

The Cauchy problem has been proven to be globally well-posed in $H^s(\mathbb{R}), s > 3/2$ by a compactness method using the various invariants[3] of the equations [1] and actually in much bigger spaces (see [94, 166, 219] and the references therein), in particular in the energy space $H^{1/2}(\mathbb{R})$, by sophisticated methods based on the dispersive properties of the equations.

[3]The existence of an infinite sequence of invariants [42, 163] is of course a consequence of the complete integrability of the Benjamin-Ono equation.

Moreover it was proven in [1] that the solution u_δ of (14) with initial data u_0 converges as $\delta \to +\infty$ to the solution of the Benjamin-Ono equation (15) with the same initial data.

Furthermore, if u_δ is a solution of (14) and setting

$$v_\delta(x,t) = \frac{3}{\delta} u_\delta(x, \frac{3}{\delta} t),$$

v_δ tends as $\delta \to 0$ to the solution u of the KdV equation

$$u_t + uu_x + u_{xxx} = 0 \tag{16}$$

with the same initial data.

Remark 4. The hierarchy of conserved quantities of the Benjamin-Ono equation leads to a hierarchy of higher order BO equations BO_n (by considering the associated Hamiltonian flow). Those equations have order $\frac{n}{2}$, $n = 4, 5, \ldots$ It was established in [173] that a family of order 3 equations containing BO_6 is globally well-posed in the energy space $H^1(\mathbb{R})$. A similar result is expected for the whole hierarchy.

Both the BO and the ILW equations are classical examples of equations solvable by IST methods. The situation is however less satisfactory than for the KdV equation. Actually the present state of the art in IST only allows for a complete solution of the Cauchy problem with small data. The small data assumption is used to insure that integral equations in the direct spectral problem for a Jost-type function $\psi(x,k)$ have solutions for all values of the complex parameter k (the spectral variable). Whether or not this obstruction is a fundamental one or merely a technical one is an interesting open question.

The formal IST theory has been given by Ablowitz and Fokas [3] (see also [12]). They found the inverse spectral problem and Beals and Coifman [15, 16] observed that it is equivalent to a nonlocal $\bar{\partial}$ problem.

Unfortunately, the direct scattering problem can be only solved for small data and a complete theory for IST (as for the KdV equation) is a challenging open problem. In particular, one does not know (but expects) that any localized initial data decompose into a train of solitary waves and a dispersive tail. The rigorous theory of the Cauchy problem for small initial data is given in [48].

As previously recalled, the BO equation has an infinite number of conserved quantities ([42], the first ones are displayed in [1]). The Hamiltonian flow of those invariants define the aforementioned Benjamin-Ono hierarchy.

The BO equation possesses explicit soliton and multi-solitons [43, 161–163]. The one soliton reads

$$Q(x) = \frac{4}{1+x^2}$$

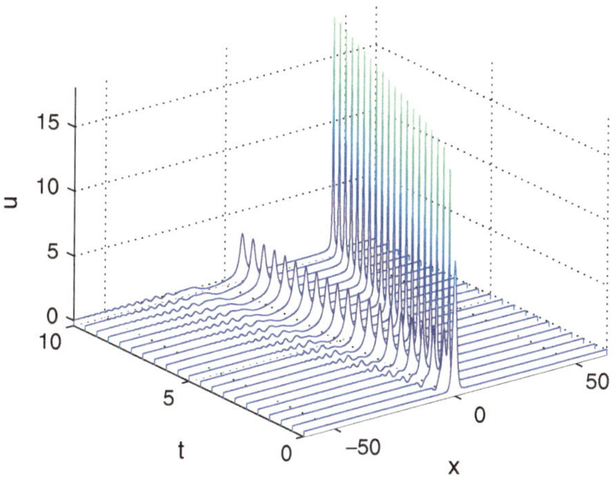

Fig. 4 Solution to the BO equation for the initial data $u_0 = 10\text{sech}^2 x$ in dependence of t

and is unique (up to translations) among all solitary wave solutions [11]. Its slow (algebraic) decay is due (by Paley-Wiener type arguments) to the fact that the BO symbol $\xi|\xi|$ is not smooth at the origin. Actually similar arguments imply that the solution to the Cauchy problem cannot decay fast at infinity see [96]. The BO solitary wave is orbitally stable (see [34, 35] and the references therein).

On the other hand Kenig and Martel [112] have proven the asymptotic stability of the BO solitary wave as well as that of the explicit multi-solitons in the energy space, a fact which reinforces the above conjecture on the long time dynamic of BO solutions. They do not use the integrability of the equation except for the exact expressions for the solitons (that help to study the spectral properties of the linearized operators).

Remark 5. The H^1 stability of the 2-soliton has been proven in [175] by variational methods.

We show the formation of solitons from localized initial data in Fig. 4. Again there is a tail of dispersive oscillations propagating to the left.

3.2.2 The ILW Equation

The formal IST theory has been given in [135, 136]. The direct scattering problem is associated with a Riemann-Hilbert problem in a strip of the complex plane. As the BO equation, the ILW equation possesses an infinite sequence of conserved quantities (see eg [146]) which leads to a ILW hierarchy. They can be used to provide the global well-posedness of the Cauchy problem in $H^s(\mathbb{R}), s > 3/2$, [1].

Moreover, the fact that for large $|\xi|$, $|\xi|\coth \xi = |\xi|(1 + O(e^{-|\xi|}))$ implies that the well-posedness results are similar to those obtained for the BO equation, for instance in the energy space $H^{1/2}(\mathbb{R})$.

On the other hand, we do not know of rigorous results for the Cauchy problem using the IST method (see however [208]). It is likely that they would require a smallness condition on the initial data.

Explicit N-soliton solutions are given in [100, 161]. Contrary to those of the Benjamin-Ono equation, they decay exponentially at infinity (the ILW symbol is smooth). For instance, when the ILW equation is written in the form

$$u_t + 2uu_x + \frac{1}{\delta}u_x - L_\delta u_x = 0, \tag{17}$$

where

$$\widehat{L_\delta u}(\xi) = (\xi \coth \xi\delta)\hat{u}(\xi),$$

the 1-soliton reads see [8, 99, 100]

$$u(x,t) = Q_{c,\delta}(x-ct), \quad Q_{c,\delta}(x) = \frac{a \sin a\delta}{\cosh ax + \cos a\delta}, \quad x \in \mathbb{R}$$

for arbitrary $c > 0$ and $\delta > 0$, and a is the unique positive solution of the transcendental equation

$$a\delta \coth a\delta = (1 - c\delta), \quad a \in (0, \pi/\delta).$$

Its uniqueness (up to translations) is proven in [8]. The orbital stability of this soliton is proven in [7, 9] (see also [34, 35]). We do not know of asymptotic stability results for the 1 or N-soliton similar to the corresponding ones for the BO equation. Those results should be in some sense easier than the corresponding ones for BO since the exponential decay of the solitons should make the spectral analysis of the linearized operators easier.

We show the decomposition of localized initial data into solitons and radiation in Figs. 5 and 6. Note that this case is numerically easier to treat with Fourier methods since the soliton solutions are more rapidly decreasing (exponentially instead of algebraically) than for the fKdV, fBBM and BO equations before. The different shape of the solitons is also noticeable in comparison to Fig. 4.

One also sees the change in the number of emerging solitons according to the size of the initial data. As in the case of BO, fKdV, fBBM equations, predicting the number of solitary waves which seem to form from a given (smooth and localized) initial data is a challenging open question. Except for the KdV case where the solitons are related to the discrete spectrum of the associated Schrödinger equation, there does not appear to be a clear characterization of the solitons emerging from given initial data for $t \to \infty$ even for integrable equations.

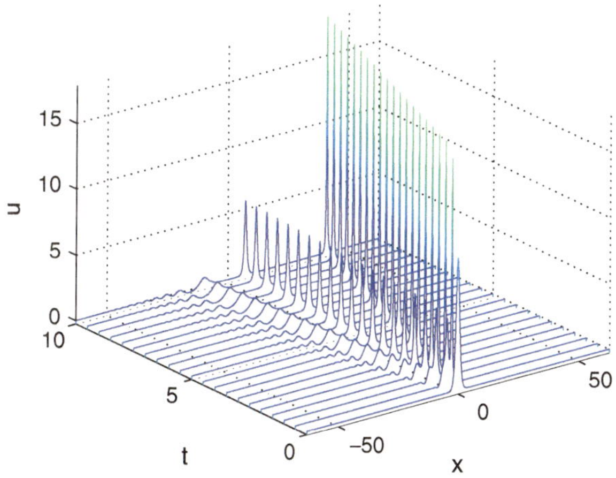

Fig. 5 Solution to the ILW equation with $\delta = 1$ for the initial data $u_0 = 10\text{sech}^2 x$ in dependence of t

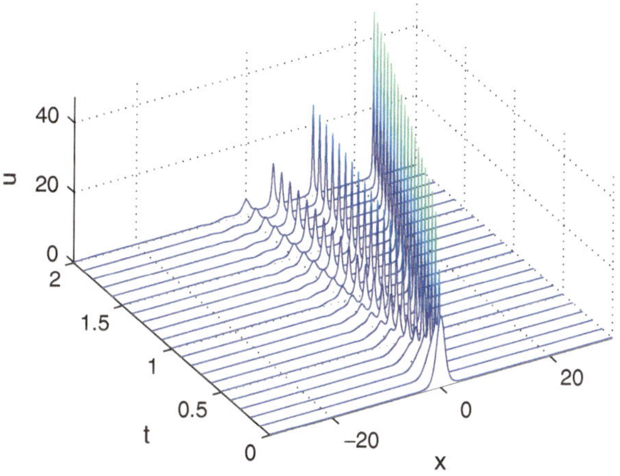

Fig. 6 Solution to the ILW equation with $\delta = 1$ for the initial data $u_0 = 20\text{sech}^2 x$ in dependence of t

3.3 The Gross-Pitaevskii Equation

The Gross-Pitaevskii equation (GP) is a version of a nonlinear Schrödinger equation (NLS), namely

$$i\partial_t \Psi = \Delta\Psi + \Psi(1 - |\Psi|^2) \quad \text{on} \quad \mathbb{R}^d \times \mathbb{R}, (d = 1, 2, 3). \tag{18}$$

It is a relevant model in nonlinear optics ("dark" and "black" solitons, optical vortices [176]), fluid mechanics (superfluidity of Helium II), Bose-Einstein condensation of ultra cold atomic gases.

At least on a formal level, the Gross-Pitaevskii equation is hamiltonian. The conserved Hamiltonian is a **Ginzburg-Landau energy**, namely

$$E(\Psi) = \frac{1}{2}\int_{\mathbb{R}^d} |\nabla\Psi|^2 + \frac{1}{4}\int_{\mathbb{R}^d} (1-|\Psi|^2)^2 \equiv \int_{\mathbb{R}^d} e(\Psi), \tag{19}$$

associated to the natural energy space

$$\mathscr{E}(\mathbb{R}^d) = \{v \in H^1_{\text{loc}}(\mathbb{R}^d), E(v) < +\infty\}.$$

In order for $E(\Psi)$ to be finite, $|\Psi|$ should in some sense tend to 1 at infinity. Actually this "non trivial" boundary condition provides (GP) with a richer dynamics than in the case of null condition at infinity which, for a defocusing NLS type equation, is essentially governed by dispersion and scattering. For instance, in nonlinear optics, the "dark solitons" are localized nonlinear waves (or "holes") which exist on a stable continuous wave background. The boundary condition $|\Psi(x,\cdot)| \to 1$ is due to this non-zero background.

Similarly to the energy, the **momentum** $\mathbf{P}(\Psi) = \frac{1}{2}\int_{\mathbb{R}^N}\langle i\nabla\Psi\ ,\Psi\rangle$, is formally conserved. We will denote by $p = Im\int_{\mathbb{R}^N} i\Psi\bar{\Psi}_x$, the first component of $\mathbf{P}$, which is hence a scalar. Justifying the momentum (and its conservation) is one of the difficulties one has to face when dealing with the (GP) equation. We will restrict here to the one-dimensional case, $d = 1$ since the GP equation is then completely integrable [231]. We just consider

$$i\partial_t u + \partial_x^2 u + (1-|u|^2)u = 0. \tag{20}$$

In fact Zakharov-Shabat [231] consider the case when $|u(x,t)|^2 \to c > 0$, $|x| \to \infty$ (propagation of waves through a condensate of constant density).

More precisely, the GP has a Lax pair (B_u, L_u), where (for $c = 1$)

$$L_u = i\begin{pmatrix} 1+\sqrt{3} & 0 \\ 0 & 1-\sqrt{3}\end{pmatrix}\partial_x + \begin{pmatrix} 0 & \bar{u} \\ u & 0\end{pmatrix} \tag{21}$$

$$B_u = -\sqrt{3}\begin{pmatrix} 1 & 0 \\ 0 & 1\end{pmatrix}\partial_x^2 + \begin{pmatrix} \frac{|u|^2-1}{\sqrt{3}+1} & i\partial_x\bar{u} \\ -i\partial_x\bar{u} & \frac{|u|^2-1}{\sqrt{3}-1}\end{pmatrix} \tag{22}$$

so that u satisfies (GP) if and only if

$$\frac{d}{dt}L_u = [L_u, B_u]. \tag{23}$$

As a consequence, the 1D Gross-Pitaevskii equation has an infinite number of (formally) conserved energies $E_k, k \in \mathbb{N}$ and momentum $P_k, k \in \mathbb{N}$. For instance,

$$I_5 = \int_{\mathbb{R}} \{2|u|^6 + 6|u|^2|u_x|^2 + (\frac{d}{dx}|u|^2)^2 + |u_{xx}|^2 - 2\}.$$

It is of course necessary to prove rigorously that the E_k and the P_k are well defined and conserved by the GP flow, in a suitable functional setting.

We do not know of a *complete* resolution of the Cauchy problem by IST methods, including a possible decomposition into solitons. We will see however that one can get a proof of stability of the solitons by using IST techniques. One can prove that the energy space in one dimension is identical to the *Zhidkov space* see [234]

$$X^1(\mathbb{R}) = \{u \in H^1_{loc}(\mathbb{R}), u_x \in L^\infty(\mathbb{R}), (1 - |u|^2) \in L^2(\mathbb{R})\}$$

and actually Zhidkov [234] proved that the Cauchy problem for (20) is globally well-posed in $X^1(\mathbb{R})$.[4]

The one-dimensional GP equation possesses two types of solitary waves of velocity c, $0 \leq c < \sqrt{2}$:

- The **"dark"** solitons : $\mathfrak{v}_c(x) \equiv \sqrt{\frac{2-c^2}{2}}\text{th}\Big(\frac{\sqrt{2-c^2}}{2}x\Big) - i\frac{c}{\sqrt{2}}$.
- The **"black"** soliton :

$$\mathfrak{v}_0(x) = \text{th}\Big(\frac{x}{\sqrt{2}}\Big).$$

Note that when $0 < c < \sqrt{2}$, $\mathfrak{v}_c(x) \neq 0, \forall x$ while the black soliton vanishes at $x = 0$. The orbital stability of the dark soliton has been proven in [27] (see also [148] for the cubic-quintic case and the numerics in [61]). This case is easier since the dark soliton does not vanish and the momentum can be defined in a straightforward way. The orbital stability of the black soliton is more delicate since it vanishes at 0. Both PDE and IST techniques provide the result, in a slightly different form though, and this is a good opportunity to compare them.

The first method is used in [29]. This is the "Hamiltonian" method (see [20, 31] for the stability of the KdV solitary wave or [44] for the stability of the focusing NLS ground states), that is one considers the black soliton as a minimizer of the energy with fixed momentum. As previously noticed, serious difficulties arise here from the momentum (definition, conservation by the flow,...) because the black soliton vanishes at 0.

Given any $A > 0$, we consider on the energy space $X^1(\mathbb{R}) = X^1$ the distance d_{A,X^1} defined by

[4]P. Gérard has proven in fact [74] that the Cauchy problem for the GP equation is globally well -posed in the energy space $\mathscr{E}(\mathbb{R}^d)$ equipped with a suitable topology when $d \leq 4$.

$$d_{A,X^1}(v_1, v_2) \equiv \|v_1 - v_2\|_{L^\infty([-A,A])} + \|v_1' - v_2'\|_{L^2(\mathbb{R})} + \||v_1| - |v_2|\|_{L^2(\mathbb{R})}.$$

One can show that for $v \in X^1$, $|v(x)| \to 1$ as $|x| \to \infty$, (but not v itself). We have the orbital stability result [29]:

Theorem 1. *Assume that $v_0 \in X^1$ and consider the global in time solution v to (GP) with initial datum v_0. Given any numbers $\varepsilon > 0$ and $A > 0$, there exists some positive number δ, such that if*

$$d_{A,X^1}(v_0, \mathfrak{v}_0) \leq \delta, \tag{24}$$

then, for any $t \in \mathbb{R}$, there exist numbers $a(t)$ and $\theta(t)$ such that

$$d_{A,X^1}\big(v(\cdot + a(t), t), \exp(i\theta(t))\, \mathfrak{v}_0(\cdot)\big) < \varepsilon. \tag{25}$$

We also have a control of the shift $a(t)$:

Theorem 2. *Given any numbers $\varepsilon > 0$, sufficiently small, and $A > 0$, there exists some constant K, only depending on A, and some positive number $\delta > 0$ such that, if v_0 and v are as in the previous Theorem, then*

$$|a(t)| \leq K\varepsilon(1 + |t|), \tag{26}$$

for any $t \in \mathbb{R}$, and for any of the points $a(t)$ as above.

We will not give a proof of the previous results see [29] but just explain how to extend the definition of the momentum $P(v) = \frac{1}{2}\int_{\mathbb{R}}\langle iv, v'\rangle$ for general functions in X^1.

- Assume $v \in X^1$ and $v = \exp(i\phi)$.

Then $\langle iv, v'\rangle = \phi' \in L^2(\mathbb{R})$ and $P(v) \equiv \frac{1}{2}[\phi(+\infty) - \phi(-\infty)]$. This is meaningful if

$$v \in Z^1 = \Big\{v \in X^1,\ \text{s.t. } v_{\pm\infty} = \lim_{x \to \pm\infty} v(x) \text{ exist}\Big\},$$

but not for any arbitrary phase ϕ whose gradient is in L^2.

Let

$$\tilde{X^1} = \{v \in X^1,\ \text{s.t. } |v(x)| > 0,\ \forall x \in \mathbb{R}\}.$$

For $v \in \tilde{X^1}$, $v = \rho\exp(i\phi)$, so that $\langle iv, v'\rangle = \rho^2\phi'$. If $v \in \tilde{Z^1} \equiv \tilde{X^1} \cap Z^1$, one has

$$P(v) = \frac{1}{2}\int_{\mathbb{R}} \rho^2\phi' = \frac{1}{2}\int_{\mathbb{R}} (\rho^2 - 1)\phi' + \frac{1}{2}\int_{\mathbb{R}} \phi' = \frac{1}{2}\int_{\mathbb{R}} (\rho^2 - 1)\phi' + \frac{1}{2}[\phi]_{-\infty}^{+\infty}. \tag{27}$$

This is well controlled since by Cauchy-Schwarz

$$|\int_{\mathbb{R}}(\rho^2-1)\phi'|\leq\frac{2}{\delta^2}E(v)$$

where

$$\delta=inf\{|v(x)|,x\in\mathbb{R}\}.$$

- The momentum for maps with zeroes:

Lemma 1. *Let $v\in Z^1$. Then, the limit*

$$\mathscr{P}(v)=\lim_{R\to+\infty}P_R(v)\equiv\lim_{R\to+\infty}\int_{-R}^{R}\langle iv,v'\rangle$$

exists. Moreover, if v belongs to $\tilde{Z}^1$, then

$$\mathscr{P}(v)=\frac{1}{2}\int_{\mathbb{R}}(\rho^2-1)\phi'+\frac{1}{2}[\phi]_{-\infty}^{+\infty}.$$

- Example is the black soliton $\mathfrak{v}_0$: since $\mathfrak{v}_0$ is real-valued, $\langle i\mathfrak{v}_0,\mathfrak{v}_0'\rangle=0$ and $\mathscr{P}(\mathfrak{v}_0)=0$.
- Elementary observation. Let $V_0\in Z^1$ and $w\in H^1(\mathbb{R})$. Then, $V_0+w\in Z^1$ and

$$\mathscr{P}(V_0+w)=\mathscr{P}(V_0)+\frac{1}{2}\int_{\mathbb{R}}\langle iw,w'\rangle+\int_{\mathbb{R}}\langle iw,V_0'\rangle.$$

- Note that besides $\mathscr{P}(V_0)$ the integrals are definite ones.
- **The renormalized and the untwisted momentum.**

  The renormalized momentum p is defined for $v\in\tilde{X}_1$ by

$$p(v)=\frac{1}{2}\int_{\mathbb{R}}(\varrho^2-1)\varphi',\tag{28}$$

as seen before, if v belongs to $\tilde{Z}^1$, then,

$$p(v)=\mathscr{P}(v)-\frac{1}{2}[\varphi]_{-\infty}^{+\infty}.$$

If $v\in Z^1\setminus\tilde{Z}^1$, the integral is a priori not well-defined since the phase φ is not globally defined. Nevertheless, the argument $\arg v$ of v is well-defined at infinity as an element of $\mathbb{R}/2\pi\mathbb{Z}$. For $v\in Z^1$, one introduces **the untwisted momentum**

$$[p](v)=\Big(\mathscr{P}(v)-\frac{1}{2}\big(\arg v(+\infty)-\arg v(-\infty)\big)\Big)\bmod\pi,$$

which is hence an element of $\mathbb{R}/\pi\mathbb{Z}$. A remarkable fact concerning $[p]$ is that its definition extends to the whole space X^1, although for arbitrary maps in X^1, the quantity $\arg v(+\infty) - \arg v(-\infty)$ may not exist.

Lemma 2. *Assume that v belongs to X^1. Then the limit*

$$[p](v) = \lim_{R\to+\infty}\left(\int_{-R}^{R}\langle iv, v'\rangle - \frac{1}{2}\big(\arg v(R) - \arg v(-R)\big)\right) \bmod \pi$$

exists. Moreover, if v belongs to $\tilde{X}^1$, then

$$[p](v) = p(v) \bmod \pi. \tag{29}$$

One has also:

- Let $V_0 \in X^1$ and $w \in H^1(\mathbb{R})$. Then, $V_0 + w \in X^1$ and

$$[p](V_0 + w) = [p](V_0) + \frac{1}{2}\int_{\mathbb{R}}\langle iw, w'\rangle + \int_{\mathbb{R}}\langle iw, V_0'\rangle \bmod \pi. \tag{30}$$

- The evolution preserves the untwisted momentum:

Lemma 3. *Assume $v_0 \in X^1$, and let v be the solution to (GP) with initial datum v_0. Then,*

$$[p](v(\cdot,t)) = [p](v_0),\ \forall t \in \mathbb{R}.$$

If moreover $v_0 \in Z^1$, then $v(t)$ belongs to Z^1 for any $t \in \mathbb{R}$, and

$$\mathscr{P}(v(\cdot,t)) = \mathscr{P}(v_0),\ \forall t \in \mathbb{R}.$$

The orbital stability of the dark soliton ($0 < c < \sqrt{2}$) is based on the minimization problem:

$$E_{\min}(\mathfrak{p}) = \inf\{E(v), v \in \tilde{X}^1 \text{ s.t. } p(v) = \mathfrak{p}\}.$$

For map having zeroes like $\mathfrak{v}_0$, one should use the untwisted momentum and consider instead

$$\mathfrak{E}_{\min}\Big(\frac{\pi}{2}\Big) \equiv \inf\Big\{E(v), v \in X^1 \text{ s.t. } [p](v) = \frac{\pi}{2} \bmod \pi\Big\}. \tag{31}$$

We now turn to the second approach using IST in [75] which avoids the renormalization by factors of modulus 1 in Theorem 1, at least for sufficiently smooth and decaying perturbations.

Theorem 3. *Assume that the initial datum of (1.1) has the form:* $u_0(x) = U_0(x) + \epsilon u_1(x), U_0(x) = tanh\left(\frac{x}{\sqrt{2}}\right)$, *where* $u_1(x)$ *satisfies the following condition,*

$$\sup_{x\in\mathbb{R}} |(1+x^2)^2 \partial^k u_1(x)| \leq 1, \quad \text{for } k \leq 3.$$

$$\forall t \in \mathbb{R}, \quad \exists\, y(t) \in \mathbb{R}, \quad |\tau_{y(t)} u(.,t) - U_0|_\infty \leq C\epsilon, \quad \text{for } 0 \leq t < +\infty.$$

Using a classical functional analytic argument, Theorem 3 easily yields the orbital stability, at least for sufficiently smooth and decaying perturbations [75]:

Corollary 1. *For every* $\delta > 0$ *there exists* $\epsilon > 0$ *such that, if*

$$\sup_{x\in\mathbb{R}} |(1+x^2)^2 \partial^k u_1(x)| \leq 1, \quad \text{for } k \leq 3,$$

then the solution u of (20) *satisfies:*

$$\forall t \in \mathbb{R},\ \exists y(t) \in \mathbb{R}, \quad d_E(\tau_{y(t)} u(t), U_0) \leq \delta.$$

Remark 6. As previously noticed Theorem 3 improves on Theorems 1, 2 since it does not require a rotation factor. On the other hand it deals with a narrower class of perturbations and does not provide an evaluation of the time shift.

One can combine PDE and IST techniques as for instance in the long wave limit of the GP equation which we describe now. The transonic (long wave, small amplitude) limit of GP to KP I in 2D or KdV in 1D is quite a generic phenomenon for nonlinear dispersive systems. A good example is that of the water wave systems (incompressible Euler with upper free surface). While the *consistency* of the approximation is relatively easy to obtain, the *stability* (and thus the *convergence*) of the approximation is much more delicate, specially if one looks for the optimal error estimates on the correct time scales.

Kuznetsov and Zakharov [229] have observed formally that the KdV equation provides a good approximation of long-wave small amplitude solutions to the 1D Gross-Pitaevskii equation. We recall here briefly a rigorous proof of this fact [25] (see also [45] for a different approach which dose not provide an error estimate and [26] for a similar analysis for two-way propagation leading to a coupled system of KdV equations). We thus start from the 1D GP equation

$$i\partial_t \Psi + \partial_x^2 \Psi = \Psi(|\Psi|^2 - 1) \quad \text{on} \quad \mathbb{R}\times\mathbb{R}, \quad \Psi(.,0) = \Psi_0, \tag{32}$$

$$|\Psi(x,t)| \to 1, \text{as} \quad |x| \to +\infty, \tag{33}$$

and recall the formal conserved quantities

$$E(\Psi) = \frac{1}{2}\int_{\mathbb{R}} |\partial_x \Psi|^2 + \frac{1}{4}\int_{\mathbb{R}} (1-|\Psi|^2)^2 \equiv \int_{\mathbb{R}} e(\Psi),$$

$$P(\Psi) = \frac{1}{2}\int_{\mathbb{R}} \langle i\partial_x \Psi, \Psi \rangle.$$

One can prove (see for instance [72]) well-posedness in Zhidkov spaces :

$$X^k(\mathbb{R}) = \{u \in L^1_{\text{loc}}(\mathbb{R}, \mathbb{C}), \text{ s.t. } E(u) < +\infty, \text{ and } \partial_x u \in H^{k-1}(\mathbb{R})\}.$$

Theorem 4. *Let $k \in \mathbb{N}^*$ and $\Psi_0 \in X^k(\mathbb{R})$. Then, there exists a unique solution $\Psi(\cdot, t)$ in $\mathscr{C}^0(\mathbb{R}, X^k(\mathbb{R}))$ to* (32) *with initial data Ψ_0. If Ψ_0 belongs to $X^{k+2}(\mathbb{R})$, then the map $t \mapsto \Psi(\cdot, t)$ belongs to $\mathscr{C}^1(\mathbb{R}, X^k(\mathbb{R}))$ and $\mathscr{C}^0(\mathbb{R}, X^{k+2}(\mathbb{R}))$. Moreover, the flow map $\Psi_0 \mapsto \Psi(\cdot, T)$ is continuous on $X^k(\mathbb{R})$ for any fixed $T \in \mathbb{R}$.*

Remark 7. One can prove conservation of energy and momentum (under suitable assumptions).

We recall that if Ψ does not vanish, one may write (Madelung transform)

$$\Psi = \sqrt{\rho} \exp i\varphi.$$

This leads to the hydrodynamic form of the equation, with $v = 2\partial_x \varphi$,

$$\begin{cases} \partial_t \rho + \partial_x(\rho v) = 0, \\ \rho(\partial_t v + v.\partial_x v) + \partial_x(\rho^2) = \rho \partial_x \left(\frac{\partial_x^2 \rho}{\rho} - \frac{|\partial_x \rho|^2}{2\rho^2} \right), \end{cases} \tag{34}$$

which can be seen as a compressible Euler system with pressure law $p(\rho) = \rho^2$ and a quantum pressure term. It is shown in [24] that (32) or (34) can be approximated by the *linear* wave equation. We justify here the long wave approximation on larger time scales $\mathscr{O}(\varepsilon^{-3})$ and, following Kuznetsov and Zakharov, introduce the slow variables :

$$X = \varepsilon(x + \sqrt{2}t), \text{ and } \tau = \frac{\varepsilon^3}{2\sqrt{2}}t. \tag{35}$$

This corresponds to a reference frame traveling to the left with velocity $\sqrt{2}$ in the original variables (x, t). In this frame the left going wave is stationary while the right going wave has a speed $8\varepsilon^{-2}$ and is appropriate to study waves traveling to the left (we need to impose additional assumptions which imply the smallness of the right going waves).

We define the rescaled functions N_ε and Θ_ε as follows

$$\begin{aligned} N_\varepsilon(X, \tau) &= \frac{6}{\varepsilon^2}\eta(x, t) = \frac{6}{\varepsilon^2}\eta\Big(\frac{X}{\varepsilon} - \frac{4\tau}{\varepsilon^3}, \frac{2\sqrt{2}\tau}{\varepsilon^3}\Big), \\ \Theta_\varepsilon(X, \tau) &= \frac{6\sqrt{2}}{\varepsilon}\varphi(x, t) = \frac{6\sqrt{2}}{\varepsilon}\varphi\Big(\frac{X}{\varepsilon} - \frac{4\tau}{\varepsilon^3}, \frac{2\sqrt{2}\tau}{\varepsilon^3}\Big), \end{aligned} \tag{36}$$

where $\Psi = \varrho \exp(i\varphi)$ and $\eta = 1 - \varrho^2 = 1 - |\Psi|^2$.

Theorem 5. *[25] Let $\varepsilon > 0$ be given and assume that the initial data $\Psi_0(\cdot) = \Psi(\cdot, 0)$ belong to $X^4(\mathbb{R})$ and satisfy the assumption*

$$\|N_\varepsilon^0\|_{H^3(\mathbb{R})} + \varepsilon\|\partial_x^4 N_\varepsilon^0\|_{L^2(\mathbb{R})} + \|\partial_x \Theta_\varepsilon^0\|_{H^3(\mathbb{R})} \leq K_0. \tag{37}$$

Let $\mathcal{N}_\varepsilon$ and $\mathcal{M}_\varepsilon$ denote the solutions to the Korteweg-de Vries equation

$$\partial_\tau N + \partial_x^3 N + N\partial_x N = 0 \tag{38}$$

with initial data N_ε^0, respectively $\partial_x \Theta_\varepsilon^0$. There exist positive constants ε_0 and K_1, depending possibly only on K_0 such that, if $\varepsilon \leq \varepsilon_0$, we have for any $\tau \in \mathbb{R}$,

$$\begin{aligned} &\|\mathcal{N}_\varepsilon(\cdot,\tau) - N_\varepsilon(\cdot,\tau)\|_{L^2(\mathbb{R})} + \|\mathcal{M}_\varepsilon(\cdot,\tau) - \partial_x\Theta_\varepsilon(\cdot,\tau)\|_{L^2(\mathbb{R})} \\ &\quad \leq K_1\big(\varepsilon + \|N_\varepsilon^0 - \partial_x\Theta_\varepsilon^0\|_{H^3(\mathbb{R})}\big)\exp(K_1|\tau|). \end{aligned} \tag{39}$$

- This is a convergence result to the KdV equation for appropriate *prepared* initial data.
- Since the norms involved are translation invariant, the KdV approximation can be only valid if the right going waves are negligible. This is the role of the term $\|N_\varepsilon^0 - \partial_x\Theta_\varepsilon^0\|_{H^3(\mathbb{R})}$.
- In particular, if $\|N_\varepsilon^0 - \partial_x\Theta_\varepsilon^0\|_{H^3(\mathbb{R})} \leq C\varepsilon^\alpha$, with $\alpha > 0$, then the approximation is valid on a time interval $t \in [0, S_\varepsilon]$ with $S_\varepsilon = o(\varepsilon^{-3}|\log(\varepsilon)|)$. Moreover, if $\|N_\varepsilon^0 - \partial_x\Theta_\varepsilon^0\|_{H^3(\mathbb{R})}$ is of order $\mathcal{O}(\varepsilon)$, then the approximation error remains of order $\mathcal{O}(\varepsilon)$ on a time interval $t \in [0, S_\varepsilon]$ with $S_\varepsilon = \mathcal{O}(\varepsilon^{-3})$.

The functions N_ε and $\partial_X\Theta_\varepsilon$ are rigidly constrained one to the other:

Theorem 6. *Let Ψ be a solution to GP in $\mathscr{C}^0(\mathbb{R}, H^4(\mathbb{R}))$ with initial data Ψ^0. Assume that* (37) *holds. Then, there exists some positive constant K, which does not depend on ε nor τ, such that*

$$\|N_\varepsilon(\cdot,\tau) \pm \partial_X\Theta_\varepsilon(\cdot,\tau)\|_{L^2(\mathbb{R})} \leq \|N_\varepsilon^0 \pm \partial_X\Theta_\varepsilon^0\|_{L^2(\mathbb{R})} + K\varepsilon^2\big(1 + |\tau|\big), \tag{40}$$

for any $\tau \in \mathbb{R}$.

The approximation errors provided by the previous theorems diverge as time increases. Concerning the weaker notion of **consistency**, we have the following result whose peculiarity is that the bounds are independent of time.

Theorem 7. *Let Ψ be a solution to GP in $\mathscr{C}^0(\mathbb{R}, H^4(\mathbb{R}))$ with initial data Ψ^0. Assume that* (37) *holds. Then, there exists some positive constant K, which does not depend on ε nor τ, such that*

$$\|\partial_\tau U_\varepsilon + \partial_X^3 U_\varepsilon + U_\varepsilon\partial_X U_\varepsilon\|_{L^2(\mathbb{R})} \leq K(\varepsilon + \|N_\varepsilon^0 - \partial_X\Theta_\varepsilon\|_{H^3(\mathbb{R})}), \tag{41}$$

for any $\tau \in \mathbb{R}$, where $U_\varepsilon = \frac{N_\varepsilon + \partial_x\Theta_\varepsilon}{2}$.

- One gets explicit bounds for the traveling wave solutions

$$\Psi(x,t) = v_c(x+ct).$$

Solutions do exist for any value of the speed c in the interval $[0, \sqrt{2})$. Next, we choose the wave-length parameter to be $\varepsilon = \sqrt{2-c^2}$, and take as initial data Ψ_ε the corresponding wave v_c. We consider the rescaled function

$$\nu_\varepsilon(X) = \frac{6}{\varepsilon^2}\eta_c\Big(\frac{x}{\varepsilon}\Big),$$

where $\eta_c \equiv 1 - |v_c|^2$. The explicit integration of the travelling wave equation for v_c leads to the formula

$$\nu_\varepsilon(X) = \nu(x) \equiv \frac{3}{\mathrm{ch}^2(\frac{X}{2})}.$$

The function ν is the classical soliton to the Korteweg-de Vries equation, which is moved by the KdV flow with constant speed equal to 1, so that

$$\mathcal{N}_\varepsilon(X,\tau) = \nu(X-\tau).$$

On the other hand, we deduce from (36) that $N_\varepsilon^0 = \nu$, so that

$$N_\varepsilon(X,\tau) = \nu\Big(X - \frac{4}{\varepsilon^2}\Big(1 - \sqrt{1-\frac{\varepsilon^2}{2}}\Big)\tau\Big).$$

Therefore, we have for any $\tau \in \mathbb{R}$,

$$\|\mathcal{N}_\varepsilon(\cdot,\tau) - N_\varepsilon(\cdot,\tau)\|_{L^2(\mathbb{R})} = \mathcal{O}(\varepsilon^2\tau).$$

This suggests that the ε error in the main theorem should be of order ε^2. This is proved in [26] at a cost of higher regularity on the initial data (and also for a two way propagation, described by a system of two KdV equations).

We now give some elements of the proofs which rely on energy methods. We first write the equations for N_ε and Θ_ε:

$$\partial_X N_\varepsilon - \partial_X^2\Theta_\varepsilon + \frac{\varepsilon^2}{2}\Big(\frac{1}{2}\partial_\tau N_\varepsilon + \frac{1}{3}N_\varepsilon\partial_X^2\Theta_\varepsilon + \frac{1}{3}\partial_X N_\varepsilon\partial_X\Theta_\varepsilon\Big) = 0, \tag{42}$$

and

$$\partial_X\Theta_\varepsilon - N_\varepsilon + \frac{\varepsilon^2}{2}\Big(\frac{1}{2}\partial_\tau\Theta_\varepsilon + \frac{\partial_X^2 N_\varepsilon}{1-\frac{\varepsilon^2}{6}N_\varepsilon} + \frac{1}{6}(\partial_X\Theta_\varepsilon)^2\Big) + \frac{\varepsilon^4}{24}\frac{(\partial_X N_\varepsilon)^2}{(1-\frac{\varepsilon^2}{6}N_\varepsilon)^2} = 0. \tag{43}$$

The leading order in this expansion is provided by $N_\varepsilon - \partial_X\Theta_\varepsilon$ and its spatial derivative, so that an important step is to keep control on this term. In view of (42) and (43) and d'Alembert decomposition, we are led to introduce the new variables U_ε and V_ε defined by

$$U_\varepsilon = \frac{N_\varepsilon + \partial_X\Theta_\varepsilon}{2}, \text{ and } V_\varepsilon = \frac{N_\varepsilon - \partial_X\Theta_\varepsilon}{2},$$

and compute the relevant equations for U_ε and V_ε,

$$\partial_\tau U_\varepsilon + \partial_X^3 U_\varepsilon + U_\varepsilon \partial_X U_\varepsilon = -\partial_X^3 V_\varepsilon + \frac{1}{3}\partial_X\big(U_\varepsilon V_\varepsilon\big) + \frac{1}{6}\partial_X\big(V_\varepsilon^2\big) - \varepsilon^2 R_\varepsilon, \tag{44}$$

and

$$\partial_\tau V_\varepsilon + \frac{8}{\varepsilon^2}\partial_x V_\varepsilon = \partial_X^3 U_\varepsilon + \partial_X^3 V_\varepsilon + \frac{1}{2}\partial_x(V_\varepsilon^2) - \frac{1}{6}\partial_x(U_\varepsilon)^2 - \frac{1}{3}\partial_X(U_\varepsilon V_\varepsilon) + \varepsilon^2 R_\varepsilon, \tag{45}$$

where the remainder term R_ε is given by the formulae

$$R_\varepsilon = \frac{N_\varepsilon \partial_X^3 N_\varepsilon}{6(1 - \frac{\varepsilon^2}{6}N_\varepsilon)} + \frac{(\partial_X N_\varepsilon)(\partial_X^2 N_\varepsilon)}{3(1 - \frac{\varepsilon^2}{6}N_\varepsilon)^2} + \frac{\varepsilon^2}{36}\frac{(\partial_X N_\varepsilon)^3}{(1 - \frac{\varepsilon^2}{6}N_\varepsilon)^3}. \tag{46}$$

The main step is to show that the RHS of the equation for U_ε is small in suitable norms. In particular one must show that V_ε which is assumed to be small at time $\tau = 0$ remains small, and that U_ε, assumed to be bounded at time $\tau = 0$, remains bounded in appropriate norms. We use in particular various conservation laws provided by the integrability of the one-dimensional Gross-Pitaevskii equation.

- For instance we use the conservation of momentum and energy to get the L^2 estimates.
- It turns out that the other conservation laws behave as higher order energies and higher order momenta. We use them to get :

Theorem 8. *Let Ψ be a solution to (GP) in $\mathscr{C}^0(\mathbb{R}, H^4(\mathbb{R}))$ with initial data Ψ^0. Assume that* (37) *holds. Then, there exists some positive constant K, which does not depend on ε nor τ, such that*

$$\|N_\varepsilon(\cdot,\tau)\|_{H^3(\mathbb{R})} + \varepsilon\|\partial_X^4 N_\varepsilon(\cdot,\tau)\|_{L^2(\mathbb{R})} + \|\partial_X\Theta_\varepsilon(\cdot,\tau)\|_{H^3(\mathbb{R})} \le K, \tag{47}$$

and

$$\|N_\varepsilon(\cdot,\tau) \pm \partial_X\Theta_\varepsilon(\cdot,\tau)\|_{H^3(\mathbb{R})} \le K\big(\|N_\varepsilon^0 \pm \partial_X\Theta_\varepsilon^0\|_{H^3(\mathbb{R})} + \varepsilon\big), \tag{48}$$

for any $\tau \in \mathbb{R}$.

The proof of the previous result led to a number of facts linked to integrability which have independent interest:

- It provides expressions for the invariant quantities of GP[5] and prove that they are well-defined in the spaces $X^k(\mathbb{R})$. Their expressions are not a straightforward consequence of the induction formula of Zakharov and Shabat since many renormalizations have to be performed to give them a sound mathematical meaning.
- It establishes rigorously that they are conserved by the GP flow in the appropriate functional spaces.
- It displays a striking relationship between the conserved quantities of the Gross-Pitaevskii equation and the KdV invariants:

$$\mathscr{E}_k(N, \partial_X \Theta) - \sqrt{2}\mathscr{P}_k(N, \partial_x \Theta) = E_k^{KdV}\Big(\frac{N - \partial_X \Theta}{2}\Big) + \mathscr{O}(\varepsilon^2).$$

It would be interesting to investigate further connections between the IST theories of the KdV and GP equations.

Remark 8. We do not know of a rigorous result by IST methods describing the qualitative behavior of a solution (solitons+radiation,...) of the GP equation corresponding to a smooth and localized initial data.

4 The Kadomtsev-Petviashvili Equation

The Kadomtsev-Petviashvili equations are universal asymptotic models for dispersive systems in the weakly nonlinear, long wave regime, when the wavelengths in the transverse direction are much larger than in the direction of propagation.

The (classical) Kadomtsev-Petviashvili (KP) equations read

$$(u_t + u_{xxx} + uu_x)_x \pm u_{yy} = 0. \tag{49}$$

Actually the (formal) analysis in [101] consists in looking for a *weakly transverse* perturbation of the one-dimensional transport equation

$$u_t + u_x = 0. \tag{50}$$

This perturbation amounts to adding a nonlocal term, leading to

$$u_t + u_x + \frac{1}{2}\partial_x^{-1} u_{yy} = 0. \tag{51}$$

---

[5]They appear in pairs: generalized energies $\mathscr{E}_k$ and generalized momenta $\mathscr{P}_k$.

Here the operator ∂_x^{-1} is defined via Fourier transform,

$$\widehat{\partial_x^{-1} f}(\xi) = \frac{i}{\xi_1}\hat{f}(\xi), \text{ where } \xi = (\xi_1, \xi_2).$$

When this same formal procedure is applied to the KdV equation written in the context of water waves (where $T \geq 0$ is the Bond number measuring the surface tension effects)

$$u_t + u_x + uu_x + (1 - \frac{1}{3}T)u_{xxx} = 0, \ x \in \mathbb{R}, \ t \geq 0, \tag{52}$$

this yields the KP equation in the form

$$u_t + u_x + uu_x + (1 - \frac{1}{3}T)u_{xxx} + \frac{1}{2}\partial_x^{-1} u_{yy} = 0. \tag{53}$$

By change of frame and scaling, (53) reduces to (49) with the + sign (KP II) when $T < \frac{1}{3}$ and the $-$ sign (KP I) when $T > \frac{1}{3}$.

Of course the same formal procedure could be applied to *any* one-dimensional weakly nonlinear dispersive equation of the form

$$u_t + u_x + f(u)_x - Lu_x = 0, \ x \in \mathbb{R}, \ t \geq 0, \tag{54}$$

where $f(u)$ is a smooth real-valued function (most of the time polynomial) and L a linear operator taking into account the dispersion and defined in Fourier variable by

$$\mathfrak{F}(Lu)(\xi) = p(\xi)\mathfrak{F}u(\xi), \tag{55}$$

where the symbol $p(\xi)$ is real-valued. The KdV equation corresponds for instance to $f(u) = \frac{1}{2}u^2$ and $p(\xi) = -\xi^2$. This leads to a class of generalized KP equations

$$u_t + u_x + f(u)_x - Lu_x + \frac{1}{2}\partial_x^{-1} u_{yy} = 0, \ x \in \mathbb{R}, \ t \geq 0. \tag{56}$$

Thus one could have KP versions of the Benjamin-Ono, Intermediate Long Wave, Kawahara, etc... equations, but only the KP I and KP II equations are completely integrable (in some sense).

Let us notice, at this point, that alternative models to KdV-type equations (54) are the equations of Benjamin–Bona–Mahony (BBM) type

$$u_t + u_x + uu_x + Lu_t = 0 \tag{57}$$

with corresponding two-dimensional "KP–BBM-type models" (in the case $p(\xi) \geq 0$)

$$u_t + u_x + uu_x + Lu_t + \partial_x^{-1}\partial_y^2 u = 0 \tag{58}$$

or, in the *derivative form*

$$(u_t + u_x + uu_x + Lu_t)_x + \partial_y^2 u = 0 \tag{59}$$

and free group

$$S(t) = e^{-t(I+L)^{-1}[\partial_x + \partial_x^{-1}\partial_y^2]} .$$

It was only after the seminal paper [101] that Kadomtsev-Petviashvili type equations have been derived as asymptotic weakly nonlinear models (under an appropriate scaling) in various physical situations (see [4] for a formal derivation in the context of water waves [142–144] for a rigorous one in the same context) and [102] in a different context.

Remark 9. In some physical contexts (not in water waves!) one could consider higher dimensional transverse perturbations, which amounts to replacing $\partial_x^{-1} u_{yy}$ in (66) by $\partial_x^{-1}\Delta^{\perp}u$, where $\Delta^{\perp}$ is the Laplace operator in the transverse variables. For instance, as in the one-dimensional case the KP I equation (in both two and three dimensions) also describes after a suitable scaling the long wave *transonic* limit of the Gross-Pitaevskii equation (see [28] for the solitary waves and [45] for the Cauchy problem).

Note again that in the classical KP equations, the distinction between KP I and KP II arises from the *sign* of the dispersive term in x.

4.1 KP by Inverse Scattering

It is usual in the Inverse Scattering community to write the Kadomtsev-Petviashvili equations as

$$\partial_x(\partial_t u + 6u\partial_x u + \partial_x^3 u) = -3\sigma^2 \partial_t^2 u, \tag{60}$$

where $\sigma^2 = 1$ corresponding to KP II and $\sigma^2 = -1$ to KP I.

4.1.1 The KP II Equation

The direct scattering problem is associated to the heat equation with the initial potential $u_0(x, y)$:

$$-\partial_y \phi + \partial_x^2 \phi + 2ik\partial_x \phi + u_0 \phi = 0, \quad \phi_{|k| \to \infty} = 1, \tag{61}$$

and the scattering data are calculated by

$$F(k) = (2\pi)^{-1}\mathrm{sign}(\mathscr{R}e\,k)\int_{\mathbb{R}^2} u_0(x,y)\phi(x,y,k)\exp\{-i(k+\bar{k})x-(k^2-\bar{k}^2)y\}dxdy.$$

The time evolution of the scattering data is trivial:

$$\mathscr{F}(k,t) = F(k)\exp(4it(k^3+\bar{k}^3)).$$

The inverse scattering problem, that is the reconstruction of the potential $u(x,y,t)$ reduces to a $\bar{\partial}$ problem:

$$\begin{cases} \partial_k\phi = \psi F(-k)\exp(itS), \\ \partial_k\psi = -\phi F(k)\exp(-itS) \end{cases} \tag{62}$$

$\begin{pmatrix}\phi \\ \psi\end{pmatrix} \to \begin{pmatrix}1 \\ 1\end{pmatrix}$ as $|k| \to \infty$, and where

$$S = 4(k^3+\bar{k}^3) + (k+\bar{k})\xi - i(k^2-\bar{k}^2)\eta,\ \xi = x/t,\ \text{and}\ \eta = y/t.$$

It turns out that, for reasons we already mentioned in the BO and ILW cases, the best known result on the direct scattering problem required that the initial data is small in spaces of type $L^1 \cap L^2$, yielding global existence of uniformly bounded (in the space of L^2 functions with bounded Fourier transform) solutions of KP II provided u_0 has small derivatives up to order 8 in $L^1 \cap L^2(\mathbb{R}^2)$ [225]. We will see that PDE methods provide a much better result.

Remark 10. In [82], Grinevich has proven that the direct spectral problem is nonsingular for real nonsingular exponentially decaying at infinity, arbitrary large potentials. Unfortunately, this does not mean that the solution of the direct scattering problem belongs to an appropriate functional class for existence of an inverse scattering problem when $t > 0$. In fact, the direct problem and inverse problem are different and the solvability of the first one does not give the automatic solvability of the second one.

Remark 11. Since KP II type equations do not have localized solitary waves [36], one expects the large time behavior of solutions to be just governed by scattering. In particular, one can conjecture safely than the global solutions of KP II (that exist by the result of Bourgain, see [40] and the discussion below) should decay in the sup-norm as $1/t$. This is also suggested by our numerical simulations [126].

A very precise asymptotics as $t \to \infty$ is given in [120] (see also [121]) for a specific class of scattering data. It differs according to different domains in the (x,y,t) space, expressed in terms of the variables $\xi = x/t$ and $\eta = y/t$. The main term of the asymptotics has order $O(1/t)$ (which is exactly the decay rate of the free linear evolution, see [202]) and rapidly oscillates. In one of the domains, the decay is $o(1/t)$. It is not clear however how the hypothesis on the scattering data translate to the space of initial data.

On the other hand, the Inverse Scattering method has been used formally in [5] and rigorously in [6] to study the Cauchy problem for the KP II equation with nondecaying data along a line, that is $u(0,x,y) = u_\infty(x - vy) + \phi(x,y)$ with $\phi(x,y) \to 0$ as $x^2 + y^2 \to \infty$ and $u_\infty(x) \to 0$ as $|x| \to \infty$. Typically, u_∞ is the profile of a traveling wave solution $U(\mathbf{k}.\mathbf{x} - \omega t)$ with its peak localized on the moving line $\mathbf{k}.\mathbf{x} = \omega t$. It is a particular case of the N- soliton of the KP II equation discovered by Satsuma [200] (see the derivation and the explicit form when $N = 1, 2$ in the Appendix of [181]). As in all results obtained for KP equations by using the Inverse Scattering method, the initial perturbation of the non-decaying solution is supposed to be small enough in a weighted L^1 space (see [6] Theorem 13).

4.1.2 The KP I Equation

The direct scattering problem for KP I is associated to the Schrödinger operator with potential (see [63])

$$i\psi_t + \psi_{xx} = -u\psi.$$

As for the KP II case, there is a restriction on the size of the initial data to solve the direct scattering problem see [152]. In [235] the nonlocal Riemann-Hilbert problem for inverse scattering is shown to have a solution leading to the global solvability of the Cauchy problem (with a smallness condition on the initial data). A formal asymptotic of small solutions is given in [155]. It would be interesting to provide a rigorous proof of this result.

It is proven in [216] that the solution constructed by the IST belongs to the Sobolev spaces $H^s(\mathbb{R}^2), s \geq 0$, provided the initial data is a small function in the Schwartz space $\mathscr{S}(\mathbb{R}^2)$, thus not assuming the zero mass constraint (see Sect. 4.2.1 below) contrary to the result in [235] (see also [67] where the IST solution is shown to be C^∞ for a small Schwartz initial data).

Finally the Cauchy problem of the background of a one-line soliton is solved formally (for small initial perturbations) in [64].

4.1.3 Conservation Laws

The KP equations being integrable have an infinite set of (formally) conserved quantities. For instance, see [232] the KP I equation has a Lax pair representation. This in turn provides an algebraic procedure generating an infinite sequence of conservation laws. More precisely, if u is a formal solution of the KP I equation then

$$\frac{d}{dt}\Big[\int \chi_n\Big] = 0,$$

where $\chi_1 = u$, $\chi_2 = u + i\partial_x^{-1}\partial_y u$ and for $n \geq 3$,

$$\chi_n = \Big(\sum_{k=1}^{n-2} \chi_k\, \chi_{n+1-k}\Big) + \partial_x \chi_{n-1} + i\partial_x^{-1}\partial_y \chi_{n-1}\ .$$

For $n = 3$, we find the conservation of the L^2 norm, $n = 5$ corresponds to the energy functional giving the Hamiltonian structure of the KP I equation, that is the following quantities are well defined and conserved by the flow (in an appropriate functional setting, see [169])

$$M(\phi) = \int_{\mathbb{R}^2} |u|^2,$$

$$E(u) = \frac{1}{2}\int_{\mathbb{R}^2} u_x^2 + \frac{1}{2}\int_{\mathbb{R}^2} (\partial_x^{-1}u_y)^2 - \frac{1}{6}\int_{\mathbb{R}^2} u^3,$$

$$F(u) = \frac{3}{2}\int_{\mathbb{R}^2} u_{xx}^2 + 5\int_{\mathbb{R}^2} u_y^2 + \frac{5}{6}\int_{\mathbb{R}^2} (\partial_x^{-2}u_{yy})^2 - \frac{5}{6}\int_{\mathbb{R}^2} u^2\partial_x^{-2}u_{yy}$$

$$-\frac{5}{6}\int_{\mathbb{R}^2} u(\partial_x^{-1}u_y)^2 + \frac{5}{4}\int_{\mathbb{R}^2} u^2\, u_{xx} + \frac{5}{24}\int_{\mathbb{R}^2} u^4.$$

As was noticed in [169], there is a serious analytical obstruction to give sense to χ_9 as far as $\mathbb{R}^2$ is considered as a spatial domain. More precisely the conservation law which controls $\|u_{3x}(t,\cdot)\|_{L^2}$ involves the L^2 norm of the term $\partial_x^{-1}\partial_y(\phi^2)$which has no sense for a nonzero function ϕ in $H^3(\mathbb{R}^2)$ say. Actually one easily checks that if $\partial_x^{-1}\partial_y(\phi^2) \in L^2(\mathbb{R}^2)$, then $\int_{\mathbb{R}^2} \partial_y(u^2)dx = \partial_y \int_{\mathbb{R}^2} u^2 dx \equiv 0,\ \forall y \in \mathbb{R}$, which, with $u \in L^2(\mathbb{R}^2)$, implies that $u \equiv 0$. Similar obstructions occur for the higher order "invariants".

The fact that the invariants ξ_n, $n \geq 9$ do not make sense for a nonzero function yields serious difficulties to define the so-called KP hierarchy.

4.2 KP by PDE Methods

The basic difference between KP I and KP II as far as PDE techniques are concerned, is that KP I is *quasilinear* while KP II is *semilinear*. We recall that this means that the Cauchy problem for KP I cannot be solved by a Picard iterative scheme implemented on the integral Duhamel formulation, for any initial data in very general spaces (that is the Sobolev spaces $H^s(\mathbb{R}^2)$, $\forall s \in \mathbb{R}$, or the anisotropic ones $H^{s_1,s_2}(\mathbb{R}^2)$, $\forall s_1, s_2 \in \mathbb{R}$) . Alternatively, this implies that the flow map $u_0 \mapsto u(t)$ cannot be *smooth* in the same spaces. Here are precise statements of those results from [168].

Theorem 9. *Let $s \in \mathbb{R}$ and T be a positive real number. Then there does not exist a space X_T continuously embedded in $C([-T,T], H^s(\mathbb{R}))$ such that there exists $C > 0$ with*

$$\|S(t)\phi\|_{X_T} \leq C\|\phi\|_{H^s(\mathbb{R})}, \quad \phi \in H^s(\mathbb{R}), \tag{63}$$

and

$$\left\| \int_0^t S(t-t')\left[u(t')u_x(t')\right] dt' \right\|_{X_T} \leq C\|u\|_{X_T}^2, \quad u \in X_T. \tag{64}$$

Note that (63) and (64) would be needed to implement a Picard iterative scheme on the integral (Duhamel) formulation of the equation in the space X_T. As a consequence of Theorem 9 we can obtain the following result.

Theorem 10. *Let $(s_1, s_2) \in \mathbb{R}^2$ (resp. $s \in \mathbb{R}$). Then there exists no $T > 0$ such that KPI admits a unique local solution defined on the interval $[-T, T]$ and such that the flow-map*

$$S_t : \phi \longmapsto u(t), \quad t \in [-T, T]$$

for (1) is C^2 differentiable at zero from $H^{s_1,s_2}(\mathbb{R}^2)$ to $H^{s_1,s_2}(\mathbb{R}^2)$, (resp. from $H^s(\mathbb{R}^2)$ to $H^s(\mathbb{R}^2)$).

Remark 12. It has been proved in [134] that the flow map cannot be uniformly continuous in the energy space.

Proof. We merely sketch it (see [168] for details). Let

$$\sigma(\tau, \xi, \eta) = \tau - \xi^3 - \frac{\eta^2}{\xi},$$

$$\sigma_1(\tau_1, \xi_1, \eta_1) = \sigma(\tau_1, \xi_1, \eta_1),$$

$$\sigma_2(\tau_1, \xi, \eta_1, \tau_1, \xi_1, \eta_1) = \sigma(\tau - \tau_1, \xi - \xi_1, \eta - \eta_1).$$

We then define

$$\chi(\xi, \xi_1, \eta, \eta_1) := 3\xi\xi_1(\xi - \xi_1) - \frac{(\eta\xi_1 - \eta_1\xi)^2}{\xi\xi_1(\xi - \xi_1)}.$$

Note that $\chi(\xi, \xi_1, \eta, \eta_1) = \sigma_1 + \sigma_2 - \sigma$. The "resonant" function $\chi(\xi, \xi_1, \eta, \eta_1)$ plays an important role in our analysis. The "large" set of zeros of $\chi(\xi, \xi_1, \eta, \eta_1)$ is responsible for the ill-posedness issues. In contrast, the corresponding resonant function for the KP II equation is

$$\chi(\xi,\xi_1,\eta,\eta_1) := 3\xi\xi_1(\xi-\xi_1) + \frac{(\eta\xi_1-\eta_1\xi)^2}{\xi\xi_1(\xi-\xi_1)}.$$

Since it is essentially the sum of two squares, its zero set is small and this is the key point to establish the crucial bilinear estimate in Bourgain's method [40].

Remark 13. It is worth noticing that the property of the resonant set of the KP II equation was used by Zakharov [228] to construct a Birkhoff formal form for the *periodic* KP II equation with small initial data. On the other hand, the fact that for the KP I equation the corresponding resonant set is non trivial is crucial in the construction of the counter-examples of [168] and is apparently an obstruction to the Zakharov construction for the periodic KP I equation.

Since the next property of KP equations is only based on the presence of the operator $\partial_x^{-1}\partial_y^2$ we will consider it in the context of *generalized* KP type equations:

$$u_t + u_x + f(u)_x - Lu_x + \frac{1}{2}\partial_x^{-1}u_{yy} = 0, \ (x,y) \in \mathbb{R}^2, \ t \geq 0, \tag{65}$$

where $\widehat{Lu}(\xi) = p(\xi)\hat{u}(\xi)$, p real and $f(u)$ is a nonlinear function, for instance $f(u) = \frac{1}{q+1}u^{q+1}$. An important particular case is the *generalized KP equation*

$$u_t + u_x + u^r u_x + u_{xxx} \pm \partial_x^{-1}u_{yy} = 0, \ (x,y) \in \mathbb{R}^2, \ t \geq 0, \tag{66}$$

where $r \in \mathbb{N}$ or $r = \frac{p}{q}$, p, q relatively prime integers, q odd.

4.2.1 The Zero Mass Constraint

In (65) or (66), it is implicitly assumed that the operator $\partial_x^{-1}\partial_y^2$ is well defined, which a priori imposes a constraint on the solution u, which, in some sense, has to be an x-derivative. This is achieved, for instance, if $u \in \mathcal{S}'(\mathbb{R}^2)$ is such that

$$\xi_1^{-1}\,\xi_2^2\,\hat{u}(t,\xi_1,\xi_2) \in \mathcal{S}'(\mathbb{R}^2)\,, \tag{67}$$

thus in particular if $\xi_1^{-1}\,\hat{u}(t,\xi_1,\xi_2) \in \mathcal{S}'(\mathbb{R}^2)$. Another possibility to fulfill the constraint is to write u as

$$u(t,x,y) = \frac{\partial}{\partial x}v(t,x,y), \tag{68}$$

where v is a continuous function having a classical derivative with respect to x, which, for any fixed y and $t \neq 0$, vanishes when $x \to \pm\infty$. Thus one has

$$\int_{-\infty}^{\infty} u(t,x,y)dx = 0, \qquad y \in \mathbb{R}, \ t \neq 0, \tag{69}$$

in the sense of generalized Riemann integrals. Of course the differentiated version of (65), (66), for instance

$$(u_t + u_x + uu_x - Lu_x)_x + \partial_y^2 u = 0, \tag{70}$$

can make sense without any constraint of type (67) or (69) on u, and so does the Duhamel integral representation of (66), (65), for instance

$$u(t) = S(t)u_0 - \int_0^t S(t-s)(u(s)u_x(s))ds, \tag{71}$$

where $S(t)$ denotes the (unitary in all Sobolev spaces $H^s(\mathbb{R}^2)$) group associated with (65),

$$S(t) = e^{-t(\partial_x - L\partial_x + \partial_x^{-1}\partial_y^2)}\ . \tag{72}$$

In particular, the results established on the Cauchy problem for KP type equations which use the Duhamel (integral) formulation (see for instance [39] for the KP II equation and [205] for the KP II BBM equation) are valid *without* any constraint on the initial data.

One has however to be careful *in which sense* the *differentiated equation* is taken. For instance let us consider the integral equation

$$u(x,y,t) = S(t)u_0(x,y) - \int_0^t S(t-t')[u(x,y,t')u_x(x,y,t)]dt', \tag{73}$$

where $S(t)$ is here the KP II group, for initial data u_0 in $H^s(\mathbb{R}^2),\ s > 2$, (thus u_0 does not satisfy any zero mass constraint), yielding a local solution $u \in C([0,T]; H^s(\mathbb{R}^2))$. By differentiating (73) first with respect to x and then with respect to t, one obtains the equation

$$\partial_t\partial_x u + \partial_x(uu_x) + \partial_x^4 u + \partial_y^2 u = 0 \quad \text{in} \quad C([0,T]; H^{s-4}(\mathbb{R}^2)).$$

However, the identity $\partial_t\partial_x u = \partial_x\partial_t u$ holds only in a very weak sense, for example in $\mathfrak{D}'((0,T)\times\mathbb{R}^2)$.

On the other hand, a constraint has to be imposed when using the Hamiltonian formulation of the equation. In fact, the Hamiltonian for (70) is

$$\frac{1}{2}\int\left[-u\,Lu + (\partial_x^{-1}u_y)^2 + u^2 + \frac{u^3}{3}\right] \tag{74}$$

and the Hamiltonian associated with (59) is

$$\frac{1}{2}\int\left[(\partial_x^{-1}u_y)^2+u^2+\frac{u^3}{3}\right]. \tag{75}$$

It has been established in [170] that, for a rather general class of KP or KP–BBM equations, the solution of the Cauchy problem obtained for (70), (59) (in an appropriate functional setting) satisfies the zero-mass constraint in x for any $t \neq 0$ (in a sense to be made precise below), even if the initial data does not. This is a manifestation of the infinite speed of propagation inherent to KP equations. Moreover, KP type equations display a striking *smoothing effect* : if the initial data belongs to the space $L^1(\mathbb{R}^2) \cap H^{2,0}(\mathbb{R}^2)$ and if $u \in C([0,T]; H^{2,0}(\mathbb{R}^2))$[6] is a solution in the sense of distributions, then, for any $t > 0$, $u(.,t)$ becomes a *continuous* function of x and y (with zero mean in x). Note that the space $L^1(\mathbb{R}^2) \cap H^{2,0}(\mathbb{R}^2)$ is not included in the space of continuous functions.

The key point when proving those results is a careful analysis of the fundamental solution of KP-type equations[7] which turns out to be an x derivative of a continuous function of x and y, C^1 with respect to x, which, for fixed $t \neq 0$ and y, tends to zero as $x \to \pm\infty$. Thus its generalized Riemann integral in x vanishes for all values of the transverse variable y and of $t \neq 0$. A similar property can be established for the solution of the nonlinear problem see [170]. Those results have been checked in the numerical simulations of [130] as can be seen in Fig. 7 taken from this reference. It can be seen that for initial data not satisfying the constraint, after an arbitrary short time some sort of infinite trench forms the integral over which just ensures that the constraint holds at all t.

We have already referred to [67, 216] for a rigorous approach to the Cauchy problem with (small) initial data which do not satisfy the zero-mass condition via the Inverse Spectral Method in the integrable case.

Nevertheless, the singularity at $\xi_1 = 0$ of the dispersion relation of KP type equations make them rather *bad* asymptotic models. First the singularity at $\xi_1 = 0$ yields a very bad approximation of the dispersion relation of the original system (for instance the water wave system) by that of the KP equation.

Another drawback is the poor error estimate between the KP solution and the solution of the original problem. This has been established clearly in the context of water waves see [142–144].

4.2.2 The Cauchy Problem by PDE Techniques

All the KP type equations can be viewed as a linear skew-adjoint perturbation of the Burgers equation. Using this structure, it is not difficult (for instance by a compactness method) to prove that the Cauchy problem is locally well-posed for

---

[6]We will see below that KP type equations (in particular the classical KP I and KP II equations) do possess solutions in this class.

[7]In the case of KP II, one can use the explicit form of the fundamental solution found in [195].

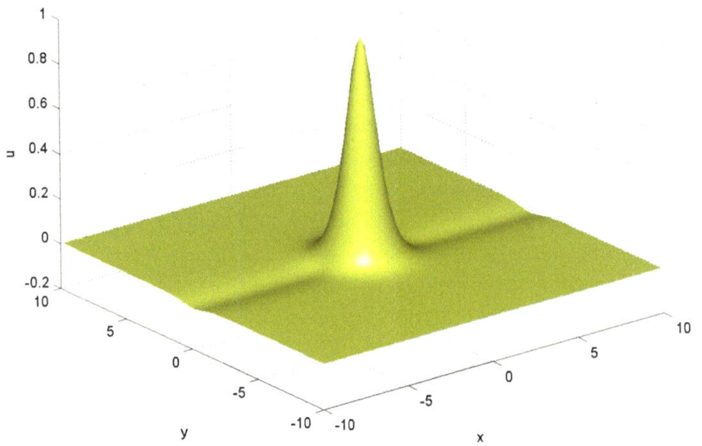

Fig. 7 Solution at time $t = 4.8 \times 10^{-4}$ to the KP-I solution with initial data $\operatorname{sech}^2\sqrt{x^2+y^2}$, which are not subject to the zero mass constraint

data in the Sobolev spaces $H^s(\mathbb{R}^2)$, $s > 2$ (see [97, 202, 222] for results in this direction).

Unfortunately, this kind of result does not suffice to obtain the *global* well-posedness of the Cauchy problem. This would need to use the *conservation laws* of the equations. For *general* KP type equations, there are only two of them, the conservation of the L^2 norm and the conservation of the energy (Hamiltonian). For the general equation (66) where $f(u) = \frac{1}{p+1}u^{p+1}$, and without the transport term u_x (which can be eliminated by a change of variable), we recall that the Hamiltonian reads

$$E(u) = \frac{1}{2}\int\left[-u\,Lu + (\partial_x^{-1}u_y)^2 + \frac{u^{p+2}}{p+2}\right], \tag{76}$$

and for the classical KP I/II equations

$$E(u) = \frac{1}{2}\int\left[u_x^2 \pm (\partial_x^{-1}u_y)^2 - \frac{u^3}{3}\right], \tag{77}$$

where the $+$ sign corresponds to KP I and the $-$ sign to KP II. We recall that the "integrable" KP I and KP II equations possess more conservation laws, but only a finite number of them make sense rigorously (see above).

In any case, it is clear that for KP II type equation, the Hamiltonian is useless to control any Sobolev norm, and to obtain the *global* well-posedness of the Cauchy problem one should consider L^2 solutions, a very difficult task. On the other hand, for KP I type equations, one may hope (for a not too strong nonlinearity) to have a global control in the *energy* space, that is

$$E = \{u \in L^2,\ uLu \in L^2,\ \partial_x^{-1} u_y \in L^2\}. \tag{78}$$

For the usual KP I equation, E reduces to

$$Y = \{u \in L^2,\ u_x \in L^2,\ \partial_x^{-1} u_y \in L^2\}.$$

The problem is thus reduced to proving the local well-posedness of the Cauchy problem in spaces of very low regularity, a difficult matter which has attracted a lot of efforts in the recent years.

Remark 14. By a standard compactness method, one obtains easily the existence of global weak finite energy solutions (without uniqueness) to the KP I equation see eg [221].

A fundamental step for KP II is made in [39] who proved that the Cauchy problem for the KP II equation is locally (thus globally in virtue of the conservation of the L^2 norm) for data in $L^2(\mathbb{R}^2)$, and even in $L^2(\mathbb{T}^2)$. This result is based on an iterative method implemented on the Duhamel formulation, in the functional framework of the *Fourier restriction* $X^{s,b}$ spaces of Bourgain (see a nice description of this framework in [80]). The basic *bilinear estimate* which aims to regain the loss of one x-derivative uses in a crucial way the fact (both in the periodic and full space case) that the dispersion relation of the KP II equation induces the triviality of a *resonant set* (the zero set of the aforementioned resonant function). With in particular the injection of various linear dispersive estimates (see for instance [18, 202]), Bourgain's result was later improved (see [87, 88, 93, 217] and the references therein) to allow the case of initial data in negative order Sobolev spaces.

We also would like to mention the paper by Kenig and Martel [113] where the Miura transform is used to prove the global well-posedness of a modified KP II equation. Moreover it was proven in [172] that the Cauchy problem for KP II is globally well-posed with initial data

$$u(0,x,y) = \phi(x,y) + \psi_c(x,y), \tag{79}$$

where ψ_c is the profile[8] of a non-localized (i.e. not decaying in all spatial directions) traveling wave of the KP II equation moving with speed $c \neq 0$.

We recall ([36]) that, contrary to the KP I equation, the KP II equation does not possess any *localized in both directions* traveling wave solution. The background solution ψ_c could be for instance the line soliton (1-soliton) of the Korteweg-de Vries (KdV) equation

[8]This means that $\psi(x - ct, y)$ solves KP II.

$$s_c(x,y) = \frac{3c}{2}\cosh^{-2}\Big(\frac{\sqrt{c}\,x}{2}\Big), \tag{80}$$

or the N-soliton solution of the KdV equation, $N \geq 2$. The KdV N-soliton is of course considered as a two dimensional (constant in y) object.

There are two suitable settings for an initial perturbation ϕ, either localized in x and periodic in y (eg $\phi \in L^2(\mathbb{R}\times\mathbb{T})$) or localized (eg $\phi \in L^2(\mathbb{R}^2)$.) Solving the Cauchy problem in both those functional settings can be viewed as a preliminary step towards the study of the dynamics of the KP II equation on the background of a non fully localized solution, in particular towards a PDE proof of the nonlinear stability of the KdV soliton or N-soliton with respect to transversal perturbations governed by the KP II flow. This has been established in [6] Proposition 17 by Inverse Scattering methods. The advantage of the PDE approach is that it can be straightforwardly applied to non integrable equations such as the higher order KP II equations see [203, 204].

We now state the main result of [172] in the two aforementioned functional settings.

Theorem 11. *The Cauchy problem associated with the KP II equation is globally well-posed in $H^s(\mathbb{R}\times\mathbb{T})$ for any $s \geq 0$.*

Theorem 12. *Let $\psi_c(x-ct,y)$ be a solution of the KP II equation such that for some $s \geq 0$,*

$$J^s\psi_c : \mathbb{R}^2 \longrightarrow \mathbb{R}$$

is bounded and belongs to $L^2_xL^\infty_y$.[9] *Then for every $\phi \in H^s(\mathbb{R}^2)$ there exists a unique global solution u of KP II with initial data (79) satisfying for all $T > 0$,*

$$[u(t,x,y) - \psi_c(x-ct,y)] \in X_T^{1/2+,s} \cap X_T^{3/4+,s} \cap C([0,T];H^s(\mathbb{R}^2)).$$

Furthermore, for all $T > 0$, the map $\phi \mapsto u$ is continuous from $H^s(\mathbb{R}^2)$ to $C([0,T];H^s(\mathbb{R}^2)))$.

Remark 15. As was previously noticed, the hypothesis on ψ_c in Theorem 12 is satisfied by the N-soliton solutions of the KdV equation, but not by a function ψ which is non-decaying along a line $\{(x,y)|x-vy = x_0\}$, as for instance the line-soliton of the KP II equation which has the form $\psi(x-vy-ct)$. However, the change of variables ($X = x, Y = x - vy$) transforms the KP II equation into

$$u_t - 2vu_Y + v^2u_X + u_{XXX} + uu_X + \partial_X^{-1}u_{YY} = 0$$

[9] The bounds can of course depend on the propagation speed c.

and the analysis applies to this equation with an initial datum which is a localized (in (X, Y)) perturbation of $\psi(X)$.

As was previously mentioned the Cauchy problem for the KP I cannot be solved by a Picard iteration on the integral formulation and one has to implement instead sophisticated *compactness methods* to obtain the local well-posedness and to use the conservation laws to get global solutions. For the classical KP I equation, the first global well-posedness result for arbitrary large initial data in a suitable function space was obtained in [169] in the space

$$Z = \{\phi \in L^2(\mathbb{R}^2) \ : \ \|\phi\|_Z < \infty\},$$

where

$$\|\phi\|_Z = |\phi|_2 + |\phi_{xxx}|_2 + |\phi_y|_2 + |\phi_{xy}|_2 + |\partial_x^{-1}\phi_y|_2 + |\partial_x^{-2}\phi_{yy}|_2.$$

By an anisotropic Sobolev embedding theorem cf. [22] if $\phi \in Z$ then $\phi, \phi_x \in L^\infty(\mathbb{R}^2)$, so the global solution is uniformly bounded in space and time. Moreover, if $\phi \in Z$ then the first three formal invariants $M(\phi), E(\phi), F(\phi)$ are well defined and conserved. Furthermore it is easily checked that any finite energy solitary waves (in particular the lumps, see below) of the KP I equation belong to Z.

The proof is based on a rather sophisticated compactness method and uses the first invariants of the KP I equation to get global in time bounds. As already mentioned, only a small number of the formal invariants make sense and in order to control $|\phi_{xxx}|_2$ one is thus led to introduce a *quasi-invariant* (by skipping the non well defined terms) which eventually will provide the desired bound. There are also serious technical difficulties to *justify* rigorously the conservation of the meaningful invariants along the flow and to control the remainder terms

The result of [169] was extended by Kenig [111] (who considered initial data in a larger space), and by Ionescu, Kenig and Tataru [95] who proved that the KP I equation is globally well-posed in the energy space Y. Moreover it is proven in [171] that the Cauchy problem for the KP I equation is globally well-posed for initial data which are localized perturbations (of arbitrary size) of a **non-localized** i.e. not decaying in all directions) traveling wave solution (e.g. the KdV line solitary wave or the Zaitsev solitary waves which are localized in x and y periodic or conversely (see Sect. 4.2 below).

4.2.3 Long Time Behavior

The results above do not give information on the behavior of the global solution for large time. Actually no result in this direction is known by PDE techniques. However, one can make precise the large time behavior of small solutions to the generalized KP equation (66) when $r \geq 2$.

Actually, it is shown in [92, 177] that for initial data small in an appropriate weighted Sobolev space, (66) for $r \geq 2$ has a unique global solution satisfying

$$|u(.,t)|_\infty \leq C(1+|t|)^{-1}(\mathrm{Log}(1+|t|))^\kappa,$$
$$|\partial_x u(.,t)|_\infty \leq C(1+|t|)^{-1},$$

where $\kappa = 1$ when $r = 2$ and $\kappa = 0$ when $r > 2$. This result does not distinguish between the KP I and KP II case since it relies on the (same) large time asymptotic of the KP I and KP II groups $S_\pm(t) = e^{it(-\partial_x^3 \pm \partial^{-1}{}_x \partial_y^2)}$, namely see [202]

$$|S_\pm(t)\phi|_\infty \leq \frac{C}{|t|}|\phi|_1.$$

Remark 16. Since this phenomena does not happen for the classical KP I and KP II equation, we do not comment on the possible *blow-up* in finite time of solutions to the generalized KP I equation (66) when $r \geq \frac{4}{3}$. We refer to Liu [151] for a theoretical study and to [122, 126] for numerical simulations.

Remark 17. Issues on the dispersionless limit of KP equations can be found in [123, 129].

4.3 Solitary Waves

We are interested here in localized solitary wave solutions to the KP equations, that is solutions of KP equations of the form

$$u(x,y,t) = \psi_c(x-ct,y),$$

where y is the transverse variable and $c > 0$ is the solitary wave velocity. The solitary wave is said to be *localized* if ψ_c tends to zero at infinity in all directions. For such solitary waves, the *energy space* Y is natural. Recall that

$$Y = Y(\mathbb{R}^2) = \{u \in L^2(\mathbb{R}^2),\ u_x \in L^2(\mathbb{R}^2),\ \partial_x^{-1}u_y \in L^2(\mathbb{R}^2)\},$$

and throughout this section we will deal only with *finite energy solitary waves*.

Due to its integrability properties, the KP I equation possesses a localized, finite energy, explicit solitary wave, the *lump*:

$$\phi_c(x-ct,y) = \frac{8c(1-\frac{c}{3}(x-ct)^2+\frac{c^2}{3}y^2)}{(1+\frac{c}{3}(x-ct)^2+\frac{c^2}{3}y^2)^2}. \tag{81}$$

The formula was found in [154] where one can also find a study of the interaction of lumps. The interactions do not result in a phase shift as in the case of line solitons (KdV solitons). More general rational solutions of the KPI equation were subsequently found [140, 187, 188, 200, 201]. These solutions were incorporated within the framework of the IST in [223] where it was observed that, in general, the spectral characterization of the potential must include, in addition to the usual information about discrete and continuous spectrum, an integer-valued topological quantity (the *index* or winding number), defined by an appropriate two-dimensional integral involving both the solution of the KP equation and the corresponding scattering eigenfunction.

Another interesting explicit solitary wave of the KP I equation which is *localized in x and periodic in y* has been found by Zaitsev [226]. It reads

$$Z_c(x,y) = 12\alpha^2 \frac{1-\beta\cosh(\alpha x)\cos(\delta y)}{[\cosh(\alpha x)-\beta\cos(\delta y)]^2}, \tag{82}$$

where

$$(\alpha,\beta) \in (0,\infty)\times(-1,+1),$$

and the propagation speed is given by

$$c = \alpha^2 \frac{4-\beta^2}{1-\beta^2}.$$

Let us observe that the transform $\alpha \to i\alpha$, $\delta \to i\delta$, $c \to ic$ produces solutions of the KP I equation which are periodic in x and localized in y. No real non-singular rational solutions are known for KP II. Moreover, it was established in [36] that no localized solitary waves exist for the KP II equation (and generalized KP II equations).

For obvious (stability) issues it is important to characterize the solitary waves by variational principles. We will consider in fact the slightly more general class of generalized KP I equations

$$u_t + u_x + u^p u_x + u_{xxx} - \partial_x^{-1} u_{yy} = 0,\ (x,y)\in\mathbb{R}^2,\ t\geq 0, \tag{83}$$

where again $p \in \mathbb{N}$ or $p = \frac{p_1}{q_1}$, p_1, q_1 relatively prime integers, q_1 odd.

Solitary waves are looked for in the energy space $Y(\mathbb{R}^2)$ which can also be defined see [36] as the closure of the space of x derivatives of smooth and compactly supported functions in $\mathbb{R}^2$ for the norm

$$\|\partial_x f\|_{Y(\mathbb{R}^2)} \equiv \left(\|\nabla f\|^2_{L^2(\mathbb{R}^2)} + \|\partial_x^2 f\|^2_{L^2(\mathbb{R}^2)}\right)^{\frac{1}{2}}.$$

The equation of a solitary wave ψ of speed c is given by

$$c\partial_x\psi - \psi\partial_x\psi - \partial_x^3\psi + \partial_x^{-1}(\partial_y^2\psi) = 0, \tag{84}$$

which implies

$$c\partial_{xx}\psi - (\psi\partial_x\psi)_x - \partial_x^4\psi + \partial_y^2\psi = 0. \tag{85}$$

When $\psi \in Y(\mathbb{R}^2)$, the function $\partial_x^{-1}\partial_y^2\psi$ is well-defined see [36], so that (84) makes sense. Given any $c > 0$, a solitary wave ψ_c of speed c is deduced from a solitary wave ψ_1 of velocity 1 by the scaling

$$\psi_c(x, y) = c\psi_1(\sqrt{c}x, cy). \tag{86}$$

We now introduce the important notion of *ground state* solitary waves. We set

$$E_{KP}(\psi) = \frac{1}{2}\int_{\mathbb{R}^2}(\partial_x\psi)^2 + \frac{1}{2}\int_{\mathbb{R}^2}(\partial_x^{-1}\partial_y\psi)^2 - \frac{1}{2(p+2)}\int_{\mathbb{R}^2}\psi^{p+2},$$

and we define the action

$$S(N) = E_{KP}(N) + \frac{c}{2}\int_{\mathbb{R}^2} N^2.$$

We call *ground state*, a solitary wave N which minimizes the action S among all finite energy non-constant solitary waves of speed c (see [36] for more details). It was proven in [36] that ground states exist if and only if $c > 0$ and $1 \leq p < 4$. Moreover, when $1 \leq p < \frac{4}{3}$, the ground states are minimizers of the Hamiltonian E_{KP} with prescribed mass (L^2 norm).

Remark 18. When $p = 1$ (the classical KP I equation), we should emphasize that it is unknown (but conjectured) whether the lump solution is a ground state. This important issue is of course related to the *uniqueness* of the ground state or of localized solitary waves, up to symmetries. A similar question stands for the focusing nonlinear Schrödinger equation but it can be solved there because the ground state is shown to be radial and uniqueness follows from (non trivial!) ODE arguments see [141]. Of course such arguments cannot work in the KP I case since the lump (or the ground states) are not radial.

It turns out that qualitative properties of solitary waves which are that of the lump solution can be established for a large class of KP type equations. Ground state solutions are shown in [37] to be *cylindrically symmetric*, that is radial with respect to the transverse variable up to a translation of the origin. On the other hand, *any* finite energy solitary wave is infinitely smooth (at least when the exponent p is an integer) and decays with an *algebraic* rate r^{-2} [37]. Actually the decay rate is sharp in the sense that a solitary wave *cannot* decay faster that r^{-2}. Moreover a precise *asymptotic expansion* of the solitary waves has been obtained by Gravejat [81].

Remark 19. Whether or not the ground states are minimizers of the Hamiltonian is strongly linked to the *orbital stability* of the set $\mathscr{G}_c$ of ground states of velocity c. We recall that the *uniqueness*, up to the obvious symmetries, of the ground state of velocity c is a difficult open problem, even for the classical KP I equation.

Saying that $\mathcal{G}_c$ is orbitally stable in Y means that for $\phi_c \in \mathcal{G}_c$, then for all $\epsilon > 0$, there exists $\delta > 0$ such that if $u_0 \in Y$[10] is such that $||u_0 - \phi_c||_Y \leq \delta$, then the solution $u(t)$ of the Cauchy problem initiating from u_0 satisfies

$$\sup_{t\geq 0} \inf_{\psi \in \mathcal{G}} ||u(t) - \psi||_Y \leq \epsilon.$$

Of course, the previous inequality makes full sense only if one knows that the Cauchy problem is globally well-posed. As we have previously seen, this is the case for the classical KP I equation [171], even in the energy space [95], but is still an open problem for the generalized KP I equation when $1 < p < 4/3$. Actually it is proved in [38] that the ground state solitary waves of the generalized KP I equations (83) are orbitally stable in dimension two if and only if $1 \leq p < \frac{4}{3}$.

4.4 Transverse Stability of the Line Soliton

The KP I and KP II equations behave quite differently with respect to the *transverse* stability of the KdV 1-soliton. Zakharov [227] has proven, by exhibiting an explicit perturbation using the integrability, that the KdV 1-soliton is *nonlinearly* unstable for the KP I flow. Rousset and Tzvetkov [197–199] have given an alternative proof of this result, which does not use the integrability, and which can thus be implemented on other problems (eg for nonlinear Schrödinger equations). The *nature* of this instability is not known (rigorously) and one has to rely on numerical simulations [126, 189].

On the other hand, Mizomachi and Tzvetkov [165] have recently proved the $L^2(\mathbb{R} \times \mathbb{T})$ orbital stability of the KdV 1-soliton for the KP II flow. The perturbation is thus localized in x and periodic in y. The proof involves in particular the Miura transform used in [113] to established the global well-posedness for a modified KP II equation. Such a result is not known (but expected) for a perturbation which is localized in x and y.

The transverse stability of the KdV soliton has been numerically studied in [126] from which the figures in this subsection are taken. We considered perturbations of the form

$$u_p = 6(x - x_1)\exp(-(x - x_1)^2)\left(\exp(-(y + L_y\pi/2)^2) + \exp(-(y - L_y\pi/2)^2)\right), \tag{87}$$

which are in the Schwartz class for both variables and satisfy the zero mass constraint. They are of the same order of magnitude as the KdV soliton, i.e., of order 0(1), and thus test the nonlinear stability of the KdV soliton. As discussed in [126], the computations are carried out in a doubly periodic setting, i.e., $\mathbb{T}^2$ and not on $\mathbb{R}^2$.

[10] Some extra regularity on u_0 is actually needed.

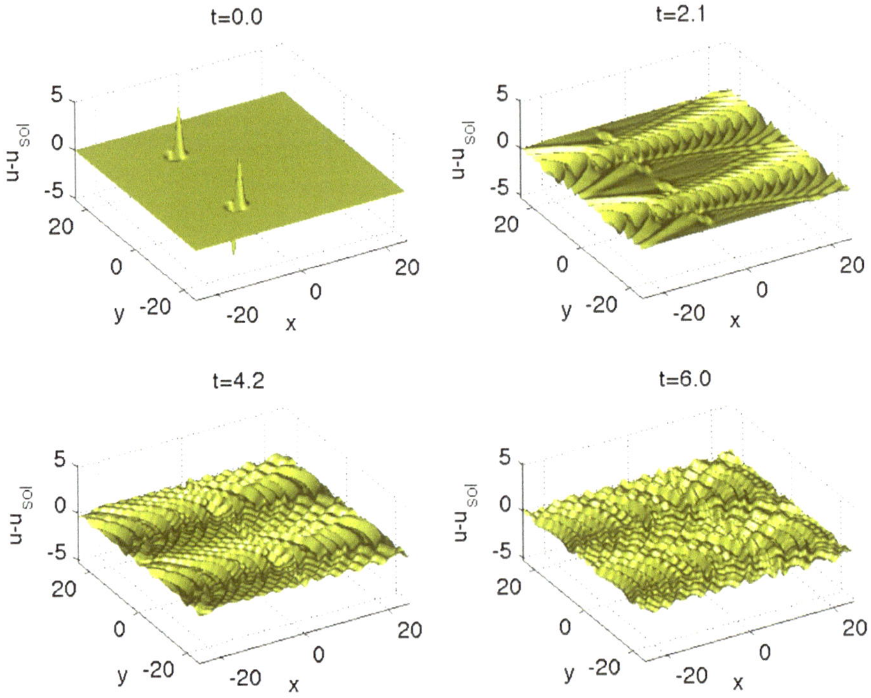

Fig. 8 Difference of the solution to the KP II equation for initial data given by the KdV soliton plus perturbation, $u(0,x,y) = u_{sol}(x + 2L_x, 0)$ and perturbation $u_p = 6(x - x_1)\exp(-(x - x_1)^2)\left(\exp(-(y + L_y\pi/2)^2) + \exp(-(y - L_y\pi/2)^2)\right)$, $x_1 = -L_x$ and the KdV soliton for various values of t

For KP II we consider the initial data $u_0(x,y) = 12\text{sech}^2(x + 2L_x) - u_p$ $(x_1 = x_0/2)$, i.e., a superposition of the KdV soliton and the not aligned perturbation which leads to the situation shown in Fig. 8. The perturbation is dispersed in the form of tails to infinity which reenter the computational domain because of the imposed periodicity. The soliton appears to be unaffected by the perturbation which eventually seems to be smeared out in the background of the soliton.

The situation is somewhat different if the perturbation and the initial soliton are centered around the same x-value initially, i.e., the same situation as above with $x_1 = x_0 = -2L_x$. In Fig. 9 we show the difference between the numerical solution and the KdV soliton u_{sol} for several times for this case. It can be seen that the initially localized perturbations spread in y-direction, i.e., orthogonally to the direction of propagation and take finally themselves the shape of a line soliton. It appears that the perturbations lead eventually to a KdV soliton of slightly higher mass. As discussed in [126], different types of perturbation all indicate the stability of the KdV soliton for KP II.

It was shown in [197–199, 227] that the KdV soliton is nonlinearly unstable against transversal perturbations in the KP I setting if its mass is above a critical

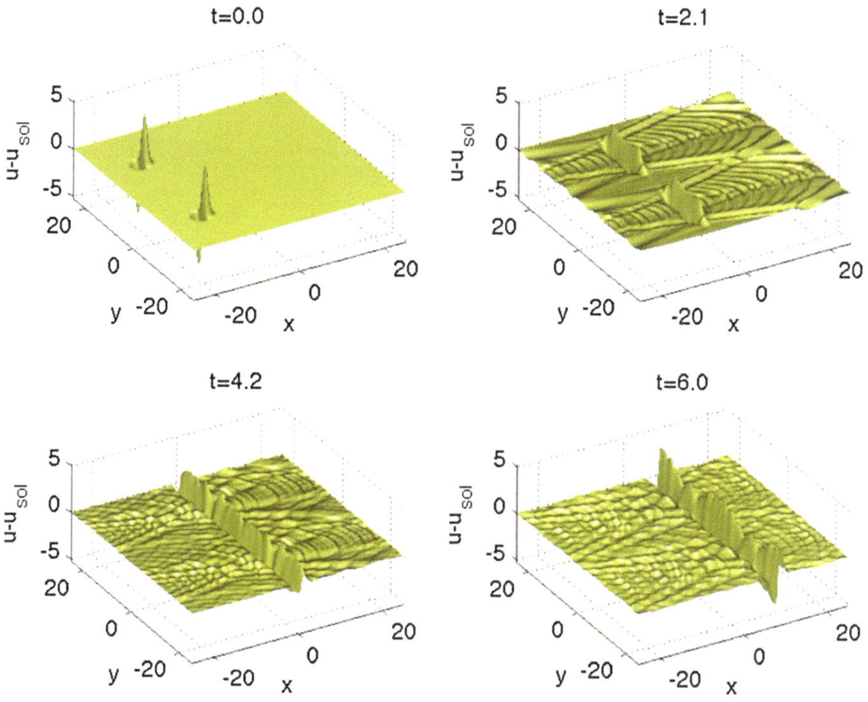

Fig. 9 Difference of the solution to the KP II equation for initial data given by the KdV soliton $u_{sol}(x + 2L_x, 0)$ plus perturbation $u_p = 6(x - x_1)\exp(-(x - x_1)^2)\left(\exp(-(y + L_y\pi/2)^2) + \exp(-(y - L_y\pi/2)^2)\right)$, $x_1 = -2L_x$, and the KdV soliton for various values of t

value. The proof in [227] relies on the integrability of the KP I equation, but the methods in [197–199] apply to general dispersive equations.

However, the type of the instability is unknown. Therefore in [126], this question was addressed numerically. In Fig. 10 we show the KP I solution for the perturbed initial data of a line soliton with the perturbation (87) and $x_1 = x_0 = -2L_x$, i.e., the same setting as studied in Fig. 9 for KP II. Here the initial perturbations develop into 2 lumps which are traveling with higher speed than the line soliton. The formation of these lumps essentially destroys the line soliton which leads to the formation of further lumps. It appears plausible that for sufficiently long times one would only be able to observe lumps and small perturbations which will be radiated to infinity if studied on $\mathbb{R}^2$.

We can give some numerical evidence for the validity of the interpretation of the peaks in Fig. 10 as lumps in an asymptotic sense. We can identify numerically a certain peak, i.e., obtain the value and the location of its maximum. With these parameters one can study the difference between the KP solution and a lump with these parameters to see how well the lump fits the peak. This is illustrated for the two peaks, which formed first and which have therefore traveled the largest

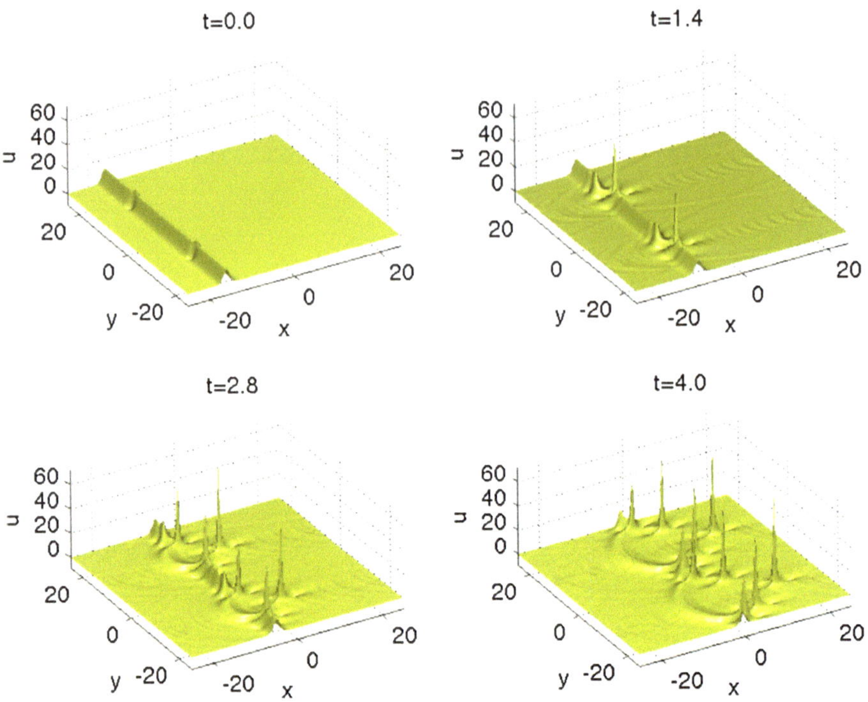

Fig. 10 Solution to the KP I equation for initial data given by the KdV soliton $u_{sol}(x + 2L_x, 0)$ plus perturbation $u_p = 6(x - x_1)\exp(-(x - x_1)^2)\left(\exp(-(y + L_y\pi/2)^2) + \exp(-(y - L_y\pi/2)^2)\right)$, $x_1 = -2L_x$, for various values of t

distance in Fig. 11. A multi-lump solution might be a better fit, but here we mainly want to illustrate the concept which obviously cannot fully apply at the studied small times. Nonetheless Fig. 11 illustrates convincingly that the observed peaks will asymptotically develop into lumps.

Further examples for the nonlinear instability of the KdV soliton in this setting are given in [126]. However, the nonlinear stability discussed in [197, 198] can be seen in Fig. 12 where the KP I solution is given for a perturbation of the line soliton as before $u_0(x, y) = 12\mathrm{sech}^2(x + 2L_x) + u_p$, but this time with $x_1 = x_0/2$, i.e., perturbation and soliton are well separated. The figure shows the difference between KP I solution and line soliton. It can be seen that the soliton is essentially stable on the shown time scales. The perturbation leads to algebraic tails towards positive x-values and to dispersive oscillations as studied in [130]. Due to the imposed periodicity both of these cannot escape the computational domain and appear on the respective other side. The important point is, however, that though the oscillations of comparatively large amplitude hit the line soliton quickly after the initial time, its shape is more or less unaffected till $t = 6$. The KdV soliton eventually decomposes into lumps once it comes close to the boundaries of the computational domain.

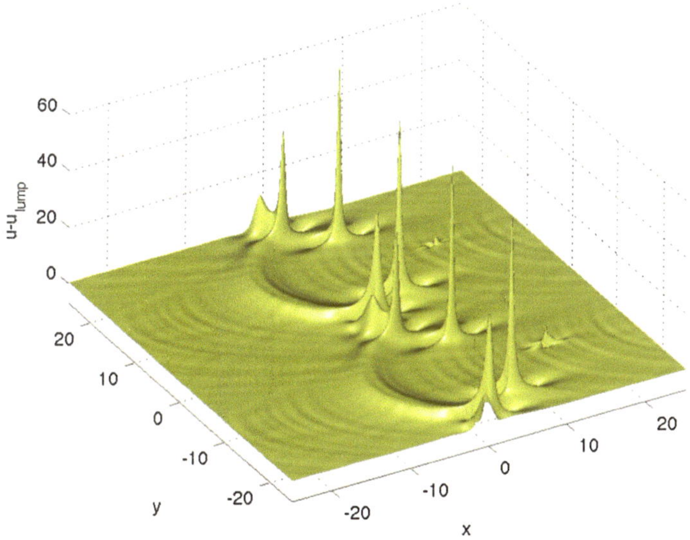

Fig. 11 Difference of the solution to the KP I equation in Fig. 10 for $t = 4$ and two lump solutions fitted at the peaks farthest to the right. Only very small peaks remain of these 'lumps' indicating that they develop asymptotically into true lumps

5 The Davey-Stewartson Systems

The Davey-Stewartson (DS) systems are derived as asymptotic models in the so-called *modulation regime* from various physical situations (water waves, plasma physics, ferromagnetism, see [4, 49, 54, 62, 78, 147, 174, 233]). They provide also a good approximate solution to general quadratic hyperbolic systems using diffractive geometric optics [49, 50]. They have the general form, where a, b, c, ν_1, ν_2 are real parameters depending on the physical context

$$\begin{aligned} i\partial_t\psi + a\partial_x^2\psi + b\partial_y^2\psi &= (\nu_1|\psi|^2 + \nu_2\partial_x\phi)\psi, \\ \partial_x^2\phi + c\partial_y^2\phi &= -\delta\partial_x|\psi|^2, \end{aligned} \tag{88}$$

where one can assume (up to a change of unknown) $a > 0$ and $\delta > 0$. Using the terminology of Ghidaglia and Saut [76], one says that (88) is

$$\begin{aligned} &\text{elliptic-elliptic if} \quad (\text{sgn}\, b, \text{sgn}\, c) = (+1, +1), \\ &\text{hyperbolic-elliptic if} \quad (\text{sgn}\, b, \text{sgn}\, c) = (-1, +1), \\ &\text{elliptic-hyperbolic if} \quad (\text{sgn}\, b, \text{sgn}\, c) = (+1, -1), \\ &\text{hyperbolic-hyperbolic if} \quad (\text{sgn}\, b, \text{sgn}\, c) = (-1, -1). \end{aligned}$$

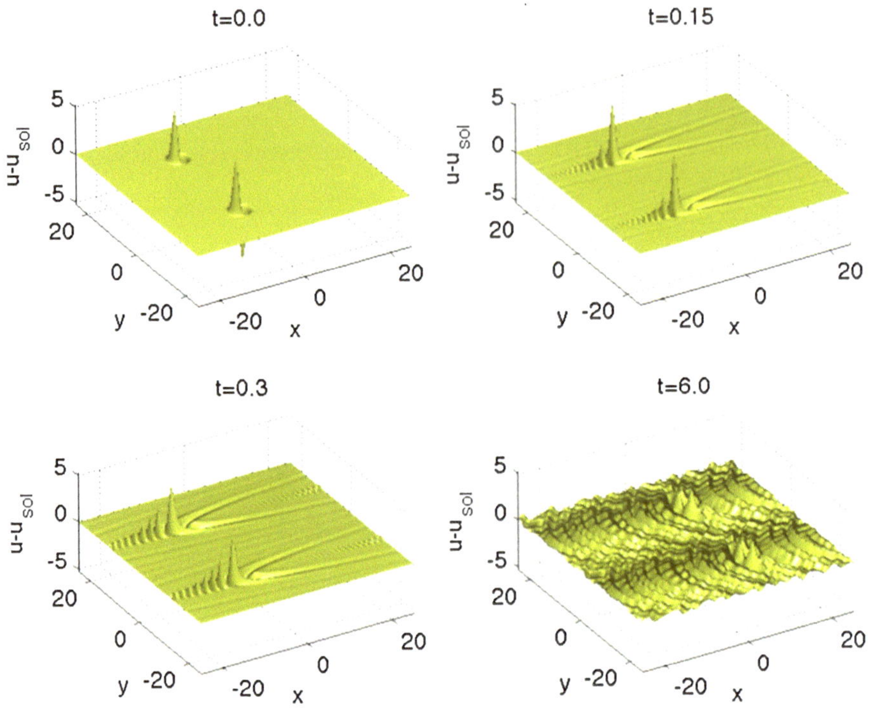

Fig. 12 Difference of the solution to the KP I equation for initial data given by the KdV soliton $u_{sol}(x + 2L_x, 0)$ plus perturbation $u_p = 6(x - x_1)\exp(-(x - x_1)^2)\left(\exp(-(y + L_y\pi/2)^2) + \exp(-(y - L_y\pi/2)^2)\right)$, $x_1 = -L_x$ and the KdV soliton for various values of t

It is worth noticing that the Davey-Stewartson systems are "degenerate" versions of a more general class of systems describing the interaction of short and long waves, the Benney-Roskes, Zakharov-Rubenchik systems [21, 230]. None of those systems is known to be integrable.

It turns out that a very special case of the hyperbolic-elliptic and the elliptic-hyperbolic DS systems are completely integrable ([206, 233]). They are then classically known respectively as the DS II and DS I systems. Since we want to compare IST and PDE methods, we will focus on the hyperbolic-elliptic and elliptic-hyperbolic cases (referred to as DS II type and DS I type). We refer to [46, 47, 76, 78, 182–184, 186] for results on the elliptic-elliptic DS systems.

We will from now on write the DS II system in the form

$$\begin{aligned} i\partial_t\psi + \partial_{xx}\psi - \partial_{yy}\psi + 2\rho\left(\beta\Phi + |\psi|^2\right)\psi &= 0, \quad \psi : \mathbb{R}^2 \times \mathbb{R} \to \mathbb{C}, \\ \partial_{xx}\Phi + \partial_{yy}\Phi + 2\partial_{xx}|\psi|^2 &= 0, \quad \Phi : \mathbb{R}^2 \times \mathbb{R} \to \mathbb{R}, \\ \psi(.,0) &= \psi_0, \end{aligned} \tag{89}$$

where the integrable DS II system corresponds to $\beta = 1$ and ρ takes the values -1 (focusing) and 1 (defocusing).

The DS II system can be viewed as a nonlocal cubic nonlinear Schrödinger equation. Actually one can solve Φ as

$$\Phi = 2[(-\Delta)^{-1}\partial_{xx}]|\psi|^2,$$

where $(-\Delta)^{-1}\partial_{xx} = R_1^2$ is a zero order operator with Fourier symbol $-\frac{\xi_1^2}{|\xi|^2}$ and is thus bounded in all $L^p(\mathbb{R}^2)$ spaces, $1 < p < \infty$ and all Sobolev spaces $H^s(\mathbb{R}^2)$, allowing to write (89) as

$$i\partial_t\psi + \partial_{xx}\psi - \partial_{yy}\psi + 2\rho\left(2\beta R_1^2(|\psi|^2) + |\psi|^2\right)\psi = 0. \tag{90}$$

One easily finds that (89) has two formal conservation laws, the L^2 norm

$$\int_{\mathbb{R}^2} |\psi(x,y,t)|^2 dxdy = \int_{\mathbb{R}^2} |\psi(x,y,0)|^2 dxdy$$

and the energy (Hamiltonian)

$$\begin{aligned} E(\psi(t)) &= \int_{\mathbb{R}^2} \left[|\partial_x\psi|^2 - |\partial_y\psi|^2 - \rho(|\psi|^2 + \beta\Phi)|\psi|^2)\right] dxdy \\ &= E(\psi(0)). \end{aligned} \tag{91}$$

Note that the integrable case $\beta = 1$ is distinguished by the fact that the same hyperbolic operator appears in the linear and in the nonlinear part. In this case the equation is invariant under the transformation $x \to y$ and $\psi \to \bar{\psi}$ and (90) can be written in a "symmetric" form as

$$i\partial_t\psi + \Box\psi - 2\rho[(\Delta^{-1}\Box)|\psi|^2]\psi = 0, \tag{92}$$

where $\Box = \partial_{xx} - \partial_{yy}$. This extra symmetry in the integrable case could be responsible for properties (existence of localized lump solutions and to blow-up phenomena in case of DS II and existence of coherent "dromion" structures in case of DS I) that exist in the integrable case and might not persist in the non integrable cases as the following discussion will suggest.

We summarize now some issues discussed in [128] where one can also find many numerical simulations. (see also [124, 125, 196])

5.1 DS II Type Systems

Systems (90) (whatever the value of ρ or β) can be seen as nonlocal variants of the *hyperbolic nonlinear Schrödinger equation* [79]

$$i\psi_t + \psi_{xx} - \psi_{yy} \pm |\psi|^2\psi = 0, \quad \psi(\cdot, 0) = \psi_0 \tag{93}$$

and actually one can obtain (using Strichartz estimates in the Duhamel formulation) exactly the same results concerning the Cauchy problem see [76]. Namely the Cauchy problem for (90) is locally well-posed for initial data ψ_0 in $L^2(\mathbb{R}^2)$ or $H^1(\mathbb{R}^2)$, and *globally* if $|\psi_0|_2$ is small enough. Nevertheless, since the existence time does not depend only on $|\psi_0|_2$ but on ψ_0 in a more complicated way, one cannot infer from the conservation of the L^2 norm that the L^2 solution is a global one. Actually, proving (or disproving) the global well-posedness of the Cauchy problem for (93) is an outstanding open problem.

As for the KP II equation, the inverse scattering problem for the (integrable) DS II is a $\bar{\partial}$ problem. It turns out that in the integrable case ($\beta = 1$) inverse scattering techniques provide far reaching results which seem out of reach of purely PDE methods. In particular, Sung [211–214] has proven the following

Theorem 13. *Assume that $\beta = 1$. Let $\psi_0 \in \mathscr{S}(\mathbb{R}^2)$. Then (89) possesses a unique global solution ψ such that the mapping $t \mapsto \psi(\cdot, t)$ belongs to $C^\infty(\mathbb{R}, \mathscr{S}(\mathbb{R}^2))$ in the two cases:*

(i) Defocusing.
(ii) Focusing and $|\widehat{\psi_0}|_1 |\widehat{\psi_0}|_\infty < C$, where C is an explicit constant.

Moreover, there exists $c_{\psi_0} > 0$ such that

$$|\psi(x,t)| \leq \frac{c_{\psi_0}}{|t|}, \quad (x,t) \in \mathbb{R}^2 \times \mathbb{R}^*.$$

Remark 20. 1. Sung obtains in fact the global well-posedness (without the decay rate) in the defocusing case under the assumption that $\hat{\psi}_0 \in L^1(\mathbb{R}^2) \cap L^\infty(\mathbb{R}^2)$ and $\psi_0 \in L^p(\mathbb{R}^2)$ for some $p \in [1, 2)$, see [214].

2. Recently, Perry [193] has given a more precise asymptotic behavior in the defocusing case for initial data in $H^{1,1}(\mathbb{R}^2) = \{f \in L^2(\mathbb{R}^2)$ such that $\nabla f, (1 + |\cdot|)f \in L^2(\mathbb{R}^2)\}$, proving that the solution obeys the asymptotic behavior in the $L^\infty(\mathbb{R}^2)$ norm:

$$\psi(x,t) = u(\cdot, t) + o(t^{-1}),$$

where u is the solution of the linearized problem.

On the other hand, using an *explicit* localized lump like solution (see below) and a pseudo-conformal transformation, Ozawa [185] has proven that the integrable focusing DS II system possesses an L^2 solution that blows up in finite time T^*. In fact the mass density $|\psi(.,t)|^2$ of the solution converges as $t \to T^*$ to a Dirac measure with total mass $|\psi(.,t)|_2^2 = |\psi_0|_2^2$ (a weak form of the conservation of the L^2 norm). Every regularity breaks down at the blow-up point but the solution persists after the blow-up time and disperses in the sup norm when $t \to \infty$ as t^{-2}. The numerical simulations in [127] suggest that a blow-up in finite time may also happen for other initial data, eg a sufficiently large Gaussian. Other numerical simulations suggest that the finite time blow-up does not persist in the *non integrable*, $\beta \neq 1$ case, both for the defocusing and focusing cases.

The family of lump solutions (solitons) to the integrable focusing DS II system reads [2, 14, 154]

$$\psi(x,y,t) = 2c\frac{\exp\left(-2i(\xi x - \eta y + 2(\xi^2 - \eta^2)t)\right)}{|x + 4\xi t + i(y + 4\eta t) + z_0|^2 + |c|^2} \tag{94}$$

where $(c, z_0) \in \mathbb{C}^2$ and $(\xi, \eta) \in \mathbb{R}^2$ are constants. The lump moves with constant velocity $(-4\xi, -4\eta)$ and decays as $(x^2 + y^2)^{-1}$ for $x, y \to \infty$.

As explained in [14], there is a one-to-one correspondence between the lumps and the pole of the matrix solution of the direct scattering problem. It is shown formally in [71] and rigorously in [121] (modulo an unproven assumption about the integral operator for the soliton potential) that the lump is *unstable* in the following sense. The soliton structure of the scattering data is unstable with respect to a small compactly supported perturbation of the soliton-like potential. It was also proven in [190, 191] that the lump is spectrally unstable. We refer to [65, 192] for the study of interaction of lumps with a line soliton.

The stability of the lump was numerically studied in [131, 164]. It was shown that the lump is both unstable against an L^∞ blow-up in finite time and against being dispersed away. In Fig. 13 taken from [131], we consider an initial condition of the form $\psi(x,y,-3) = 0.9\psi_l$, where ψ_l is the lump solution (94) with $c = 1$. The solution travels at the same speed as before, but its amplitude varies, growing and decreasing successively.

If instead the initial data $\psi(x,y,-3) = 0.9\psi_l$, is considered, the solution appears to blow up in finite time as can be seen in Fig. 14. Note that also the Ozawa solution [185] was in [131] numerically shown to be unstable against both an earlier blow-up and being dispersed away. In [128] it was shown that the blow-up of the Ozawa solution is not generic.

5.1.1 Localized Solitary Waves of DS II Type Systems

It is well known that hyperbolic NLS equations such as (93) do not possess solitary waves of the form $e^{i\omega t}\psi(x)$ where ψ is localized see [77]. It was furthermore proven in [77] that non trivial solitary waves may exist for DS II type systems only when $\rho = -1$ (focusing case) and $\beta \in (0, 2)$. Note that the (focusing)

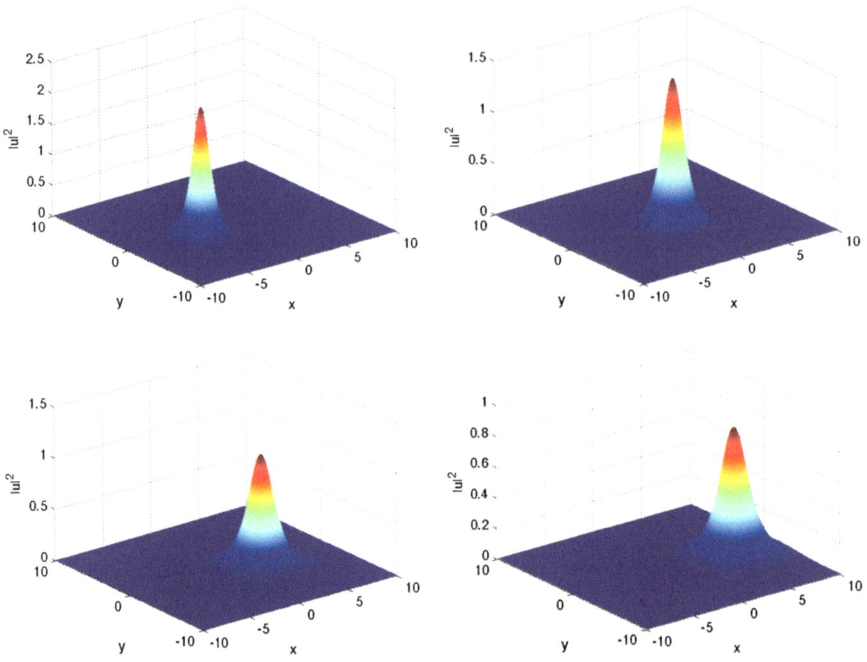

Fig. 13 Solution to the focusing DS II equation (89) for an initial condition of the form $\psi(x, y, -3) = 0.9\psi_l$

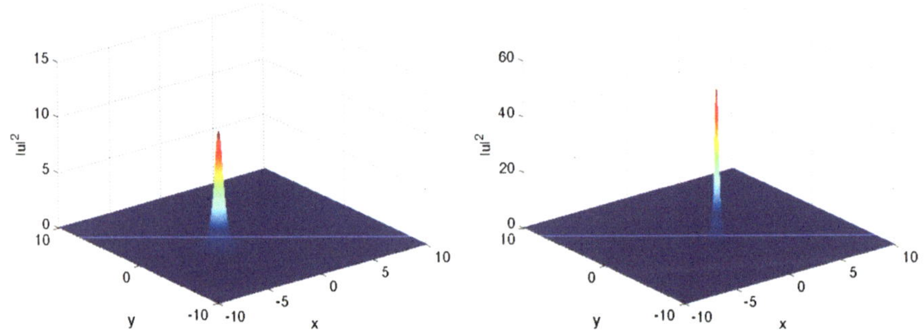

Fig. 14 Solution to the focusing DS II equation (89) for an initial condition of the form $\psi(x, y, -3) = 1.1\psi_l$

integrable case corresponds to $\beta = 1$. Moreover solitary waves with *radial* (up to translation) profiles can exist only when $\rho = -1$ and $\beta = 1$, that is in the focusing integrable case. Those results (and the numerical simulations in [127, 131]) suggest that localized solitary waves for the focusing DS II systems exist only in the integrable case and this might be due to the new symmetry of the system we were alluding to above in this case.

To summarize, one is led to conjecture that neither the existence of the lump nor the associated Ozawa blow-up persist in the focusing DS II non integrable case. One can also conjecture that the solution of DS II type solution is global and decays in the sup norm as $1/t$ as was shown by Sung and Perry in the (integrable) DS II case.

Remark 21. The previous results and conjectures do not of course exclude the existence of nontrivial, non localized, traveling waves. In the context of the hyperbolic cubic NLS, see for instance Remark 2.1 in [77] or [117]. We do not know of existence of similar solutions for DS II type systems.

5.2 DS I Type Systems

The DS I type systems are quite different from the other DS systems. Actually, solving the hyperbolic equation for ϕ (with suitable conditions at infinity) yields a *loss of one derivative* in the nonlinear term and the resulting NLS type equation is no more semilinear. Even proving the rigorous conservation of the Hamiltonian leads to serious problems. We describe now how to solve the equation for ϕ in a H^1 framework see [76]. The elliptic-hyperbolic DS system can be written after scaling as

$$\begin{cases} i\partial_t\psi + \Delta\psi = \chi|\psi|^2\psi + b\psi\phi_x \\ \phi_{xx} - c^2\phi_{yy} = \dfrac{\partial}{\partial x}|\psi|^2. \end{cases}$$

The integrable DS I system corresponds to $\chi + \frac{b}{2} = 0$.

We now solve the equation for ϕ. Let $c > 0$. Consider the equation

$$\frac{\partial^2\phi}{\partial x^2} - c^2\frac{\partial^2\phi}{\partial y^2} = f \quad \text{in } \mathbb{R}^2 \tag{95}$$

with the boundary condition

$$\lim_{\xi\to+\infty}\phi(x,y) = \lim_{\eta\to+\infty}\phi(x,y) = 0 \tag{96}$$

where $\xi = cx - y$ and $\eta = cx + y$. Let $K = K_c$ the kernel

$$K(x,y;x_1,y_1) = \frac{1}{2}H(c(x_1 - x) + y - y_1)H(c(x_1 - x) + y_1 - y)$$

where H is the usual Heaviside function.

Lemma 4 ([76]). *For every $f \in L^1(\mathbb{R}^2)$, the function $\phi = \mathscr{K}(f)$ defined by*

$$\phi(x,y) = \int_{\mathbb{R}^2} K(x,y;x_1,y_1)f(x_1,y_1)dx_1dy_1 \tag{97}$$

is continuous on $\mathbb{R}^2$ *and satisfies* (95) *in the sense of distributions. Moreover,* $\phi \in L^\infty(\mathbb{R}^2)$, $(\partial\phi/\partial x)^2 - c^2(\partial\phi/\partial y)^2 \in L^1(\mathbb{R}^2)$ *and we have the following estimates*

$$\sup_{(x,y)\in\mathbb{R}^2} |\phi(x,y)| \leq \int_{\mathbb{R}^2} |f| dxdy \tag{98}$$

$$\int_{\mathbb{R}^2} \left| \left(\frac{\partial\phi}{\partial x}\right)^2 - c^2 \left(\frac{\partial\phi}{\partial y}\right)^2 \right| dxdy \leq \frac{1}{2c} \left(\int_{\mathbb{R}^2} |f| dxdy \right)^2. \tag{99}$$

Remark 22. 1. No condition is required as ξ or η tends to $-\infty$.
2. In general, $\nabla\phi \notin L^2(\mathbb{R}^2)$ even if $f \in C_0^\infty(\mathbb{R}^2)$, but Lemma 4 allows to solve the ϕ equation as soon as $\psi \in H^1(\mathbb{R}^2)$ for instance.

The DS I type system possesses the formal Hamiltonian

$$E(t) = \int_{\mathbb{R}^2} \left[|\nabla\psi|^2 + \frac{\chi}{2}|\psi|^4 + \frac{b}{2}(\phi_x^2 - c^2\phi_y^2) \right] dxdy.$$

Lemma 4 allows to prove that this Hamiltonian makes sense in an H^1 setting for ψ see [76]. Proving its conservation on the time interval of the solution is an open problem as far as we know (this would lead to global existence of a weak H^1 solution).

5.2.1 DS I Type by PDE Methods

The first local well-posedness result is due to Linares-Ponce [149] and we summarize below the best known results, due to Hayashi-Hirata [90, 91] and Hayashi [89]. After rotation, one can write the DS-I type systems as

$$\begin{cases} i\partial_t\psi + \Delta\psi = i(c_1 + \frac{c_2}{2})|\psi|^2\psi - \frac{c_2}{4}\left(\int_x^\infty \partial_y|\psi|^2 dx' + \int_y^\infty \partial_x|\psi|^2 dy'\right)\psi \\ + \frac{c_2}{\sqrt{2}}\left((\partial_x\phi_1) + \partial_y\phi_2)\right)\psi, \end{cases}$$

where $c_1, c_2 \in \mathbb{R}$ and ϕ satisfies the radiation conditions

$$\lim_{y\to\infty} \phi(x,y,t) = \phi_1(x,t), \quad \lim_{x\to\infty} \phi(x,y,t) = \phi_2(y,t).$$

Theorem 14 ([89]). *Assume* $\psi_0 \in H^2(\mathbb{R}^2)$, $\phi_1 \in C(\mathbb{R}; H_x^2)$ *and* $\phi_2 \in C(\mathbb{R}; H_y^2)$. *Then there exist* $T > 0$ *and a unique solution* $\psi \in C([0,T]; H^1) \cap L^\infty(0,T; H^2)$ *with initial data* ψ_0.

- The proof uses in a crucial way the *smoothing properties* of the Schrödinger group.

The next result concerns global existence and scattering of small solutions in the weighted Sobolev space

$$H^{m,l} = \{f \in L^2(\mathbb{R}^2); |(1 - \partial_x^2 - \partial_y^2)^{m/2}(1 + x^2 + y^2)^{l/2} f|_{L^2} < \infty\}.$$

Theorem 15 ([91]). *Let* $\psi_0 \in H^{3,0} \cap H^{0,3}$, $\partial_x^{j+1}\phi_1 \in C(\mathbb{R}; L_x^\infty)$, $\partial_y^{j+1}\phi_2 \in C(\mathbb{R}; L_y^\infty)$, $(0 \le j \le 3)$, *"small enough". Then*

- *There exists a solution* $\psi \in L^\infty_{loc}(\mathbb{R}; H^{3,0} \cap H^{0,3}) \cap C(\mathbb{R}; H^{2,0} \cap H^{0,2})$.
- *Moreover*

$$||\psi(\cdot, t)||_{L^\infty} \le C(1 + |t|)^{-1}(||\psi||_{H^{3,0}} + ||\psi||_{H^{0,3}}).$$

There exist $u^\pm$ *such that*

$$||\psi(t) - U(t)u^\pm||_{H^{2,0}} \to 0, \quad as \to \pm\infty,$$

where $U(t) = e^{it(\partial_x^2 + \partial_y^2)}$.

5.2.2 DS I by IST: Comparison with Elliptic-Hyperbolic DS

Contrary to the results of the previous subsection which were valid for arbitrary values of β we focus here on the integrable case, $\beta = 1$. The first set of results concerns coherent structures (dromions) for DS I with nontrivial conditions (on ϕ) at infinity. The existence of dromions is established in [30, 66] and the perturbations of the dromion are investigated in [119] (see also Sect. 7 of [121]). We do not know of any study of dromions by PDE techniques or of existence of similar structures in the non-integrable case. Actually they might have no physical relevance.

Concerning the Cauchy problem, the global existence and uniqueness of a solution $\psi \in C(\mathbb{R}; \mathcal{S}(\mathbb{R}^2))$ of DS I for data $\psi_0 \in \mathcal{S}(\mathbb{R}^2)$, $\phi_1, \phi_2 \in C(\mathbb{R}; \mathcal{S}(\mathbb{R}))$ is proven in [68]. Under a smallness condition, the solutions with trivial boundary conditions $\phi_1 = \phi_2 = 0$ disperse as $1/t$ (Kiselev [119], see also [121]). A precise asymptotics is also given. The numerical simulations in [23] for general DS I type systems confirm the dispersion of solutions of DS I with trivial boundary conditions and suggest that the dromion is not stable with respect to the coefficients, that is it does not persist in the non-integrable case.

6 Final Comments

We briefly comment here on two other integrable equations.

6.1 The Ishimori System

The Ishimori systems were introduced in [98] as two-dimensional generalizations of the Heisenberg equation in ferromagnetism. They read

$$\begin{aligned} S_t &= S \wedge (S_{xx} - S_{yy}) + b(\phi_x S_y + \phi_y S_x) \\ \phi_{xx} + \phi_{yy} &= 2S \cdot (S_x \wedge S_y) \\ S(\cdot, 0) &= S_0, \end{aligned} \tag{100}$$

$$\begin{aligned} S_t &= S \wedge (S_{xx} + S_{yy}) + b(\phi_x S_y + \phi_y S_x) \\ \phi_{xx} - \phi_{yy} &= -2S \cdot (S_x \wedge S_y) \\ S(\cdot, 0) &= S_0, \end{aligned} \tag{101}$$

where S is the spin, $S(\cdot, t) : \mathbb{R}^2 \to \mathbb{R}^3$, $|S|^2 = 1$, $S \to (0, 0, 1)$ as $|(x, y)| \to \infty$ and $\wedge$ is the wedge product in $\mathbb{R}^3$. The coupling potential ϕ is a scalar unknown related to the topological charge density $q(S) = 2S \cdot (S_x \wedge S_y)$. b is a real coupling constant. When $b = 1$, (100), (101) are completely integrable [137, 138]. Note the formal analogy[11] of (100), (101) with DS II and DS I type systems.

It is proven in [210] that the Cauchy problem for (100) (for arbitrary values of b) is locally well-posed in $H^m(\mathbb{R}^2)$, $m \geq 3$ provided the initial spins are almost parallel. Under a stronger regularity assumption on the initial data it is furthermore proven that the solution is global and converges to a solution of the linear "hyperbolic" Schrödinger equation as t goes to infinity.

The idea is to reduce (100), (101) to a nonlinear (hyperbolic) Schrödinger type equation by the stereographic projection

$$u : \mathbb{R}^2 \to \mathbb{C}, \ u = \frac{S^1 + iS^2}{1 + S^3},$$

reducing (100) to

$$\begin{cases} iu_t + u_{xx} - u_{yy} = \frac{2\bar{u}}{1+|u|^2}(u_x^2 - u_y^2) + ib(\phi_x u_y + \phi_x u_x) \\ \Delta\phi = 4i\frac{u_x\bar{u}_y - \bar{u}_x u_y}{(1+|u|^2)^2} \\ u(\cdot, 0) = u_0, \end{cases} \tag{102}$$

[11] Which is clearer after the stereographic projection below.

with the condition $|u(x,y)| \to 0$ as $|(x,y)| \to \infty$. The initial condition on S becomes thus a smallness condition on $u_{|t=0}$.

The local well-posedness of the Cauchy problem for (100) for *arbitrary* initial data in $H^m(\mathbb{R}^2)$, $m \geq 4$ is proven in [114]. This result is improved in [17] where local well-posedness is proven for arbitrary large initial data in $H^s(\mathbb{R}^2)$, $s > \frac{3}{2}$ having a range that avoids a neighborhood of the north pole.

Concerning the IST method for (100), the rigorous justification of the procedure in [137, 138] is not trivial and Sung [215] used instead a gauge transform which relates (100) (when $b = 1$) to the focusing integrable DS II system. This allows to prove the global well-posedness of the Cauchy problem for small initial data (in a different functional setting than [210]).

The connection between the (integrable) Ishimori system (100) is nicely used in [17] to prove the *global* well-posedness of the Cauchy problem for (100) in the *defocusing* case, that is when the target of S is no more the sphere $\mathbb{S}^2$, but the hyperbolic space $\mathbb{H}^2 = \{(x,y,z) \in \mathbb{R}^3; x^2 - y^2 - z^2 = 1, x > 0\}$. The gauge transform relates in this case the *defocusing* Ishimori system (100) to the *defocusing* DS II system. Such a result is not known in the non integrable case, $b \neq 1$.

We do not know of any finite time blow-up for (100) or of any rigorous result (by PDE or IST methods) for (101).

6.2 The Novikov-Veselov Equation

The Novikov-Veselov system

$$\begin{cases} v_t = 4\mathrm{Re}\,(4\partial_z^3 v + \partial_z(vw) - E\partial_z w), \\ \partial_{\bar{z}} w = -3\partial_z v, \quad v = \bar{v} \end{cases} \tag{103}$$

where $\partial_z = \frac{1}{2}\left(\partial_x - i\partial_y\right)$, $\partial_{\bar{z}} = \frac{1}{2}\left(\partial_x + i\partial_y\right)$, was introduced in [179, 180] as a two dimensional analog of the Korteweg-de Vries equation, integrable via the inverse scattering transform for the following 2-dimensional stationary Schrödinger equation at a fixed energy E:

$$L\psi = E\psi, \quad L = -\Delta + v(x,y,t), \tag{104}$$

where $\Delta = \frac{\partial^2}{\partial x^2} + \frac{\partial^2}{\partial y^2}$ and E is a fixed real constant.

The Novikov-Veselov equation has the Manakov triple representation [153]

$$\dot{\mathscr{L}} = [A, \mathscr{L}] - B\mathscr{L}, \tag{105}$$

where

$$\mathscr{L} = -\Delta + v + E,$$

$$A = 8(\partial^3 + \bar{\partial}^3) + 2(w\partial + \bar{w}\bar{\partial}),$$
$$\bar{B} = -2(\partial w + \bar{\partial}\bar{w}),$$

that is (v, w) solves (107) if and only if (105) holds.

Remark 23. There is an interesting formal limit of (103) to KP I (resp. KP II) as $E \to +\infty$ (resp. $E \to -\infty$), under an appropriate scaling, see [83, 139], assuming that the wavelengths in y are much larger than those in x.

Remark 24. As far as we know, and contrary to the integrable equations studied above, the Novikov-Veselov equation does not seem be derived as an asymptotic model from a more general system.

To make the equation more "PDE like", we first write the equation in a slightly different form. Setting $w = w_1 + iw_2$, (103) becomes:

$$\begin{cases} v_t = 2v_{xxx} - 6v_{xyy} + 2[(vw_1)_x + (vw_2)_y] - 2E(\partial_x w_1 + \partial_y w_2), \\ \qquad (x, y) \in \mathbb{R}^2,\ t > 0, \\ \partial_x w_1 - \partial_y w_2 = -3v_x \\ \partial_x w_2 + \partial_y w_1 = 3v_y. \end{cases} \tag{106}$$

Note that the last two equations imply

$$(\partial_{xx} - \partial_{yy})w_2 + 2\partial_{xy}w_1 = 0$$

and that w_1 et w_2 are defined up to an additive constant. One can express $w = (w_1, w_2)$ in dependence of v, in a unique way via the Fourier transform

$$\hat{w_1} = \frac{3(\xi_2^2 - \xi_1^2)}{|\xi|^2}\hat{v}, \quad \hat{w_2} = \frac{6\xi_1\xi_2}{|\xi|^2}\hat{v}.$$

We denote L_1 and L_2 the zero order corresponding operators,

$$w_1 = L_1 v, \quad w_2 = L_2 v.$$

With these notations, the equation reads

$$v_t - 2v_{xxx} + 6v_{xyy} + 2E(L_1 v_x + L_2 v_y) - 2[(vL_1 v)_x + (vL_2 v)_y] = 0. \tag{107}$$

One remarks that the dispersive part in (107) is reminiscent of that of the Zakharov-Kuznetzov equation [230]

$$u_t + u_x + \Delta u_x + uu_x = 0,$$

another two-dimensional extension of the KdV equation, which is not integrable though. We refer to [19] for a systematic study of the dispersive properties of general third order (local) operators in two-dimensions.

6.2.1 The Zero Energy Case $E = 0$

We refer to the excellent survey [52] for a rather complete account of what is known in this case and which comprises some interesting numerics and a rich bibliography. We extract from [52] the following definition :

Definition 1. The operator $L = -\Delta + v$ is said to be:

(i) subcritical if the operator L has a positive Green's function and the equation $L = 0$ has a strictly positive distributional solution,
(ii) critical if $L = 0$ has a bounded strictly positive solution but no positive Green's function, and
(iii) supercritical otherwise.

The following conjecture is made in [52].

Conjecture. The Novikov-Veselov equation with zero energy has a global solution for critical and subcritical initial data, but its solution may blow up in finite time for supercritical initial data.[12] We refer to [52] for some partial results toward its resolution.

Added in proofs. This conjecture is partially solved in [173] (see also [194]) where global well-posedness results with initial data of arbitrary size in suitable functional spaces are proved.

6.2.2 The Case $E \neq 0$

For nonzero energy E and potentials v which vanish at infinity, the scattering transform and inverse scattering method was developed by P. Grinevich, R. G. Novikov, and S.-P. Novikov (see Kazeykina's thesis [103] for an excellent survey and [83, 85, 86] for the original papers).

We also refer to [103] and to the papers [104–106, 108–110, 178] for many interesting results on the soliton solutions (absence, decay properties,..) and the long time asymptotics. More precisely it is proven that the rational, non singular solutions introduced in [83, 84] at positive energy are multi-solitons. It is also proven that solitons cannot decay faster than $O(|x|^{-3})$ when $E \neq 0$ and than $O(|x|^{-2})$ when $E = 0$. Finally, the evolution solutions corresponding to non singular scattering data are shown to decay asymptotically as $O(t^{-1})$ when $E > 0$ and as $O(t^{-3/2})$ when $E < 0$.

[12] By the result in [13] mentioned below, the Novikov-Veselov equation with zero energy has a *local* solution for a very general class of initial potentials.

6.2.3 Novikov-Veselov by PDE Techniques

There are few results on Novikov-Veselov by PDE techniques. Angelopoulos [13] has proven the local well-posedness in $H^{1/2}(\mathbb{R}^2)$ of the Cauchy problem in the zero energy case ($E = 0$) case following the method used in [173]. It is very likely that the methods in [132] for the Dysthe type systems[13] would also lead to a local well-posedness result when $E = 0$, not in the "optimal" space though. The case $E \neq 0$ is more delicate and is considered in [107] where local well-posedness is obtained in $H^s(\mathbb{R}^2), s > 1/2$.

Note than one cannot perform simple energy estimates on (107) (as in the KP or Zakharov-Kuznetsov case) leading "for free" to a local result in $H^s(\mathbb{R}^2), s > 2$ (actually even the L^2 estimate on (107) fails).

Also PDE methods might solve the aforementioned problem of the "long wave" KP limit when $E \to \pm\infty$. Actually similar questions are already solved for the Gross-Pitaevskii equation see [25, 26, 45].

Finally we do not know of any stability results (by PDE methods) for the Novikov-Veselov solitary waves.

Acknowledgements The authors thank Anne de Bouard, Anna Kazeykina, Thanasis Fokas, Oleg Kiselev and Peter Perry for fruitful conversations. They also thank an anonymous referee for his/her constructive remarks and suggestions. J.-C. S. was partially supported by the project GEODISP of the Agence Nationale de la Recherche.

References

1. Abdelouhab, L., Bona, J., Felland, M., Saut, J.-C.: Non local models for nonlinear dispersive waves. Physica D **40**, 360–392 (1989)
2. Ablowitz, M.J., Clarkson, P.A.: Solitons, Nonlinear Evolution Equations and Inverse Scattering. London Mathematical Society Lecture Notes Series, vol. 149, Cambridge University Press, Cambridge (1991)
3. Ablowitz, M.J., Fokas, A.S.: The inverse scattering transform for the Benjamin-Ono equation: a pivot to multidimensional problems. Stud. Appl. Math. **68**, 1–10 (1983)
4. Ablowitz, M.J., Segur, H.: On the evolution of packets of water waves. J. Fluid Mech. **92**, 691–715 (1979)
5. Ablowitz, M.J., Villarroel, J.: The Cauchy problem for the Kadomtsev-Petviashvili II equation with with nondecaying data along a line. Stud. Appl. Math. **109**, 151–162 (2002)
6. Ablowitz, M.J., Villarroel, J.: The Cauchy problem for the Kadomtsev-Petviashvili II equation with data that do not decay along a line. Nonlinearity **17**, 1843–1866 (2004)
7. Albert, J.P., Bona, J.L.: Total positivity and the stability of internal waves in stratified fluids of finite depth. IMA J. Appl. Math. **46**(1–2), 1–19 (1991)
8. Albert, J.P., Toland, J.F.: On the exact solutions of the intermediate long-wave equation. Differ. Integral Equ. **7**(3–4), 601–612 (1994)

[13]Mainly based on the smoothing properties of the linear group generated by third order partial differential operators. Such a property holds for the operator $-2\partial_x^3 + 6\partial_{xyy}^3$ which appears in the Novikov-Veselov equation when $E = 0$.

9. Albert, J.P., Bona, J.L., Henry, D.: Sufficient conditions for stability of solitary-wave solutions of model equations for long waves. Physica D **24**, 343–366 (1987)
10. Albert, J.P., Bona, J.L., Nguyen, N.V.: On the stability of KdV multi-solitons. Differ. Integral Equ. **20**(8), 841–878 (2007)
11. Amick, C.J., Toland, J.: Uniqueness and related analytic properties for the Benjamin-Ono equation-a nonlinear Neumann problem in the plane. Acta Math. **167**, 107–126 (1991)
12. Anderson, R.L., Taflin, E.: The Benjamin-Ono equation-Recursivity of linearization maps-Lax pairs. Lett. Math. Phys. **9**, 299–311 (1985)
13. Angelopoulos, Y.: Well-posedness and ill-posedness results for the Novikov-Veselov equation. (2013). arXiv 1307.4110
14. Arkadiev, V.A., Pogrebkov, A.K., Polivanov, M.C.: Inverse scattering transform and soliton solution for Davey-Stewartson II equation. Physica D **36**, 188–197 (1989)
15. Beals, R., Coifman, R.R.: Scattering, transformations spectrales et équations d'évolution non linéaires I,II. Séminaire Goulaouic-Meyer-Schwartz 1980/81, exposé XXII, and 1981/1982, exposé XXI, Ecole Polytechnique, Palaiseau
16. Beals, R., Coifman, R.R.: Scattering and inverse scattering for first-orde systems. Commun. Pure Appl. Math. **37**, 39–90 (1984)
17. Bejenearu, A.D., Ionescu, A.D., Kenig, C.E.: On the stability of certain spin models in $2+1$ dimensions. J. Geom. Anal. **21**, 1–39 (2011)
18. Ben-Artzi, M., Saut, J.-C.: Uniform decay estimates for a class of oscillatory integrals and applications. Differ. Integral Equ. **12**(2), 137–145 (1999)
19. Ben-Artzi, M., Koch, H., Saut, J.-C.: Dispersion estimates for third order dispersive equations in 2 dimensions. Commun. Partial Differ. Equ. **28**(11–12) 1943–1974 (2003)
20. Benjamin, T.B.: The stability of solitary waves. Proc. R. Soc. (Lond.) Ser. A **328**, 153–183 (1972)
21. Benney, D.J., Roskes, G.J.: Waves instabilities. Stud. Appl. Math. **48**, 377–385 (1969)
22. Besov, O., Ilin, V., Nikolski, S.: Integral Representation of Functions and Embedding Theorems. Wiley, New York (1978)
23. Besse, C., Bruneau, C.H.: Numerical study of elliptic-hyperbolic Davey-Stewartson system: dromions simulation and blow-up. Math. Models Methods Appl. Sci. **8**(8), 1363–1386 (1998)
24. Béthuel, F., Danchin, R., Smets, D.: On the linear regime of the Gross-Pitaevskii equation. J. d'Analyse Mathématique. **110**(1), 297–338 (2010)
25. Béthuel, F., Gravejat, P., Saut, J.-C., Smets, D.: On the Korteweg-de Vries long-wave approximation of the Gross-Pitaevskii equation I. Int. Math. Res. Not. **14**, 2700–2748 (2009)
26. Béthuel, F., Gravejat, P., Saut, J.-C., Smets, D.: On the Korteweg-de Vries long-wave approximation of the Gross-Pitaevskii equation II. Commun. Partial Differ. Equ. **35**(1), 113–164 (2010)
27. Béthuel, F., Gravejat, P., Saut, J.-C.: Existence and properties of traveling waves for the Gross-Pitaevskii equation. In: Farina, A., Saut, J.C. (eds.) Stationary and Time Dependent Gross-Pitaevskii Equations. Contemporary Mathematics, vol. 473, pp. 55–103. American Mathematical Society, Providence, RI (2008)
28. Béthuel, F., Gravejat, P., Saut, J.-C.: On the KP-I transonic limit of two-dimensional Gross-Pitaevskii travelling waves. Dyn. Partial Differ. Equ. **5**(3), 241–280 (2008)
29. Béthuel, F., Gravejat, P., Saut, J.-C., Smets, D.: Orbital stability of the black soliton to the Gross-Pitaevskii equation. Indiana J. Math. **57**, 2611–2642 (2008)
30. Boiti, M., Leon, J.J., Martina, L., Pempinelli, F.: Scattering of localized solitons in the plane. Phys. Lett. **A32**, 432–439 (1988)
31. Bona, J.L.: On the stability theory of solitary waves. Proc. R. Soc. Lond. Ser. A **349**, 363–374 (1975)
32. Bona, J.L., Chen, M., Saut, J.-C.: Boussinesq equations and other systems for small-amplitude long waves in nonlinear dispersive media I : derivation and the linear theory. J. Nonlinear Sci. **12**, 283–318 (2002)
33. Bona, J.L., Lannes, D., Saut, J.-C.: Asymptotic models for internal waves. J. Math. Pures Appl. **89**, 538–566 (2008)

34. Bona, J.L., Soyeur, A.: On the stability of solitary-wave solutions of model equations for long-waves. J. Nonlinear Sci. **4**, 449–470 (1994)
35. Bona, J.L., Souganidis, P.E., Strauss, W.A.: Stability and instability of solitary waves of KdV type. Proc. R. Soc. Lond. **A 411**, 395–412 (1987)
36. de Bouard, A., Saut, J.-C.: Solitary waves of the generalized KP equations. Ann. IHP Analyse Non Linéaire **14**(2), 211–236 (1997)
37. de Bouard, A., Saut, J.-C.: Symmetry and decay of the generalized Kadomtsev-Petviashvili solitary waves. SIAM J. Math. Anal. **28**(5), 104–1085 (1997)
38. de Bouard, A., Saut, J.-C.: Remarks on the stability of the generalized Kadomtsev-Petviashvili solitary waves. In: Dias, F., Ghidaglia, J.-M., Saut, J.-C. (eds.) Mathematical Problems in the Theory of Water Waves. Contemporary Mathematics, vol. 200, 75–84. American Mathematical Society, Providence, RI (1996)
39. Bourgain, J.: Fourier transform restriction phenomena for certain lattice subsets and applications to nonlinear evolution equations. II. The KdV equation. Geom. Funct. Anal. **3**(3), 209–262 (1993)
40. Bourgain, J.: On the Cauchy problem for the Kadomtsev-Petviashvili equation. Geom. Funct. Anal. GAFA **3**, 315–341 (1993)
41. Calogero, F., Degasperis, A.: Spectral Transforms and Solitons. North-Holland, Amsterdam, New-York (1982)
42. Case, K.M.: Properties of the Benjamin-Ono equation. J. Math. Phys. **20**, 972–977 (1979)
43. Case, K.M.: The N-soliton solution of the Benjamin-Ono equation. Proc. Natl. Acad. Sci. **75**, 3562–3563 (1978)
44. Cazenave, T., Lions, P.-L.: Orbital stabiliy of standing waves for some nonlinear Schrödinger equations. Commun. Math. Phys. **85**(4), 549–561 (1982)
45. Chiron, D., Rousset, F.: The KdV/KP limit of the nonlinear Schrödinger equation. SIAM J. Math. Anal. **42** (1), 64–96 (2010)
46. Cipolatti, R.: On the existence of standing waves for a Davey-Stewartson system. Commun. Partial Differ. Equ. **17**(5–6), 967–988 (1992)
47. Cipolatti, R.: On the instability of ground states for a Davey-Stewartson system. Ann. Inst. H. Poincaré, Phys.Théor. **58**, 85–104 (1993)
48. Coifman, R.R., Wickerhauser, M.V.: The scattering transform for the Benjamin-Ono equation. Inverse Prob. **6** (5), 825–862 (1990)
49. Colin, T.: Rigorous derivation of the nonlinear Schrödinger equation and Davey-Stewartson systems from quadratic hyperbolic systems. Asymptot. Anal. **31**, 69–91 (2002)
50. Colin, T., Lannes, D.: Justification of and long-wave correction to Davey-Stewartson systems from quadratic hyperbolic systems. Discrete Contin. Dyn. Syst. **11**(1), 83–100 (2004)
51. Craig, W.L.: An existence theory for water waves and the Boussinesq and Korteweg-de Vries scaling limits. Commun. Partial Differ. Equ. **10**(8), 787–1003 (1985)
52. Croke, R., Mueller, J.L., Music, M., Perry, P., Siltanen, S., Stahel, A.: The Novikov-Veselov equation: theory and computation. (2013). arXiv:1312.5427v1 [math.AP]
53. Darrigol, O.: Worlds of Flow. A History of Hydrodynamics from the Bernoullis to Prandtl. Oxford University Press, Oxford (2005)
54. Davey, A., Stewartson, K.: One three-dimensional packets of water waves. Proc. R. Soc. Lond. A **338**, 101–110 (1974)
55. Deift, P.A., Its, A.R., Zhou, X.: Long-time asymptotics for integrable nonliner wave equations. In: Fokas, A.S., Zakharov, V.E., (eds.) Important Developments in Soliton Theory. Springer series in Nonlinear Dynamics, pp. 181–204. Springer, Berlin (1993)
56. Deift, P.A., Venakides, S., Zhou, X.: New results in small dispersion KdV by an extension of the steepest descent method for Riemann-Hilbert problems. Int. Math. Res. Not. **6**, 286–299 (1997)
57. Deift, P.A., Venakides, S., Zhou, X.: An extension of the steepest descent method for Riemann-Hilbert problems: the small dispersion limit of the Korteweg-de Vries equation. Proc. Natl. Acad. Sci. USA **95**(2), 450–454 (1998)

58. Deift, P.A., Zhou, X.: A steepest descent method for oscillatory Riemann-Hilbert problems. Asymptotics for the MKdV equation. Ann. Math. **137**(2), 295–398 (1993)
59. Deift, P.A., Zhou, X.: Long-time asymptotics for integrable systems. Higher order theory. Commun. Math. Phys. **165** (1), 175–191 (1994)
60. Deift, P.A., Zhou, X.: Long-time asymptotics for solutions of the NLS equation with initial data in a weighted Sobolev space. Commun. Pure Appl. Math. **58**(8), 1029–1077 (2003)
61. Di Menza, L., Gallo, C.: The black solitons of one-dimensional NLS equations. Nonlinearity **20**, 461–496 (2007)
62. Djordjevic, V.D., Redekopp, L.G.: On two-dimensional packets of capillary-gravity waves. J. Fluid Mech. **79**, 703–714 (1977)
63. Fokas, A.S.: On the inverse scattering for the time dependent Schrödinger equation and the associated Kadomtsev-Petviashvilii equation. Stud. Appl. Math. **69** (8), 211–222 (1983)
64. Fokas, A.S., Pogorobkov, A.K.: Inverse scattering transform for the KP-I equation on the background of a one- line soliton. Nonlinearity **16**, 771–783 (2003)
65. Fokas, A.S., Pelinovsky, D., Sulem, C.: Interaction of lumps with a line soliton for the Davey-Stewartson II equation. Physica D, **152–153**, 189–198 (2001)
66. Fokas, A.S., Santini, P.M.: Dromions and a boundary-value problem for the Davey-Stewartson I equation. Physica D **44**(1–2), 99–130 (1990)
67. Fokas, A.S., Sung, L.Y.: The inverse spectral method for the KP I equation without the zero mass constraint. Math. Proc. Camb. Phil. Soc. **125**, 113–138 (1999)
68. Fokas, A.S., Sung, L.Y.: On the solvability of the N-wave, Davey-Stewartson and Kadomtsev-Petviashvili equations. Inverse prob. **8**, 673–708 (1992)
69. Fonseca, G., Linares, F., Ponce, G.: The well-posedness results for the dispersion generalized Benjamin-Ono equation via the contraction principle. (2012). arXiv:1205.540v1
70. Fonseca, G., Linares, F., Ponce, G.: The IVP for the dispersion generalized Benjamin-Ono equation in weighted spaces. Annales IHP Non Linear Anal. **30**(5), 763–790 (2013)
71. Gadyl'shin, R.R., Kiselev, O.M.: On lump instability of Davey-Stewartson II equation. Teor. Mat. Fiz. **118**(3), 354–361 (1999)
72. Gallo, C.: The Cauchy problem for defocusing nonlinear Schrödinger equations with non vanishing initial data at infinity. Commun. Partial Differ. Equ. **33**(4–6), 729–771 (2008)
73. Gardner, C.S., Greene, J.M., Kruskal, M.D., Miura, R.M.: Method for solving the Korteweg-de Vries equation. Phys. Rev. Lett. **19**, 1095–1097 (1967)
74. Gérard, P.: The Cauchy Problem for the Gross-Pitaevskii equation. Ann. Inst. H. Poincaré Anal. Non Linéaire **23**(5), 765–779 (2006)
75. Gérard, P., Zhang, Z.: Orbital stability of traveling waves for the one-dimensional Gross-Pitaevskii Equation. J. Math. Pures Appl. **91** (2), 178–210 (2009)
76. Ghidaglia, J.-M., Saut, J.-C.: On the initial value problem for the Davey-Stewartson systems. Nonlinearity **3**, 475–506 (1990)
77. Ghidaglia, J.-M., Saut, J.-C.: Non existence of traveling wave solutions to nonelliptic nonlinear Schrödinger equations. J. Nonlinear Sci. **6**, 139–145 (1996)
78. Ghidaglia, J.-M., Saut, J.-C.: On the Zakharov-Schulman equations. In: Debnath, L. (ed.) Non-linear Dispersive Waves, pp. 83–97. World Scientific, Singapore (1992)
79. Ghidaglia, J.-M., Saut, J.-C.: Nonelliptic Schrödinger evolution equations. J. Nonlinear Sci. **3**, 169–195 (1993)
80. Ginibre, J.: Le problème de Cauchy pour des EDP semi linéaires périodiques en variables d'espace. Sém. N. Bourbaki **37**706, 163–187 (1994–1995)
81. Gravejat, P.: Asymptotics of the solitary waves for the generalized Kadomtsev-Petviashvili equations. Discrete Contin. Dyn. Syst. **21**(3), 835–882 (2008)
82. Grinevich, P.G.: Non singularity of the direct scattering transform for the KP II equation with a real exponentially decaying-at-infinity potential. Lett. Math. Phys. **40**, 59–73 (1997)
83. Grinevich, P.G.: The scattering transform for the two-dimensional Schrödinger operator with a potential that decreases at infinity at fixed nonzero energy. Russ. Math. Surv. **55**(6), 1015–1083 (2000)

84. Grinevich, P.G.: Rational solutions of the Veselov-Novikov equation are reflectionless potentials at fixed energy. Theor. Math. Phys. **69**, 1170–1172 (1986)
85. Grinevich, P.G., Manakov, S.V.: Inverse problem of scattering theory for the two-dimensional Schrödinger operator, the $\bar{\partial}$- method and nonlinear equations. Funct. Anal. Appl. **20**(2), 94–103 (1986)
86. Grinevich, P.G., Novikov, S.P.: A two-dimensional inverse scattering problem for negative energies, and generalized analytic functions. I. Energies lower than the ground state. Funct. Anal. Appl. **22**(1), 19–27 (1988)
87. Hadac, M.: Well-posedness for the Kadomtsev-Petviashvili II equation and generalisations. Trans. Am. Math. Soc. **360**(12), 6555–6572 (2008)
88. Hadac, M., Herr, S., Koch, H.: Well-posedness and scattering for the KP II equation in a critical space. Ann. IHP Anal. Non Linéaire **26**(3), 917–941 (2009)
89. Hayashi, N.: Local existence in time of solutions to the elliptic-hyperbolic Davey-Stewartson system without smallness condition on the data. J. Anal. Math. **73**, 133–164 (1997)
90. Hayashi, N., Hirota, H.: Local existence in time of small solutions to the elliptic-hyperbolic Davey-Stewartson system in the usual Sobolev space. Proc. Edinb. Math. Soc. **40**, 563–581 (1997)
91. Hayashi, N., Hirota, H.: Global existence and asymptotic behavior in time of small solutions to the elliptic-hyperbolic Davey-Stewartson system. Nonlinearity **9**, 1387–1409 (1996)
92. Hayashi, N., Naumkin, P., Saut, J.-C.: Asymptotics for large time of global solutions to the generalized Kadomtsev-Petviashvili equation. Commun. Math. Phys. **201**, 577–590 (1999)
93. Isaza, P., Mejia, J.: Local and global Cauchy problems for the Kadomtsev-Petviashvili (KP-II) equation in Sobolev spaces of negative indices. Commun. Partial Differ. Equ. **26**, 1027–1057 (2001)
94. Ionescu, A.D., Kenig, C.E.: Global well-posedness of the Benjamin-Ono equation in low-regularity spaces. J. Am. Math. Soc., **20**, 753–798 (2007)
95. Ionescu, A.D., Kenig, C.E., Tataru, D.: Global well-posedness of the initial value problem for the KP I equation in the energy space. Invent. Math. **173**(2), 265–304 (2008)
96. Iório Jr., R.J.: On the Cauchy problem for the Benjamin-Ono equation. Commun. Partial Differ. Equ. **11** (10), 1031–1081 (1986)
97. Iório Jr., R.J., Nunes, W.V.L.: On equations of KP-type. Proc. R. Soc. Edinb. **128**, 725–743 (1998)
98. Ishimori, Y.: Multivortex solutions of a two dimensional nonlinear wave equation. Prog. Theor. Phys. **72** (1), 33–37 (1984)
99. Joseph, R.I.: Solitary waves in a finite depth fluid. J. Phys. A: Math. Gen. **10**(12), L225–L228 (1977)
100. Joseph, R.I., Egri, R.: Multi-soliton solutions in a finite depth fluid. J. Phys. A: Math. Gen. **11**(5), L97–L102 (1978)
101. Kadomtsev, B.B., Petviashvili, V.I.: On the stability of solitary waves in weakly dispersing media. Sov. Phys. Dokl. **15**, 539–541 (1970)
102. Kalyakin, L.A.: Long -wave asymptotics. Integrable equations as the asymptotic limit of nonlinear systems. Russ. Math. Surv. **44**(1), 3–42 (1989)
103. Kazeykina, A.V.: Solitons and large time asymptotics of solutions for the Novikov-Veselov equation. PhD Thesis at Ecole Polytechnique (2012)
104. Kazeykina, A.V.: A large time asymptotics for the solution of the Cauchy problem for the Novikov-Veselov equation at negative energy with non-singular scattering data. Inverse Prob. **28**(5), 055017 (2012)
105. Kazeykina, A.V.: Absence of solitons with sufficient algebraic localization for the Novikov-Veselov equation at nonzero energy. Funct. Anal. Appl. (2012). arXiv :1201.2758. (to appear)
106. Kazeykina, A.V.: Absence of traveling wave solutions of conductivity type for the Novikov-Veselov equation at zero energy. Funct. Anal. Appl. (2012). arXiv :1106.5639. (to appear)
107. Kazeykina, A.V., Munoz, C.: Dispersive estimates for rational symbols and local well-posedness of the nonzero energy NV equation. (2015). arXiv:1502.00968v1. (Submitted)

108. Kazeykina, A.V., Novikov, R.G.: Large time asymptotics for the Grinevich-Zakharov potentials. Bull. des Sci. Math. **135**, 374–382 (2011)
109. Kazeykina, A.V., Novikov, R.G.: Absence of exponentially localized solitons for the Novikov-Veselov equation at negative energy. Nonlinearity **24**, 1821–1830 (2011)
110. Kazeykina, A.V., Novikov, R.G.: A large time asymptotics for the solution of the Cauchy problem for the Novikov-Veselov equation at negative energy with nonsingular scattering data. Inverse Prob. **28**(5), 055017 (2012)
111. Kenig, C.E.: On the local and global well-posedness for the KP-I equation. Annales IHP Analyse Non Linéaire **21**, 827–838 (2004)
112. Kenig, C.E., Martel, Y.: Asymptotic stability of solitons for the Benjamin-Ono equation. Revista Matematica Iberoamericana **25**, 909–970 (2009)
113. Kenig, C.E., Martel, Y.: Global well-posedness in the energy space for a modified KP II equation via the Miura transform. TAMS **358**, 2447–2488 (2006)
114. Kenig, C.E., Nahmod, A.: The Cauchy problem for the hyperbolic-elliptic Ishimori system and Schrödinger maps. Nonlinearity **18**, 1987–2005 (2005)
115. Kenig, C.E., Ponce, G., Vega, L.: Well-posedness of the initial value problem for the Korteweg- de Vries equation. J. Am. Math. Soc. **4**, 323–346 (1991)
116. Kenig, C.E., Ponce, G., Vega, L.: Well-posedness and scattering results for the generalized Korteweg- de Vries equation via the contraction principle. Comm. Pure Appl. Math., **46**, 527–620 (1993)
117. Kevrekidis, P., Nahmod, A.R., Zeng, C.: Radial standing and self-similar waves for the hyperbolic cubic NLS in 2D. Nonlinearity **24**, 1523–1538 (2011)
118. Kiselev, O.M.: Asymptotics of solution of the Cauchy problem for the Davey-Stewartson I equation. Teor. Mat. Fiz. **114** (1), 104–114 (1998)
119. Kiselev, O.M.: Dromion perturbation for the Davey-Stewartson I equation. J. Nonlinear Math. Phys. **7**(4), 411–422 (2000)
120. Kiselev, O.M.: Asymptotics of a solution of the Kadomstev-Petviashvili II equation. Proc. Steklov Math. Inst. Suppl. **7**(1), S107–S139 (2001)
121. Kiselev, O.M.: Asymptotics of solutions of higher-dimensional integrable equations and their perturbations. J. Math. Sci. **138**(6), 6067–6230 (2006)
122. Klein, C., Peter, R.: Numerical study of blow-up in solutions to generalized Kadomtsev-Petviashvili equations. Discrete Contin. Dyn. Syst. B **19**(6), (2014). doi:10.3934/dcdsb.2014.19.1689
123. Klein, C., Roidot, K.: Numerical study of shock formation in the dispersionless Kadomtsev-Petviashvili equation and dispersive regularizations. Physica D **265**, 1–25 (2013)
124. Klein, C., Roidot, K.: Numerical Study of the semiclassical limit of the Davey-Stewartson II equations. Nonlinearity **27**, 2177–2214 (2014)
125. Klein, C., Roidot, K.: Fourth order time-stepping for Kadomtsev-Petviashvili and Davey-Stewartson equations. SIAM J. Sci. Comput. **33**(6), (2011) doi: 10.1137/100816663
126. Klein, C., Saut, J.-C.: Numerical study of blow-up and stability of solutions to generalized Kadomtsev-Petviashvili equations. J. Nonlinear Sci. **22**(5), 763–811 (2012)
127. Klein, C., Saut, J.-C.: A numerical approach to blow-up issues for dispersive perturbations of Burgers equation. Physica D 295–296 (2015), 46–65
128. Klein, C., Saut, J.-C.: A numerical approach to Blow-up issues for Davey-Stewartson II systems. Commun. Pure Appl. Anal. **14**(4) (2015), 1449–1467
129. Klein, C., Sparber, C.: Numerical simulation of generalized KP type equations with small dispersion. In: Liu, W.-B., Ng, M., Shi, Z.-C. (eds.) Recent Progress in Scientific Computing. Science Press, Beijing (2007)
130. Klein, C., Sparber, C., Markowich, P.: Numerical study of oscillatory regimes in the Kadomtsev-Petviashvili equation. J. Nonlinear Sci. **17**(5), 429–470 (2007)
131. Klein, C., Muite, B., Roidot, K.: Numerical Study of Blowup in the Davey-Stewartson System. Discrete Contin. Dyn. Syst. B **18**(5), 1361–1387 (2013)
132. Koch, H., Saut, J.-C.: Local smoothing and local solvability for third order dispersive equations. SIAM J. Math. Anal. **38**, 5, 1528–1541 (2007)

133. Koch, H., Tzvetkov, N.: Nonlinear wave interactions for the Benjamin-Ono equation. Int. Math. Res. Not. **30**, 1833–1847 (2005)
134. Koch, H., Tzvetkov, N.: On finite energy solutions of the KP-I equation. Math. Z. **258**(1), 55–68 (2008)
135. Kodama, Y., Satsuma, J., Ablowitz, M.J.: Nonlinear intermediate long-wave equation: analysis and method of solution. Phys. Rev. Lett. **46**, 687–690 (1981)
136. Kodama, Y., Ablowitz, M.J., Satsuma, J.: Direct and inverse scattering problems of the nonlinear intermediate long wave equation. J. Math. Phys. **23**, 564–576 (1982)
137. Konopelchenko, B.G., Matkarimov, B.T.: On the inverse scattering transform for the Ishimori equations. Phys. Lett. **A 135**(3), 183–189 (1989)
138. Konopelchenko, B.G., Matkarimov, B.T.: Inverse spectral transform for the nonlinear evolution equation generated by the Davey-Stewartson and Ishimori equations. Stud. Appl. Math. **82**, 319–359 (1990)
139. Konopelchenko, B.G., Moro, A.: Integrable equations in nonlinear geometrical optics. Stud. Appl. Math. **113** (4), 325–352 (2004)
140. Krichever, I.M.: Rational solutions of the Kadomtsev-Petviashvili equation and the integrable systems of N particles on a line. (Russian) Funkcional. Anal. i Prilozhen **12**, 76–78 (1978)
141. Kwong, M.K.: Uniqueness of positive solutions of $\Delta u - u + u^p = 0$ in $\mathbb{R}^n$. Arch. Ration. Mech. Anal. **65**, 243–266 (1989)
142. Lannes, D.: Water Waves: Mathematical Theory and Asymptotics. Mathematical Surveys and Monographs, vol. 188. American Mathematical Society (AMS), Providence, RI (2013)
143. Lannes, D.: Consistency of the KP approximation. In: Proceedings of the 4th International Conference on Dynamical Systems and Differential Equations, pp. 517–525. Wilmington, NC, USA, May 24–27 2002
144. Lannes, D., Saut, J.-C.: Weakly transverse Boussinesq systems and the KP approximation. Nonlinearity **19**, 2853–2875 (2006)
145. Lax, P.D.: Integral of nonlinear equations of evolution and solitary waves. Commun. Pure Appl. Math. **21**, 467–490 (1968)
146. Lebedev, D.R., Radul, A.O.: Generalized internal long waves equations, construction, Hamiltonian structure, and conservation laws. Commun. Math. Phys. **91**, 543–555 (1983)
147. Leblond, H.: Electromagnetic waves in ferromagnets. J. Phys. A **32**(45), 7907–7932 (1999)
148. Lin, Z.: Stability and instability of traveling solitonic bubbles. Adv. Differ. Equ. **7**(8), 897–918 (2002)
149. Linares, F., Ponce, G.: On the Davey-Stewartson systems. Ann. Inst. H. Poincaré Anal. Non Linéaire **10**, 523–548 (1993)
150. Linares, F., Pilod, D., Saut, J.-C.: Dispersive perturbations of Burgers and hyperbolic equations I : local theory. SIAM J. Math. Anal. **46**(2), 1505–1537 (2014)
151. Liu, Y.: Blow-up and instability of solitary wave solutions to a generalized Kadomtsev-Petviashvilii equation. Trans. AMS **353**(1), 191–208 (2000)
152. Manakov, S.V.: The inverse scattering transform for the time-dependent Schrödinger equation and Kadomtsev-Petviashvilii equation. Physica D **3**(1–2), 420–427 (1981)
153. Manakov, S.V.: The inverse scattering method and two-dimensional evolution equations. Uspekhi Mat. Nauk **31**(5), 245–246 (1976)
154. Manakov, S.V., Zakharov, V.E., Bordag, L.A., Its, A.R., Matveev, V.B.: Two-dimensional solitons of the Kadomtsev- Petviashvili equation and their interaction. Phys. Lett. A **63**, 205–206 (1977)
155. Manakov, S.V., Santini, P.M., Takchtadzhyan, L.A.: An asymptotic behavior of the solutions of the Kadomtsev-Petviashvili equations. Phys. Rev. Lett. A **75**, 451–454 (1980)
156. Marchenko, V.A.: Sturm-Liouville operators and applications. Revised edn., vol. 373 American Mathematical Society Chelsea Publishing, Providence, RI (2001)
157. Martel, Y., Merle, F.: Asymptotics stability of solitons of the subcritical generalized KdV equations revisited. Nonlinearity **18**, 55–80 (2005)
158. Martel, Y., Merle, F.: Asymptotic stability of solitons of the gKdV equations with a general nonlinearity. Math. Ann. **341**, 391–427 (2008)

159. Martel, Y., Merle, F.: Inelastic interaction of nearly equal solitons for the BBM equation. Discrete Contin. Dyn. Syst. **27** (2), 487–532 (2010)
160. Martel, Y., Merle, F., Mizumachi, T.: Description of the inelastic collision of two solitary waves for the BBM equation. Arch. Ration. Mech. Anal. **196**(2), 517–574 (2010)
161. Matsuno, Y.: Exact multi-solitons of the Benjamin-Ono equation. J. Phys. A Math. Gen. **12**(4), 619–662 (1979)
162. Matsuno, Y.: Interaction of the Benjamin-Ono solitons. J. Phys. A Math. Gen. **13**(5), 1519–1536 (1980)
163. Matsuno, Y.: Bilinear Transformation Method. Academic Press, New York (1984)
164. McConnell, M., Fokas, A., Pelloni, B.: Localized coherent Solutions of the DSI and DSII equations. Numer. Study. Math. Comput. Simul. **69** (5–6), 424–438 (2005)
165. Mizomachi, T., Tzvetkov, N.: Stability of the line soliton of the KP II equation under periodic transverse perturbations. arXiv: 1008.0812v1
166. Molinet, L., Pilod, D.: Global well-posedness and limit behavior for a higher-order Benjamin-Ono equation. Commun. Partial Differ. Equ. **37**, 2050–2080 (2012)
167. Molinet, L., Saut, J.-C., Tzvetkov, N.: Ill-posedness issues for the Benjamin-Ono and related equations. SIAM J. Math. Anal. **33**(4), 982–988 (2001)
168. Molinet, L., Saut, J.-C., Tzvetkov, N.: Well-posedness and ill-posedness results for the Kadomtsev-Petviashvili-I equation. Duke Math. J. **115**(2), 353–384 (2002)
169. Molinet, L., Saut, J.-C., Tzvetkov, N.: Global well-posedness for the KP-I equation. Math. Annalen **324**, 255–275 (2002). Correction: Math. Ann. **328**, 707–710 (2004)
170. Molinet, L., Saut, J.-C., Tzvetkov, N.: Remarks on the mass constraint for KP type equations. SIAM J. Math. Anal. **39**(2), 627–641 (2007)
171. Molinet, L., Saut, J.-C., Tzvetkov, N.: Global well-posedness for the KP-I equation on the background of a non localized solution. Commun. Math. Phys. **272**, 775–810 (2007)
172. Molinet, L., Saut, J.-C., Tzvetkov, N.: Global well-posedness for the KP-II equation on the background of a non localized solution. Annales IHP, Analyse Non Linéaire, **28**(5), 653–676 (2011)
173. Music, M., Perry, P.: Global solutions for the zero energy Novikov-Veselov equation by inverse scattering, arXiv:1502.02632v1. 9 Feb (2015)
174. Musher, S.L., Rubenchik, A.M., Zakharov, V.E.: Hamiltonian approach to the description of nonlinear plasma phenomena. Phys. Rep. **129** (5), 285–366 (1985)
175. Neves, A., Lopes, O.: Orbital stability of double solitons for the Benjamin-Ono equation. Commun. Math. Phys. **262**, 757–791 (2006)
176. Newell, A., Moloney, J.V.: Nonlinear Optics. Addison-Wesley, New York (1992)
177. Niizaki, T.: Large time behavior for the generalized Kadomtsev-Petviashvilii equations. Differ. Equ. Appl. **3** (2), 299–308 (2011)
178. Novikov, R.G.: Absence of exponentially localized solitons for the Novikov-Veselov equation at positive energy. Phys. Lett. **A 375**, 1233–1235 (2011)
179. Novikov, S.P., Veselov, A.P.: Finite-zone, two-dimensional, potential Schrödinger operators. Explicit formula and evolutions equations. Dokl. Akad. Nauk SSSR **279** 20–24 (1984); Translation in Sov. Math. Dokl. **30**, 588–591 (1984)
180. Novikov, S.P., Veselov, A.P.: Finite-zone, two-dimensional, potential Schrödinger operators. Potential operators. Dokl. Akad. Nauk SSSR **279**, 784–788 (1984); Translation in Sov. Math. Dokl. **30**, 705–708 (1984)
181. Novikov, S.P., Manakov, S.V., Pitaevskii, L.P., Zakharov, V.E.: Theory of solitons. The inverse scattering method. Contemporary Soviet Mathematics, Consultant Bureau, New York and London (1984)
182. Ohta, M.: Stability and instability of standing waves for the generalized Davey-Stewartson system. Differ. Integral Equ. **8**, 1775–1788 (1995)
183. Ohta, M.: Instability of standing waves for the generalized Davey-Stewartson system. Ann. Inst. H. Poincaré, Phys. Théor. **62**, 69–80 (1995)
184. Ohta, M.: Blow-up solutions and strong instability of standing waves for the generalized Davey-Stewartson system. Ann. Inst. H. Poincaré, Phys. Théor. **63**, 111–117 (1995)

185. Ozawa, T.: Exact blow-up solutions to the Cauchy problem for the Davey-Stewartson systems. Proc. R. Soc. Lond. A **436**, 345–349 (1992)
186. Papanicolaou, G., Sulem, C., Sulem, P.-L., Wang, X.P.: The focusing singularity of the Davey-Stewartson equations for gravity-capillary waves. Physica D **72**, 61–86 (1994)
187. Pelinovsky, D.E.: Rational solutions of the Kadomtsev-Petviashvili hierarchy and the dynamics of their poles. I. New form of a general rational solution. J. Math. Phys. **35**, 5820–5830 (1994)
188. Pelinovsky, D.E., Stepanyants, Yu.A.: New multisolitons of the Kadomtsev-Petviashvili equation. Sov. Phys. JETP Lett. **57**, 24–28 (1993)
189. Pelinovsky, D.E., Stepanyants, Yu.A.: Self-focusing instability of plane solitons and chains of two-dimensional solitons in positive-dispersion media. Sov. Phys. JETP **77**(4), 602–608 (1993)
190. Pelinovsky, D.E., Sulem, C.: Eigenfunctions and Eigenvalues for a Scalar Riemann-Hilbert Problem Associated to Inverse Scattering. Commun. Math. Phys. **208**, 713–760 (2000)
191. Pelinovsky, D.E., Sulem, C.: Spectral decomposition for the Dirac system associated to the DS II equation. Inverse Prob. **16**, 59–74 (2000)
192. Pelinovsky, D.E., Sulem, C.: Embedded solitons of the Davey-Stewartson II equation. CRM In: Sulem, C., Sigal, I.M. (eds.) Proceedings and Lecture Notes, vol. 27, pp. 135–145 (2001)
193. Perry, P.A.: Global well-posedness and long time asymptotics for the defocussing Davey-Stewartson II equation in $H^{1,1}(\mathbb{R}^2)$. (2012). arXiv:1110.5589v2
194. Perry, P.A.: Miura maps and inverse scattering for the Novikov-Veselov equation. Analysis Partial Differ. Equ. **7**(2), 311–343 (2014)
195. Redekopp, L.: Similarity solutions of some two-dimensional nonlocal wave evolution equations. Stud. Appl. Math. **63**, 185–207 (1980)
196. Roidot, K., Mauser, N.: Numerical study of the transverse stability of NLS soliton solutions in several classes of NLS type equations. arXiv:1401.5349v1 [math-ph] 21 Jan 2014
197. Rousset, F., Tzvetkov, N.: Transverse nonlinear instability for some Hamiltonian PDE's. J. Math. Pures Appl. **90**, 550–590 (2008)
198. Rousset, F., Tzvetkov, N.: Transverse nonlinear instability for two-dimensional dispersive models. Ann. IHP, Analyse Non Linéaire **26**, 477–496 (2009)
199. Rousset, F., Tzvetkov, N.: A simple criterion of transverse linear instability for solitary waves. Math. Res. Lett. **17**, 157–169 (2010)
200. Satsuma, J.: N-soliton solution of the two-dimensional Kotreweg-de Vries equation. J. Phys. Soc. Jpn. **40**, 286–290 (1976)
201. Satsuma, J., Ablowitz, M.J.: Two-dimensional lumps in nonlinear dispersive systems. J. Math. Phys. **20**, 1496–1503 (1979)
202. Saut, J.-C.: Remarks on the generalized Kadomtsev-Petviashvili equations. Indiana Univ. Math. J. **42**, 1011–1026 (1993)
203. Saut, J.-C., Tzvetkov, N.: The Cauchy problem for higher order KP equations. J. Differ. Equ. **153**(1), 196–222 (1999)
204. Saut, J.-C., Tzvetkov, N.: The Cauchy problem for the fifth order KP equation. J. Math. Pures Appl. **79** (4), 307–338 (2000)
205. Saut, J.-C., Tzvetkov, N.: Global well-posedness for the KP-BBM equations. AMRX Appl. Math. Res. Express **1**, 1–16 (2004)
206. Schulman, E.I.: On the integrability of equations of Davey-Stewartson type. Theor. Math. Phys. **56**, 131–136 (1983)
207. Schuur, P.C.: Asymptotic Analysis of Soliton problems. An Inverse scattering Approach. Lecture Notes in Mathematics, vol 1232, Springer, Berlin (1986)
208. Scoufis, G., Cosgrove, C.M.: An application of the inverse scattering transform to the modified long wave equation. J. Math. Phys. **46** (10), 103501 (2005)
209. Segur, H.: Who cares about integrability. Physica D **51**, 343–359 (1991)
210. Soyeur, A.: The Cauchy problem for the Ishimori equations. J. Funct. Anal. **105**, 233–255 (1992)

211. Sung, L.Y.: An inverse scattering transform for the Davey-Stewartson equations. I. J. Math. Anal. Appl. **183**(1), 121–154 (1994)
212. Sung, L.Y.: An inverse scattering transform for the Davey-Stewartson equations. II. J. Math. Anal. Appl. **183**(2), 289–325 (1994)
213. Sung, L.Y.: An inverse scattering transform for the Davey-Stewartson equations. III. J. Math. Anal. Appl. **183**(3), 477–494 (1994)
214. Sung, L.Y.: Long-Time Decay of the Solutions of the Davey-Stewartson II Equations. J. Nonlinear Sci. **5**, 433–452 (1995)
215. Sung, L.Y.: The Cauchy problem for the Ishimori equation. J. Funct. Anal. **139**, 29–67 (1996)
216. Sung, L.Y.: Square integrability and uniqueness of the solutions of the Kadomtsev-Petviashvili-I equation. Math. Phys. Anal. Geom. **2**, 1–24 (1999)
217. Takaoka, H., Tzvetkov, N.: On the local regularity of Kadomtsev-Petviashvili-II equation. IMRN **8**, 77–114 (2001)
218. Tanaka, S.: On the N-tuple wave solutions of the Korteweg-de Vries equation. Publ. R.I.M.S. Kyoto Univ. **8**, 419–427 (1972)
219. Tao, T.: Global well-posedness of the Benjamin-Ono equation in H^1. J. Hyperbolic Differ. Equ. **1**, 27–49 (2004)
220. Tao, T.: Why are solitons stable? Bull. AMS **46** (1), 1–33 (2009)
221. Tom, M.M.: On a generalized Kadomtsev-Petviashvili equation. Contemp. Math. AMS, **200**, 193–210 (1996)
222. Ukaï, S.: Local solutions of the Kadomtsev-Petviashvili equation. J. Fac. Sci. Univ. Tokyo Sect. IA Math. **36**, 193–209 (1989)
223. Villarroel, J., Ablowitz, M.J.: On the discrete spectrum of the nonstationary Schrödinger equation and multipole lumps of the Kadomtsev-Petviashvili I equation. Commun. Math. Phys. **207**, 1–42 (1999)
224. Whitham, G.B.: Linear and Nonlinear Waves. Wiley, New York (1974)
225. Wickerhauser, M.V.: Inverse scattering for the heat equation and evolutions in $(2+1)$ variables. Commun. Math. Phys. **108**, 67–89 (1987)
226. Zaitsev, A.A.: Formation of stationary waves by superposition of solitons. Sov. Phys. Dokl. **28** (9), 720–722 (1983)
227. Zakharov, V.E.: Instability and nonlinear oscillations of solitons. JETP Lett. **22**, 172–173 (1975)
228. Zakharov, V.E.: Stability of periodic waves of finite amplitude on the surface of a deep fluid. J. Appl. Mech. Tech. Phys. **2**, 190–194 (1968)
229. Zakharov, V.E., Kuznetsov, A.: Multi-scale expansion in the theory of systems integrable by the inverse scattering transform. Physica D **18**(1–3), 455–463 (1986)
230. Zakharov, V.E., Rubenchik, A.M.: Nonlinear interaction of high-frequency and low frequency waves. Prikl. Mat. Techn. Phys., **5**, 84–98 (1972)
231. Zakharov, V.E., Shabat, A.B.: Interaction between solitons in a stable medium. Sov. Phys. JETP **37**, 823–828 (1973)
232. Zakharov, V.E., Schulman, E.I.: Degenerate dispersion laws, motion invariants and kinetic equations. Physica **1D**, 192–202 (1980)
233. Zakharov, V.E., Schulman, E.I.: Integrability of nonlinear systems and perturbation theory. In: Zakharov, V.E. (ed.) What is Integrability? Springer Series on Nonlinear Dynamics, pp. 185–250. Springer, Berlin (1991)
234. Zhidkov, P.E.: Korteweg-de Vries and Nonlinear Schrödinger Equations: Qualitative Theory. Lecture Notes in Mathematics, vol. 1756. Springer, Berlin (2001)
235. Zhou, X.: Inverse scattering transform for the time dependent Schrödinger equation with applications to the KP-I equation. Commun. Math. Phys. **128**, 551–564 (1990)

Erratum to: IST Versus PDE: A Comparative Study

Christian Klein and Jean-Claude Saut

P. Guyenne et al. (eds.), *Hamiltonian Partial Differential Equations and Applications*,
Fields Institute Communications 75, DOI 10.1007/978-1-4939-2950-4_14,

DOI 10.1007/978-1-4939-2950-4_15

Beginning on page 403, in Chapter 14 (IST Versus PDE: A Comparative Study), the equations were incorrectly numbered. The numbering has been corrected in the current version of the chapter

The online version of the original chapter can be found at
http://dx.doi.org/10.1007/978-1-4939-2950-4_14

P. Guyenne et al. (eds.), *Hamiltonian Partial Differential Equations and Applications*, Fields Institute Communications 75,
DOI 10.1007/978-1-4939-2950-4_15